I0044130

ENGAGING COMMUNITIES FOR HIGH-IMPACT
THREATS TO CRITICAL INFRASTRUCTURE

Conference Proceedings of the InfraGard National EMP SIG Sessions

at the

Dupont Summit 2015

Video-recorded sessions at
http://www.ipsonet.org/conferences/the-dupont-summit/infragard-videos-2015

Edited by Charles L. Manto and Stephanie A. Lokmer

Friday, December 4, 2015
Whittemore House
1526 New Hampshire Ave, NW
Washington, DC

Engaging Communities for High-Impact Threats to Critical Infrastructure

Conference Proceedings

InfraGard National EMP SIG Sessions at the 2015 Dupont Summit

December 4, 2015

At the Whittemore House
1526 New Hampshire, NW, Washington, DC 20005
Hosted and Published by the Policy Studies Organization

Edited by Charles L. Manto and Stephanie A. Lokmer

November 2016

ISBN-10: 1-63391-429-1
ISBN-13: 978-1-63391-429-2

Cover and interior design by Jeffrey Barnes
jbarnesbook.design

Daniel Gutierrez-Sandoval, Executive Director
PSO and Westphalia Press

Updated material and comments on this edition can be found at the Westphalia Press website:
www.westphaliapress.org

Westphalia Press
An imprint of Policy Studies Organization
1527 New Hampshire Ave., NW
Washington, DC. 20036
info@ipsonet.org

Table of Contents

|

Preface.....3

2015 Program Guide for Dupont Summit EMP SIG Sessions.....5

Presenter Biographies.....8

FBI's Roles and Perspectives on the EMP Special Interest Group and the 2015 Dupont Summit Sessions.....23

- Chuck Manto, CEO, Instant Access Networks (IAN) and Founder/Infragard EMP SIG (Electromagnetic Pulse Special Interest Group) Chair
- Jerry Bowman, InfraGard National Members Alliance Board President
- John Pi, Unit Chief , Supervisory Special Agent, Federal Bureau of Investigation (FBI)
- Eric Sporre, Acting Deputy Assistant Director of the FBI's Cyber Division

Coordinating the DHS Protective Security Advisor Program and InfraGard's EMP SIG.....29

- Jerry Bowman, InfraGard National Members Alliance Board President
- Scott Breor, Director of the Protective Security Coordination Division, DHS

Congressional Updates on EMP and High-Impact Threat Planning and Protection for Communities.....31

- Mr. Jerry Bowman introduces Session Chair
- Dr. Wallace E. Boston, President and CEO, American Public University System introduces
- Congressman Roscoe Bartlett, former Congressman(Maryland)

Private Sector Finance and Capital Markets for Resilient Community Infrastructure.....35

- Mr. Jeff Weiss, Cochairman and Managing Director, Distributed Sun

The Cyber Threat to Critical Infrastructure.....42

- Paul Joyal, former InfraGard National Board member
- Brett Leatherman, Assistant Chief of the FBI Cyber Division

Cyber, EMP, Space Weather Complications to Complex Systems Engineering.....52

- Mr. Chuck Manto, Moderating
- Mr. Thomas Popik, Founder, Foundation for Resilient Societies, Nexus of engineering and organizational issues
- Mr. Mark Walker, INCOSE and Mr. Michael deLamare, Bechtel and INCOSE, reviews the joint INCOSE & EMP SIG Planning Models to Manage Complexity of High Impact Threats

- Mr. Joseph Weiss, Author, Protecting Industrial Control Systems from Electronic Threats
- Cyber complications

FEMA Perspective on the New Space Weather Strategy and Action Plan..59

- Hon. Craig Fugate, FEMA Administrator

State Plans and Regional EMP SIGs: Workshops and Table Top Exercises..71

- Ms. Mary Lasky, National EMP SIG
- Mr. Steve Pappas, MW EMP SIG and IN EMP Planning
- Mr. Steve Volandt, SE EMP SIG and the NC EMP Plan
- Prof. Mel Lewis, NE EMP SIG, Updates on NYC area planning
- Hon. Andrea Boland, National EMP SIG Policy Adv Panel, Maine update

Transformer Protection Test Data...80

- Dr. Fred Faxvog, Emprimus

The National Space Weather Strategy and Action Plan Impact on the Whole of Community...........85
and Preparedness (also accepts presentation of the EMP SIG Triple Threat Power Grid Exercise Program)

- Hon. Caitlin Durkovich, Assistant Secretary, DHS; SWORM Cochair

Key Operational Space Weather Facts and the Action Plan...93

- Mr. William (Bill) Murtagh, Assistant Director for Space Weather at the White House Office of Science and Technology Policy

Community Action in Light of High-impact threats Reviewing the Major Culture Shift of Emergency Management Engaging Space Weather, EMP and Cyber Threats.................................101

- Hon. Andrea Boland
- Mr. William Harris
- Mr. Thomas Popik

Role of Protected Micro-grids for Community Protection...108

- Hon. R. James Woolsey, Chairman of the Foundation for Defense of Democracies and a Venture Partner with Lux Capital, former CIA Director
- Mr. David Geary, DC Fusion, Direct Current Microgrids
- Mr. Terrance (Terry) Hill, Passive House Instituteand Community Microgrids
- Congressman Roscoe Bartlett
- Thomas (Tom) Popik
- Charles (Chuck) Manto

DTRA EMP Program and SBIR for EMP Protected Defense Critical Infrastructure......................120
How does DCI include communities for geographically broad long-term events? What is required to protect them? What is underway now?

- MajGen Robert Newman, USAF Ret and former Adjutant General of Virginia, Cochair EMP SIG Civilian Military Liaison Panel
- Mr. Kevin Briggs, National Cybersecurity & Communications Integration Center (NCCIC), DHS staff performing EMP analysis.
- Dr. George Baker, Previous DTRA staff, current contractor)
- Amb. Henry (Hank) F. Cooper, Chairman, High Frontier; Cochair EMP SIG Civilian Military Liaison Panel
- Charles (Chuck) Manto, CEO, Instant Access Networks, LLC

Industry Best Practices for EMP, Space Weather and other High-Impact Events...........................128

- Mr. Charles (Chuck) Manto, EMP SIG Chair; CEO, Instant Access Networks, LLC
- Mr. Michael deLamare, INCOSE and Bechtel
- Mr. Gale Nordling, CEO, EMPrimus, MW Transformer protection users
- Mr. Jack Pressman, Cyber Innovation Labs, NE Insurance Industry Data Center
- Mr. William (Bill) Harris, Foundation for Resilient Societies

Key EMP/GMD Next Steps..139

- Dr. Peter Vincent Pry, EMP Task Force on National and Homeland Security.
- Dr. George Baker, Professor Emeritus, James Madison University, EMP SIG EMP Advisory Panel Chair

II

Grid Vulnerabilities to Cyber Attack..149
Analysis of the Cyber Attack on the Ukrainian Power Grid
Defense Use Case
March 18, 2016
By SANS, and E-ISAC

FERC on Cyber Threats to Electric Grids..178
FEDERAL ENERGY REGULATORY COMMISSION
18 CFR Part 40
[Docket No. RM15-14-000]
Revised Critical Infrastructure Protection Reliability Standards
(Issued January 21, 2016)

FERC on Supply Chain Risk..241
FEDERAL ENERGY REGULATORY COMMISSION
18 CFR Part 40

[Docket No. RM15-14-002; Order No. 829]
Revised Critical Infrastructure Protection Reliability Standards
(Issued July 21, 2016)

FERC Ruling on GMD......**334**
FEDERAL ENERGY REGULATORY COMMISSION
18 CFR Part 40; [Docket No. RM15-11-000; Order No. 830]
Reliability Standard for Transmission System Planned Performance for Geomagnetic Disturbance Events (Issued September 22, 2016)

FERC-NERC Joint on Restoration......**406**
FEDERAL ENERGY REGULATORY COMMISSION
18 CFR Part 40 Docket No. RM15-11-000; Order No. 830]
Reliability Standard for Transmission System Planned Performance for Geomagnetic Disturbance Events (Issued September 22, 2016)

Challenge to Supply Chain Protection Standards......**536**
Critical Infrastructure Protection Docket No. RM15-14-000 Reliability Standards
JOINT COMMENTS OF ISOLOGIC, LLC AND THE FOUNDATION FOR RESILIENT SOCIETIES (Submitted to FERC on September 21, 2015)

III

EMP SIG Sessions at the NOAA Space Weather Workshop April 27, 2016

Balancing Risk—Understanding & Preparing for Catastrophes......**555**
By Catherine L. Feinman

Space Weather—A Historic Shift in Emergency Preparedness......**560**
By Charles (Chuck) Manto

Preparing for Everything Under the Sun......**564**
By Josh Sparber

Space Weather & Electrical Grid—GPS the Weakest Link......**568**
By Dana Goward

Electrical Systems & 21st Century Threats......**570**
By Benjamin Dancer

Cascadia Catastrophe—Not If, But When......**573**
By Arthur Glynn

Orientation to the Society of Disaster Medicine and Public Health

About the Society of Disaster Medicine and Public Health.........576

Disaster Medicine and Public Health Preparedness: A Discipline for All Health Professionals.........577
A reprint by: James, J.J., Benjamin, G.C., Burkle, F.M., Gebbie, K.M., Kelen, G. and Subbarao, I. (2013), *Disaster Medicine and Public Health Preparedness*, 4(2), pp. 102–107. doi: 10.1001/dmp.v4n2.hed10005,

Overview of the *Powering Through* Planning Guide.........583

Bibliography.........586

Resilience Assessment for Cyber Industry by DHS.........611

Preface and Acknowledgements

This conference proceedings of the InfraGard EMP SIG (Electromagnetic Pulse Special Interest Group) sessions at the Dupont Summit 2015 provides written presentations and background material for the video recordings available at http://www.ipsonet.org/conferences/the-dupont-summit/infragard-videos-2015.

The sessions covered high-impact threats to critical infrastructure with a special emphasis on geomagnetic disturbance (GMD), a topic of the sessions provided by the EMP SIG at each Dupont Summit since 2011 and the contingency planning workshops and exercises with the National Defense University and the Maryland Emergency Management Agency in October 2011. On the pre-conference day of Dec. 3, 2015, the EMP SIG held a workshop and discussions at the National Guard Association of the US headquarters in order to develop a planning guide so that the whole of community might improve their business continuity and disaster mitigation and recovery plans.

Subsequent to the Summit, the White House Office of Science and Technology Policy published their final National Space Weather Strategy and Action Plan that has been added to the conference proceedings along with other notable items such as a related press release from the Defense Department in June 2016. This was a small business innovation research (SBIR) program request for strategies to protect defense critical infrastructure and the civilian infrastructure they need from the effects of long-term regional and national blackouts due to high-altitude nuclear burst EMP or drive-by directed energy weapons. Both of these documents were ground breaking initiatives showing that local communities need to be more resilient since the time to rescue might not be four days, but, possibly 40 or 400 days.

The National Space Weather Strategy has six goals, the second of which is to have the "whole-of-community" conduct planning and exercises focusing on long-term regional and national power outages. This is precisely what the EMP SIG has been doing and continues to expand through its regional activities of its regional EMP SIGs. The 2014 Summit continued to set the stage for local planning that will be developed further in the 2015 Summit. The significance of this will become more apparent as the strategy becomes a coordinated federal action plan late in 2015.

EMP SIG participation at the NOAA Space Weather Prediction Center's Space Weather Workshop in April 27, 2016 is documented at the end of these proceedings in the form of articles covering discussions there.

This is the fourth year that we have published conference proceedings. Each transcript of the prior year's presentation is hyperlinked to YouTube videos. Other essays and items in the bibliography section are also hyperlinked. This year, the EMP SIG is also publishing the first edition of a planning guide entitled, *Powering Through*. It is available at Westphalia Press and Amazon along with its companion, the *Triple Threat Power Grid Exercise* that was published last year. These materials will assist local communities and businesses to better prepare for long-term power outages as long as a year. In turn, that work can lead to greater investment and development of sustainable local infrastructure. An overview of the *Powering*

Through planning guide is included near the end of this year's conference proceedings with program procurement information. Material will be made available to National Guard units, the National Governors Association, local InfraGard chapters and the DHS Protective Security Advisor program that is coordinating these planning activities with the EMP SIG.

The EMP SIG wishes to thank the Policy Studies Organization (PSO) for its generous support of the conference and the assistance of many who helped with editing. We also wish to thank the National Guard Association of the US who provided use of their facilities for workshops and tabletop exercises. Of course, the EMP SIG, as a nationwide special interest group of InfraGard, appreciates the strong support of the national board and staff, local chapters across the country, its members across all the 50 states and territories, and the Federal Bureau of Investigation who provides significant support to its InfraGard program.

As the chairman of the InfraGard EMP SIG and its conference session organizer, I welcome you to contact me for more information about InfraGard's EMP SIG and ways to participate in future activities. For information on InfraGard and how to join, see www.infragard.org.

Charles Leo Manto
EMP SIG Chairman, InfraGard National
cmanto@stop-EMP.com

National InfraGard EMP Special Interest Group

Program Details for
Engaging Communities for High Impact Threats to Critical Infrastructure

Historic Whittemore House, Dupont Circle
1526 New Hampshire Avenue, Washington DC

Main Auditorium Schedule
Friday, December 4, 2015

For secure DHS WEBCAST access on the day of event, go to the following web address and register as a Guest, preferably 15 minutes ahead of time: https://share.dhs.gov/empsig
HSIN requires Flash Player to be installed.
For a mere broadcast of the events on Friday Dec 4, a livestream can be accessed here:
https://livestream.com/futureview/EMPSIGDupont
Phone only conference bridge is also available at:
Dial in: 540-667-8701
When prompted, enter the session ID: 8602#
When prompted, enter the access code: 93231#
Each caller will then have to state their name followed by the # to be added to the call.

8:00–8:30 **Registration**

8:30–8:35 **Introduction**

Mr. Charles (Chuck) Manto, InfraGard National EMP SIG Chairman, provides a brief overview of the 2015 EMP SIG Dupont Summit, the purpose of the InfraGard National EMP SIG and introduces **Mr. Jerry Bowman,** INMA President who introduces **FBI Unit Chief John Pi.**

8:35–8:50 **InfraGard and High-Impact Threats Planning with DHS PSA Program**

FBI Unit Chief John Pi introducing **FBI Deputy Assistant Director Don Good (Cyber Division)** who provides a program overview of the InfraGard EMP SIG "Triple Threat Power Grid Exercise" program and planning with the DHS PSA program.

DHS PSA Program Director Scott Breor, Overviews coordination of DHS PSA Program and EMP SIG.

8:55–9:30 **Mr. Jerry Bowman** introduces **Session Chair, Dr. Wallace E. Boston,** President and CEO, American Public University System, introduces **Congressman Roscoe Bartlett**
"Congressional Updates on EMP and High-Impact Threat Planning and Protection for

Communities"
Hon. Roscoe Bartlett, former Congressman(Maryland)

9:35–9:50 **"Private Sector Finance and Capital Markets for Resilient Community Infrastructure"**
Mr. Jeff Weiss, Cochairman and Managing Director, Distributed Sun

10:00–10:30 **"The Cyber Threat to Critical Infrastructure": Mr. Brett Leatherman,** Assistant Section Chief (ASC) of FBI Cyber Division, Cyber Outreach Section

10:30–10:55 **Panel: "Cyber, EMP, Space Weather Complications to Complex Systems Engineering"**
Mr. Chuck Manto, Moderating
Mr. Joseph Weiss, Author, Protecting Industrial Control Systems from Electronic Threats Cyber complications
Mr. Thomas Popik, Founder, Foundation for Resilient Societies, Nexus of engineering and organizational issues
Mr. Mark Walker, INCOSE and **Mr. Michael deLamare,** Bechtel and INCOSE,reviews the joint INCOSE & EMP SIG Planning Models to Manage Complexity of High Impact Threats

11:00–11:40 **"FEMA Perspective on the New Space Weather Strategy and Action Plan"**
Hon. Craig Fugate, FEMA Administrator

11:45–12:10 **Panel: "State Plans and Regional EMP SIGs—Workshops and Table Top Exercises"** covers the current and emerging activities of regional EMP SIGs and the development of a preplanning framework for high-impact disasters
Ms. Mary Lasky, National EMP SIG
Mr. Steve Pappas, MW EMP SIG and IN EMP Planning
Mr. Steve Volandt, SE EMP SIG and the NC EMP Plan
Prof. Mel Lewis, NE EMP SIG, Updates on NYC area planning
Hon. Andrea Boland, National EMP SIG Policy Adv Panel, Maine update

12:10–12:55 **Lunch Break, Transformer Protection Test Data, Dr. Fred Faxvog, Emprimus** (Meal provided on site)

12:55–1:00 **Welcome Back by INMA Chairman Gary Gardner and President Gerry Bowman/ Cameo Presentations**

1:00–1:30 **Key Note Presentation: "The National Space Weather Strategy and Action Plan Impact on the Whole of Community and Preparedness"**
Hon. Caitlin Durkovich, Assistant Secretary, DHS; SWORM Cochair

1:35–2:00 **"Key Operational Space Weather Facts and the Action Plan"**
Mr. William (Bill) Murtagh, Assistant Director for Space Weather at the White House Office of Science and Technology Policy

2:05–2:30 **Panel: Community Action in Light of High-impact threats Reviewing the Major Culture Shift of Emergency Management Engaging Space Weather, EMP and Cyber Threats**
(**Hon. Andrea Boland** and various presenters)

2:35–3:25 **Panel: Presentation: "Role of Protected Micro-grids for Community Protection"**

Hon. R. James Woolsey, Chairman of the Foundation for Defense of Democracies and a Venture Partner with Lux Capital, former CIA Director

Mr. Scott Sklar, President of The Stella Group, Ltd, and Adjunct Professor at The George Washington University (GWU)

Mr. David Geary, DC Fusion, Direct Current Microgrids

Mr. Terry Hill, Passive House Instituteand Community Microgrids

Utility Microgrids, panelist TBA

3:30–3:55 **Panel: DTRA EMP Program and SBIR for EMP Protected Defense Critical Infrastructure.** How does DCI include communities for broad long-term event? What is required to protect them? What is underway now?

MajGen Robert Newman, USAF Ret and former Adjutant General of Virginia, Cochair EMP SIG Civilian Military Liaison Panel

Mr. Kevin Briggs, DHS staff performing EMP analysis

Dr. George Baker, previous DTRA staff, (current contractor)

Amb. Henry (Hank) F. Cooper, Chairman,High Frontier; Cochair EMP SIG Civilian Military Liaison Panel

4:00–4:40 **Panel: "Industry Best Practices for EMP, Space Weather and other High-impact Events"**

Dr. Paul Stockton, Panel Moderator, Managing Director Sonecon, LLC. And former Assistant Secretary of Defense

Mr. Michael deLamare, INCOSE and Bechtel

Mr. Gale Nordling, CEO, EMPrimus, MW Transformer protection users

Mr. Jack Pressman, Cyber Innovation Labs, NE Insurance Industry Data Center

Mr. William Harris, Foundation for Resilient Societies

4:45–5:10 **Panel: "Key EMP/GMD Next Steps"**

Dr. Peter Vincent Pry, EMP Task Force on National and Homeland Security

Dr. George Baker, Professor Emeritus, James Madison University, EMP SIG Advisory Panel Chair

5:15-5:30 **Next Steps for EMP SIG Working Groups and Concluding Remarks.** Overview of next steps with emerging regional EMP SIGs and the workshops and table top exercises expected over the next year. A reminder of the EMP SIG meeting in April at the Space Weather Workshop (April 15) in Boulder, CO.

Mr. Chuck Manto, EMP SIG Chair

FBI Coordinators

Biographies of Presenters

Baker, Dr. George

Email: BakerGH@JMU.edu

Dr. Baker is emeritus professor of applied science at James Madison University (JMU). In addition to teaching graduate and undergraduate applied science courses from 2001-2012, he organized and directed the university's Institute for Infrastructure and Information Assurance (IIIA). From 1987-1994 Baker led the Defense Nuclear Agency's EMP R&D program protecting strategic systems against electromagnetic pulse (EMP) effects and developing DoD's EMP guidelines and standards. From 1994-1996 Baker directed DNA's Innovative Concepts Division, managing advanced technology development and demonstration including electric guns, high density energy storage, radio-frequency directed energy weapon technology, and space nuclear power. A primary research interest stems from his experience as Director of DTRA's Springfield Research Facility, a national center for critical system vulnerability assessment and protection guidance, during 1996-1999. His organization was responsible for developing and implementing the Joint Chiefs of Staff Force Protection assessment methodology. From 1999-2001 he served as a senior scientist at Northrop-Grumman, advising DTRA nuclear effects and test programs. He now applies lessons-learned from DoD experience to critical national infrastructure assurance and community resilience. He consults in the areas of critical infrastructure protection, EMP and geomagnetic disturbance (GMD) protection, nuclear and directed energy weapon effects, and risk assessment for customers including DoD, DOE, DHS, the White House, National Guard units, the National Park Service, SAIC, and Defense Group Inc. He also serves on the Congressional EMP Commission, the Board of Directors of the Foundation for Resilient Societies, the Board of Advisors for the Congressional Task Force on National and Homeland Security and the JMU Research and Public Service Advisory Board. Degrees include M.S., Physics (University of Virginia) and Ph.D., Engineering Physics (U.S. Air Force Institute of Technology).

Bartlett, Congressman Roscoe G.

Email: roscoegbartlett@gmail.com

Elected to serve his tenth term in the United States House of Representatives, Roscoe G. Bartlett considers himself a citizen-legislator, not a politician. Prior to his election to Congress, he pursued successful careers as a professor, research scientist and inventor, small business owner, and farmer. He was first elected in 1992 to represent Maryland's Sixth District.

In the 112th Congress, Bartlett serves as Chairman of the Tactical Air and Land Forces Subcommittee of the House Armed Services Committee. Owing to his 10 years of experience as a small business owner, he also serves on the Small Business Committee. One of three scientists in the Congress, Dr. Bartlett is also a senior member of the Science, Space and Technology Committee.

Prior to his election to the Congress, Dr. Bartlett worked for more than 20 years as a scientist and engineer on research and development programs for the military and NASA. Nineteen of his 20 patents are held by the U.S. Government for his inventions of life support equipment used by military pilots, astronauts, search and rescue personnel, and firefighters.

In 2008, *Slate* magazine applauded him as "an advocate for reducing dependency on fossil fuels." The Association for the Study of Peak Oil (ASPO-USA) created the Roscoe G. Bartlett "Speak Truth to Power" Award in his honor in 2008. It had previously awarded him the M. King Hubbert Award in 2006 for his leadership in the Congress to promote efficiency and conservation and alternative renewable sources of domestic energy to enable the United States to overcome the challenges to national security and economic prosperity of global peak oil. Congressman Bartlett is the cofounder and cochairman of the Congressional Peak Oil Caucus. He is also the cochairman of the House Renewable Energy and Energy Efficiency Caucus and Defense Energy Security Caucus. He is also a member of the Oil and National Security Caucus.

Boland, Hon. Andrea

Email: sixwings@metrocast.net

State Representative Andrea Boland recently completed 8 years (or 4 terms) in the Maine legislature. She is considered a leader in safety issues of electromagnetic radiation, especially from cellphones and smart meters. She became involved in electric grid protection against electromagnetic pulse and geomagnetic solar storms (GMD) at the suggestion of her regular scientific advisor. Her work is supported by several national experts. She has a B.A. degree from Elmira College and an MBA from Northeastern University, and studied at the Sorbonne and Institute of Political Studies in Paris. She was awarded the 2011 Health Freedom Hero Award by the National Health Federation for her work on health freedom and safety. Her legislative work has led to confronting major corporate interests on matters of transparency and regulatory capture, and public protections.

Breor, Mr. Scott

Email: scott.breor@hq.dhs.gov

Scott Breor serves as the Director for the Protective Security Coordination Division (PSCD) within the Office of Infrastructure Protection at the Department of Homeland Security (DHS). Mr. Breor is responsible for the day-to-day business and management of infrastructure security specialists known as Protective Security Advisors (PSAs) and the Office for Bombing Prevention. These programs provide communities with effective vulnerability and security gap analyses, training, support to Special Events, and improvised explosive device (IED) awareness and risk mitigation training. Mr. Breor has over thirty years of military and senior executive experience in the United States government. Prior to DHS, Mr. Breor was a Naval Aviator and had served as the Senior Policy Advisor for the Chief of Naval Operations on all Homeland Security matters. While assigned to the Office of the Chief of Naval Operations (CNO) he led a division that supported the CNO on key warfare and Homeland Security and Defense policy decisions, which included: interagency coordination, incident management, and Department of Homeland Security/Department of Defense integration. In addition, he directed and managed the Navy Joint Doctrine program. For his work for the CNO and his efforts following the tragic events of September 11, 2001 at the Pentagon, he was awarded the Legion of Merit. As a Naval Aviator he supported operations in Iceland, Greenland, Adriatic, Mediterranean, Azores, and South America. Mr. Breor was a Senior Executive Fellow at the John F. Kennedy School of Government, Harvard University. He received a

Masters of Arts in National Security Studies and in Homeland Security and Defense from the Naval Post Graduate School, and received a Masters of Business Administration from the University of Oklahoma. In addition, he earned a Bachelor of Science in Physics from The Citadel.

Briggs, Mr. Kevin

Email: kevin.briggs@hq.dhs.gov

Mr. Briggs is the DHS Team Chief, NCCIC (National Cyber Security and Communications Integration Center) and in that capacity leads various strategic and tactical activities regarding emergency communications including the SHARES program and provided technical guidance for the EMP protection of federal civilian communications systems.

Cooper, Ambassador Henry F.

Email: hcooper@ara.com

Ambassador Henry F. Cooper is Chairman of High Frontier, Chairman Emeritus of Applied Research Associates, and a Director on the Boards of The Foundation for Resilient Societies, the London Center for Policy Research, and the EMP Task Force for Homeland and National Security. He previously served as Senior Associate of the National Institute for Public Policy and Visiting Fellow at the Heritage Foundation. At High Frontier, he is working with local, state, and federal authorities to provide effective defenses against ballistic missiles, particularly those that pose an existential EMP threat to all Americans. Since 1979, he has been appointed by the President to serve as Deputy Assistant Secretary of the Air Force with oversight responsibility for Air Force strategic and space systems (1979-81); Assistant Director of the U.S. Arms Control and Disarmament Agency, backstopping all bilateral negotiations with the Soviet Union (1983-85); Ambassador and Chief U.S. Negotiator at the Geneva Defense and Space Talks with the USSR (1985-90); and Director of the Strategic Defense Initiative (SDI, 1990-93). He served on numerous technical working groups and high level advisory boards--including the Defense Science Board, the Air Force Scientific Advisory Board, U.S. Strategic Command's Strategic Advisory Group, the Defense Nuclear Agency's Scientific Advisory Group on Effects, and the Congressional Commission to Assess the U.S. Government's Organization and Programs to Combat the Proliferation of Weapons of Mass Destruction. He received the Defense Department's Distinguished Public Service Medal, the Defense Special Weapons Agency Lifetime Achievement Award, the U.S. Missile Defense Agency's Ronald Reagan Award, the U.S. Navy Aegis BMD Pathfinder Award, and Clemson University's Distinguished Service Medal. Ambassador Cooper taught at Clemson University and worked at Bell Telephone Labs, the Air Force Weapons Lab, R&D Associates, and JAYCOR. He holds BS and MS degrees from Clemson University and a Ph.D. from New York University, all in Mechanical Engineering.

deLamare, Mr. Michael

Email: MADELAMA@bechtel.com

Michael deLamare has practiced systems engineering for 33 years. His experiences include design and manufacture of satellite components (Hughes Aircraft); development, modification and sustainment of Intercontinental Ballistic Missiles (TRW); and requirements, engineering process development and configuration management of a nuclear waste vitrification plant and other complex, one-of-a-kind facilities (Bechtel National Inc). He currently serves as the corporate Systems Engineering Manager for Bechtel's

Nuclear, Security and Environmental business unit. His responsibilities include defining systems engineering processes for the business line; developing the SE culture and training; and establishing and overseeing SE implementation on projects for the business line. Mr. deLamare holds a B.S. in Physics from the University of California at Irvine, an M.S. in Systems Engineering from Johns Hopkins University, and is certified as an Expert Systems Engineering Professional (ESEP) by the International Council On Systems Engineering (INCOSE). Within INCOSE, Mr. deLamare represents Bechtel on the Corporate Advisory Board, chairs the Critical Infrastructure Protection and Recovery (CIPR) working group, and is the co-chair on the Infrastructure working group.

Faxvog, Dr. Frederick R.

Email: FFaxvog@emprimus.com

Frederick R. Faxvog, is a Senior Program Director and member of the Emprimus technical staff. He received his BS, MS, and PhD in Electrical Engineering from the University of Minnesota. He currently leads the development and marketing of the SolidGround™ transformer neutral grounding system for transformers and power grid protection against induced quasi- DC currents from either geomagnetic disturbances (GMD) or nuclear EMP weapons.

Dr. Faxvog has over 40 years of engineering, research and marketing experience with General Motors, Honeywell and Emprimus. He is the author of thirty-four publications in refereed journals and twenty three patents.

Gardner, Mr. Gary L.

Email:gary.gardner@totaleaccess.com

Gary Gardner is the President & CEO of TOTALeACCESS a security consulting firm. He brings over forty years of investigative, security, protection, intelligence, analysis, forensic, technology, consulting, management and teaching experience to the national and international business and law enforcement communities. His career spanned more than thirty years of diverse service with the Federal Bureau of Investigation in various capacities. His experience covers numerous areas of concern to today's business and sports world. While serving in the FBI, he held many positions in records management, laboratory technician, a Special Agent, investigating a wide variety of Federal offenses. In the New York City Office, he served as Supervisor/Co-Commander of both the renowned Bank Robbery and Terrorism Task Forces. Then, he was assigned to FBI Headquarters where he oversaw cutting-edge investigative information technology and crisis management. He led the FBI's investigative computer support systems, directed investigative support and training for crisis situations, special events and major cases. He also designed, managed development and directed the FBI's Law Enforcement OnLine (LEO), an international interactive computer communications capability and information service, used exclusively for the law enforcement/criminal justice/public safety community. He served as a consultant to *Lowe's Motor Speedway* providing Security & Safety Coordination. He then he served as NASCAR's Director of Security, overseeing all aspects of NASCAR's corporate and event security, encompassing investigations; physical and cyber security; international security, employment screening and executive protection, implementing many new collaborative risk mitigation approaches. He collaborated with the US Department of Homeland Security to establish the first evacuation guide for large sporting events. Currently, Mr. Gardner established TOTALeACCESS specializing in state-of-the–art security, risk/crisis management, intelligence collection/analysis, simulation/modeling, business continuity/disaster recovery, information security/cybercrime, biometrics, training, tracking and e-commerce solutions. Mr. Gardner is a decorated U.S. Army Ranger veteran with service in Vietnam and a graduate of the American University in Washington, D.C. Mr. Gardner previously served as the President of the Charlotte InfraGard Chapter; a public/private sector organization for the protection of our Nation's Critical Infrastructures, then on the national InfraGard Board of Directors and now is Chairman of the Board. He is also a member of the National Center for Spectator Sports Security and Safety (NCS4) Advisory Board and their Conference & Summit Director.

Geary, Mr. David

Email: dgeary@poweranalytics.com

David Geary is the Vice President of Power Delivery at Power Analytics. Dave joined the company in 2015 when his company, dcFUSION, merged with Power Analytics. Dave is a licensed professional electrical engineer with a distinguished career spanning more than 30 years. Throughout his career Dave has brought innovation and leadership to a variety of electrical engineering projects in many markets including: government and private infrastructure engineering, design, construction, and operations. Dave is considered one of the pioneers and experts in the new evolution of higher voltage direct current (dc) power distribution for telecom, data centers, and dc microgrids—specifically, for the evolution of alternative energy integration with energy storage in dc microgrids. In his role as co-founder, principal engineer, at dcFUSION, Dave provided technical engineering innovation, vision and leadership for the use and development of alternative energy solutions, as well as higher voltage (380/400V dc) electrical system topologies for data centers and technology facilities.

Hill, Mr. Terry

Email: tjh3@me.com

Terry Hill trained as an elementary school teacher for two years in Newcastle, Australia, 1960-61. Taught for three years in the NSW public school system, received certification and then spent a year teaching in the British Columbia system in Canada.

January 1, 1967 Terry entered the States, making my way to DC where he got a job in the Australian Embassy as a file clerk. While there he began flying lessons and in 1968 returned to GW to complete a 4 year degree in Education so he could join the US Air Force as a pilot. Subsequently, in order finish his degree he got a job in the IMF and became the beneficiary of it's tuition repayment scheme, graduating GW in 1972 with a MBA in International Business. Terry then spent the next 30 years at the IMF in various capacities with the majority of the time in the Budget and Planning section. He served on the boards of PHIUS and the Emerge Alliance and I represent PHIUS as member of the High Performance Building Congressional Caucus Coalition (HPBCCC) and DC Nexus. Upon retirement, he undertook the rebuilding of his house in Alexandria, VA and, as a direct consequence of that, was introduced to the Passive House concept. In 2008, he took the Passive House training. He was subsequently able to introduce the Passive House principles to many sections of government in DC. Terry attended his first EMPSIG in 2013 and immediately made the connection between Passive House and the central role it could play in dramatically reducing load on the utility grid. When coupled with direct current (DC) further utility load reductions were possible and, with the addition of DC microgrids, the potential to mitigate many of the concerns of the EMPSIG readily became apparent.

Lasky, Ms. Mary

Email: mary.lasky@jhuapl.edu

Ms. Lasky is a Certified Business Continuity Professional (CBCP). She has been the Program Manager for Business Continuity Planning for the Johns Hopkins University Applied Physics Laboratory (JHU/APL), and also coordinated the APL Incident Command System Team. She is the immediate Past President of the Community Emergency Response Network Inc. (CERN) in Howard County, Maryland as well as the immediate Past President of the Central Maryland Chapter of the Association of Contingency Planners (ACP). She is a member of InfraGard and the vice chair of InfraGard EMP-SIG. She is a member of the FEMA Nuclear - Radiation Communications Working Group. She has held a variety of supervisory positions in Information Technology and in business services. For many years, she has been on the adjunct faculty of the Johns Hopkins University Whiting School of Engineering, teaching in the graduate degree program in Technical Management. Ms. Lasky is the President of the Board of Directors of Grassroots Crisis Intervention Center in Howard County, MD. She served on the Finance Committee for Leadership Howard County and is co-chair of the Steering Committee for the Leadership Premier Program. Her consulting work has included helping non-profit organizations create and implement their business continuity plans.

Leatherman, Mr. Brett E. CISSP, GISP, GSEC

Email: Brett.Leatherman@ic.fbi.gov

Brett Leatherman is Assistant Section Chief of FBI Cyber Division, Operational Engagement Section, managing FBI engagement initiatives to private sector agencies, other government agencies, and critical infrastructure in the realm of both national security and criminal based cyber threats. In this role, Mr. Leatherman manages teams responsible for key partner engagement, FBI field office driven engagement, and private sector outreach programs and initiatives at the national level.

Prior to this role, Mr. Leatherman worked as a Supervisory Special Agent in an FBI national security unit overseeing investigation and intelligence work related to cyber actors targeting U.S. companies and interests using sophisticated cyber network exploitation tools. Mr. Leatherman led FBI efforts to identify and pursue these actors using all FBI authorities. Mr. Leatherman has worked extensively with United States Intelligence Community partners, private sector agencies, and international law enforcement and intelligence partners to mitigate sophisticated cyber threats targeting the United States.

Mr. Leatherman entered on duty as a Special Agent with the FBI in 2003. Prior to his current assignment at FBI Headquarters, he served in the Cleveland and Detroit field offices. Mr. Leatherman has also served as an FBI pilot and hostage negotiator. Mr. Leatherman has received a number of awards over his career, including the FBI's Exceptional Performance Award, Medal of Excellence, the National Intelligence Meritorious Unit Citation, and the FBI Director's High Impact Leadership Award.

Prior to joining the FBI, Mr. Leatherman was the Director of Information Technology for a non-profit organization, and graduated from Cornerstone University with an undergraduate degree in Computer Information Systems. Mr. Leatherman holds several widely recognized information security certifications and speaks regularly on cyber threats facing the United States and its interests.

Manto, Mr. Charles

Email: cmanto@stop-EMP.com

Mr. Manto is CEO of Instant Access Networks, LLC a consulting and R&D firm that produced independently tested solutions for EMP protected micro-grids and won the competitive R&D 2016 contract from DTRA for EMP-protected microgrids. He received six patents in telecommunications and computer mass storage and EMP protection and assisted other entrepreneurs and investors with their intellectual property strategies. Developed valuation methodology accepted by the U.S. DOD, countries, and companies participating in industrial defense conversion. Facilitated due diligence of over 200 deals, managed a venture capital service, a revolving loan fund, an economic development corporation, a computer mass storage manufacturer, and broadband CLEC. Mr. Manto has also founded and leads InfraGard National's EMP SIG. He received his B.A. and M.A. from the University of IL at Urbana/Champaign.

Newman, Major General Robert Newman (USAF, Ret.)

Email: robertnewmanjr1@gmail.com

Major General Robert B. Newman, Jr., USAF (retired), has over thirty years of business, military, and homeland security expertise in both government and private sectors in the fields of infrastructure protection, financial services, energy, and information security.

As Senior Vice President and Director of Strategic Partnerships for Sera-Brynn, Mr. Newman assists the CEO with the company's strategic planning effort and is responsible for the strategic marketing and industry awareness programs. He works with professionals both in and out of the information security field to develop new partnerships to expand the company's client base and with other senior leaders in the company to develop new products and analyze their impact on the company's clients and the market.

Long having recognized the rapidly evolving cyber challenges that face our country, Mr. Newman worked as a consultant prior to joining Sera-Brynn with both public and private entities to help secure our nation's cyber, financial, and energy infrastructures and to ensure their reliability for the American economy. He has also worked with the modeling and simulation industry in Virginia to identify opportunities and develop a strategic way ahead to ensure the industry's continued success across the business spectrum in the commonwealth.

Mr. Newman graduated as a distinguished aerospace graduate from the Virginia Military Institute. While a cadet at VMI, Newman was regimental commander of the Corps of Cadets, president of the honor court, and co-captain of the soccer team being twice named to the Virginia All-State soccer team. He received a Master of Arts in Management and Public Administration from Webster University.

Following graduation from VMI, Mr. Newman attended US Air Force pilot training at Reese AFB, TX. Following graduation, he was assigned to the 23 Tactical Fighter Wing flying the A-7D Corsair II fighter aircraft. During his active duty Air Force career, he was an instructor pilot, a flight examiner, and was named to an 8-man USAF aerial gunnery team that flew against 10 Royal Air Force teams and won the RAF annual tactical fighter competition.

Mr. Newman separated from active duty and returned to Virginia where he began a career as an institutional bond trader working for both Wall Street and regional brokerage firms. He also joined the Virginia Air National Guard flying the A-7D fighter later transitioning to the F-16C Fighting Falcon.

Following the attacks on September 11, 2001, Mr. Newman was mobilized and assigned to the National Guard Bureau in Washington where he helped stand up homeland security missions assigned to the Guard, including the infrastructure protection mission, and directed the standup of the National Guard's initial element at US Northern Command. He was later promoted to brigadier general serving as the deputy J3/4 for operations and logistics at US Joint Forces Command.

Following a two year tour of duty at the National Guard Bureau, Newman returned to Richmond where he served in the cabinet of Governor Mark Warner as deputy homeland security advisor continuing work to secure Virginia's critical infrastructure and directing homeland security efforts that established the Virginia fusion center.

In 2006, Governor Tim Kaine selected Newman to be the Adjutant General of Virginia commanding 10,000 Virginia Army and Air Guardsmen and promoting him to the rank of major general. Newman served on many national committees while serving as adjutant general including the Reserve Forces Policy Board and, drawing on his business experience, as a member of the Board of Directors of the Army and Air Force Exchange Services.

Mr. Newman continues to be active in the community having served on many local and national boards to include the Board of Advisors for Linxx Global Solutions, the Board of Advisors of the Center for American Studies at Christopher Newport University, the Board of Directors of the Virginia War Memorial Educational Foundation, and the Board of Directors of the Congressional Award.

Nordling, Mr. Gale K.

Email: gnordling@emprimus.com

Mr. Nordling, President and CEO of Emprimus, has 35 years experience as an engineer, risk manager, risk management and insurance consultant, and expert witness. He has been involved with the preparation, negotiation, settlement, litigation, arbitration, mediation, and insurance coverage of over $500 million of claims and contract disputes for engineers, contractors, suppliers and owners including universities, hospitals, states, airlines, casinos, and utilities.

Mr. Nordling has been employed by a nuclear utility, a disaster recovery company, national construction company, and international risk management firm. Mr. Nordling served on a national committee to create a national pooled inventory and management of safety related spare equipment for all nuclear plants. The disaster recovery company included some of the largest upper Midwest companies including ConAgra, Cargill, Northwest Airlines, National Car, Gelco, Minnegasco, Northern States Power, and various insurance and banking institutions.

Pappas, Mr. Steve

Email: scpappas@comcast.net

Steve Pappas has over 30 years of experience on active duty in the U.S. Army, higher education at Indiana University and in system safety and security positions in the public and private sectors. Steve Pappas is a partner with 4 Star a homeland security consulting firm based in Indianapolis, Indiana. He served as the G 3 – Operations, Training and Plans officer and the G-4 Logistics Officer for the Indiana Guard Reserve, Indiana's State Defense Force from 2008 to 2012, retiring as a Colonel (O-6). He has written a variety of USEPA, FEMA and OSHA regulatory compliant programs and has lectured on security and safety related topics at the state and national level. In a DHS funded grant program, he revised several Indiana county level Comprehensive Emergency Management Plans and their state agency emergency support function annexes. During this same period, he developed and facilitated a number of Homeland Security Exercise Evaluation Programs (HSEEP) from workshops to full scale exercises. From 2009 through 2012, he designed and implemented HSEEP compliant exercises for Indiana's State Defense Force. In 2008, he was awarded grant funding from the United States Department of Homeland Security and the FBI to conduct security related workshops for several InfraGard chapters. For the past six years Steve Pappas has conducted NIMS workshops, HSEEP table top scenario based exercises and functional exercises for the public and private sectors. Steve Pappas has extensive experience with local government emergency management agencies. He served both as a Deputy Director for the Johnson County, Indiana Emergency Management Agency and concurrently the Local Emergency Planning Committee chair for over eight years.

Pi, Mr. John, M.D.

Email: John.Pi@ic.fbi.gov

Dr. Pi graduated from Columbia University School of Engineering with a BS Degree in Computer Science and subsequently worked at Poughkeepsie IBM as a project manager for IBM mainframe security and resource management applications. After IBM, Dr. Pi proceeded to attend New York University School of Medicine and completed Emergency Medicine Residency at UCLA School of Medicine. He worked as Assistant Associate Clinical Professor at UCLA and served as clinical faculty, Emergency Department Medical Director, and Chief of Staff at various academic and community hospitals.

In 1997 Dr. Pi joined the FBI as a Special Agent assigned to Los Angeles Division for 14 years including five years in Asian Organized Crimes/drug Trafficking, five years in National Security Cyber Intrusion, and four years in crticial incident response. He also served as the Medical Director for Los Angeles SWAT Tactical Medical Program, a tactical crew chief for CIRG Field Helicopter program, testified as court experts, conducted undercover operations as a certified undercover agent, deployed to special events, such as 911 World Trade Center Bombing, Africa Embassy Bombing, Juarez Mexico Mass Grave Investigation, and Beijing Olympic. In 2004, Dr. Pi received the FBI Director's Award for Special Achievement for his accomplishments to include being the case agent for the landmark prosecution Title 18 U.S.C. § 842(p)(2)(A) for a domestic terrorism case.

In 2010, Dr. Pi transferred from LA to FBI HQ at Quantico as a team leader for the FBI CIRG Render Safe Mission which is the national asset for the counterterrorism response for all domestic WMD incidents to include biological, chemical, radiological, and nuclear weapons and IED. In 2012, Dr. Pi joined the FBI Cyber Division Operation Section as a Threat Manager for national security cyber intrusions. In 2014, Dr. Pi became the Unit Chief for Cyber Division NIPU I which has the national-level management responsibility for the InfraGard program.

Popik, Mr. Thomas (Tom)

Email: thomasp@resilientsocieties.org

Thomas Popik is chairman of the Foundation for Resilient Societies, a nonprofit group dedicated to the protection of critical infrastructure against infrequently occurring natural and manmade disasters. He is principal author of a Petition for Rulemaking submitted to the Nuclear Regulatory Commission that would require backup power sources for spent fuel pools at nuclear power plants. Previously, as a U.S. Air Force officer, Mr. Popik investigated unattended power systems for remote military installations. Mr. Popik graduated from MIT with a B.S. in mechanical engineering and from Harvard Business School with an M.B.A.

Pry, Dr. Peter Vincent

Email: Peterpry@verizon.net

Dr. Peter Vincent Pry is Chief of Staff of the newly re-established Congressional Electromagnetic Pulse (EMP) Commission, and Executive Director of the Task Force on National and Homeland Security, a Congressional Advisory Board dedicated to achieving protection of the United States from electromagnetic pulse (EMP), Cyber Warfare, mass destruction terrorism and other threats to civilian critical infrastructures, on an accelerated basis. Dr. Pry also is Director of the United States Nuclear Strategy Forum, an advisory board to Congress on policies to counter Weapons of Mass Destruction. Foreign governments, including the United Kingdom, Israel, Canada, and Kazakhstan consult with Dr. Pry on EMP, Cyber, and other strategic threats.

Dr. Pry served on the staffs of the Congressional Commission on the Strategic Posture of the United States (2008-2009); the Commission on the New Strategic Posture of the United States (2006-2008); and the Commission to Assess the Threat to the United States from Electromagnetic Pulse (EMP) Attack (2001-2008).

Dr. Pry served as Professional Staff on the House Armed Services Committee (HASC) of the U.S. Congress, with portfolios in nuclear strategy, WMD, Russia, China, NATO, the Middle East, Intelligence, and Terrorism (1995-2001). While serving on the HASC, Dr. Pry was chief advisor to the Vice Chairman of the House Armed Services Committee and the Vice Chairman of the House Homeland Security Committee, and to the Chairman of the Terrorism Panel. Dr. Pry played a key role: running hearings in Congress that warned terrorists and rogue states could pose EMP and Cyber threats, establishing the Congressional EMP Commission, helping the Commission develop plans to protect the United States from EMP and Cyber Warfare, and working closely with senior scientists and the nation's top experts on critical infrastructures, EMP and Cyber Warfare.

Dr. Pry was an Intelligence Officer with the Central Intelligence Agency responsible for analyzing Soviet and Russian nuclear strategy, operational plans, military doctrine, threat perceptions, and developing U.S. paradigms for strategic warning (1985-1995). He also served as a Verification Analyst at the U.S. Arms Control and Disarmament Agency responsible for assessing Soviet arms control treaty compliance (1984-1985).

Dr. Pry has written numerous books on national security issues, including *Blackout Wars*; *Apocalypse Unknown: The Struggle To Protect America From An Electromagnetic Pulse Catastrophe*; *Electric Armageddon: Civil-Military Preparedness For An Electromagnetic Pulse Catastrophe*; *War Scare: Russia and America on the Nuclear Brink*; *Nuclear Wars: Exchanges and Outcomes*; *The Strategic Nuclear Balance: And Why It Matters*; and *Israel's Nuclear Arsenal*. Dr. Pry often appears on TV and radio as an expert on national security issues. The BBC made his book *War Scare* into a two-hour TV documentary *Soviet War Scare 1983* and his book *Electric Armageddon* was the basis for another TV documentary *Electronic Armageddon* made by the National Geographic.

Sklar, Mr. Scott

Email: solarsklar@aol.com

Scott Sklar runs a clean energy technology optimization and strategic policy firm, The Stella Group, Ltd, since 2000, which facilitates clean distributed energy. specializing on blending technologies and financing for projects, assisting companies to scale-up market penetration, and facilitating federal and state polices to expand markets.

Previously, Sklar served as Executive Director for 15 years of two national trade associations concurrently, the Solar Energy Industries Association and the National BioEnergy Industries Association. Prior of running trade associations, Sklar was Political Director of The Solar Lobby for two years, —a renewable energy advocacy group founded by the big nine US environmental organizations. And for three years previous to joining the advocacy organization, served as Washington Director for two years and Acting RD&D Director for one year of the National Center for Appropriate Technology (NCAT), a federally-funded applied technology institution. Sklar started his energy career serving as a military and energy aide to Senator Jacob K Javits (NY) on his Washington personal and Committee staff for nine years, and cofounded the Congressional Solar caucus in the mid-1970s.

He serves on the Boards of Directors of three national non-profits: Business Council for Sustainable Energy (climate change), and The Solar Foundation. He also serves as Steering Committee Chairman of the Sustainable Energy Coalition. Sklar has coauthored two books, *The Forbidden Fuel: A History of Power Alcohol,* published in 1985 which is updated and was updated and re-released in 2010 by University of Nebraska Press, and a *Consumer Guide to Solar Energy* first published in 1998 and is in its third publishing.

Sklar is an Adjunct Professor at The George Washington University teaching two unique interdisciplinary sustainable energy course, and is an Affiliate Professor at CATIE, a United Nations-founded sustainable development university based in Costa Rica. Scott Sklar was re-appointed in September 2016 for a two year term on the US Department of Commerce Renewable Energy & Energy Efficiency Advisory Committee, where he previously served as its Chairman. He can be reached at 202-347-2214 or at solarsklar@aol.com.

Stockton, Dr. Paul N.

Email: pstockton@cloudpeak.sonecon.com

Dr. Paul N. Stockton is Managing Director of Sonecon, LLC, and directs the firm's infrastructure resilience and security-related practice. Before joining Sonecon, he served as the Assistant Secretary of Defense for Homeland Defense and Americas' Security Affairs (May 2009-January 2013). In that position, Dr. Stockton oversaw the Defense Critical Infrastructure Protection program. He led the development of the Department's *Mission Assurance Strategy* (2012), which provides analytic methodologies and policy initiatives to address DOD's dependence on civilian-owned critical infrastructure. Dr. Stockton also served as the principal civilian advisor to the Secretary of Defense on the use of the military to support civil authorities following catastrophic events such as Superstorm Sandy, including infrastructure restoration operations. Dr. Stockton was twice awarded the Department of Defense Medal for Distinguished Public Service, DOD's highest civilian award, and received the Department of Homeland Security's Distinguished Public Service medal. Dr. Stockton holds a Ph.D. from Harvard University and a B.A. from Dartmouth College. He is widely published in the area of critical infrastructure resilience, including: *Superstorm Sandy: Implications For Designing A Post-Cyber Attack Power Restoration System,* Johns Hopkins Applied Physics Laboratory, 2016. http://www.jhuapl.edu/ourwork/nsa/papers/PostCyberAttack.pdf

Volandt, Mr. Stephen

Email: Stephen Volandt svolandt@aurorosinc.com

Stephen Volandt is Vice President of Auroros, Inc., a contracting and management-consulting firm based in Raleigh, NC. He specializes in successfully connecting strategic purpose, risk management, decision-making, enterprise project portfolio management, operational user requirements and the technology that supports them. He has been instrumental in the establishment of governance structures for global and nation-wide organizations. Mr. Volandt co-authored the DoD CIO Executive board governance charter and participated in establishing enterprise-wide portfolio rationalization, harmonization, and transformation governance for the DoD technology portfolio. Mr. Volandt served as the lead architect for transforming the multi-billion-dollar United States Marine Corps business enterprise to better support combat operations and readiness cycles. He provided policy, operations modeling, IT and communications modeling, planning, and budget justification for a global US Army weapon of mass destruction response capability; and was a principal operations planner and architect for joint US Army and National Guard response to smuggled nuclear weapon ground burst terrorism in the homeland. Mr. Volandt has also supported the Joint Requirements and Integration Office (JR&IO), NRO, and served as the governance team leader for the FBI's CJIS Division. Other areas of expertise include cyber business-risk assessments, technology portfolio transformation management, disaster recovery planning and operations, and military operations. Prior to transitioning to management and technology consulting, Mr. Volandt performed or managed thousands of environmental assessments and remediation projects, to include radioactive and complexly contaminated industrial and military SUPERFUND sites, and supported state government and FEMA during several hurricane recoveries. He is a former US Marine Corps reserve officer who served at the infantry battalion, brigade, expeditionary force, and headquarters levels; specializing in readiness, decision support, exercise management, logistics, sustainment, and global operations in austere remote environments. He graduated from The Citadel with a Bachelor of Arts in Mathematics. He currently volunteers as the 2nd Vice President for the Eastern North Carolina InfraGard Chapter, as the Deputy Director, InfraGard SE Region EMP-SIG, and as the National InfraGard EMP-SIG Administrative Officer. Mr. Volandt authored the exercise scenario, exercise process, provided the maturity model for the 2015 EMP SIG annual workshop and conference. His current passion is the design, funding, and creation of resilient communities.

Walker, Mr. Loren Mark

Email: lmw107@bct-llc.com

Mark Walker was the BCT VP, Systems Engineering Programs (retired). He is a co-lead for the INCOSE Critical Infrastructure Protection & Recovery (CIPR) WG. His degrees include a BS Electrical Engineering (Bucknell) and MS Systems Management (USC) and is an INCOSE Certified Expert Systems Engineering Professional (ESEP). Mr. Walker has over 49 years of systems engineering and leadership positions in the USAF, DOD Agencies and Contractor communities (BCT LLC, Lockheed, BAH and TASC). He helped develop & has taught Systems Engineer Courses and led SE working group since 2000. His experience includes several systems design, development, integration/testing and deployments and has been a lead systems engineer and architect on several multiyear System of Systems programs for over 30+ years. He has held leadership positions on many high priority/visibility programs and organizations. He is a leader in SE development concepts, Model Based Systems Engineering (MBSE) processes/methods, DoD architecture/design and their implementation on many customer RF & IT systems developments. He has published numerous articles on systems engineering and architecture, presented papers at INCOSE and other Symposia, Conferences, Chapter meetings and customer conferences. He also helped found and was President twice (1994 and 2000) of the INCOSE Chesapeake Chapter and the INCOSE Object Oriented Systems Engineering Method (OOSEM) WG since 2000.

Weiss, Mr. Jeff

Email: jeff@distributedsun.com

Jeff Weiss is co-founder and Managing Director of Distributed Sun's Solar Energy Investment Companies (SEIC's). Mr. Weiss leads capital formation activities for SEIC's, the entities that own and operate the company's solar assets. He also plays an active management and oversight role at D-SUN. Mr. Weiss has founded, managed, and led many companies as General Manager, CFO, CMO, Board Member and venture investor. Among them are Trust Strategy Group, a $10MM strategic intelligence firm, Picture Network International (sold to Kodak in 1997), CDx (Certificate of Deposit Exchange), and Vista Information Technologies (a $100MM network services firm).

Weiss, Mr. Joseph

Email: joeweiss16@yahoo.com

Joseph Weiss is an industry expert on control systems and electronic security of control systems, with more than 40 years of experience in the energy industry. Mr. Weiss spent more than 14 years at the Electric Power Research Institute (EPRI) where he led a variety of programs including the Nuclear Plant Instrumentation and Diagnostics Program, the Fossil Plant Instrumentation & Controls Program, the Y2K Embedded Systems Program and, the cyber security for digital control systems. As Technical Manager, Enterprise Infrastructure Security (EIS) Program, he provided technical and outreach leadership for the energy industry's critical infrastructure protection (CIP) program. He was responsible for developing many utility industry security primers and implementation guidelines. He was also the EPRI Exploratory Research lead on instrumentation, controls, and communications. Mr. Weiss serves as a member of numerous organizations related to control system security. These include the North American Electric Reliability Corporation (NERC) Control Systems Security Working Group (CSSWG), the International Electrotechnical Commission (IEC) Technical Committee (TC) 57 Working Group 15 - Data and Communication Security, the Process Controls Security Requirements Forum, CIGRÉ WG D2.22 - Treatment of Information Security for Electric Power Utilities (EPUs), and other industry working groups. He served as the Task Force Lead for review of information security impacts on IEEE standards. He is also a Director on ISA's Standards and Practices Board. He has provided oral and written testimony to three House subcommittees, one Senate Committee, and a formal statement for the record to another House Committee. He has also responded to numerous Government Accountability Office (GAO) information requests on cyber security and Smart Grid issues. He is also an invited speaker at many industry and vendor user group security conferences, has chaired numerous panel sessions on control system security, and is often quoted throughout the industry. He has published over 80 papers on instrumentation, controls, and diagnostics including chapters on cyber security for Electric Power Substations Engineering and Securing Water and Wastewater Systems. He coauthored Cyber Security Policy Guidebook and authored Protecting Industrial Control Systems from Electronic Threats. He supported MITRE and NIST in extending NIST SP800-53 to include control systems and the development of NIST SP800-82. He was tasked to write the White Paper on Industrial Control Systems Security for the Center for Strategic and International Studies Blue Ribbon Panel preparing cyber security recommendations for the Obama administration. In February 2016, Mr. Weiss gave the keynote to the National

Academy of Science, Engineering, and Medicine on control system cyber security. Mr. Weiss has conducted SCADA, substation, plant control system, and water systems vulnerability and risk assessments and conducted short courses on control system security. He has amassed a database of more than 750 actual control system cyber incidents. He is a member of Transportation Safety Board Committee on Cyber Security for Mass Transit. He also established the annual Industrial Control System (ICS) Cyber Security Conference. Mr. Weiss has received numerous industry awards, including the EPRI Presidents Award (2002) and is an ISA Fellow, Managing Director of ISA Fossil Plant Standards, ISA Nuclear Plant Standards, ISA Industrial Automation and Control System Security (ISA99), a Ponemon Institute Fellow, and an IEEE Senior Member. He has been identified as a Smart Grid Pioneer by Smart Grid Today. He is a Voting Member of the TC65 TAG and a US Expert to TC65 WG10, Security for industrial process measurement and control – network and system security and IEC TC45A Nuclear Plant Cyber Security. He has two patents on instrumentation and control systems, is a registered professional engineer in the State of California, a Certified Information Security Manager (CISM) and Certified in Risk and Information Systems Control (CRISC).

Ambassador R. James Woolsey

Email: jim@woolseypartners.com

Ambassador R. James Woolsey is Chairman of the Foundation for Defense of Democracies Venture Partner with Lux Capital. He is the former Director Central Intelligence. Mr. Woolsey previously served in the U.S. Government on five different occasions, where he held Presidential appointments in two Republican and two Democratic administrations, most recently (1993-1995) as Director of Central Intelligence. From July 2002 to March 2008. Mr. Woolsey was a Vice President and officer of Booz Allen Hamilton, and then a Venture Partner with Vantage Point Venture Partners of San Bruno, California until January 2011. He was also previously a partner at the law firm of Shea & Gardner in Washington, DC, now Goodwin Procter, where he practiced for 22 years in the fields of civil litigation, arbitration, and mediation. During his 12 years of government service, in addition to heading the CIA and the Intelligence Community, Mr. Woolsey was: Ambassador to the Negotiation on Conventional Armed Forces in Europe (CFE), Vienna, 1989–1991; Under Secretary of the Navy, 1977–1979; and General Counsel to the U.S. Senate Committee on Armed Services, 1970–1973. He was also appointed by the President to serve on a part-time basis in Geneva, Switzerland, 1983–1986, as Delegate at Large to the U.S.–Soviet Strategic Arms Reduction Talks (START) and Nuclear and Space Arms Talks (NST). As an officer in the U.S. Army, he was an adviser on the U.S. Delegation to the Strategic Arms Limitation Talks (SALT I), Helsinki and Vienna, 1969–1970. Mr. Woolsey serves on a range of government, corporate, and non-profit advisory boards and chairs several, including that of the Washington firm, Executive Action LLC. He serves on the National Commission on Energy Policy. He is currently Co-Chairman of the Committee on the Present Danger. He is Chairman of the Advisory Boards of the Clean Fuels Foundation and the New Uses Council,and a Trustee of the Center for Strategic & Budgetary Assessments. Previously he was Chairman of the Executive Committee of the Board of Regents of The Smithsonian Institution, and a trustee of Stanford University. He has also been a member of The National Commission on Terrorism, 1999–2000; The Commission to Assess the Ballistic Missile Threat to the U.S. (Rumsfeld Commission), 1998; The President's Commission on Federal Ethics Law Reform, 1989; The President's Blue Ribbon Commission on Defense Management (Packard Commission), 1985–1986; and The President's Commission on Strategic Forces (Scowcroft Commission), 1983. Mr. Woolsey has served in the past as a member of boards of directors of a number of publicly and privately held companies, generally in fields related to technology and security, including Martin Marietta; British Aerospace, Inc.; Fairchild Industries; and Yurie Systems, Inc. In 2009, he was the

Annenberg Distinguished Visiting Fellow at the Hoover Institution at Stanford University and in 2010–2011 he was a Senior Fellow at Yale University's Jackson Institute for Global Affairs. Mr. Woolsey was born in Tulsa, Oklahoma, and attended Tulsa public schools, graduating from Tulsa Central High School. He received his B.A. degree from Stanford University (1963, With Great Distinction, Phi Beta Kappa), an M.A. from Oxford University (Rhodes Scholar 1963–1965), and an LL.B from Yale Law School (1968, Managing Editor of the Yale Law Journal). Mr. Woolsey is a frequent contributor of articles to major publications, and from time to time gives public speeches and media interviews on the subjects of energy, foreign affairs, defense, and intelligence. He is married to Suzanne Haley Woolsey and they have three sons, Robert, Daniel, and Benjamin.

The FBI's Roles and Perspectives on the EMP Special Interest Group and the 2015 Dupont Summit Sessions

Corresponding Video:

https://www.youtube.com/watch?v=ThAeXBLIohU&feature=em-share_video_user

Panelists:

- **Chuck Manto,** CEO, Instant Access Networks (IAN) and Founder/Infragard EMP SIG (Electromagnetic Pulse Special Interest Group) Chair
- **Jerry Bowman,** InfraGard National Members Alliance Board President
- **John Pi,** Unit Chief , Supervisory Special Agent, Federal Bureau of Investigation (FBI)
- **Eric Sporre,** Acting Deputy Assistant Director of the FBI's Cyber Division

(Charles Manto introduces the InfraGard National EMP SIG and InfraGard National Members Alliance Board President Jerry Bowman, master of ceremonies)

CHUCK MANTO: So we'd like to begin as soon as we can, now. And I'm going to introduce our emcee for the morning in just a little bit. But before I do, I just want to make certain you have the chance to grab your seats as I tell you where we are and what we're doing here today.

My name is Chuck Manto. I head up a group within InfraGard called The Electromagnetic Pulse Special Interest Group. And, as you know, InfraGard is a 501c3 organization of professionals in any critical infrastructure area. And they sign nondisclosure agreements with the FBI, and get background checks, so that we can hold trusted conversations between private sector and the public sector on security issues. And for example, we had a great meeting yesterday, where we were doing a lot of planning work taking our recent Triple Threat Power Grid Exercise book that we developed last year, and we're now beginning to make a cookbook of all the things you might need if you are a planner and want to up your game and improve your plan to account for high-impact threats that could impact the country nationwide for more than a month.

And that's what the EMP SIG does as its program. We look at any threat that could impact the entire country for a month or longer, chief among them is electromagnetic pulse because of its unique characteristics. But we also look at Space Weather. We look at cyber. And today we're going to have top leaders on the technology and policy side who will talk about all of those. A substantial theme of the day, today, will be presentations from administration officials and others on the new National Space Weather Strategy, which is significant, because for the first time in a long time, the Government is now saying, "We might not be there to rescue you at day four after a disaster. It may be Day 44, or it may be Day 404."

This is a substantial cultural shift that could make it more likely that more people will think about their resilience, their readiness, and their ability to prepare. So today we're going to have an opportunity to hear from a number of people. And you already have, if you've come in here this morning, you should have a copy of this, Planning Resilience for High Impact Threats. Some of the people who are speaking today have articles in this book, including organizations like the International Council on Systems Engineering who we'll hear from today, as well.

So, as we get started, I'd like to call up to the front the InfraGard National Members Alliance Board President Jerry Bowman, who will say a word of greeting and introduce our speakers from the FBI this morning. And he will be here to be the emcee throughout the day, so you won't have to see me as much. Thank you very much. Jerry, come on up.

[Applause]

JERRY BOWMAN: Thanks Chuck, and Good Morning. Just a couple of housekeeping notes before we get started. The acoustics here are very, very, very good. And I could hear conversations in the back of the room, even as Chuck was speaking. So we've got the networking area. Just be aware of the fact that people up front can hear what you're saying. It might be distracting to the speakers. And to the gentleman in the brown suit back there, yes your wife does still love you, okay. [laughter] So just be aware of that fact, we do have the networking area if you need to talk.

Number Two—cell phones, make sure they're on stun. [laughter] No pun intended. Actually, Chuck's got a portable EMP device. And anybody's phone that rings won't work after today. [laughter]

We do want to focus on networking. We do want to encourage you to meet with the folks around you, be provided opportunities during the day. And obviously, you can go back and talk during the sessions if you need to.

Just a note, our presenter schedules, our presenters are on fairly tight schedules, some of them have to come and go. We may change up the agenda a little bit. Don't get confused or upset if it doesn't match what you've got, because we may reshuffle things if we get out of sync too much, just out of courtesy to our presenters.

Oh, Kelly handed me a note before I came up. In the coat room back there, we found a roll of $100 dollar bills wrapped in a rubber band. You can see Kelly to claim your rubber band. [laughter]

So let's go ahead and get started. The first presenter or introduction that is going to be made is John Pi. John is one of those people that you work with every day. And, just like family and other people close to you, you tend to take for granted. And some days I think he prefers it that way, just because he doesn't want the credit for the hard work that he does. John studied Engineering and Computer Science at Columbia, where the New York Police Department recruited him out of Columbia.

Instead of going there, he was hired by IBM to write code. And just as he was about to be promoted, he was accepted into medical school, where he completed medical school and became a medical doctor. He completed his Residency at USC-LA, and then ended up working in a Cleveland Emergency Room where he was picked as the most likely to become the Chief of Staff.

However, the FBI recruited him a year later, and he went to FBI Academy. Last year, we were very fortunate to have John accept the position of Unit Chief for NIPU1 who manages the InfraGard program for The Bureau.

I can't tell you how happy we are to be working with John. The relationship is good. He's a very strong leader and a visionary. And it's hard for us some days just to keep up with his ideas. And so now, here is a man who needs no introduction, and yet insists on one, John Pi.

[Applause]

JOHN PI: Thanks Jerry. My wife just said I can't hold a steady job, so that's the—that's where I stand on my home front. Good to see everyone again. We had a great day yesterday. We worked hard, we played hard. We had to get kicked out. People were turning out the lights to kick us out at 9:30 after the dinner. But we got a lot of stuff done.

One of the things I want to emphasize is that Triple Threat Workbook. That's a great workbook. If you have not gotten a copy, please pick a copy up today. It covers three important scenarios. This is actually a framework that the EMP SIG leadership has developed over the years. It covers—first set of scenarios will be Cyber Threats with the large scale ICS SCADA compromise. the second set, it will be a natural disaster, where you have geomagnetic storm, a solar storm; and then the last part, it's a weaponized EMP, it's nation-state sponsored.

Each of the scenarios is different. Virtual world, natural disaster, a nation-state sponsored event, where you can have anywhere from a military—heavy military involvement, all the way down to a virtual environment. It's a very, very good workbook to have. It's really the full digest of all of the important things that you need to know to maintain that resiliency, to know how to do a response in this kind of, you know, large-scale disasters.

Today's program is fantastic. You have many, many great speakers who will cover across the board as a whole government approach, and also to include the local regional response, as well as the private sector. The private sector owns 90 percent of the critical infrastructure. Without you being part of the response, we cannot possibly achieve the level of response that we need.

With that being said, I would like to introduce the FBI Cyber Division Deputy Assistant Director Eric Sporre. He is in charge of all of the Operations Section in the Cyber Division. And you can use your imagination, the type of technology, the sophistication in the degree of collaboration that we have within the Cyber Division, and under his leadership in advancement of the National Defense. Eric.

[Applause]

ERIC SPORRE: Thanks and good morning. John is one of our great recruitment successes that we try to duplicate all the time. He's an extremely hard person to follow when you're coming up behind that great background. But I always feel comforted that, if I collapse on stage today, his medical training can kick in and he'll save me. So it's a great thing that he's here in the room.

As I was prepared to come and speak today, I'll tell you a little bit about my background, and I'll tell you a little bit about what I know about this threat and what I don't know about this threat. But I was sitting up in my office last night in FBI Headquarters, looking at all the traffic down below, as they were about to do the lighting of the National Christmas Tree. And I don't have a particularly good joke about it, but I thought, "What if they flipped the switch and the Christmas tree doesn't go on?" That's a big event for me, and I know that's going to be all over the news.

And, after reading all the materials that John had fed me and briefed me on regarding these threats, I'm like, you know, that's nothing. But you just think about how much we take for granted, our electrical power and the power Grid. I participated in Grid X a few weeks ago, and again, an eye-opening experience for me.

So let me just start off by saying, thanks for having me here today. Thanks to the organizers, thanks to the InfraGard Board, thanks to Chuck Manto for chairing this important Special Interest Group. But, most importantly, thanks to all of you. Thanks to you for your expertise. Thanks to you for being here and spending your time.

My current role in the Cyber Division looks at computer intrusions, right. It's all the computer intrusion threats that we face as a nation, and really throughout the world, because it's a partnership. And I've been involved in cyber since around 2003. And I tell the story often, I was in the Miami Field Office at the time. I was a white collar agent. And it was as we were developing cyber squads. They were just starting in the Bureau. We had the—what we called NIPC (National Infrastructure Protection Center), at the time.

And my boss was walking around, our Assistant Special Agent-in-Charge. And he came over to our squad and told us we were now the Cyber Squad. And, as agents do, you then start figuring it out. You started attending classes. We had some folks on the squad that had some computer background, and that's kind of why they put it there, but I didn't. I was a cop and a lawyer. And we started taking all those NIPC classes. And again, back then, they were—I think they were not FireEye, but we had courses through a group called Cytex and Red Cliff, and just now, it's the SANS courses and everything else. And you got yourself up to speed. And then it's really the recruiting efforts, as you heard about with John and others, and bringing the right people into the organization to look at these threats.

But, even being involved in cyber since 2003, I have to be honest, I didn't really have an in depth understanding of this EMP issue. And I say in depth, I still don't have an in depth understanding like all of you do, but really, an understanding of the EMP issue until my current position. So I'm working cyber, but I guess I was focused on my cases, and working my particular threats, and really hadn't thought about this or learned about this.

And, as I learned more, both through the reading and the briefing, quite frankly, I initially found it to be quite overwhelming. I mean this is huge, right? I mean this goes across so much more than I have an expertise in. It's so broad. And when you talk about high impact threats to critical infrastructure, it's tremendously important and an interesting area of work, but an area that really takes me out of my comfort zone.

So I looked through the list that John gave me, and I kind of read through some of the books. And I saw—so computer intrusions and industrial control systems. I feel like I can talk about that. I'm pretty good. I don't feel really nervous about getting up and speaking with groups and interacting. Terrorist events, nation-state man-made events, same thing. Worked in that area for a number of years.

Then I started reading about the natural disasters, never really had much, you know, I see them on the news like everybody else, but really, as a—as a citizen and not really in my professional capacity. And then I started reading about weather and space and started to go, now here is an area that I know very little. So today's agenda, I started looking through that, and talk about an intimidating agenda, a really highly qualified—and I know, FBI people shouldn't use the term "intimidated," but an intimidating schedule of really qualified, you know, highly-skilled, accomplished presenters, attendees across the board. And I was quickly overtaken, so I kind of had that initial thought of intimidation. And then I was overtaken by kind of a sense of pride, that the FBI and the InfraGard program is involved in this endeavor with all of you, whether it's my DHS colleagues, FEMA, our Congressional leadership, the White House, our private industry partners, academia—all of us in the same room, all of us talking it together.

And it was clear that this issue has really done that. And, under the leadership of this group, you know, you see all those people in one place, sharing what you know—and just like you did with me reading this stuff, taking each of us out of our comfort zones and learning a little bit about each other's expertise, how we can integrate and do things better.

And, quite frankly, you're accomplishing many of the things we are continuing to progress on in the Cyber Division at the FBI. And that's integration. You'll hear it everywhere you go, cyber touches all our different programs. And it really does. And we continue to evolve as an organization in how do we work in an integrated fashion with our Terrorism Division, with our Counterintelligence Division, with our Criminal Division, so that all those expertises, how do we work with the other Government Agencies? How do we work with our private industry? And that's going to continue. It's getting better and better and better. And it's going to be a continual evolution, and it's one that we're committed to.

One other area that I like to hit at every opportunity when I stand up in front of folks is engagement with your local Field Offices. So when Director Comey came in, I can almost remember the actual speech, where he said he expects his Field Commanders, his Special Agents-in-Charge in the field, to have those personal relationships with the industry, academia and other stakeholders in their areas of responsibility.

I would guess, through the participation in InfraGard, that many of you already have that, and that I'm probably telling you something that you don't necessarily need to know. But, for those of you that don't, please develop those relationships, reach out to those folks, know the leadership. Much like we've seen in the last few days out in California, before we have a significant event, regardless of how it occurred or why it occurred, we'd like to know you. We'd like our leadership not to be seeing you and meeting you for the first time when a significant event occurs. And we'd like you to get to the point where the discussions are not always about big events, they're about small, incremental things, questions that you have about how we work or how we do certain things, how you can give us the information that you have. So please, if you don't have those relationships, know that we want to have those relationships with you, and that it's a goal of ours to have that.

With that, I have—I'll thank you all for the opportunity to be here today. You have a tremendous lineup of speakers which I know are going to give you a great day. But I did have one more obligation before I—before I relinquish the microphone. And that's if I could ask Janet to step up. She's going to say a few words about Mr. Manto. And I'd like to present him an award from Director Comey.

JANET: I have had the pleasure of working with Chuck over almost a year, now. And he is so significant in EMP SIG; he is EMP SIG. He's brought this group to a new height and a new energy. He has designed testing facilities. He's just brought this to such new levels. And he has done so much for this community and the knowledge and brought us all here together, which is just across the community, and bringing all of these Agencies together in this room today is just amazing to see this, to see all these people here today. It's just wonderful. And so we want to present him with this. (Presented a signed and sealed certificate of appreciation from Federal Bureau of Investigation Director James B. Comey to Charles Manto, Chair, Electro-Magnetic Pulse Special Interest Group.)

ERIC SPORRE: Chuck, on behalf of Director Comey, thanks very much for your leadership. And we look forward to a continued partnership.

[Applause]

CHUCK MANTO: Very much a surprise and another attempt to make me blush. But it's not going to work. [laughter] Thank you very much.

ERIC SPORRE: Thanks. Have a great day.

Coordinating the DHS Protective Security Advisor Program and InfraGard's EMP SIG

Corresponding Video:

https://www.youtube.com/watch?v=NUkZOFsEphk&feature=em-share_video_user

Panelists:

- Jerry Bowman, InfraGard National Members Alliance Board President
- Scott Breor, Director of the Protective Security Coordination Division, DHS

JERRY BOWMAN: Our next presenter for today is DHS PSA Program Director, Scott Breor. Scott currently serves as the Director for the Protective Security Coordination Division (PSCD) within the Office of Infrastructure Protection at the Department of Homeland Security. Mr. Breor is responsible for the day-to-day business and management of infrastructure security specialists known as Protective Security Advisors, or PSAs, and the Office for Bombing Prevention.

Prior to PSCD, Mr. Breor served as the Deputy Director for the Infrastructure Security Coordination Division (ISCD) which was responsible for the execution of the Chemical Facility Anti-Terrorism Standard that imposes comprehensive Federal security regulations for high-risk chemical facilities. Please welcome to the podium Mr. Scott Breor.

[Applause]

SCOTT BREOR: Please note I have no notes with me. Chuck said keep it short, and I promised him I would. But I have to—I'm retired Navy. I used to fly P3s. And so I have to tell a little sea story. And this morning, I decided I'm going to come out of the closet. I am, by training, a physicist. And back in 1983, my Senior research project was on the "Effects of Solar Flares on The Magnetosphere of the Earth". And this was back in 1983. So I had to do this on a typewriter. [laughter]

And I had to take the data from Pioneer satellites, I believe it was 15 and 17, you know; and back then, we thought the Pioneer satellites were the greatest thing since sliced bread. I had to take that data and manually put it on graph paper, connect that, you know, with a nice ballpoint pen, go to a copy machine. It's not like back then you had a Kinko's at every corner. And try to make nice, like, looking little graphs on standard 8-by-11 ½ paper, and then put that in a typewriter and type around it, and you know, address it as a Table. So that was painstaking. But back then, I knew what the effects of a solar flare event could do. So I have to ask Chuck, where have you been for the last 30 years? [Laughter]

So Chuck asked me, because we actually met this summer, Chuck got involved with a Regional Resiliency Assessment Program that we are doing now out at Ashburn. It's one of the projects that our program gets involved with. And he thought it'd be a good idea for us to come and speak about our Program, just so folks have awareness, but then to build on the relationship that we are definitely going to continue, not only with Chuck's group, but with all the aspects of InfraGard.

I came to this Program back in January. I have 102 Protective Security Advisors across the U.S. and one in Puerto Rico. And in your high Metropolitan areas, we have a couple PSAs there. So you can see that's not many folks to cover the U.S. And some States, like Colorado, I have one PSA. And that person is responsible not only for having the conduit to the critical infrastructure, but also the lifeline sectors. And, as we're seeing now with recent events in Paris and events prior to that, like down in Charleston, really places where people gather. So, needless to say, they have a lot on their plate.

But it's also for the representatives of InfraGard and area. You know, they're facing the same thing every day. And back in my day in DoD, yes we had the Goldwater Nichols Act, and we were supposed to be Joint when we worked with the other services. But I'll be honest, it was not a Coordination Operations. It was De-conflict Operations. And I think, in a lot of places and a lot of areas, we have been sort of like working like that. We have good relationships with some InfraGard Chapters. Others, it's more, you know, they know them.

So since this summer, you know, with Chuck's help and working with other folks at InfraGard, we're going to make it a more concerted effort going into '16 to build this relationship stronger. Whole of Government was already mentioned today. But if we have an event like EMP, it's going to take that Whole of Government, and really, that's—you know, at the end, it's the communities. And the communities need to be safe and secure. And we need to do what we can do to ensure that they know what is out there to understand and build resilience.

So with our program, we primarily focus on physical security assessments. We do have Cyber Security Analysts within DHS. We have recently started working with them to cross-train our Protective Security Advisors, so they would also be able to do cyber security assessments. So, going into '16, that's a capability and skill set that we're going to be asking of our PSAs.

And we also do training. We do training, whether it's Cyber Security or whether it's Counter-IED (improvised explosive device), or whether it's Active Shooter. And now, with any aspects of EMP that a facility would like to know, any products that come out of InfraGard, those sources and materials are known to the PSAs, and they can help facilitate that.

So I know I only had a couple minutes. I want to introduce Kyle Wolf. So Kyle, if you could stand up. Kyle, could you stand up? Oh sorry. [Laughter] So Kyle is our Protection Security Advisor for the National Capital Region. So I actually have to run to go to another engagement, and Kyle will be here all day. He can speak further about our program. But, more importantly, he's the touch point for what we offer here in NCR. So with that, thank you.

[Applause]

"Congressional Updates on EMP and High-Impact Threat Planning and Protection for Communities"

Corresponding Video:

https://www.youtube.com/watch?v=NUkZOFsEphk&feature=em-share_video_user

Panelists:

- Mr. Jerry Bowman introduces Session Chair,
- Dr. Wallace E. Boston, President and CEO, American Public University System introduces
- Congressman Roscoe Bartlett, former Congressman(Maryland)

BEGIN PART 2 OF INTRODUCTION

JERRY BOWMAN: Thanks, Scott. Introducing our next presenter is Dr. Wallace E. Boston, President and CEO of the American Public University. Dr. Boston was appointed President and CEO of American Public University System and its parent company, American Public Education Incorporated in July of 2004. He joined APUS as its Executive Vice-President and Chief Financial Officer in 2002. Now, Dr. Boston if you would come.

DR. WALLACE BOSTON: Thank you. It's my pleasure today to introduce Congressman Roscoe Bartlett. Congressman Bartlett served the State of Maryland's Sixth District for 20 years, from 1992-to-2002. A graduate of Washington Adventist University, he earned a Masters and PhD from the University of Maryland in Physiology. I'm proud to say that my parents were and still are residents of the area that he served, although their District was gerrymandered to the Eighth District in 2011.

Congressman, today I'm presenting you with the American Public University and Policy Studies Organization Sword of Leadership in recognition of your efforts to focus attention on the National Power Grid and the threats to it. No one has really worked longer and harder than you to make those of us face up to the challenges that our dependence on the Grid presents. We hope you will continue to lead efforts to make the Grid secure, and that this Crusader's Sword is recognition of your dedication to this cause. Thank you.

[Applause]

HON. ROSCOE BARTLETT: I came to the Congress when Bill Clinton came to the Presidency. And at that time, we were pretty aggressively downsizing the military. And one of the things that they did in downsizing the military was to waive chemical and EMP hardening of all of our new weapons systems. I have no idea how I first learned of EMP, but I mentioned it to the Committee. At least two-thirds of the Armed Services Committee at that time had never even heard of EMP.

So each time that one of the military would come forward, I would ask them, "Why are you waiving chemical and EMP hardening?" because in any war with a peer, and it's in all of their open literatures and all of their war games, one of the first things they would do is an aggressive EMP laydown, which denies the use of all of our non-hardened equipment, which now is all of our new equipment. The only thing that was hardened was the old, ancient equipment.

And about the year 2000, I was sufficiently concerned about this. And in working with Dr. Peter Pry, who was then a Senior Staff Member in the Armed Services Committee, we drafted legislation to set up an EMP Commission. Neither the Chairman of the Committee, nor certainly the Staff Director, wanted this legislation. So they told us, "Gee, we can't do that because it would require a sequential referral." And nothing in the Armed Services Committee Bill requires sequential referral. What that means is, it would have to go to other Committees because they have some jurisdiction there. And it would never get through those Committees in time for the very early, as you noticed, the Armed Services Committee Bill is one of the first Bills that's passed in the Congress.

So I asked them, "What other Committees might it be referred to?" And they gave me the names of those Committees. And so I went to the Chairman of those Committees, and I got a letter from each of the Chairmen saying that they waived jurisdiction of this. So I brought those letters back, and now they had to put our legislation in. And so we set up the EMP Commission. It functioned, I think, for what, six or eight years? And I'm really, really gratified now that they are reestablishing that Commission. Finally they recognize that this is really something important to us.

Not only is the Commission being reestablished, but as you all know, the two important pieces of legislation are active in the current Congress—the Grid Infrastructure Protection Act, which has passed the House, and the big H.R.8, which is a big bill, includes many things, I understand includes the Shield Act, has also passed. And now it awaits its future in the Senate.

I've been gratified that I see that a number of States are not waiting for the Federal Government, which has been pretty slow, and I think will continue to be pretty slow. And so these States are now setting up their own programs to do something meaningful in preparing for an EMP attack.

There's other legislation that relates to this, that's not—EMP will not even be mentioned in the legislation. One of these is all the Missile Defense Acts. I was opposed, and vocally opposed, to the way we were doing Ballistic Missile Defense. We were setting up systems in, what?, California? and Alaska? And these were to intercept weapons that were coming, essentially, over the Pole. The only nation today, and maybe tomorrow if China would be added to that, but today the only nation that would come over the Pole is Russia. And they have thousands of missiles. And we might if we were lucky shoot down the first, what?, half dozen or so. And so our Defenses, as far as Russia is concerned, would be fairly meaningless at the Poles.

We have been watching Iran and North Korea to see when they would have a missile—a Continental Ballistic Missile which would reach us. I always got somewhat confused by our focus on that. These people may be evil, they are not idiots. They are not going to launch over the Pole. We would know where it came from. If they launched a missile at us, it's going to come from the ocean. Probably be from a tramp steamer off our coast. And then the tramp steamer will be sunk, and there will be no fingerprints about who did it. So whose country would we bomb if the missile came from the water somewhere? And I'm sure that that's where it would come from.

Very interesting little anecdote, and I can't remember the year, I'm not very good at remembering years. But you all know the date because it was when we were involved with the war in Europe. And we had three—was it three members?, who were held by Slobodan Milosevic. And I was sitting in a hotel room in Vienna, Austria with eight other members of Congress, three members of the Russian Duma, and a personal representative of Slobodan Milosevic.

By the way, they were released to Jesse Jackson, you remember there were pictures of him holding hands in a prayer circle with Slobodan Milosevic, which really disturbed our Secretary of State at that time, Madeline Albright.

One of the members of the Russian Duma who was there had been the Ambassador to this country. And he sat there with his arms folded for three days in that hotel room, very angry. And finally he said, "If we really wanted to hurt you, with no fear of retaliation, we'd launch an SLBM, we'd detonate a nuclear weapon high above your country, and shut down your power Grid and your communications for six months or so."

The third-ranking Communist in Russia at that time was Alexander Shurbanov, a tall blonde, handsome fellow. And he smiled and said, "And if one weapon wouldn't do it we have some spares." At that time, I think it was about 10,000 spares that they had.

Currently, I don't believe we have—we have limited ability to detect a launch off our coast. And I think essentially no ability to intercept a missile that comes from off our coast. I can't imagine either North Korea or Iran launching a missile from their soil, which we would certainly identify, and we would know who it came from, when they can so easily launch a much shorter range missile from off our coast where we have little detection capability. And, as far as I know, no Defense capability.

Another major threat to our Grid, of course, in addition to EMP, is the cyber threat. Ted Koppel just out with a new book, Lights Out, which I'm sure you've all heard about. He says that both Russia and China now sufficiently hacked into our Grid, that at any time they wished, they could shut down the Grid. And countries like Iran and North Korea are gaining capability. And who knows when they might be able to do that.

Of course the third thing that could shut down our grid is a big solar storm. The granddaddy of all those was in, what, 1859, described as a Carrington Event after Dr. Carrington in England, who identified what it was. It was a near miss, what, just in '12, missed us by just a few days. And had it hit our world, and had it been active for 24 hours in the Northern Hemisphere, who knows what it might have done to civilization.

The dinosaurs all disappeared a while ago, and it's conceivable that humans might follow them with an event like that. That would have been really, really devastating, would it not? That event, by the way, was as big as the Carrington Event in 1859—would shut down all of the microelectronics in all of the Northern Hemisphere, if it hit the Northern Hemisphere and stayed there for 24 hours span of the Earth.

You know, it's very interesting. I was wondering sometime ago why the aggressive use of organisms like the Small Pox organism was not more addressed. Only two countries are supposed to have the Small Pox virus. That's the United States and Russia. I've been assured, with some certainty, that China now has it. If an enemy had that, why are they not releasing this virus in our country? It would kill, what?, a third of our people. It's very aggressive, it could easily do this. Just come here dressed in garbs so only your eyes are showing, and you're in an effective stage of it, and visit major public events, and it would just spread like wildfire through our country.

And one of the people in this discussion mentioned that maybe they weren't doing this because we were not sufficiently prepared to address this challenge. And it might come back to bite them. That's very interesting—which means that you would have less chance of an attack if you didn't prepare for the attack, because you're inviting the attack if you prepare for it.

A similar argument might be made for defense against EMP. If we are aggressively—if a potential enemy has the ability to inflict an EMP event on us, and if we are now preparing for it, would not our preparation perhaps invite an earlier attack because if they waited it might not be effective? Kind of interesting, isn't it?

Just recently, I was reading a book in which they mentioned something that I've been familiar with for a very long time. And that's Yamantau Mountain. How many of you have heard of Yamantau Mountain? Yamantau Mountain - nobody heard of Yamantau? Okay, a couple of hands back there. Yamantau Mountain is about 600 miles—about due east of Moscow. It is in the Ural Mountains. It has the largest underground nuclear secure facility in the world. It is as large as inside the beltway, the Washington Beltway. We've had two defectors from there, and it has train tracks going in two different directions, so they're prepared to move huge amounts of material around there.

I've had several briefings while I was in the Congress on Yamantau Mountain. It was started during the Age of Brezhnev. It still continues. The current leaders in Russia ski fairly close to that. It was supported by a city that does not even appear on their maps. And they have cities in Russia that don't appear on their maps. Mezhgorye it had—they can't hide it from our satellites, of course, where 60,000 people in Mezhgorye, which means about 20,000 people working on Yamantau Mountain. Today I think there's about maybe 20,000 people in Mezhgorye, which means about 6,000 people working on Yamantau. What in the heck are they doing there?

The book I was reading said that they were doing three things there. They were aggressively building biological weapons, aggressively building chemical weapons, and they were preparing a site there that they could take the cream of their civilization so that they would have a nucleus to re-populate with after a catastrophic event.

Whatever they are doing in Yamantau Mountain, it is very disquieting, is it not? It's a very secret program. It doesn't appear in any of the budgets, obviously costs a great deal of money. Imagine something as big as inside the Washington Beltway, huge nuclear secure underground facility. What in the heck is it for?

The last briefing I had was that third use, they're going to bring the cream of their civilization there to re-populate after a catastrophic event. Whatever, whatever the intended use of Yamantau Mountain is, it is somewhat disquieting, is it not?

Well, I couldn't be more gratified. When I started this journey, two-thirds of the Members of the Armed Services Committee had never heard of EMP. And now we have a room full of people committed to making sure that we do something to prepare for this. Thank you very much.

[Applause]

Private Sector Finance and Capital Markets for Resilient Community Infrastructure

Corresponding Video:

https://www.youtube.com/watch?v=5ERDXHCI5N4&feature=em-share_video_user

Panelist:

- **Mr. Jeff Weiss,** Cochairman and Managing Director, Distributed Sun

JERRY BOWMAN: Mr. Jeff Weiss is our next presenter. He serves as the Managing Director of SEICs (solar investment energy companies) at Distributed Sun, LLC. Mr. Weiss is the Cofounder and Managing Director at ASAP Ventures, LLC, in their Investment arm. Prior to forming ASAP Ventures in June of 2000, he was President of the ComQuest Group, helping e-Business and Information Technology Entrepreneurs to establish valuable market positions by providing Strategy, Marketing, and Capital Formation advice. Please welcome to the podium, Jeff Weiss.

[Applause]

JEFF WEISS: Thank you very much. Good morning. So this is the Cheerios and School of Hard Knocks part of the conversation, because it's about the folks who give the party. So, a long time ago in business, I was told that it's the customer who gives the party, so it's not the CEO. It's the people who pay. And in the energy business, I learned that, in the energy business, the person who gives the party is the person with the money, because we are in the infrastructure business, and it costs a lot of money to get started.

So when we're talking about resilience and reliability and EMP and everything we're talking about, there is a layer of it which is my role for the next 15 minutes to talk about, which is about where the money comes from and how it gets funded, which of course—which is, of course, complex.

So I have a couple of questions I want to ask, just to get a sense of folks who are here. So how many people feel they have, at home, a week's supply of food? Okay, how many people feel, at home, they have a month's supply of food? There's a very prepared group. How about six months? Very—I'm going to come to your house. How many people have a personal generator that they feel would work for a month? Okay, a smaller group. And how many people have their own renewable generation that they feel would last indefinitely? Great.

So part of the reason I'm asking that, it's from an energy and electricity point-of-view. Sitting here in 2015, while we're talking about resilience and reliability and potential risk and threats, at the same time, with the Grid and with where we get all the electricity that we use at our homes, at our businesses, in our governments, we're at the very front edge, right?, of the distributed generation revolution.

So, just like Nicholas Negroponte at MIT said 25 years ago, there was going to be what then became known as the "Negroponte Inversion," which was that anything that was wired became wireless—he was talking about the telephone—and anything that was wireless was going to become wired. And, of course, these are no longer wired. And that which we do with them is radically different than that which we thought about when it used to be what we called Plain Old Telephone Service.

Well, I will tell you that that which is wired is going to become wireless, and the opposite for electricity. What do I mean by that? What do I mean by that, because this is very meaningful from a resilience and a reliability point-of-view. What I mean by that is- virtually all of the electricity that's distributed and used everywhere is centralized today. It's managed by the Grid. It's produced centrally. And the Utility Model and the way that the Public Service Commissions regulate Utilities generally have a common structure where generation transmission and distribution are all commonly regulated and threaded together.

As we go to the distributed generation revolution and work towards what will be a more resilient set of infrastructure, we're also at the front-edge of what you would think of and can start to call Energy Democracy. So the great thing about energy democracy, right?, is that the capital markets will finance different subsets of that differently if they're unfettered and allowed to do that. And the reason that didn't happen ten years ago, 50 years ago, or 100 years ago, is because the cost and the infrastructure needed to build a large Grid was truly enormous and truly something you only want to do one time and have one group do.

So we set up, as a country, Public Service Commissions and Utilities in an oligopolus [sic] style, where they provide for our power and our generation. Now, with the revolution that's going on, from a technology and a cost point-of-view, we're at the front-edge of being able to create our own locally. And do the folks who are doing this believe that, 20 years from now, 100 percent of the Grid will be distributed? No they don't. But, if we get to 20 percent of the Grid being distributed, the impact from a resilience and reliability point-of-view, well managed, is truly enormous, is truly enormous.

Which gets back to the question that I asked as to how many people have how much food and how much power at their own home? It's the same concept. Most of you, I would wager, who raised your hands are getting nearly all your electricity from the Grid. And yet, you're investing in and spending money and time and thinking about your backup, and how you're doing it personally. And, as you're doing that personally, it's the same decision that's being made—needs to be made for schools, for hospitals, for military, for our critical infrastructure and our society.

So renewables, as a subset of that, right, have benefited—everyone knows this—from very sharp declines in component prices, right? Solar panels, which are more than two-thirds of all renewable investments, have come down 80-to-90-percent in price. Panels which used to be $5- or $10-dollars-a-watt are now 55-cents-a-watt. And inverters and everything else have come way down.

So a way of measuring that, just to give you a piece of data, is that in 2008, right?, if all of us in the room were to install a bunch of Solar on average, anywhere in the United States, if you were an investor, you would ask the question, how many cents-per-kilowatt hour do I need to be paid every month in order to recoup my investment and give me an above-hurdle return on capital? And those are the words that investors use.

Well the answer to that question, given the very high cost, just in the middle of last decade, would have been about 50-cents-a-kilowatt hour. That's a very high number. And no one pays 50-cents-a-kilowatt hour unless they're maybe in parts of Hawaii. So, by definition, you know, when I tell you that number, that every solar panel that was put up in that period of time was put up with a deep subsidy. Someone was subsidizing it, because investors are only going to invest or they're going to get a return-on-their-capital. So if you needed 50 cents—and by the way, at that time, the average retail electricity price in the United States was just over 12-½ cents-a- kilowatt hour, again, average across all 50 states.

Today, if you asked the same question, where the prices have plummeted, two things have happened. Number One, the Average Price of Retail Power has increased, right. It's just shy of 14-cents-a-kilowatt hour, average nationwide, at retail. At the same time, the average price you need as an investor, average nationwide is about 15-cents-a-kilowatt hour. So we're very close to Total-Grid Parity.

And I told some people last night, that in the middle of Texas, solar is now being banked, meaning successfully financed at an acceptable market rate of return to investors, where they're selling the power for 3.8-cents-a-kilowatt hour. 3.8-cents-a-kilowatt hour. So solar, in that part of Texas, in that example, beats natural gas.

Well, that's the leading edge data point, right?, about this inversion, about the prospect of energy democracy. Because if you can do that-- it's just unlocking a tremendous potential. So what's happening from the capital markets point-of-view? It turns out that investors are really mean. They're mean because they're very narrow-minded. Sorry, and I—they're narrow-minded in that they don't really care about what you care about. They don't care about resilience. They don't care about reliability. And they really don't care about things that are extra backup, things that are not needed every day. They only care about one thing. They care about return on their money, right?

And the capital markets, I'll tell you a secret. The capital markets have an infinite supply of money, infinite, for one thing, for long-duration, low-risk investments. Did I say anything about resilience in that sentence? Did I say anything about electricity in that sentence? No. I mean I could be talking about buying tables and chairs. As long as you have something to offer to the capital markets, that's long-duration and low-risk, there's infinite money for it.

So our challenge in Industry, our challenge in this room, our challenge in thinking about resilience, is not to force the investors to think like we think, because they'll never do that. They'll never change. They never have changed, and they won't change. Our challenge is to understand their thinking, right?, and then to understand what we do and what we need, which is different. We know what we need. But to fit it into their framework, right? And their framework is actually one that's not so far off from our framework. Because in our framework, what do we need? We need a lot of money, right?, and we need some sort of contracts attached to that money, which represents some sort of a monetary return on it. There needs to be a way to pay for it, right? People don't invest a dollar without getting a dollar back. And it's just a matter of what their cost to capital and what their time is.

The great thing about the capital markets is, they don't need it back today. They don't need it back tomorrow. They can have it back in a reasonable way over decades, right? That's okay. And by the way, if you think about it, that really is what our society created, right?, with the Incumbent Utility Business Model, right? You know, back a century or a century-and-a-half ago when we needed to start and build infrastructure from scratch, we needed an enormous amount of money. But that money, which is called and which is all about the building of the Grid, was paid for upfront, and monetized, paid back, every day and every month, by all the consumers and all the businesses and all the Government who pay a little bit in cents per kilowatt hour.

So, at the individual level, we're getting a good value proposition, right?, because we just turn on our lights, and we get a bill at the end-of-the-month, and give-or-take, we feel it's a fair trade. We don't think about the billions of dollars invested, we really just think about the fact that we have electricity. Now, what we're talking about is another layer on that, right?, which is all the negative outcomes that can happen. So, what we have to do as a group and as an industry, from an EMP and a resilience point-of-view, is we have to understand how to translate what we need, which is a lot of money, and a lot of change in infrastructure, into what the investment community wants to give us, which is money for long-duration, low-risk investments. And once we can productize and structure it in that way and figure out how to create the outcomes, we're in good shape because there's unlimited money.

It's happening in distributed generation, right? So the leading edge of this charge, the leading edge of the charge is this distributed generation revolution. And by itself, *by itself*, the advent of energy democracy, the advent of five—which is what we have now, and then 10-, and then 20-percent of the power in the country being created from distributed, resilient, low-carbon or no-carbon resources, will create a more resilient Grid, in general, *in general*. And it will be integrated and work better because the second part—the First part is Generation, and the Second part is Transmission, and the Third part is Distribution—those two parts can interoperate better when they know that it's distributed, as long as the engineer set it up to work that way, right?

So the next phase of the Grid operation is to integrate that and anticipate that we've got a lot of distributed resource, and figure out how to work on it that way—which is not unlike my knowing that Chuck just told me he's got a lot of food at home, so I don't have to have it anymore, because I know where he lives, and he's my friend, right? So I'm going to go be in his—in his friend group as soon as—as soon as this happens, as long as I can wander over to his house.

So I want to go a little bit more into what's happening from a business point-of-view and a finance market point-of-view, between Public Service Commissions, investors, utilities, and how that—and how that's working—because, there's a lot of friction right now in the market. And the friction is holding back investment.

So the great thing about our country is we, from an energy markets point-of-view, we're not the United States of America, right, we're 50 States. So we have 50 State Public Utility Commissions, and we have a couple of hundred utilities, with different regulatory structures. And everyone in the room knows that—and if one wants to understand what an opportunity is in North Carolina or New York or California or Hawaii or Idaho, you have to actually know it at the State level in order to know the answer. It's not about—there aren't Federal—there are very few Federal laws that apply to that.

So interestingly, because they're 50 States-- that's pushing us closer to choice and democracy than having only the Federal Government apply these decisions. So New York State is taking the lead, which some people know with their structure, which is called "Reforming the Energy Vision." And they talk about energy democracy a lot, and they use that word. And they are expressly, through the Public Service

Commission, deregulating and decoupling, which goes beyond deregulating, the nature of who owns and who provisions generation-versus- transmission-versus-distribution. Because those three elements require different capital, right? They have different time horizons, different forms of investment return. And, in fact, functionally and engineering and from a technology point-of-view, they're very different businesses.

We understand why, in the past, they've been managed as the same business, because there was no way around it, right? We now—and New York State is among the leaders in determining how to re-change their mix—we now have the opportunity to bring not only energy democracy but energy capitalism in and allow Investors to separately and from a competitive point-of-view, invest in, own, and provision those different elements. And the investors will decide, and they're deciding all these things, whether they want to do it for a coal plant or whether they want to do it for a hydro-plant or whether they want to do it for a nuclear plant or whether they want to do it for a solar plant. And in the fullness of time, all those things are happening and will happen.

Another word from a capital Markets point-of-view about how this is growing, and where it's going. So in 2008, just to benchmark the number, in solar, which as I told you is about two-thirds of all the renewable energy invested in America, there was $5 billion dollars in the whole United States of America invested in solar. In the capital markets, that's a really, really small number. In the incumbent energy markets, that's a small number. And obviously, $5 billion in the whole country was widely distributed to where it was, although a lot of it was in California in that year.

This year it's about $25 billion. So between 2008 and now, we've grown five times. And I just told you that the costs have come down by 80 percent, so obviously, the number of units in the $25 billion is a lot bigger than the number of units in the $5 billion. So it's growing really nicely and really astronomically.

All of that money, the $25 billion, is what I would tell you to think of as primary market investment, meaning that's the cost of building a new plant somewhere. We're going to put Solar on the roof of this building. We're going to put a 100-megawatts out in Loudoun County. Whatever it is, it's the direct cost of the installation and provisioning of the new plant. So let's just call that the Primary Market.

A lot of people probably know that the capital markets work very efficiently when they can figure out how to lower the cost-of-capital, right? So if I come to Gerry and say, "Have I got a deal for you, and here is what it looks like," and he says, "That's great. I've got a 10-percent cost of capital. Therefore, I need this return. Well, I'll have to work it into your equation." Remember, we can't change the investor's point-of-view, we have to understand their point-of-view and fit what we're doing into their equation.

By contrast, if I go to Gerry with the same-of-everything, and I say I've got—and he says, "I've got a six-percent cost of capital," Gerry is going to finance a lot more of my deals because a lot more will bank if he's got a lower cost of capital. So the trick that's going on through the distributed generation revolution is to structure and productize the investment so it's less bespoke, more productized, so that the more simplified it becomes, the more the capital markets can finance at a lower cost of capital.

The capital markets have a trick as to how they do that, it's called the secondary market. So again, most people in this room, and most people in industry, when they think of secondary markets, they don't think about any form of energy, right? You think of the mortgage-backed security market, you think about the asset-backed securities markets, you may know that their secondary markets for credit card loans, you probably know that, in the old days, if you went and got a home loan or mortgage on your house, a 30-year mortgage, the bank that you went to, you'd sit down with the bank, and they'd actually not only underwrite you, but they'd own that loan on their balance sheet.

Well that's very old fashioned, and that hasn't happened in 10 or 20 years. Today, if you go and you apply for a loan, the person who's giving you the loan and underwriting you for the loan is not the organization or institution who's going to own the loan, they're going to keep it for a very short amount of time, and they're going to sell it into the secondary markets. And that, for our society, has brought—has allowed a broadening of home ownership, huge broadening of home ownership, because it's brought down the cost of capital for residential mortgages.

So that's an example of what has also happened for a couple of trillion dollars of investments through a broader asset classes, which are called asset-backed securities. So we're just at the front-edge in energy of applying asset-backed securities, financial technology, to allow us to resell our assets in the secondary market.

What does that mean? The capital markets, as I have said three times now, love long-duration low-risk investments, nothing about energy. Well what is an energy investment? Fundamentally, energy investments are pretty nice. You put in a lot of money upfront, you get a contract with someone if they're going to pay you a long-term stream of cash flows, whether it's a lease or a power purchase agreement. But pretty much no one invests in a large energy anything, right?, without someone—without knowing what they're going to get paid for. And it's usually by contract.

Those contracts can be resold and structured as cash-flow payments. So, just like the mortgage market, right?, allows—has facilitated a movement so there are originators, and then there are secondary market holders, the energy market is at the front-edge of that. And in total, in the United States, there's been about $15 billion dollars of secondary market transactions in renewable energy.

So that's going to be pretty good. I would say we're in our very early adolescence. It's not attractive, it's not yet productized, but it's happening. And the fact that there is a forming secondary market is absolutely critical to the growth. Because I will tell you that, while we've grown in the primary market, from $5 billion-to-$25 billion, between 2008 and now, and there are tons and tons and tons of studies by large well-known companies who are saying that this is going to be a trillion-dollar market, or a $250 billion dollar market, it's going to be, you know, like Crazy Eddie said, "it's going to be big, really big". But, however big it's going to be, I will tell you, we can't get much beyond the $25 billion without a well-formed secondary market.

So the next thing to watch for, as we go from $25-to-$50-to-$100 billion dollars in annual new investment in distributed generation in the United States, and in renewable energy, whether it's distributed or centralized, is the formation of the secondary market to support and finance these transactions. And, what they do is, they allow other people to build them and invest in them. And then, those who build them and invest take their contracted amounts and resell them to other people, which is just—which is a way of getting some of their money back, right.

So again, if Gerry and I have to put $100 million dollars out of our own pocket, and he looks like he's got it, we'd probably—we'd probably want to have someone else pay us back at some point. And the good—A great way to do that, and a cool way to do that, is in fact for us to put our $100 million dollars in and resell the cash flows. Because when we do that, we still own the underlying asset, right? And we greatly leverage our capital position.

And that's happened in every other industry. What's cool about the distributed generation revolution is we're at the front-end of taking that finance technology and applying it to energy, in many cases for the first time. So this is all going to happen, and is happening around us, right? It's happening, it's happening across energy. It's happening in large part around Clean Energy. It's led by solar. It's led by wind. There's a lot of biofuel stuff going on. There's a lot of—there's just a lot of financial structuring happening which can happen once the technology and the product part is de-risked.

So, now that the technology and the product of Clean Energy is de-risked, the capital markets are now able to come in and apply finance technology, which is the way which provides the kind of booster rockets to get it to the tens of billions, and then the hundreds of billions of dollars. So that's where it is.

So what I would advise this group, as we think slightly differently about the Grid, and we think about everything that's being talked about from an EMP-, and a resilience-, and a potential disaster point-of-view, think about how to bring more money in. And the way to bring more money in is to fit in, right?, and to fit into the way the capital markets' crowds look at things. The good thing about microgrids, the good thing about distributed generation, the good thing about having backup, the good thing about having some form of alternate sourcing, is, increasingly, that's financeable on a capital markets basis.

So again, if Chuck and I were on the Board of a hospital ten years ago, and someone asked the CEO of the hospital, "What's our plan in the case of a disaster? What are we going to do if the power goes out?" Well nearly 100 percent of the hospitals in the country would have answered the question the same way. They would have said, "Well, sir/ma'am, we have a backup plan. We have invested in generators. There they are. This is what they cost. We fix them every year. We bring enough oil in. And the generators are going to take care of our problem."

What we know as a society today is that two things have happened. Number One, that's a terrible answer, right?, because the generators really are not resilient, they don't provide backup, and they don't work for very long. So it's really not an acceptable answer. It was an okay answer a decade ago, not because it was a great answer, it was the only answer. It was the only answer.

And I'm telling you, in 2015, and from now forward, is we now have Technology-at-a-cost and Infrastructure-at-a-cost that there are much better answers. And by investing and in creating local generation, and having the local generation be enough of a subset, with storage, that it can actually not only be paid for day-in-and-day-out by being a big subset of the energy that you're actually using, but then by also being there to generate the power in the case of the power going out, that's a huge win. And that is going to help drive energy democracy as we move along.

So Chuck, thanks so much for inviting me. And, good luck today for everybody.

[Applause]

The Cyber Threat to Critical Infrastructure

Corresponding Video:

https://www.youtube.com/watch?v=Uk0rGxaPTYc&feature=em-share_video_user

Panelists:

- **Paul Joyal,** former InfraGard National Board member
- **Brett Leatherman,** Assistant Chief of the FBI Cyber Division

JERRY BOWMAN: So our next presentation is "The Cyber Threat to Critical Infrastructure." I want to introduce the presenter, and then we've got some additional information that's going to be presented by Paul Joyal, who just came off of the National Board and has been a very strong leader in this area for InfraGard, and he's got a few words to say, as well.

So our next presenter is Mr. Brett Leatherman. Brett is the Assistant Section Chief for the FBI Cyber Division. He manages all FBI Public-Private Sector Engagement. He previously served as a Supervisory Special Agent in Eurasia for Cyber Operations for that Unit, managing some of the most sophisticated nation-state cyber campaigns.

Prior to the FBI, he was the Director of Information Technology. He holds undergraduate degree in Computer Information Systems and holds several Cyber Certifications. He's also an FBI Pilot and a Hostage Negotiator. So, if you want to get out early today, he's your guy to talk to.

So I'm going to go ahead and bring Paul Joyal up to do some further introductions for our speaker Brett Leatherman.

PAUL JOYAL: Thanks. Thank you, Jerry. It's a great pleasure to introduce Brett. Brett is a true professional, not only in the FBI, but within the entire U.S. Intelligence Community and is well known internationally.

I want to emphasize one point. I think what he brings to the table today is a unique understanding of the type of combined threat that we face. While Brett was Supervisory Special Agent in that Eurasian Division, he came into contact, very closely, with the nexus that we fight against, which is the criminal organization and the state-sponsored intelligence services. And how these two groups interplay, especially in Russia.

EMP and EMP-attack philosophy falls under the rubric of Information Warfare in Russian Military Doctrine. That also includes cyber. It also includes Information Ops. But they have a holistic understanding of this. The First Step as the Chinese say comes well before Military Operations. It's in this area. The EMP, the cyber operations, to cripple the will of the opponent to resist. This is the essential nature. Brett has seen these capabilities close up. You're going to see a fascinating presentation. Obviously, SCADA and the electrical network is on—is in the crosshairs of these services. It's a great pleasure to introduce Brett to you today. Thank you.

[Applause]

BRETT LEATHERMAN: Well thanks for the great, unexpected introduction, Paul. And I appreciate you guys' time and allowing the FBI to come out and talk to you about kind of the cyber threat landscape, both as we see it from an investigative agency, and a USIC, U.S. Intelligence Community Agency, as well as from our Incident Response capabilities.

So I always like to give a brief introduction of how I joined the FBI in 2003, because it kind of shows also where the FBI is developed in our cyber capabilities. I joined in 2003. My wife and I lost a good friend in the World Trade Centers on 9/11. And like many, that was kind of my call to action. I had always wanted to become an FBI agent, and I applied at that point and was accepted.

I had a computer background. I came in under the cyber background for the FBI as an Agent. So I went to the FBI Academy, expecting that they were going to utilize my cyber expertise as terrorists, criminals, spies were more utilizing technology more and more often to commit their acts, their criminal acts. So, of course, I get to the FBI Academy. Six months later I graduate. And I was thrown for ten years into the Violent Crime Program. So I did get a great chance to work violent-crime and other criminal acts over the course of ten years, and really started to learn the degree to which these actors, again, whether they're Intelligence Officers, terrorists, criminals, child predators, some of the worst folks that we face here in society, who want to do us harm and hurt us, they are utilizing technology more and more. And that was in 2003. That is certainly true of what we face in 2015.

So what I want to do is kind of give you an overview of the threat as we see it in general, as well as to ICS SCADA. When I was in the Eurasia Cyber Ops Unit, I got to work a lot of these nation-state campaigns against the United States. And the implications are real of these—of cyber warfare. And so I'm hoping to give you an optic here and then maybe take a few questions if we have time at the end.

So, as we all know, those who are in risk management know, those net defenders know, what is risk? It really comes down to the equation, threat times vulnerability, right? Without a Threat Actor, if you can have all the vulnerabilities on your network in the world, but you're safe; without a vulnerability on your network, you can have all the threat actors in the world, and, by and large, you're safe.

However, it's when that threat meets that vulnerability that you have some kind of exploitation. That's what we learned working these cases. Unfortunately, most folks look at this from a prevention standpoint. Ninety percent of the money we put into this effort is on prevention, ten percent on detection. And what we're trying to encourage is that you really have to balance that out a little bit more. The adage right now in cyber security is prevention when able, but detection is a must. You have to be able to detect that threat actor rapidly when that exploitation happens. And I'll talk a little bit more about why.

So threat times vulnerability, let's take a look at who some of the threat actors are. And really, as I move from category-to-category, although you see these go from kind of least sophisticated in their capabilities to most sophisticated, the implications are still significant. So the first category is the hacktivists. The hacktivists' desire to perpetuate their cause, their belief system, based on their capabilities to affect change. Social injustice right now is a big cause for them.

In fact, I read this morning that over a thousand officials attending a Climate Summit in Paris had their personal identifiable information released by Anonymous. They allegedly released the information out there. They're not monetizing that, they're doing that for a reason. They're trying to out those individuals to perpetuate their cause. Unfortunately, we've seen that a lot in our dialogue domestically with U.S. law enforcement officers. A lot of people say, "Hey, these hacktivists are just defacing web pages. What does that really do?" And that, in fact, has gone well beyond defacing Web pages.

Hacktivists now are changing the way we police our communities, and they're really changing the dialogue here, even though their techniques aren't as sophisticated as some of the other actors. The implications are significant. If you look at the fact that the FBI put out a publication saying that hacktivists were targeting U.S. Law Enforcement personnel, this was back earlier this year. It still holds true today. We have Police Officers in communities right now—Baltimore, Ferguson and elsewhere, where we have a significant dialogue in our communities about how the Police police their communities, those that they serve. There's a significant dialogue going on.

But what effect happens when you have a Police Officer serving on one of those Departments whose information is obtained by a hacktivist group, who had no involvement in a particular incident, other than he serves on that Department. This hacktivist group is able to obtain his information, the information on his family, specifically a daughter who's in little league soccer, and releases that information. In today's day-and-age, when we're dealing with ISIS threats and everything else, that information is released on the Internet. And now, the entire Internet knows where this Police Officer's daughter goes to school, plays soccer. And, by the way, she's going to be on this field on this day wearing this jersey number. All that can be mined via Social Media, school websites, athletic websites.

When these hacktivists pull all that together, and they release that, they're really changing the dialogue and the way we police our communities. Same thing with the airlines right now. A lot of the airlines will tell you, you've got a lot of Hacktivists engaged in Twitter and Social Media campaigns to actually ground aircraft in-flight based on threats that they can perpetuate, realistically perpetuate, by mining data publicly available. So this is a significant concern for us.

The next category is the criminals out there who just who just take information for the sole purpose of monetizing it for their own benefit. They want to commit this exfiltration of data, whether it's PII, PHI, take that information and either sell it on the black web, or monetize it in some other way, shape or form. So we had a recent intrusion a few months ago that sparked a lot of dialogue around dinner tables here in America. Anybody heard of the Ashley Madison attack? So you had a lot of conversation around dinner tables because of that.

Let me ask you. When we look at these threat actors, we look at the intent behind them. Would a hack like that fall under a criminal attack? Certainly there are criminal implications there. Or would it fall under a hacktivist attack? Their goal was to shut down the website. Their goal was to take that information, use it as blackmail, and shut the website down. That portion of the attack, because of their intent—they weren't looking at selling the information, blackmailing folks, that part of a hack was a hacktivist attack, based on the intent behind it.

So we always look at the intent behind the actors. Now, plenty of criminal hackers jumped on the bandwagon after that, because now they've released the information of these folks who were out there doing this—engaged in this activity. So then the criminal actors are able to get that information, pull additional information, and go after individuals on the website to blackmail them. That's an example of a criminal attack. It's the intent behind it.

Now the insider threat, the reason I put that in kind of the middle of all these threat actors, is we have two things. We have "knowing" and "unknowing" insiders affecting cyber network defense within our networks. So, but still, the number one way networks are compromised, whether it's in ICS SCADA, energy networks, many times classified networks, unclassified networks, theft- of-trade secrets, is through spear phishing or social engineering. So still, that trusted insider is still the means by which these threat actors exploit the vulnerability and get into the network, because we're not managing access, privileged access to our networks appropriately. And I'll talk about that in a little bit as well.

But it's that trusted insider. Now the insider provides the keys to the Kingdom. Your data, the data you hold in that network, the capabilities that those control systems have to manipulate controls and other things, are all held within the network. And you have users who have access to that, and the threat actors know it. So, if it's a nation-state, a cyber-espionage campaign, a criminal actor, they're looking at those Insiders for access into the network.

In the old days, you used to have the old spy-versus-spy days. You had Robert Hanson, one of the most notorious spies, unfortunately coming from our Organization, who spied for the Russians. And the Russians worked very hard, in his case and every other case, to develop a relationship with him because of the information, the knowledge and the data that he had, ultimately. They wanted the data, which would hurt us from a national security standpoint, and give them a national security advantage over us.

So they spent months, weeks, years, cultivating that relationship. And it was hit-or-miss. What I'm going to tell you today is that, what we've seen over the last few years, is these same threat actors who can sit in Moscow or elsewhere, will spend the same amount of time identifying who in your networks have that access that the old human intelligence sources had. And they're going to target them in the same way. They're going to look at their LinkedIn. They're going to look at their Facebook. We see them doing it every day. They're going to craft very believable social engineering campaigns. And they're going to get them eventually to click on a link, which could provide access to malware to exploit your network. So the actors are doing it. They just don't have the cost-of-business of doing it, because they can sit right in their Agency's network and still target U.S. personnel.

We get to the cyber-espionage campaigns. Right now, this is crippling a lot of our private sector businesses who hold critical information, trade secret information, that's going overseas in all the research and development dollars that we've put in a lot recently, within our energy networks, are now going over and are freely obtained by these other competitors overseas. So the espionage campaigns, those actors are very involved in the advanced persistent threat. This is where you go from just exploiting a network, pop in somebody's email account, getting in and exfiltrating information, to actually get it in the network and setting up persistence within that network.

The goal of an APT is to get in there, set up alternate communications with the threat actor, get in there quietly. And again, once they breach the network, it's not like they're going to go after your data within days or weeks. They're going to spend time within that network, enumerating what the hosts are, what the other servers in the network. Am I in an ICS SCADA system? Am I in a Point-of-Sale system? Am I in an email server? And how does this network connect to the keys to the Kingdom? How do I get, if I'm in a business network, how do I get to the SCADA devices? If I'm a cyber-espionage actor or worse, a terrorist, or worse, a nation state looking to preposition themselves for warfare, how do I get into that network and preposition myself for illicit activity?

And so, when we talk about prevention in detection in the cyber threat, you cannot prevent. And that's just the truth. And I go out, and I talk to a lot of CEOs and Board of Directors and folks who say, "How can the FBI help us prevent this? What can my IT personnel do to prevent a breach?" And don't get me wrong, prevention is still important. We can prevent a significant amount of low-hanging fruit attacks. But if an APT Actor wants to get into your network, based on my experience working these cases and responding to incidents, they're going to get into your network.

The key is detecting it. I mentioned that I used to work criminal violations for the FBI. One of the violations I used to work for a number of years was kidnappings. Very, very difficult cases to work. But you knew, as soon as you got that call that a child went missing, you knew that it was—that you weren't playing around. You had to bring the troops in, en masse, and you had to find that child. You had 48-to-72 hours. You guys hear that a lot on the TV shows, 48-to-72 hours to find that missing child. Because what happens beyond that timeframe?

Beyond that timeframe, evidence degrades. Physical evidence will degrade. Surveillance tape will become overwritten, and you might not see something that you should have seen. Witness recollection degrades over time. So beyond those 72 hours, the chance of finding that child becomes more and more difficult the more time that goes on.

The same is true, now that I'm working in the APT world and this sophisticated cyber network exploitation world, the same is true here. Your networks here are going to get breached. If you don't detect that breach before they escalate privilege, enumerate your network and find where they are and how to get into cross-domains platforms. Get into your SCADA system, get into your Point-of-Sale network, and then I mentioned escalate privilege. It's at that point that you have really those first few weeks to identify that they're there. What are the changes that happened to my system to find them and kick them out of my network? And that timeframe could mean the difference between your company staying in business or going-out-of-business, or networks here in the U.S., in the energy sector, staying up or going down.

We just did a major exercise. The FBI was part of it. Many of you were probably part of it in the last two weeks called Grid X. Have you guys heard of that, the Grid X Exercise? It brings all energy sector partners from the private sector together, across the country, thousands of participants, everybody from the Government, to partner together and really sit there and game out what happens if there's a cyber network attack, a physical attack, or a terrorist attack on an energy sector system. And it's the recognition of that threat, early on, that's going to keep the lights on or keep the lights out, unfortunately.

Then we look at kind of the vulnerability. So we looked at the threat actor standpoint, the intent behind what these threat actors want to do. Again, when we're talking about intent, what's scary to me is an FBI Agent working in the national security realm. And we're kind of blessed and cursed at the same time, because we get to work the criminal standpoint of these cases, and put these guys in jail. And where we can't put them in jail, we have Title 50 authorities that allow us to utilize our national security authorities.

What's scary to me is the recognition that these actors are intent, in some cases, to sit on an energy sector network and not exfiltrate any information. And what does that mean when you have a sophisticated APT sitting on a PLC or an HMI within the energy sector, and there's no data being exfiltrated? Why is that piece of malware there? That's obviously a concern to us as well. Could that mean prepositioning of malware on that device for potential wartime capability? That's, from our standpoint, that's what we have to mitigate, protect, and defend against. And so that's what keeps us busy in the FBI, within DHS and the rest of our Partners, in DoD and elsewhere, is how do we ensure we find that malware that has complete remote capability to manipulate the system and work with our private sector partners to kick it out of the network?

So, some current and emerging attack vectors, as we see it right now. Really, the supply chain has become huge. If you guys see on here, I've got Energetic Bear, Black Energy campaigns, two major campaigns last year affecting ICS SCADA devices. Black Energy had the ability to manipulate the human-machine interface on all these energy sector networks—Siemens, WinCC, as well as two other HMIs, within the energy sector. It allowed them to basically pop those computers, obtain the HMIs, and manipulate controls within the ICS SCADA system.

And what's interesting is Dragonfly, Energetic Bear, another campaign, was one of the first campaigns to actually weaponize a software update to PLC devices within the ICS SCADA networks. Now, as network defenders, and I used to work in network defense in the private sector, as network defenders, we're taught to do what when we identify a vulnerability? Patch it, right? Patch, patch, patch, that's our mantra. Let's get in there, patch the vulnerability as soon as we can.

Well what happens, when you have a threat actor who knows there's a vulnerability on PLCs within the energy sector network, they know that there's a patch available for it, and they know the energy sector is going to rush to get this patch, to patch the PLCs. Why not, if you're sophisticated enough, weaponize that patch, so that when they patch a legitimate vulnerability, they get a prepositioned piece of malware on there, as well. So the Energetic Bear campaign was one of the first campaigns to do that.

I did also put on there an old Cessna control panel, because I'm a pilot as well. I learned on steam gauges. A lot of these new platforms, aircraft platforms, that carry over 500 passengers have over 1,000 pieces of unique firmware and software on that airframe. So how do you validate and test that firmware and software to ensure its integrity? And then, how do you validate the updates to every individual piece of firmware and software on that platform, to ensure that there's no Energetic Bear malware or otherwise on there as well? Scary stuff. So that's what we're trying to talk to private sector. You have the ingenuity in private sector. You guys are the innovators. Let's work together to try to determine how we standardize patch management, and we ensure integrity is there during the patch management process.

Internet-of-Things—everybody is aware of what this is, right, the wearable devices. I just looked at my Apple Watch. You have PANs now, instead of LANs. You have PANs, Personal Area Networks. You have the house that CNET is developing right now that is completely connected to the Internet, so that if you are driving home from work, your car is talking to your home server. And, as soon as you cross a certain point in town, the refrigerator is telling your vehicle to tell you that you only have four items, so you can create these five dishes for dinner. But, if you stop at the grocery store that's two miles away, and you pick up these additional three items, you're going to have a dozen more dishes that you can create.

And then the thermostat is going to start warming up certain parts of the house as you get close to the home, so that when you get there, you're comfortable. And, as you walk from your vehicle into the house, your Pandora is going to transition into the house so that you hear that music in non-stop fashion. The Internet-of-Things is creating an expanded attack vector within businesses, within the energy sector, and within ICS SCADA systems. True stories.

We've had private sector businesses that have put LED displays or the LCD displays on conference rooms throughout their campuses to show events like this, meeting times, who the speaker is, so that participants can go to the different events, look at the LED screen, the LCD screen, and determine where they want to go. And they're centrally connected to the IPv4 space. They're connected to the production network, so that the secretary or the receptionist can push, to every one of those devices, what the updates are.

Why do we do that? We do that for convenience, right? Remember this—convenience for us is convenience for the bad guys as well. These devices were running Android Operating Systems that were not being patched because they just thought these are LCD screens. Why worry about this in the grand scheme of things? Well we had Threat actors exploit the Android Operating System within those LCDs, and they were able to then enumerate hosts within the production network and move laterally. Now, how do you want to be that CISO? Do you want to be that CISO or that Security Officer who explains that we weren't patching the LCDs, and that's how they stole this critical information within our network?

So we also had another instance of a burger kiosk within a university campus that was running the kiosk so that people, students could order burgers within the cafeteria, connected, again, to the IPv4 space. And we had the threat actors exploit that kiosk and be able to move laterally through the network. So it's recognizing what devices we're connecting. Do we need to connect these things? And what are we doing to secure these devices if we're connecting them to our production networks?

Trusted third-party has become huge. Who has access to our network? A lot of people now have partnerships where they have direct-access into other people's data centers. So how are we creating Service Level Agreements that protect us from a partner who may not have the security requirements that we have? The energy sector has done a pretty good job in the last probably two years, increasing their defenses against cyber adversaries. What we've seen is some of the third-party intermediaries have not done such a good job. So, even though the companies are spending significant amounts of money and dollars on network defense for the corporation, the threat actors still want to exploit the path-of-least-resistance. So, if there's a small Mom-and-Pop shop that's managing their email services and have connectivity to their network, we're seeing them exploited, and they then have trusted access into the corporate network, and they're able to then move into the ICS SCADA environment, or, in some cases, the business networks.

And then, of course, we have the expanded attack surface of this idea of "Bring Your Own Device," the Cloud environment -- the Infrastructure-as-a-Service, Services-as-a-Service, and Software-as-a-Service. The whole movement to the Cloud, we've gone from data centers and energy sector networks that are brick-and-mortar and hardly defined, to this robust dispersed group of people carrying laptops, tablets, cell phones, all of which are Personal Devices and connect to the network without two-factor authentication, which allows additional footprint and additional avenues of compromise.

So risk mitigation—I know I'm running low on time, so I'm going to hit some of the top points on here, so I can maybe take a few questions if you have them. And I'm really going to focus on Preparing, Prevention, and Detection, because my understanding is, there's not a lot of net defenders here. And once we get into isolate-contain-respond-and-recover, we're talking about Incident Response.

So, when it comes to preparing, the biggest thing I can say is develop plans beforehand, develop your Incident Response plan beforehand, and include all lines of business. I can't tell you how many times we go to an electric company and say, "Hey, we've got Intel from a national security source that your network has a problem." And we get sent away—or this could be any corporation—we get sent away for the next three-to-four days, as the lawyers and the CEOs and the outside counsel and the net defenders and the IT Department talk about what information they can provide to Law Enforcement, how they're going to redact the information.

And what did I tell you about the importance of detection and ex-filtrating that compromise early on? If we're out of the picture for three-to-five days because you guys are worrying about an Incident Response plan, there's no telling what the threat actors are doing. Unfortunately, a lot of corporations then, with in-band, what we call in-band, will then send email communications all over the place on that same compromised network. "The FBI was just here. We've got an issue on the SCADA network or the Point-of-Sale network. What are we going to do? What can we share?" And the threat actors are going, "All right, let's preposition here, let's go here." And they're just kind of following everything around. So really, having that Incident Response plan early on, and including all lines of business, is critical in mitigating that threat.

Prevention. I'll tell you, these are the top ways you can prevent. So many companies we go to spend millions of dollars on software that help them do all sorts of neat things, take out encryption, look at encryption, re-package encrypted technology and send it back out, communications. This is what you need to do, because they're spending all this money, but they're not looking at the fundamental ways to protect their networks.

Patch management. You've got to ensure that your Operating Systems and your software are up-to-date, especially as we talk about the amount of devices in our networks that are now utilizing software, it's important that we are centralizing the management process and ensuring the integrity of the patches that we're putting out there.

Classifying, segregating data. There's too many businesses out there that have email servers on the same enclave as their customer's PII. You've got to classify and segregate that data.

Privileged Access Management. If you get nothing else out of this presentation, if your organization is not using two-factor authentication for remote access and admin access to your network, you've got to implement that. I talked earlier about how the threat actors will escalate privilege. And too many companies have network admins, CEOs and others, who surf the Internet from the same device they use to manage critical systems within their networks. You've got to be able to deescalate privileges, only use those privileges where needed, and use two-factor when you employ it. And then application white-listing.

Detection. It comes down to basically getting an eye on your network and creating a baseline so that your log files are telling you what they need to tell you. You need to know, if you're in the energy sector, you need to have a baseline of what that ICS SCADA system, what kind of traffic is coming from the HMI to the PLCs and back, so that if there's any anomaly, you can detect that anomaly. There's two ways to do it.

We can do it through signature-based analysis, which is important, but you have to have behavior-based analysis in anomalies, as well.

This is a report from ICS CERT. This is the importance of detection. In reports last year, ICS CERT had 245 incidents reported to them in the ICS SCADA community, in the energy control system community. Of those reports, 38 percent had an unknown vector of compromise. So we can't say how those attacks got into the network. That tells me there's no detection technology in place. We're not detecting that threat actor, which means, from a Law Enforcement perspective, we might have an idea of who it is, but we can never prosecute them, we can never attribute this to a nation-state actor, because there's no detection in place. Log files are probably being overwritten, and there's not much we can do with it.

And then this all revolves around partnerships. I talk about Grid X. And one of the programs I've had the privilege to manage over the last year is the InfraGard program. And I'm very proud of what our folks do here in InfraGard with Private-Sector/Public-Sector partnerships. I can't encourage you in the private sector more. And I know I'm preaching to the choir with the group here, to engage with your Local FBI Field Office, your DHS personnel, and your other partners within the various ISACs in other organizations, to share threat Intel, best practices, and to talk about the Threat Landscape, as you see it.

So, I think I'm hitting up against my window. I appreciate your time. I think I've got a few minutes for questions if anybody has any. Got one in front and center here.

Q: [off microphone] When I buy or get food, I see USDA-certified Organic. Is there a way, when I go to a bank, I can have it healthier, and it's FBI-certified data protection? That is to say, what can the FBI do to give the Consumer the information I need to select what banks, what utilities, whatever I want to give my business to, or the bank, what cloud do I want to give my business to?

BRETT LEATHERMAN: So let me—Go ahead.

Q: And then, related to that, what Legislation would you want to see happen to make sure that the things you want to prevent are easier for you to prevent or detect?

BRETT LEATHERMAN: So the first question for those, I think, who are remote, is what recommendations can the FBI provide, based on what we see towards products or services? The answer to that, in short, is let me tell you, although we see this from a threat landscape perspective, our primary mission in our organization is Investigation and attribution. We only have, within the FBI, approximately 14,000 Special Agents assigned across the world to work Counterterrorism and every other program, 35-to-40,000 personnel total.

New York Police Department has about 35,000 people in New York City alone. So we've got a little bit of a task. So that's not part of our primary mission, is to actually vet products and provide some kind of Seal of Approval for it, unfortunately. I'd love to have a personal Blog where maybe I did that, based on my experience, but I'd probably get in trouble.

The legislation is the other question. Right now, legislation that we're focused on, from our standpoint, is the Information Sharing legislation that's going through our ability and private sector's ability to share information with us from a net defense standpoint, as well as a cyber defense standpoint, without regulatory or other sanctions or implications.

Q: [off microphone]

BRETT LEATHERMAN: The question is—Is it one of the solutions that not every critical system needs to be on a network? Tell that to the Millennials. I mean it's like they want to connect everything to the Internet and tell everybody where they're at. I'm sorry if you're a Millennial. No, I get it. Air-gapping is the best solution, in my opinion. There are probably some devices that need to be air-gapped from any other network. Unfortunately, people make business decisions outside of security as well, occasionally. And that business decision is what is impacting us. One more? All right, up front here.

Q: [off microphone] FBI does a great job ... How does FBI engage politically?

BRETT LEATHERMAN: So the question is- How does the FBI engage, really, from a political standpoint? Let me answer that in two ways. Number One, under the Hatch Act, we can't. We're a U.S. Intelligence Community Agency and a Law Enforcement Agency, so we can't engage in the political conversation, really. But we do provide classified—I've been to Congress, the White House, and elsewhere, providing classified briefings, so that they're aware of the threats we face at the National Security Level. The Candidates get many similar briefings to what others get within our political landscape.

What we do in engaging with private sector, though, and if I had more time, I could spend half an hour talking about encryption. And the implications of encryption, from a privacy standpoint, and a national security standpoint, and where that's going, because it's a dangerous area that's going right now. It's a big concern to us in the FBI. It's getting with our private sector partners, in the ISACs, within other communities, to talk to them about the implications. And we do that regularly. Not really through commercials, but let me tell you, I think I was—three weeks ago, I was in seven States in eight days or nine days, I think. I mean, we are engaging it with education, critical infrastructure, and elsewhere, to make sure they know what we're seeing from an incident response, a Law Enforcement, and an intelligence standpoint, and trying to make—really, bringing in an awareness campaign, so that Private Sector can then look at encryption and other implications, and come up with some of those solutions, or engage in that dialogue as well.

So, before I get kicked off, I want to thank you guys for your time and your partnership as well. And I look forward to talking to you guys offline, maybe, about some of this stuff later.

[Applause]

END

Cyber, EMP, Space Weather Complications to Complex Systems Engineering

Corresponding Video:

https://www.youtube.com/watch?v=XSnKbdI7cRc&feature=em-share_video_user

Panelists:

- **Mr. Chuck Manto,** Moderating
- **Mr. Thomas Popik,** Founder, Foundation for Resilient Societies; Nexus of engineering and organizational issues
- **Mr. Mark Walker**, INCOSE and **Mr. Michael deLamare,** Bechtel and INCOSE, review the joint INCOSE & EMP SIG Planning Models to Manage Complexity of High Impact Threats
- **Mr. Joseph Weiss,** Author, *Protecting Industrial Control Systems from Electronic Threats;* Cyber complications

CHUCK MANTO: And at this point, what I thought I'd do is introduce the topic of the panel. And that's looking at all the unexpected complications to our very large systems and very often these large systems that are connected to other large systems in environments that we don't anticipate fully. And it brings up a lot of issues that may go undetected, even when you're doing your own job really well.

So today, we have a couple of folks from the International Council on Systems Engineering, Mark Walker and Michael deLamare, who will talk a little bit about the role of that Council and that group of engineers. And when they explain it, you'll see why that's a significant opportunity for us to bring national-level, actually World-Class-level, talent to this issue.

Tom Popik is here from the Foundation for Resilient Societies, and has been—he is an MIT engineer himself, but he's been very active in looking at the organizational complexity between organizations like NERC and FERC and State organizations who are trying to put their hands around all of this.

And then Joe Weiss is interesting because he's been looking at this as an engineer looking at a lot of the issues regarding cyber threats that are sort of obscure to most of us. And he'll give us two or three of the reasons why that is, and how that sneaks up on us in unexpected ways in complex systems, as well.

So, what I'm going to do is start by maybe having Tom Popik discuss from his point of view—and by the way, we'll have a few minutes for each of the panelists. And then you have some opportunities for questions. But this is a very brief panel, so everyone's going to keep it to about two or three minutes now. You might hear from some of these folks later, as well. And we'll have opportunities in between some sessions and lunch and so on to grab them for some more information. But we just need a general orientation. And Tom, maybe you can start us off.

TOM POPIK: Well, certainly. Thank you, Chuck. I'm going to be talking about the nexus of engineering and organizational issues, but I'd like to expand it a little bit more to—also the political, economic, legal and scientific arenas. So the first point I'd like to make is that reliable infrastructure is essential to human populations—we can all agree on that. But I'd also like to make a second, very important point, which is countries and societies under stress have trouble prioritizing infrastructure reliability in advance of catastrophic events.

And so we see this not only in the United States, but actually in other countries in the world right now. So for example, if we look at South Africa, South Africa is currently having rolling blackouts because their society is literally having trouble keeping the lights on. If we go to another society, the United Kingdom, the reserve margin, reserve capacity margin for electric generation in the United Kingdom this winter will be approximately one percent. In November, the United Kingdom had to enact what's called Demand Response. They asked major consumers of electricity to shut down, such as factories, so they wouldn't have to have rolling blackouts.

And then I'll even give an example here closer to home. In the winter of 2014, January 2014, we had the Polar Vortex, very cold weather snap. PJM came with—I should say with PJM, it's the largest reliability coordinator and transmission operator here in the United States. They cover from the Midwest all the way to here in the Washington, D. C. area. So in January of 2014, they came within two gigawatts of having rolling blackouts and that's about two percent of reserve margin.

That's some of the societal stress that we're under. And then let's look at how we have reliability regulations set here in the United States. It's set through a Public/Private Partnership with the North American Electric Reliability Corporation setting Standards through representatives of the electric utility industry. And then a Federal regulatory body, the Federal Energy Regulatory Commission approving these standards. And there's a lot of friction, dispute in the standard-setting process.

This is really a symptom of the overall societal stress and how we have trouble coming to grips with some of these very, very large issues of how we prioritize infrastructure reliability and protection.

CHUCK MANTO: Thank you, Tom. That's a good first overview. From another systems perspective, perhaps Mark and Michael—and it doesn't matter to me which may go first—just give us a little overview of INCOSE, and why these complex issues, or systems, are of interest to you in light of these very high-impact threats. And you might just mention a little bit about your interest, how you got interested in it, and what your roles are in INCOSE, and so on, so people can get an idea of who you are and what you're about to do.

MICHAEL DELAMARE: Thank you, Chuck. Mark here won the coin toss, so I go first. I get to kick off here. So, let me just start by introducing INCOSE. What is INCOSE? You might not have head of that before. It is an acronym standing for the International Council on Systems Engineering. It is the premier Professional Society of Systems Engineers. It's a 25-year-old organization with about 10,000 members in 62 countries. And it is a certifying body for SE professionals.

It has about 100 companies on the Corporate Advisory Board, and in the Government and with university representation, as well. And it exists to push the state-of-the-art in the topic of systems engineering; applying systems engineering, if you're not familiar with it, is a discipline that was established primarily in Space and Defense and has been pushed out into other domains such as healthcare, automobile, industrial facilities where I spend a lot of my time right now. As far as the interest in this, I am on the Corporate Advisory Board for INCOSE. I also co-chair the Infrastructure Working Group. And some, well, it was about January, Mark and Chuck were conspiring together to bring INCOSE into this because of the abilities that INCOSE has, the Intellectual Properties and the skill set that the members have in being able to model complex systems, for example.

So, we established in June of this year the Critical Infrastructure Protection and Recovery Working Group which is parallel in our charter to what the EMP SIG is doing. So we are working on supporting the activity of how do we protect our critical infrastructure, how do we recover from these events? And the way that INCOSE is doing this as a Professional Society of all Volunteers is by bringing in a systems-approach to things and making that kind of skill set available to people in InfraGard and in the related organizations trying to build partnerships with universities, with InfraGard, with Public/Private sector, with other professional societies such as the IEEE or the International Society of Automation, or EIA, Electronic Industry Association, ISO, who are bodies that create standards that we'd be able to influence and try to work this all together from not just bringing in system principles to systems themselves, but to the more complex system of policy and human systems, in that regard.

So that's part of the way that we can provide expert assistance in dealing with complex systems, system-to-systems, model-based support, which I think Mark will address a little bit more about. System reliability, system resilience, principles and architectures, standards development, life-cycle approaches, information distribution, and areas along that side. So with that, let me allow Mark to take this further.

MARK WALKER: Okay, as Mike has mentioned, INCOSE's, really, main objective is to establish a well-defined and well-formulated high-level system engineering in all different areas of system engineering, in engineering-developed systems. So there's like 46 different Working Groups in INCOSE. A lot of them are process-oriented, like requirements architecture, system-of-systems, complex systems, these are Working Groups. In addition to applying it, and that's where we're getting a little more into that area. I've been in INCOSE for 23 years or so.

And so the "Super Working Group" is kind of the, okay, now let's figure out how we can really apply the high-level thinking of system engineering to a real problem. And obviously, all engineers are interested in challenging problems and how you solve them. So when we get into the EMP, the solar, as well as the cyber world of the problems, you start thinking heavily on the negative sides of developing systems, is how do you address those problems?

So, from a systems point of view, we use a term, "system thinking". It's really trying to understand the holistic perspective of the problem and understand what the customers, and that can be good or bad if you have obviously users you don't want using your system. That directly applies to failure modes, effects of analysis, stuff like that, which is very valuable to determine how you design and develop the system from a complete point of view. It can also be multiple layers. Quickly thinking of this, I look at this problem as a five-to-six level of a system-of- systems point-of-view of how are we going to manage it from the many organizations that are involved that are very high level and all the engineering you got to do to connect systems at very high levels, as well as you get down into the—all the way down to the components that have to be fixed to protect them from a hit or something like that.

So model-based system engineering is a new concept we're implementing over the last 10-, 15- years. It's based on models, but maybe not the dynamic modeling necessarily. It's more of a—I can't probably go into too much more of this. But anyhow, it's key requirements. I just want to throw out one requirement, I mentioned this yesterday. There's one requirement that really applies to the problem we're looking at here, and it's called Operational Availability. Had to design a system one time that was a very high Operational Availability requirement. It was the combination of the design of the system for all failure moves as well as what you want it to do.

In addition to the life-cycle support aspects of it, if you have failures, how do you address them? So it's a one requirement that is extremely applicable to this problem that we really need to think and look at.

We worry about all users, good and bad. You got to really look through that. Enabling capabilities, databases, all the information you want to save and pass and how you do that, obviously cryptology, stuff like that related to how do you make sure that stuff is safe.

So, anyhow, diagnostic systems. I'm thinking them from the point-of-view of identifying the problem, like what Brett was mentioning about quickly and what's your reaction to that, is it very rapid, or is it more of a long-term, response? So those are the kinds of things you think about.

CHUCK MANTO: So in a moment, we'll make certain we have time for each of you to ask questions of each other. I think we have a slide or two that Joe had for a quick illustration if it's ready. There we go.

JOE WEISS: Number One, I wanted to thank Chuck and InfraGard for inviting me and allowing me to speak. I'm a Control System Engineer. I'm also the Managing Director of the International Society of Automation's Control System and Automation Cyber Security Standards. Those are the International Cyber Security Standards for electric, water, oil, gas, chemicals, pipeline, transportation, you name it.

This is my little slide to explain why we're different and what needs to be done. When you're over here, this is what you just heard from the FBI and what they're looking at; that is Windows, that's Internet Protocol, it's what you're used to. Stuxnet affected what was called the Programmable Logic Controllers. They don't look like IT, they don't look like anything else. What's actually inside those controllers, there really isn't any forensics for. It's part of the reason Stuxnet wasn't found because—it would change in logic.

As you move further down, valves, motor control sensors, there is no security there. There is also, effectively, no cyber forensics there. This is in any industry worldwide. This is also why we need to train engineers, because they're the only ones who are going to understand what's going on at that point. That is not what the IT world is going to be able to see. This is also where you have catastrophic damage occurs. This is where you would have two-or-three day outages. That's where you have 9-to-18 month outages.

The other thing I wanted to point out, what we don't have, for anybody who saw "Diehard 4," there are no screens that say cyber attack. [laughter] Unfortunately. This next one basically tells you this is real. I've been able to amass a database of over 750 actual control systems cyber incidents, all industries worldwide. Almost none identified as cyber. This is both malicious and unintentional. We've already had more than 50 control systems cyber incidents that have killed people. There's been over a thousand deaths. There's been five major cyber related electric outages in the United States already, none identified as cyber. So when you look at this, this is real.

The other thing I wanted to point out real quick and show you one other, real quick, set of slides, people normally think of interdependencies between electric and everything else. There's another set of interdependencies that hasn't been considered; that's our vendors. What Stuxnet did was to exploit a vulnerability in the design of every single Siemens controller that was ever made. That means that same vulnerability crossed every single industry that used it. That's a totally different interdependency than has been addressed, to date.

One of the biggest issues we've got is sharing information. And the policy associated to allow people to share information with a "Get Out of Jail Free" card. Right now, there's a disincentive to share any of this information.

Last three slides, just so people know what they are. Aurora is a vulnerability that was demonstrated by the Idaho National Lab back in 2007. It's where they physically destroyed a generator. DHS, for reasons only they can answer, declassified 843 pages on Aurora. What I'm doing is showing three pages that are on hacker websites today. The first identifies by name, the Pacific Gas and Electric substations you would use to destroy the rotating equipment in the Chevron Richmond refinery.

This is every large water-pumping station in the State of California, and how you would use the substations to destroy or damage the motors for every single large pumping station in California.

That's the natural gas compressor stations in the Northeast. This is where you would use the substations to destroy the natural gas critical infrastructure. So this is backwards. Everybody is wanting "the grid" to stay there. Here's where "the grid" can be a threat to industry.

Just the very last thing—Aurora occurs in less than one minute. It needs no other software. This is not a virus or a worm. This is effectively a mechanical, if you will, deficiency in the overall system that cyber can exploit. It is totally different; it's there. The hacker world knows it, and it's a direct threat to essentially all, of not just American, North American, but any industrial infrastructure. Thank you.

CHUCK MANTO: Thank you, Joe. So, we're going to hear more about Space Weather today, and we're also going to hear more about EMP. But, all of you have a good background in each of them. For just maybe 30 seconds, each from the end of the panel up, what are some of the hope for opportunities by addressing these systems? And I'm going to ask the question this way: Many organizations I see, including businesses, look at something like this, and they throw their hands up and say, "Too complicated. I don't have time. I'm just going to hope for the best." Or, "It's too expensive," or, "I don't have the talent." Is there hope that if people do a really thorough systems review of all of these interdependencies and get their hands around it, that they might actually save and make their companies money while avoiding a major disaster? Or is it so expensive to do, that only a few companies can do it and the rest of us are just going to suffer as helpless victims to the complicated world we've inherited?

TOM POPIK: That's a really interesting question, Chuck, and let me just say I'm not going to claim to be the expert on being able to answer all those elements in there, particularly the economic ones. But from what I have seen, though, is that the triage approach has been discussed a lot over the last couple days, and it has also been written about with the idea that if you can protect a smaller number, say 20 percent, of the grid, say, and the large transformers, you can result in the ability to protect 80 percent of the economic loss that would otherwise have occurred if you did nothing at all.

Now, let me just say one of the things that I would hope would come out of the INCOSE engagement in this. One of the things that I think would be beneficial in this is that being able to model complex systems, being able to look at the behaviors, being able to look at the interactions and identify the high-leverage points. By the high leverage points, I'm looking for something that for the least amount of cost, you can make the greatest amount of effort, or the greatest amount of result. Those kinds of interactions are not always obvious and it does take some good analysis to be able to pinpoint those.

But once that work is done, I think it would save some work with respect to where we put our money and to provide an improvement.

CHUCK MANTO: Mark, anything to add? We have less than a minute, so go ahead.

MICHAEL DeLAMARE: Okay. I've been involved in a lot of big programs, a lot of mini systems. And the one thing I think I've learned over that, and I think it relates to the System-of- Systems Working Group we have, and the others, is that in order to take on a very big problem, very big interrelated system of systems, and typically we will not address down to detail levels like Joe's talking about. But if you break it down to where you start understanding the whole problem and then start breaking it up with individual pieces of work in one area and another, you can work through that.

And obviously, interface and everything else is related to that. But you have to understand the whole problem in order to know where your priority's got to be, where the big issues are. And obviously when you focus on the three we're talking about—the solar, the EMP, as well as the cyber, you would then work on each one of those and where you go from there. So it can be done. It's just a matter—it does take time, a lot of effort, and I think we probably all understand that part.

CHUCK MANTO: Thank you. Tom, you have the benefit of having gone first and last. In this case, the first shall be last because he's one in the same.

TOM POPIK: Well, certainly. I think this is an area where economic analysis can be very helpful. Let's look at opportunities where there are high-impact events that can be cost effectively protected against. For example, a solar storm impacts on the electric grid, we can protect all the electric grid for less than a billion dollars, and then we can get a double benefit if it's done correctly. We also could get a nuclear deterrent against high-altitude EMP attacks. We wouldn't get 100 percent protection, but we'd get a deterrent.

So we need to look at these opportunities which are very cost effective and, as a society, decides strategically we're going to take care of these opportunities and then move on to other things which are more complex.

CHUCK MANTO: I think later today during our lunch break, we'll hear some recently released data that's been done at Idaho National Labs and other places that reflect that there are day-to-day wear-and-tear experience of equipment such as the grid and consumers, or commercial equipment that's dependent on it, that if they were taking a certain amount of protective measures against these issues, they would actually save wear-and-tear and actually save themselves money day-to-day, while being better prepared for a larger event. And I think we'll hear more about that this afternoon.

But I just wanted to thank the panel for being here today, and thank all of you for spending your time with us today doing this. Thank you.

[Applause]

END

FEMA Perspective on the New Space Weather Strategy and Action Plan

Corresponding Video:

https://www.youtube.com/watch?v=Q1Ob1PIOQNw&feature=em-share_video_user

Panelist:

- **Hon. Craig Fugate,** FEMA Administrator

JERRY BOWMAN: Our next presenter is going to talk to us on the FEMA perspective on the new Space Weather Strategy and Action Plan. The presenter is the Honorable CRAIG FUGATE, the FEMA Administrator. He was confirmed by the U.S. Senate and began his service as Administrator of the Federal Emergency Management Agency in May of 2009. He has promulgated the Whole Community Approach to the Emergency Management, emphasizing and improving collaboration in all levels of Government, that's Federal, Tribal, State and Local, and external partners, including voluntary Agencies, faith-based organizations, the Private Sector, and Citizens.

Prior to coming to FEMA, he served as Director of the Florida Division of Emergency Management. He served as the Florida State Coordinating Officer for 11 Presidentially-declared disasters, including the management of the $4.5 billion-dollar in Federal Disaster Assistance. In 2004 he managed the largest Federal Disaster Response in Florida history.

Would you please join me in welcoming The Honorable CRAIG FUGATE.

[Applause]

HON. CRAIG FUGATE: Well, good morning. When you talk about various scary scenarios, there's always the—what I call the Science Fiction worst-case, everything's horrible, everything's going to go wrong. And then, what really can happen. And so, when you talk about risk and hazards, you always have to balance between the extreme views of, "It will never happen" to, "If it happens, there's nothing we can do about it." And that's—that tends to be where hazards fall, it's either so bad everybody goes, "We can't do anything about it. It's such an infrequent event." We need to plan for things that happen more frequently and are more likely to happen.

Well, when you talk about events like Space Weather, it falls into one of these categories that it may be so bad, who wants to think about it? And it doesn't happen very frequently. So therefore, let's focus on the floods, tornadoes, hurricanes, and things we know are going to happen more often.

But, when you look at hazard analysis, you have to look at two things. You have to look at the probability and frequency of events, but you have to look at the consequences of the events. And oftentimes, the sweet spot tends to be where people focus is the moderate-to-high frequency of events that have moderate-to-low consequence.

We tend to look a little bit differently. At the Federal level, we have to look at not just what can happen, because many of the things people focus on every day actually are handled by state and local responders. We have to look at those events that have the ability or reach a threshold that requires significant Federal response or have impacts on national security. And that's where you start looking at, oftentimes, what may be a low-consequence/high—a low-probability/high-consequence event.

So when I came to FEMA in 2009, first thing I asked them was, "what are plans for a G-5 event and a Space Weather storm?". And they just looked at me and go, "What's a G-5? What's Space Weather?" And I had began this in Florida, because we were looking at the various things that could impact as emerging threats. And what we found is, it's not so much emerging threats as existing threats, but technology may have changed the impacts.

And as we began this process when I came to FEMA, it became clear that there were very few people outside of those at the Space Weather Prediction Center and a few people in the industry that really understood what Space Weather was and what it was not. Oftentimes I hear two things that are confused, and they say, you know, "Space weather is the same thing as an EMP." They're not. They're two separate events. They may have similar consequences—they're two different types of events. And, in fact, what you would do is different in both of them. There are some similarities, but they're not the same.

So I'm going to focus on Space Weather, understanding that many things we talk about would be similar systems impacted by electromagnetic pulse, particularly those generated by nuclear detonation and Near Space. But that's an Act of War. I want to start with one that's a naturally occurring phenomenon that historically has happened and has been documented. If you go back to 1859, the Carrington Event was a coronial mass ejection that occurred, that produced all kinds of observations across the world. It was one of the most well-documented events. But, at the time, the technology that was most severely affected was a new network that nobody really understood was a network, and that was telegraph systems.

But, what's interesting about this event is, from the time the event occurred until it began impacting earth, was estimated to be about 16-to-17 hours. Many of these coronial mass ejections move much slower. So we oftentimes have days to determine that an event has occurred, potentially it is coming our way, and have some chance to begin prepping for it. But take an event like the Carrington Event, and you got maybe 15-to-16 hours, from the time the event is detected to impact. And working in those timeframes, understanding that the impacts won't be known until it gets closer, but the timeframes it takes to take protective actions oftentimes will fall outside of the decision-making cycle as it exists today.

And what happens in one of these coronial mass ejections that comes towards earth? We deal with several major phenomena, some of which are just caused by solar flares; but in a coronial mass ejection, we deal with some additional things in the geomagnetic world that get very interesting, as far as impacts on the Infrastructure.

But the Space Weather Prediction Center, if you're not familiar with it, we actually have a Center that forecasts Space Weather, 24-hours-a-day, seven-days-a-week, they are one of the Centers for Environmental

Predictions, just like the National Hurricane Center, the Storm Prediction Center. It is a Joint process that the Air Force and the National Weather Service, through observation platforms and forecasting. And they forecast essentially the three hazards of Space Weather: radiation, HF radio propagation and blackouts, and geomagnetic storms.

And they have scales assigned to them. And those scales indicate severity. And part of this was in place when I came onboard in 2009. What was not in place was an understanding, at the Federal level, of what these impacts were. There was no strategy. We had a detection system, and not so much even a warning system, as much as we could get information out, but a warning is a little bit different than an advisory. Because when you talk about warnings—warnings are, "This is an event. You need to do something differently."

With a lot of the Space Weather Advisories, there was not specific understanding of, "What do we need to do differently to protect? What were the impacts? How would you minimize those?" And so, let's talk about some of the impacts of a geomagnetic storm, of a G-5, or a Carrington-style event, some of which are going to be essentially Space Impacts, others which will become terrestrial impacts.

First one up, and the fastest one to get here, oftentimes, is radiation. Generally, this will not affect terrestrial, but it does affect space, humans in space—this is one of the things that the International Space Station has to monitor. That's one of the reasons why they always keep one of the modules on the Space Station for escape, because it's the only place that has enough shielding to offer realistic protection in a radiation storm. It affects high-flying aircraft, going particularly Polar routes. And it impacts satellites, because it can produce all kinds of static discharge and impacts, particularly on non-shielded satellites.

Then you have the impacts of the HF. HF radio propagation is still used by many Agencies as one of the backup—and this is what's kind of interesting. In communication or you take a cyber event, where you have massive disruptions in communications, HF radio is still one of those systems that is used as a backup, but in a geomagnetic storm can be rendered useless. And so those systems that use HF radio, including a lot of aircraft that are transiting the Pacific and Atlantic, still use HF radio for commercial aircraft to get and update information. Those systems can be degraded, and anybody that's using them basically from one-to-30 megahertz is at risk from disruptions.

But the one that tends to get the most attention are the terrestrial impacts and some of the impacts and satellites from the geomagnetic storm itself. What we have observed, and what happened, and why Carrington was so interesting, was the first time we saw a wired network that could actually build up energy from a coronial mass ejection in a geomagnetic storm, and produce large voltages across the system. The reports of the telegraph operators at the time were sparking, arcing, and getting shocked. When they disconnected their batteries, they could still transmit, there was that much energy in the lines. But at that time, the only thing that was networked was the telegraph system. And there wasn't that many places for it to really build up and produce surges beyond what they were observing.

Fast-forward to later years, we began seeing in some of these events, particularly the higher latitudes, impacts to the electrical Grid. And what we found was, those networks, of a conducting material over long distances, amplified the impacts. And primarily, those are going to be pipelines and electrical Grid, particularly the transmission lines moving power across long distances.

And that process that we have is kind of one of the things that, within industry, has produced a lot of interesting risk. And that is, we tend to generate power at very large quantities away from where it's consumed. And then we transport it long distances to get to where the demand is. And because demand is never going to be static, that ability to move power long distances is not always in the same direction. It can be moving power across the country to where, oftentimes, where is the cheapest power production at, and where is the highest consumption—and retailing power across these grids. So these systems are very complex. They move a lot of power. They're not always carrying the same loads. It's a very dynamic system that constantly changes, depending upon demand and production.

But the characteristic is, you have large generation being transmitted long distances. That requires transformers. And this is where we found some of the first terrestrial impacts of geomagnetic storms, was in building up the electrical charges and the system, overloading, in what is referred to in the Industry as very large transformers, or large transformers. There's no standard definition. I think the Department of Energy uses, I think it's a 100-megawatt ampere as a kind of a threshold for what is a large transformer.

But these are very important to the ability to move power across your Grid long distances, of stepping-up and stepping-down power. And it is that that has been seen as the most vulnerable, because it's happened. Hydro Quebec had a situation back where one of their storms produced enough energy in the system that it overloaded and began melting the core of the transformers. Now if transformers were a commodity that was relatively inexpensive and easy to replace, this would be, you know, okay, it's a lot, but we change it. They're not. These tend to be very specialized, custom-built, and very expensive along lead times to build.

There's not a lot in reserve, although the National Electrical Liability Council has been working with their members to build an inventory. But, from the timeframe of ordering one to having one delivered, oftentimes is measured in years. These are extremely large, oftentimes require specialized transport. And, we also have an aging infrastructure. The United States has some of the oldest of the large transformers that are reaching the end of their cycles' lifetime, and are going to be expected to be replaced anyway.

So when you look at an aging Infrastructure, you look at one piece of it. Now there are other pieces that are affected, but the one that people tend to focus on are these large transformers. The fact that it has a limited manufacturing base and doesn't really have a lot sitting on the shelf because it's got to be custom-built, is a huge operation just transporting them. If you've ever seen them, they're the ones that have the specialized railcars or trucks, where they have to go in and literally shut down the roads, and plan from where you're going to manufacture it to where it's going before you even place the order, because it doesn't do you any good to buy one that you can't get to where it goes.

And any geomagnetic storm, if no actions are taken to protect those systems, the power surge within the Grid could potentially short out, melt, and destroy those transformers, which would essentially cripple the Grid because the ability, now, to get power from the generation to the consumption will be severely hampered or disrupted. And there is not much within the system, as designed, to offset those losses. And this could result in prolonged debilitating power shortfalls.

People might say power outages, doesn't always mean you're going to have power outages. But it can mean that you could have severely reduced levels of power. And that may now require further reductions in consumptions. And that timeframe of recovery could be measured, not just in hours-to-days, but in how long it takes to rebuild, reroute, or redirect the Grid to get power where it needs to go.

So, within the world of the geomagnetic storms, these large-scale events, which basically they call them Century Events, they're about once-every-100-years, could get potentially to the level that would not just be a disruption in the flicker to the system, but could actually put enough energy in, that if nothing was done, you could have significant impacts to those transformers.

Other parts of this don't have to get to those 100-year events to see problems. One of the things that will happen in some of these geomagnetic storms is the upper atmosphere expands, increasing drag on low earth-orbiting satellites, where much of the GPS and other constellations sit, can be impacted. These systems would then be affected by the drag, shortened life frame time of those, increased fuel consumption, lowering the lifetime of those, and in some cases, actually causing enough disruption that satellites are lost.

That also plays into radiation, because very few commercial satellites are hardened for an EMP, much less radiation, from these storms. Military satellites and some commercials are—but again, the vulnerabilities are there. And what you'll start seeing, too, is in the disruption in the ionosphere—things like GPS signals become very erratic and can actually go off, not just in meters, but in tens-of-meters, particularly those that are just single channel. Dual channels can still be affected.

And this is more pronounced the further South you are, because once you get to the Equator, we already see disruption in GPS, it's not as good as it is in the Northern Latitudes. So you get satellite's impact from the—what's going on in the ionosphere jamming—not jamming, but distorting signals and, now, degrading GPS accuracy. You get increased drag on satellites. And in some cases, you can actually have satellite failures because of radiation overload. And that's oftentimes going to be in the commercial satellite industry, communications for those disruptions. And it has happened.

And, as the industry builds for that, they try to address it, but it becomes a weight issue, and a factor of lifetime-of-the-satellite, versus what's the risk and what's my return on investment? So the White House began looking at this in 2009-2010, going, "Wait a minute, we do not have a strategy. We have bits and pieces. Department of Energy, FEMA is doing some exercises. There's not really a clear-cut answer of what we do."

And in many cases, it was nobody, really, sat down and said, "What are the risks across the enterprise which become a risk that become something of national significance, versus localized or regionalized impacts?" And so the White House began bringing in the Subject Matter Experts, both industry, as well as science, as well as those Federal Agencies that either had regulatory or dealt with consequences; and in this October, released National Strategy for Space Weather and an Action Plan on things that Agencies are going to do.

Because one of our challenges has been, even though you may get a forecast, and they may have a scale of a geomagnetic or radiation or HF, there's no corresponding action plans by Agencies, "What do you do when you have a G-5? Who makes the call? Who makes—What decisions have to be made? What do you advise the President may require actions to take beyond which Agencies do every day?" We know the answer is that you turn off the Grid. That was somebody's, you know, early on, was the thought, well if you isolate and turn off the Grid, you minimize the impacts.

But there are a lot of things that Industry can do, as well as consumers can do, from the standpoint of minimizing impacts. But it still required Agencies to know, if you have a specific level event, this is what your plan says you're doing, in concert with the rest of the Federal Agencies. And again, these will not be re-occurring Space Weather Events that occur with some frequency and have some impacts. They don't rise to the level of changing how we do business. But when we see events that either could have a triggering Event that requires that, the Space Weather Strategy says each of the Agencies enumerated will develop specific plans and guidelines, based upon triggering events, as well as preparing for consequences of that.

In addition, the longer term, it's to look at how we build more resiliency into the systems and look at what the potential single points of failures are. And are there ways for the Federal Government to mitigate that? Challenge is, as you well know, most of this is owned by the private sector. So again, is this something that we can work with the Industries, either through a regulatory framework, through a cooperative framework, through others—and essentially, my joke is, in the Federal Government, we essentially have two tools in dealing with the private sector, what I call Bribes-and-Extortion. You know, is it tax credits? Or is it regulation? You know, what's the best tools? And part of that is, is getting a consistent answer on what's the best things and best strategies to use to build resiliency.

And one of the things that I think people are talking about is an outgrowth of things like smart grid. As long as we are dependent upon very long distance movement of power across systems to meet electrical demand in a very fluid situation, that in itself is one of the greatest vulnerabilities we have to geomagnetic storms. If our systems are designed to produce and consume power on a more local scale, not dependent upon that as much, or we have the ability for brief periods of time to reduce our dependency on the long distance transmissions, and use other maybe less-efficient, less-cost-effective, but could be less-vulnerable during that timeframe, it means that we could then isolate those Grids and reduce the risks totally.

Again, as we work with the various industries, what are the best strategies? And the focus is primarily on the areas of the Grid itself, particularly the long distance transmission lines and the very large transformers, and those satellite systems which, either through degradation or through damage, could affect Strategic Interests. Communications, GPS, GPS timing signals, and those systems that are used frequently by both industry and Government to perform our essential tasks.

So at FEMA, we deal with consequences. I get asked all the time, in fact Ted Koppel came in my office, I thought that was pretty interesting. And here's Ted Koppel in my office, I'm like, "What'd I do now?" [laughter] He says, "No, I'm retired, I'm writing this book." I said, "What's it about?" He said, "Well, when the lights go out." And he was writing about cyber war and other things. And I said, "Well, I'm actually more worried about geomagnetic storms."

And this tendency that people always get fixated on what causes it, is FEMA deals with consequences. So what we look at in Space Weather is, although it's helpful to understand the risk and it's also helpful to know what our Agencies are doing to work with Industry to minimize that, our job is always coming back to the default of, what if it goes out? What do you do then? And the reality is, is like I like to point out to people, you have to sometimes put things in context. As much as we talk about cyber, or as much as we talk about EMP, and as much as we talk about geomagnetic storms, the squirrels still are the most devastating things to the grid. [laughter] We have seen more power outages because of those varmints than we have from the others. It doesn't mean that it can't happen, but we see power outages on a frequent basis.

The general effect of a large-scale power outage that we deal with at FEMA is usually the result of a hurricane. Of all the natural hazards we deal with outside of Space Weather, hurricanes tend to be the most debilitating to the Grid over long distances. Even tornadoes and earthquakes tend to be more isolated. So we've had experience of what it's like to go under. And we start to see the interdependencies and the types of issues that will curb both immediately and downrange.

And Sandy is a great example of that. When you look at Hurricane Sandy, and you look at the numbers, we had about 350,000 residents and businesses that were actually physically damaged by the storm. Yet what drove the response for the first two weeks was not the damages from the storm, to those structures, it was the power outage. And that power outage drove everything from impacts to the basic electrical, water and sewage systems, to gas, pharmacies, grocery stores, and the role of the Federal Government filling those gaps. And again, we're not designed to replace the private sector distribution infrastructure en masse. We can fill gaps.

And so once the power came back on, the response smoothed out very quickly. But here is the numbers. It's estimated that, when the power was out, we were impacting over 25 million people. Now that's not 25 million customers, and it didn't mean 25 million people didn't have power. It meant that transportation, jobs, work, that whole sector. That's one of the most densely populated areas in the U.S. And so, when you looked at what happened, when you had power outages going for weeks, versus what could occur in a geomagnetic storm, if the system—if everything we know did not get applied, if we did lose those very large transformers, if we did do significant damages to the Grid, now you're looking at months-to-years.

And we can tell you that our experience in the short run—and the longest we've really dealt with power outages in this country has been a couple of weeks, to large populations. Once you move past a couple weeks, dynamics change very quickly, because systems and their ability to cope degrade fast. And the first couple days, nobody is really concerned that the gas stations weren't open. By the end of the week, that was the major crisis we were dealing with. We couldn't get fuel out of the ports, we couldn't get fuel delivered. And, what you could get delivered you couldn't get in the cars.

And so as you start looking at consequences for us, this is where this raises geomagnetic storms to an entirely different level of threat. It may be a low-frequency event, it has an extremely high consequence and now becomes a national security risk. Because, in that situation, not only are we dealing with the public and the businesses that are directly impacted by the power outages, it will start pulling so many resources to begin to affect our economy in a way that starts having national implications versus the localized impacts of a disaster such as a hurricane.

And remember, Sandy, just on the Federal side of direct funding by the Federal Government, was over $60 billion dollars. That does not count private insurance, lost businesses, and all the other costs associated with it. That would be dwarfed by a geomagnetic storm if we don't take action. We don't have plans, and we don't protect the systems that are most vulnerable.

And so, as that debate goes forward, the White House has got out, for the first time now, both the strategy of what the Federal Government will focus on, and a template for the actions of Agencies to begin those initial steps. And they may be very simplistic, but think about that. We need to correlate what actions you take based upon the level of these triggering events. And we generally look at triggering events being in the G-4 to G-5 range, most likely for most systems, it's a G-5. A geomagnetic storm of that size, a Carrington Event, what they call the Century Storm, the once-every-100-years, doesn't mean it'll only happen once every 100 years, it just means there's a one-percent risk.

But they're of such severity that, if you don't have an Action Plan, and you only have 15-to-17 hours to make a decision, and that decision is not a singular decision that says there's a button you push that says, "In case of geomagnetic storm, press here," there will be a lot of cascading impacts that have to be coordinated across various agencies who have 50 Governors who will definitely be interested in what those impacts might be to them, whether or not they need to be looking at enhancing any of their capabilities, including activating or calling to Active Duty their National Guard.

And, if none of this costs any money, it's very easy to do. But there would be significant costs. There's going to be significant costs to industry who may have to make decisions about saving systems at the expense of lost revenue, on the projection of an event, which may or may not occur, as devastating as initially thought. Because, again, if you got 15-to-17 hours, we won't know for sure once we detect that coronial mass ejection how it's going to hit the earth and whether it will hit us.

And the other thing we're going to face is there's very few hazards outside of war that affects more than one nation. Well this will not just be a national event, this will be a hemispheric event. And we've actually done exercises with our counterparts in the EU, within Sweden and others, because it is likely that in a Carrington-style event, we could have impacts across most of the Northern Hemisphere as the Earth rotates, because these effects went for about 24-to-36 hours. So it's not just whatever is facing the sun at that moment, it will be as the earth rotates. And we may find ourselves—and this is where it really gets interesting on the international scale—if impacts across Europe, Canada, United States, and many of the resources we may be competing for, we're all competing for in a market that's very limited, has limited capacity, and has tremendous lead times to buy, source, and install the types of systems that may be at risk.

So you can look at a lot of things that are generally going to be nation-specific. Geomagnetic storms and Space Weather of the Carrington-style could be hemispheric, involving multiple nations, who are both finding themselves disrupted at the same time, but also competing for resources. And if you're—if anybody is here from the electrical industry, and you talk to the Northern tier states, one of our biggest importers of power is from Hydro Quebec. And they have been impacted, and they are more vulnerable, because this, again, the closer you are to the Arctic, the more the impacts are.

And we've already had events where Hydro Quebec has had impacts. They have lost transformers, there have been disruptions. And again, if we're in a situation where it's the dead of winter, we have extremely high-consumption rates for heating, we get a disruption, and the system cannot balance out with that loss, even if it's just limited to Canada, it still impacts us. Correspondently, summertime, again, if we have these triggering events during extreme power consumptions, there's even less resiliency in the system.

Our best hope is it happens when we have moderate temperatures, low consumption, and we have enough backup and ability to take systems down to protect them without shutting down or turning out lights in major Metropolitan areas.

So the last thing I want to leave, in case you want to ask any questions, is this. New York City was kind of an interesting place during Sandy, because the haves and have-nots were basically the demarcation line of who had power and who didn't. What if it wasn't just those areas in New York that were without power? What if it was New York, Boston, Philly, Trenton, that whole corridor that was without power?

And it now wasn't measured in days-to-weeks, it was measured over a Time Scale much greater. Again,

I'm not into the doomsday scenarios. I look at consequences. And the planning for that event, and what we would do, is vastly different than what we would do in response to a hurricane. And it won't be just that area, it would be across much of the U.S. That's why I'm glad I live in Florida. Not only does it have mild winters, the two States that are least of impact from the impacts of these storms is actually Texas and Florida because of the way the interconnects work and how far South they are.

But they would not be immune because of the other distribution disruptions that would occur, because think of this. As we saw in Sandy, how many different supply chains require power to be effective? And how much capacity you lose when you have long-term disruptions? And it is quite startling how vertically integrated and how many single points of failure we start seeing when you turn the lights off for more than a couple of days.

So with that, I'll take any questions you have. Sir?

Q: [off microphone] One quick question. If there were a Carrington-like event tomorrow, the President says "I'm not convinced that we're protected, I want this whole country be dark. *(implying turn off power grid to minimize damage to equipment during a coronal mass ejection headed to earth)* Does he have the authority to declare that kind of emergency?

HON CRAIG FUGATE: It doesn't really matter. There's no way to turn it off. [laughter]

Q: [off microphone] [video 30:55]

HON. CRAIG FUGATE: Well, I mean again, it's one thing for the President to say, "Turn it off", and then you talk to industry and it says, "Well turn it off," and they go, "Turn off what? What part are you turning off? How are you turning it off?" And again, this is not an EMP. So, in some cases, if I can get away from the large very long transmission lines and isolate those very large generators, that's all I need to do. So in many cases, just going off of commercial power and going on your generator pretty well isolates you from the event.

If you have generation near the area, and you're not doing long transmission lines, there may not be a need to go there either. So there's no simple—, you just turn the system off. Because the other problem is, if you turn the system off, think about this. You go dark across the country. You know how long it's going to take to turn it back on? It's no simple—. This is the thing that we keep getting into, is the more we looked at this, the more the complexity and what Industry began telling us, there's no off-switch. And if you do manage to turn things off, it's not as easy as you turn it off and you bring it back on.

Q: [off microphone] I thought that is what happened when Hurricane Sandy blacked out lower Manhattan.

HON. CRAIG FUGATE: Yep. And that's not an uncommon approach. A utility can shut down parts of their Grid. That's not what's the greatest risk in this. It's the transmission between the grids, and the fact that those systems are live and dynamic and always moving power. And because we're moving them over long distances, you may be able to isolate out. And what Con Ed could do, if they have enough—and this is one of the scenarios people look at—if Con Ed has enough residual power generation close enough, where they could isolate themselves from the rest of the grid, they may be able to stay lit and minimize the risk from the impact. Because the greatest impact, again, is to the long transmission lines. And there are some lengths there that they start looking at increase your vulnerabilities.

And so you may be able to do what they call islanding. If you can island yourself off from the grid, you can significantly reduce the impacts and stay lit. So that's why, again, there's not like an off-switch. Sir?

Q: [off microphone]

HON. CRAIG FUGATE: Well, you're not going to go leap from, "We don't have a good plan" to "We have all the answers." And that's why I would recommend you go back, and there's two pieces to the President's strategy. There's the strategy itself, and there's the implementation plan. We've got to take some baby steps. Because within the regulatory frameworks, and what authorities Agencies have, we have not really laid out, "Okay, who's got the authority in the Department of Energy to make that decision?"

If you remember on 9/11, the Transportation Secretary made the decision to ground all the aircraft. He had no legal authority to do it, he just did it. So part of this is to go back and go, by Agency, what are the actionable steps based upon a century-style event that each Agency would have? And now start running through, where did the attorneys tell us we don't have authority or we're not exercising the authority we have?

Part of this may be going back to Congress and going, "We lack clear authority here. And either we need to have legislative language, or we need to go back and look at how do we get industry to address these issues?" And again, it comes back in a regulatory framework, what's the best way? Because for industry, there is costs associated with this which are not going to be returned on the investments, unless everybody has to follow the same regulations. That's been one of the challenges, is industry says, "We know this is a problem. But you got to understand, we have a fiduciary responsibility to shareholders. And when you're talking about investor on the utilities, this is no small matter." And many of them say, "Look, we know what our challenges are. But, unless everybody is compelled to do it, where is our incentive? And how do we go back and show our investors that a very unlikely event is causing us to pull resources and money into areas that they're not getting back in shareholder returns?"

And so, you know, I like this term "resiliency," because I've really come to the conclusion, the more efficiency we build in the systems, the less resilient we become. And that's one of the things we keep finding as we continue to streamline and modernize systems, we introduce variables that we don't quite understand of single points of failure, that, until you have a hazard like a geomagnetic storm to expose them, nobody was really thinking about what was happening as we went through electrical deregulation, we went in the merchant plants, we continue to go with, the bigger they are, the more efficient they are, but the further away we build them, because nobody wants to have smokestacks in their backyard.

And then we'd have to build power lines. And the industry can tell you, one of the worst things right now is just trying to get power lines built so that they can reduce the dependency on those lines and increase carrying capacity. People don't want them, the "NIMBY"-ism. So this is not something that, again, if there was a switch you could press, and the President could press it, yeah, we'd do it. The system is not even built that way. And that may not be the best answer.

That's why, again, we have got to take baby steps. And so for the first time, we have a National Strategy, the White House Policy on Space Weather. For the first time—and again, you may look at him and go, "This isn't enough." Trust me, you've got to start somewhere. It starts enumerating what Agencies are now required to do to begin planning specifically those initial steps of a triggering event and what those actions are going to be.

And, as we go through that, some of the things that we are doing and have done is continue to do exercises of these events and work, both with the Industry, as well as with the Federal Agencies on how we respond, what are the impacts, and the likelihood of that. And again, as industry has been doing this, they're starting to see some solutions of what they have internally to themselves that can better reduce their vulnerabilities, because nobody has a vested interest in letting these systems go, losing those transformers, and losing all the revenue and costs associated with it. So there's a vested interest on everybody to get the best answer.

But there is no easy answer, because the system was not built that way—in many cases deregulation and how that industry evolved actually set up increased vulnerabilities. And it's not just to this, if you look at when InfraGard got started by cyber, SCADA systems, and the ability to control and load balance very complex systems and a mercantile market, if those systems are disrupted, we end up with significant impacts to the system that we either do one of two things. Either the power goes out, or power consumption becomes very expensive, and we end up having to use very different types of fuels that may increase, again, pollution and other activities or costs to manufacture.

So again, we know that, as we look at systems, and you keep looking at new risks, we didn't deal with cyber 20 years ago. Now, what does this impact? So, just staying onto the geomagnetic storms, it's the first steps. And it took us a while to get it there. And I think, again, when the President starts asking about, "What's this thing called Space Weather?" a lot of people that were saying this needed to be addressed finally had a voice in the White House asking the very question that needed to be asked.

And so, from the time the President says, "Hey, what is Space Weather, and what are we doing about it?", we've been able to move from almost like voices in the wilderness, because, when you think about it, you're going to the White House saying, you know, "A Carrington Event in 1859 would take out today's Grid," and they're like going, "Well, that was in 1859. I've got all these other things to deal with. Go away."

From 2009, there was a handful of people at my Level that even knew what Space Weather was. Now we have a policy, and we have an action plan. Again, they're baby steps, but it's a move in the right direction. You want to know what the foundation and what you can do, it's putting pressure. Because we will not complete it in this term, I can just tell you, what has to be done cannot be completed in a year. So you need to make sure this is an Administration priority going forward, and that Congress holds the Administration going forward accountable to implementing that plan. And, as we learn what we need to do, that we may not have Authorities or resources to do, that we apply that or request those Authorities to execute and build that resiliency.

So with that, I thank you very much.

[Applause]

CHUCK MANTO: Thank you, Administrator. Before we go, I'd like to have John Pi come up and Gary Gardner come up. Gary Gardner is the National Chairman of InfraGard. John Pi is responsible for the program at the FBI. And first of all, we want to say, it's because of folks like you who care about this. Back in 2011, InfraGard created a nation-wide program to address Space Weather and any high impact threat that could impact the country for a month or longer. And these are the folks who were involved in it from around the country. It's called the Electromagnetic Pulse Special Interest Group.

But, as you can see from our conference proceedings every year, there is a big sun. And there's a picture of an actual solar storm taken by someone on the International Space Station. But what we've done this last year, because of the work of the Administration and the Strategy and in the Action Plan, we organized a lot of topics for some industry and Government to look at this. And we've noticed that you were calling for the Whole of Communities to respond and to planning and exercises.

So we put together a rich bibliography and a Triple Threat Exercise that looked at Space Weather, EMP, and cyber, with either one-, three-, or 12-month nation-wide consequence. And what we would like to do is share this with you, even though FEMA is the King of planning, we learned from you folks. And, in fact, we appealed to the HC Process and everything else in this work. But we would like to present this to you as the American public's response to a Whole of Community effort. And this is our Whole of Community effort gift to FEMA.

HON. CRAIG FUGATE: Thanks.

CHUCK MANTO: Thank you very much.

CRAIG FUGATE: Thank you. I've got one thing to add.

[Applause]

HON. CRAIG FUGATE: Have you found yourselves talking to a lot of business folks, and they kind of look at you when you start talking about this stuff, and they shake their head and blow you off? We were presented this in NATO. And it was basically, we had a crowd like you, except they were all about to fall asleep, until the representative, Swiss Re got up. And this is something that's very powerful, because it gets people's attention. Representative Swiss Re says, "We've done our analysis and determined that a Carrington-style event is an uninsurable risk. And we're advising all of our clients to begin exempting geomagnetic storms from business continuity and business interruption and damages in their policies."

Now think about that. Swiss Re, who would insure just about anything, has said, a geomagnetic storm, under current technology and current practices in industry, is not an insurable risk. And they will no longer reinsure the industry forward. That may get a few people to pay attention.

[Applause]

END

State Plans and Regional EMP SIGs: Workshops and Table Top Exercises

The current and emerging activities of regional EMP SIGs and the development of a preplanning framework for high-impact disasters.

Corresponding Video:

https://www.youtube.com/watch?v=EqNGn-3koKg&feature=em-share_video_user

Panelists:

- **Ms. Mary Lasky**, National EMP SIG
- **Mr. Steve Pappas,** MW EMP SIG and IN EMP Planning
- **Mr. Steve Volandt,** SE EMP SIG and the NC EMP Plan
- **Prof. Mel Lewis,** NE EMP SIG, Updates on NYC area planning
- **Hon. Andrea Boland**, National EMP SIG Policy Adv Panel, Maine update

MARY LASKY: It's a pleasure to be able to talk about the kinds of things we've been doing in this last year with the EMP SIG. And all of you are here today because of the EMP SIG. We are happy that we can be part of the Dupont Summit, and we thank you all for coming.

Last year, as Chuck mentioned, we had a similar event, and in fact we've been doing this since 2011. In 2011, it was very small, and so it's great that it's grown every year, and that's been exciting.

And since 2012, we have published the proceedings for each of those events, and we will do that again this year for this event. And what you received today is the proceedings from last year's event. So we hope that you will all take those copies and read them and absorb the information that is there.

Last year, as we just mentioned, and a copy of this was given to the Honorable Fugate was that we held a tabletop exercise to look at three kinds of events. We looked at cyber, solar storms, and then EMP. And that was very beneficial, and there are books that you can purchase to see the things that we've done.

This year, however, we decided that we would take a slightly different approach and say, "Okay, something bad can really happen and what kind of plans are we going to make to see what we can do about this?" So we started planning yesterday. We had the morning time, was looking at various things that we could do, what kind of event—what would happen in those events, what were we left with, what kind of communications would we have, and something called Ponderosa, based on the Ponderosa tree that is resilient. And so we said, what kind of things will still survive, and what would we have if something really bad happens, either cyber or solar storm or with an EMP?

So we looked at all of that and we looked at—we had a panel look at what happens with water and food and medical and our supply chain and fuel, all five important things. And then we had the afternoon of planning with breakout sessions looking at the Executive Branch, the Legislative Branch, the National Guard, Civil Defense, what happens with volunteers in our community, and then power and financial.

And the results of all of that will be compiled into a start of planning. And what we're hoping with the planning is that we will be able to take this to the next steps of having this go out to the regions, and that it's a start for the year and that next year, we can come back and see what of all the planning and where have we gone, and what have we done.

So in doing that, one of the important things is that any of you who would like to come and work with us on what are these plans? And what are the next steps? And what should we be doing? We would be very happy to have you come and help us. So, please talk to me if you want to get involved. That would be great.

The EMP SIG has done a couple of other really significant things, I think, during this last year. And one of them was that we've involved the International Council on System Engineering, or INCOSE, and you've heard that talked about this morning. And so you have a feel—I mean, I think that it's exciting that we're branching out and that we're working collaboratively with others. And so we're very grateful that we have done that.

The other really major step that we have taken is that we've been working with regions across the country. And that is a start that you're going to hear about from our regions right now, about the kind of activities that we've been doing. So not only are we organizing things like here at the Dupont Summit, but now we have regional groups that are taking this the next step and doing it across the country.

So, let's turn now to Steve Pappas, who's going to tell us about what he's been doing there in the kind of Midwest.

STEVE PAPPAS: Thank you, Mary. So, the Midwest EMP SIG consists of Indiana, Ohio, Michigan, Illinois, Minnesota, Iowa and, I believe, Missouri. It's configured in the same makeup of the regional configuration for all of the InfraGard chapters. One of the things that we've been doing in our planning process this year is put together a framework of how we're going to go and move forward with exercises and training in the EMP SIG. And Steve Volandt will cover a little bit more of that in a few minutes.

I wanted to touch on why this is important to everybody and how you can take the products that we have for sale here today and implement these activities back in your home location. My background is in the water sector, the water sector and emergency management. In Johnson County, Indiana, I was a Deputy Director for a number of years, and I was able to work at the grass roots level. In every county across the country, you have a County Emergency Management Agency, and you have a County level, comprehensive Emergency Management Plan with annexes called Emergency Support Annexes. In my case, the water annex was ESF3. ESFs address transportation to energy to law enforcement and, ultimately, there's 15 of these that address every one of the critical infrastructures.

So this involvement with the EMA, and my background going way back was in the military, plans, training, exercises. One of the opportunities I had as a Deputy Director was to take in a number of different courses at the Center of Domestic Preparedness on the National Incident Management System Doctrine, and the Homeland Security Exercise Evaluation Program Guidelines.

Now, all of our first responder agencies across the country adhere to that, NIMS Doctrine and Homeland Security Exercise Program Guidelines. In our role as a critical infrastructure and our interface at the County, State and Federal levels, with the emergency support functions, means that we've got to interact with State and Federal agencies. We need to learn their language, and that language is NIMS Doctrine and HSEP Guidelines.

Now, yesterday in a panel discussion, we had a gentleman from Public Health that left one with the impression that NIMS gets really—you can get stuck in the weeds, and that's true. But you've got to tailor it specifically to your needs and requirements. And I think that's important. The manuals that the Federal Government has produced are voluminous. Tailor to your needs and requirements. Use the checklists that are in there. I think that's the most valuable tool in all of that documentation, are the tools for incident command or for operations, planning logistics.

So the documentation that we've given you utilize NIMS Doctrine, use the HC protocols, tie in with your County levels and communicate, coordinate, train and exercise with your partners at the County level and at the State level, and then share information with sister chapters and bring them along, as well.

The last thing I want to say is simplify. That's what I've heard throughout this week and up at the INCOSE conference in Cleveland. Simplify, simplify, simplify. NIMS can do that. There's one thing that they try to teach the responder in NIMS is smart objectives, and creating smart objectives—specific, measurable, action-oriented, realistic and time-sensitive. And if you boil that down so that those people on the front lines understand exactly what needs to happen, that's exactly what you'll be able to do, is make things happen.

So let me pass that on to the next panelist.

MARY LASKY: Okay, Steve Volandt from North Carolina, will you talk to us now about what's happening there?

STEVE VOLANDT: Sure. Thanks, Mary and Steve Pappas. Hi, I'm Steven Volandt. I'm in North Carolina. We've got two InfraGard Chapters there, one in Raleigh and one in Charlotte. The Charlotte one stood up about a year ago and raised their hand to lead the effort to create a Southeast Regional EMP SIG Interest Group, which I've gotten involved with. And Tory Crafts, if you could stand up just so everybody knows—if you're here, they can recognize you and find you? I don't see him right now. I'll call him out later.

What we decided to do was we needed to get organized. We had lots of smart people talking about this problem from many different directions. So the first thing that we put together was a Charter. We've got some governance built into that Charter that's working group-based, and it's organized into interest areas, and then further segregated into domains. The Charter was written to be reusable by other regions and other entities that are interested in this problem space. So please contact one of us if you're interested in a copy, or Chuck Manto, or Mary.

We realized as we tried to get our heads around an initial exercise that we needed to do more than just work with the people that we had, at hand, in order to cover the entire Southeast. So, we settled on focusing on North Carolina, and then we've learned that the state of North Carolina was getting organized for this themselves. So we've been able to leverage their work as well. I'll talk about that in a second.

The other thing that we built for the Southeast, in addition to getting organized from a governance perspective, was we built a collaboration portal. And the Charlotte Chapter lead that, and Tory himself was in charge of that. And it was also designed to be reusable and a template for others. It's scalable. It could be scaled to handle all of the EMP efforts across the country, really. It's in a secure, redundant location and includes features like a document library and a collaboration Wiki, if you've worked with those before. It allows volunteers to register. They can submit their contact info, their industry sector, their activities of interest that they'd like to sign up for, the domains that they're interested in.

And we're also wanting to screen for people with prior military experience. If an EMP or a solar storm were to happen, we would have situations that would approximate a failed nation-state and a lot of our military veterans have experience fixing things and operating in failed nation-state environments. And then lastly asks for who your InfraGard reference is.

Tory, I just saw you walk in. Could you stand up? This is Tory Crafts, and please reach out to him or myself if you want to learn more about what we're doing in the Southeast.

Now I'm going to turn to the state of North Carolina and ask Taylor Smith to stand up. Now, she's been really the person in the middle of facilitating a lot of the work—thanks Taylor—that happened with the State of North Carolina. So please ask her if you want some detailed answers or contact information.

Now, what the State of North Carolina did is they put together a scenario that's not as severe as what we're used to. It's more of a limited EMP attack, which fades out over North Carolina. So parts of northern North Carolina have really the full blast effect. And then by the time you hit the South Carolina border, it's fairly minimal. And they put together an annex to the State Emergency Management Plan. They focused on five critical sectors that were prioritized, plus public health and safety.

So the sectors they focused on were electricity, communications, transportation, water waste water and fuel being the core for any kind of planning and resiliency preparedness for the State. And their goals were to review the scenario with the participants and what would be the expected conditions after an EMP event.

They also worked to identify existing response structures, so how can we leverage what's already designed to handle emergencies versus create something new, and establish measures that we could take if there was some advanced warning, particularly with a solar storm. And in particular review public information messaging. How do we get the word out and communicate to the public-at-large?

And I've sat in on several of those meetings, not all of them, and one thing that struck me was, like I mentioned earlier, lots of very smart people with lots of opinions and lots of differing understandings of the problem which led us to think, rethink, what we really needed in the Southeast Region. And one effort to mitigate that lack of harmony in people's opinion was to come up with a Maturity Model. And the Maturity Model is based on starting from a Level Zero Maturity where you're unaware, or participants are unaware or don't believe in the problem, would be Level Zero. And Level Five would be you've got an organization—whichever organization you're assessing is prepared to be safely resilient and sustainable for 400 days after the loss of all the infrastructure that we normally count on. So One-through-Four, and between there show different Levels of Planning and then preparation activity to achieve that State of Full Maturity. And I'm happy to share that with anybody that reaches out.

We've also been reaching out a little bit, taking baby steps, to talk to legislators in the State and other folks at the Executive Level, so we're socializing with them, I guess you could call it that description, with the aim of drafting some legislation next year. And one of our goals in the Southeast is to promote components for resilient buildings and local communities. If you think about the problem that we have with all of our just-in-time food production and manufacturing, there's no buffer. If things stop for more than a few weeks, all the supply chains are halted. And to reconstitute any single one of those is an enormous task.

And I grew up, and I think most of us grew up, hearing stories from family about how they handled the Great Depression. Almost all of us had a family farm somewhere that some relative had, and nobody was worried about starving, for example. And those buffers don't exist anymore. When I go to the grocery store, I'm eating salad from California or from Florida. I'm not eating salad from North Carolina.

So, that's a very long-term goal, but I think it's worth discussing. And if we were inoculated with that kind of a buffer, this problem would be a lot less alarming.

Ways to accomplish that that we've explored are passive solar-heated and cooled buildings with attached greenhouses and hydroponics that require no electricity, no power, just natural heat convection of the air circulating through them combined with rain catchment systems. How many of those would you need? Not everybody needs those. What percentage of a community would need that kind of a fallback for this problem to be a lot less severe?

The other thing we've got to do now that we've got our governance in place and this Maturity Model in place is we need to do some recruiting. So if you're from the Southeast, please reach out to us. We would like your participation. And we know that Tory has been talking to folks in South Carolina, so we're expanding into South Carolina, but the other States in the region, please reach out to Tory or myself. Thank you.

MARY LASKY: Thank you, Steve. Now, we'll hear from Mel Lewis, who is in the New York area, and actually was our first of our Regional Groups. So, Mel?

MEL LEWIS: Yes, hi. I'm the Chair of the InfraGard Northeast Region Committee for EMP SIG. And we started our activities at the Regional Level in May of 2014 and since that time, we've been, after creating a Charter for Chuck to approve making us official, we've been doing things like sharing Open Source Information with our committee members. We have about 40 committee members in the New York, New Jersey, Connecticut area, plus another 25 that are spread out more geographically that want to participate and be involved.

In addition to sharing—and that's how we started sharing EMP-related information on the sources and vulnerabilities—we then started doing things like holding Awareness Briefings for industry and the infrastructure people and Law Enforcement, Government agencies. We would have actual half-day meetings at a central location and there would be people coming in to become aware of things like that. We bring in people from, in addition to Chuck and others, we would bring in people from the EMP SIG Community to help us with that.

Then we had Planning Meetings, and these are HSEEP *(FEMA standard)* compliant in order to facilitate the consistency from one group to another, and also to make sure we didn't forget anything. Because it's difficult to prepare for a Tabletop Exercise, which we're planning to do in January of 2016. It'll be held in Westchester County in New York, which is just North of New York City. It's being held at the New York Power Authority facility. We've met there before, we like that facility. The people that are working for the New York Power Authority are very supportive and helpful.

So we're going to be doing that Tabletop Exercise. We're inviting people from Law Enforcement, Emergency Medical Services, Fire Service, Government Organizations, the OEMs, of course, and Non-Government Organizations, such as the Red Cross and Faith-based Organizations, FEMA, DHS. These have already been coming-- these people have already been coming to our meetings, but we're going to bring them back and encourage them to bring others to our Tabletop Exercise.

We're also very focused on communications. Loss of communications, as has been mentioned, is serious where we lose situational awareness when we don't have communications. So we're bringing in Amateur Radio People, ARRL (Amateur Radio Relay Leg). Several of us on our committee are Ham Radio operators. I got my license when I was 13, so we understand the utilization of that. Just to give you an idea how old I am, I was involved with RACIs during the Cuban Missile Crisis, okay? So that was the real deal, all right? We thought we were really going to be nuked, we really did.

So, we're bringing in the utility companies such as energy, water, sewage treatment, natural gas. Healthcare, of course, hospitals, nursing homes. Other communications entities, such as the phone companies, wireless and wired land line. Our Tabletop Exercise is not going to be a doomsday scenario where, as mentioned earlier, there's no point in preparing, it's hopeless. There isn't much we can do. Rather, it'll be serious, but not devastating so that we can address doable actions and come up with gap analyses that are useful and we can assign action items and the different Agencies will assign themselves action items to close those gaps.

In closing, I'd like to recommend, maybe all of you have done this already, I'd like to recommend that you read the book *One Second After*—I have nothing to do with the author or the publisher—and also *Lights Out*, which used to be an Internet-published book, but now it's available in paper. These are eye-opener documents that will help you to understand the possible consequences of loss of the grid for a year or more. Thank you.

MARY LASKY: Thank you, Mel. Now I'd like to turn to Andrea Boland. She has been just a wonderful light in Maine working there tirelessly. So, Andrea, give us an update.

HON. ANDREA BOLAND: Thank you very much, folks, for your attention. I actually have PowerPoint, which I've cut down, Chuck, quite a bit, try to accommodate the time. But there were a few things that I wanted to share with you. As a Legislator in Maine, I brought the first piece of legislation on EMP and GMD protections that passed. So, I got a lot of support for that from many international experts and national experts, and several people here in the room who came to testify and helped to get that passed. Because, of course, it's not an issue that any one legislator can do by himself or herself.

Just to try to jump along through this as quickly as I can, some of the main points that I want to make is—that it's important to remember that States have regulatory authority. We're inclined to always look to the top, the Federal level, and they've really had a hard time moving things forward for years. And so the States now have gotten more and more engaged. And, of course, they're closer to the people and closer for you to reach. So I suggest that you pay attention to what may be going on in your own States. They've really been leading the way.

The main problem that we have is a problem of politics and regulatory capture. And that's something that I'll talk about a little bit as I go along. The main points of the Maine legislation, which was LD 131, which passed in 2013, was—I wanted them to just do something. I wasn't interested in too much in baby steps, I wanted long strides. And, of course, as a legislator and a non-technical person, I can enjoy that kind of position and just push them. So I would just ask for them to just get protections on the Grid and give them the benefit of the doubt that they'd know how to do that.

And, of course, they protested that it wasn't needed, there wasn't a problem. And then as the testimony of experts became stronger from meeting to meeting, they finally got to the point of admitting that there's an awful lot they didn't know, and I guess it was okay to do a study. So the study was really kind of a model piece of legislation that other States have also followed in one way or another.

It just asked for the Public Utilities Commission to look at the most vulnerable components of the Maine system, the potential mitigation measures, their estimated costs—low-, moderate- and high-cost, a time-frame for adoption, the policy implications, and give us a report back by next January. And everybody agreed that that was fine, good thing to do. It was interesting, the Chair of the committee polled all the stakeholders in the room and got them all to nod their assent. It was a long way from starting out saying I was just totally nuts about this whole thing.

We ended up with two reports, one that was authored by Imprimis, who was an R&D firm that licenses some technology to block surges into the big transformers. The other was authored by Central Maine Power Company, even though as you may remember, it was supposed to be done by the POC. It eventually got shifted to the power companies.

But, you know, without the good help of experts, particularly the Foundation for Resilient Societies is just very strong on helping us, John Popik and Bill Harris. We got to something that at least gave us opportunities. So the following year, I was no longer in the Legislature, I was termed out, and a new piece of legislation was brought to say let's use some of these mitigation measures that we came up with.

And that, the utilities really fought hard. There was a high-level of lobbying going on, and they were saying it wasn't needed and they appealed very much to the Republicans in the Senate. The Senate's the easiest place to work because there are fewer people there. And they really appealed to them hard that that would be unfriendly to business. And needless to say, our side was presenting that it was quite the opposite. But anyway, the result was that action piece of legislation resulted from the studies failed by a single vote in the Senate.

So one of the issues that was brought up was cost. So I wanted to just show you this, and I don't know if you can read it from where you are, but my layman's way of trying to simplify what the two different studies gave as priorities.

Central Maine Power Company authored one that essentially as you look at it, their priorities amounted to costs of about $42 million. If we use the neutral ground blockers that, of course, Imprimis knew all about, and that generally acknowledged around the country to be effective, the alternative would have been eight-to-twelve million dollars to accomplish very much the same things, and more, for less money. So the question is, why wouldn't you want the lower cost piece?

And that's something that you need to look at. The utility companies get a guaranteed rate-of-return. In Maine, they get approximately a 12 percent rate-of-return. So, it's actually more profitable to them to spend more money rather than less. There was also the fact that they would question whether they could get cost recovery. They could get the customers to be able to offset their expense. They had a great deal of support in feeling that FERC would approve that.

So that didn't seem to be a real big problem. And then the actual manufacturer of the products then said, "Well, if you want to protect the whole Maine Grid, the biggest transformers, the 345k transformers, they would—", and they did it within a certain amount of time to make that commitment, they would halve the price. So for about $4 million, as compared to 40 that has burgeoned to more like probably, well, $70-million-plus, at this point, the choice was to go with the more expensive solutions—and basically, less protection because the neutral ground blockers work like a surge protector on your laptop or your computer. And they've been tested by National Labs, and they're actually a very good one, also working already. And they're reporting great results.

But there's a real funny thing going on in the Industry around the utility companies, and NERC, just resistance to this thing. And I still don't understand what it is.

Here's another problem that we've got, because when you look at legislation, you got to look at more about what works technically and what it costs. And I'd like to read it to you, if you don't mind.

"In no event Central Maine, which is Central Maine Power, in no event shall Central Maine be liable for any incidental consequential multiple punitive special exemplary or indirect damages or loss of revenue or profits, attorneys fees or costs arising out of or connected in any way with the performance or non-performance of this, basically what's in the Service Agreement. Even if such damages are foreseeable or the damaged party has advised Central Maine of the possibility of such damages, and regardless of whether any such damages are deemed to result from the failure or inadequacy of any exclusive or other remedy."

We've got that in all the States so what they can be assured of is that if there's a big problem, the customers and the taxpayers will foot the bill. CMP, I guess I mentioned to you, has the 11.74 percent guaranteed Return on Investment, too. My point here is that we need to link the guaranteed Rate of Return on Investments to higher protective standards, and I think that's reasonable to ask.

Actually in Maine, we have a statute that says it will do that, but it hasn't been applied. There's such a problem with regulatory capture that we've got the PUC turning to the industry to ask its questions. The Public Utilities Commission was as surprised as the committee members when they heard the testimony of the experts. And when the CMP did the study to look at all these problems, it refused to even use its own recorded historical data on the events on the Maine Grid, which has shown tripping and actually a fire resulting at our nuclear power plant following a major solar storm.

So, just to report at this point, the study group that was formed eventually that the power company who was directing included some of our national experts, foremost I must say again the Foundation for Resilient Societies have been—they're right in the next State and it's great to be able to call and just haul them on up to Maine and they do all their work and happily present it and generally leave the utility company speechless.

But eventually, we have made progress. And that's another message I want you to hear is—the States have regulatory authority, legislation is not easy, but it's valuable, if only the opportunity for these big experts to come and testify and let the word out so that legislators and others start to know more. And also because it starts to push the industry to do more, the utility company.

So, what they're doing now is installing series capacitors at some high-voltage transmission lines. They're adding reactive power capacity against voltage drops, which are a big risk in Maine. And they will continue to look at the neutral ground blockers, but they won't be taking it up until some time after the middle of next year.

But the most important thing they're doing is they have a big—an excellent development with the CMPs, is that they will install a pilot synchrophasor* unit to monitor the Grid, so for better management options and to know, really, truly what's going on. Which is terrific because they haven't had that before and they were resisting it when it was suggested in legislation.

It will include monitoring of harmonics and DC flows which is something that the neutral ground blocker would do on its own. And the really best thing about it is not only will they be collecting data because the data has been collected, an independent outside firm is actually going to be doing analytics on the data. So we know what it means and how to respond to what it says. So they'll be capturing data at high frequency, hundreds of reports-per-second, and it will allow for a cumulative look at what is actually going on.

So, the initial findings were due the end of November. I'm a member of the taskforce, we'll be meeting on December 14th, not long after we get back, and this will have an ongoing look at this whole issue and how we can respond to it. I was going to report on some other States, but I think I've used up my time pretty well. But in other legislatures, we're seeing actions trying to go forward, the resistance there and some people coming back, but some people getting discouraged. So thank you very much for listening to this report.

[Applause]

MARY LASKY: Thank you, Andrea. And there was work in other States modeled on what you've been doing. So, it's really helpful, so thank you. And thank you all, panel, for the work that you're doing across the country.

[Applause]

---

* A phasor measurement unit (PMU) is a device which measures the electrical waves on an electricity grid using a common time source for synchronization. Time synchronization allows synchronized real-time measurements of multiple remote measurement points on the grid. The resulting measurement is known as a **synchrophasor**. Phasor measurement unit: Wikipedia, the free encyclopedia, https://en.wikipedia.org/wiki/Phasor_measurement_unitWikipedia

Transformer Protection Test Data

Corresponding Video:

https://www.youtube.com/watch?v=QrFdcTYsbBk&feature=em-share_video_user

Panelist:

- **Dr. Fred Faxvog,** Emprimus

FRED FAXVOG: Okay, I've got a question for everyone. Hey, I've got a question for everyone. How many have heard of Transformer Neutral Blocking? Pretty good. How many have not? One, two—Okay, I got a few—I've got a lot of slides. I'm going to try to roll through this quickly.

Okay, we've talked about low-frequency, high-impact events. I don't call this low frequency anymore. It's at least moderate frequency, and it's extremely high consequence. So this is really important. What I'm going to talk about is the vulnerability of the grid and mitigation solution. So we're going to get to the solutions and real data.

Okay, there's a long history here of both Solar Weather studies and EMP related to the electric Grid. The Solar Weather papers—we can find them dated back to 1930; there's probably people here who have probably gone back further than that and have seen papers even further back than that. The EMP experiments in the Pacific were done in the late '50s, early '60s. There were two EMP commission reports, 2004, 2008. There was the Grid Act, the SHIELD Act. They both died in Congress several years ago. Now we heard yesterday that the SHIELD Act is going to be part of the Energy bill. Great news, absolutely great news.

There's IEEE power grid papers related to solar storms that date back to 1972. Our work started in about 2009 on this subject. And the Idaho National Labs started maybe a year before us, of really getting into this. And I just confirmed that with Professor Baker this morning.

Here are the images from the high-altitude nuclear tests in the late '50s and early '60s over the Pacific. And this is when they first realized the EMP problem, the EMP threat. They actually disabled some electronics in Hawaii, which was many, many hundreds of miles away, and all of a sudden it was a realization that a nuclear bomb above the atmosphere propagates this electromagnetic wave down upon us. That's what was discovered a long time ago.

And then the solar understanding came in the early '70s, started to be put together in the '70s, where you get coronal mass ejections, solar flares coming off the sun and they go off in all directions. But when they come our way, we know about it. And they interact with our magnetic field of our Earth and they jiggle it, and they cause this so-called E3 wave that impacts our power grid. And we see them. If you live in the northern states or Canada or Scandinavian countries, or the South, towards the South, South Africa, you see the effects of solar flares hitting the Earth.

Now, just a little perspective here, here's an image—boy, it's just barely over the top of the table here—here we superimposed Jupiter next to the sun. The sun is a big, big ball of fusion going on up there. And you can see a solar flare coming off the sun there. Well, below Jupiter is the Earth. Look at the size of the Earth compared to a solar flare, even when it's superimposed right next to the sun. When it gets to us, it's much bigger after it's traveled the 93 million miles.

Now, there have been two solar Super Storms in our history that we know of: the Carrington Event, which has been mentioned several times, the New York Railroad Storm in 1921. Both of these occurred- they occurred before we built our Power Grid. If you look at the evidence of electrical power in our country, we built the Grid after World War II. And the power went up during the years of late '40s, '50s, '60s, '70s, is when we built the Grid. We have not witnessed a solar Super Storm since we've had the Grid. Do we know what it'll do? We sure do. It's easy, very easy, to model and calculate. And it's been done by several companies now and their software, powerful software.

So we have not witnessed a big Super Storm while we've had the Grid. Now, the probability of a big storm hitting, a Super Storm, is 12 percent within the next decade, or 25 percent within the next 20 years, two decades. That's a big number. That's a big number when you consider the consequences of the power grid going down and you can't get it up for months, if not years. So this is a big number.

Now, this probability was calculated by three independent scientists (all totally independent), three independent papers and the number comes out just the same. It's right on, all three of them agree perfectly.

Now, one of the consequences of these storms is you saturate the big transformers. You have what they call half-cycle saturation, when the DC current, or quasi-DC current, is in the transmission lines and in the transformers. So when you saturate them, you generate all these harmonics. And harmonics is bad quality power. And the standard, the IEEE standard, is 1½ percent for total harmonic distortion. And we exceed this all the time, whenever the solar storms are hitting us, even for the small storms.

It only takes a small amount of current, and that's been shown in both calculations by Ray Walling on the left there, and in tests that we did out at Idaho National Labs. And just a small amount of current, anywhere from 4-to-32 amps of DC in these transformers gets you well above the IEEE Standard for Harmonic.

And in addition to that, there's been a study done, it's a statistical study by four very, very well known scientists, one of which spoke here yesterday and will be speaking this afternoon, Bill Murtagh. And, what this statistical study showed is they took all the insurance claims, and they threw out the ones where there was a known cause, what caused the damage that was being reported to the insurance companies. They threw those out, and then they looked and they found these insurance claims would peak like a month after we had a Solar Storm. So they attributed those to the Solar Storms. The correlation is very high, and it looks like ordinary storms, storms each and every year, results in a $2 billion-plus dollars annually for—and this was done over—data was looked at over ten years, 2000-to-2010. So everyday storms are causing a lot of damage, and nobody knows what it is. This was a statistical correlation that they found and published. And actually, there are three papers on it now.

Now, somebody mentioned that we had a big coronal mass ejection from the sun in 2012 that missed us. Here it is. We happen to have a little video that NOAA put together on this, the data from satellites shows the solar flare going out the back side of the sun—did not come our direction. When they looked at this one, they said this was the big one, this was one of the big ones. And you can see the Earth is over to the right side there, the arrow points to the little yellow dot. That's supposed to be the Earth. And the CME, the coronal mass ejection, comes out pretty much the backside. You can see how large it is by the time it gets out 93 million miles.

Now, if you look at the data for these very, very large ejections, on average they come off about every seven years. This isn't a one-in-a-hundred-year deal. It's a one-every-seven years, statistically. But they go different directions. And when I look at this, I say, Wow, have we been lucky or what? If the last one was 1921, the one before that was 1859. Now, you can't say that it's due or anything like that. You know, the time clock starts now. The probability is 12 percent within the next decade, or 25 percent within the next 20 years. To me, that's as big number.

Now, we looked at some data that wa submitted by EIS to Maine, to the Maine Public Utility Commission, and what they found was data collected up at Chester, Maine, at a transformer up there, and they found in one case here, 100 amps of current, and they correlated it to the magnetic field, the rate-of-change of magnetic field of 250 nanotesla per minute, a hundred amps. We know the big solar super storm is 5,000 nanotesla, it's 20 times larger. Multiply 100 amps by 20, you're looking at 2,000 amps that's possible at a substation up in Maine.

There's a scatter in the data here because they didn't have the data that told you what the electric field direction was. Doesn't always line up with the power line; but when it does, you get the higher currents. So that's why there's a max on this, and a min and the median. But the possibility is 500-to-1,000-to-2,000 amps. And that will fry a transformer. There is no doubt. You can talk to the manufacturers and they will verify that.

So, our unit is a neutral blocking device that we put in the neutral connection, not on the high- voltage side. We're on the neutral side of the transformer. And when we detect the current, we put in a blocking element. It's a capacitor bank that still effectively grounds the transformer; the power stays on, everything operates like normal. We're just grounding the transformer differently. We're grounding it AC-wise, not DC. And it blocks the DC so you've mitigated the problem.

And here what we find in our modeling that we do with power world—is, as we add transformer neutral blockers, the collapse of the voltage moves to the right. So we can go to much higher geo-electric fields, much higher EMP E3 fields, E3 and GMD are essentially the same, both the quasi-DC phenomenon.

So here we raise the collapse-point out considerably with putting 25 neutral blockers in the system. And this was done for Wisconsin. This was a Wisconsin run. We do have one unit up and running. It is in Wisconsin, it's at Morgan. When you put one in, it doesn't change the collapse.

Also, another benefit is we reduced the reactive power, what they call the VARS. And when you put in 20 percent blockers in the system, you drop the reactive power demand by over 40 percent. So there are multiple benefits.

Now, people talk about procedures that they can take care of this or mitigate with procedures. We found in our modeling, and this was done for Maine, that if you follow the ISO formula of reducing power transmission by 10 percent under a storm condition, we found no change in the voltage—in the collapse voltage, essentially none. So, procedures, moving power around, unless you move a lot—if you move it 90 percent down. But up in Maine, in the high-demand times, they're moving thousand megawatts in the summer, or in the winter.

The transformers at risk are essentially 5,000 of the big ones across the country. And in our modeling, we find that we can do a very effective job by putting neutral blocking on a thousand of the 5,000, or 20 percent. And that could go for a State, it could go for the whole country. It's a pretty good number, 20 percent. And what does that relate to? Our unit is $400,000 per copy. I had a President of a utility up in Canada come up to me and says, "How much does that run?" I said, "It's $400,000 installed." He said, "In our industry, that's peanuts. Peanuts." He said it three times, "That's peanuts."

Well, if you do all thousand in the United States, that's $400 million. It's a dollar-and-a-quarter per person amortized over ten years. It's pennies, it's peanuts. I mean, it really is. And they're a lot less costly than a $10 million Series Capacitor Bank. And we have series capacitor banks in the systems, especially out in Western states, the really long lines.

Now, the transformers with highest risk are the really big ones, the 345, 765, and if you got ones up towards a million volts, yes. That's what we find in our modeling. And Frank Koza told me this morning, or yesterday, that's what they're seeing in their modeling, as well. It's the big ones, and the modeling shows that the highest currents in these systems that we've studied, the highest geomagnetic currents, induced currents, are at the generation sites. When we start blocking, we find that's where you want to block first. It ends up in the first couple iterations of our Blocking modeling.

You can also look, and many times these transformers are single-phase. Well, single-phase transformers have more reactive demand power than other types; seven times higher than a three-phase, three-legged transformer. So it's a single-phase, big transformers. And then if you're along a coastline, there's an enhancement factor in generating the geo-electric field. We know it's a factor of two, it could be as high as seven, enhancement of the geo-electric field at the boundary. And that goes in about 50 kilometers inland.

So where do we have our big cities? They tend to be along a coastline or along the Great Lakes. And earlier, the gentleman from FEMA was saying we lost a transformer in Canada. We lost a couple in Canada. We lost one in New Jersey, Salem, New Jersey, for that same storm. And we've lost several in South Africa for these storms.

This is our unit, what it looks like—eight-foot-by-eight-foot-by-12-foot pole. We've tested it, we've modeled at the University of Manitoba. We've tested it at a lab outside of Philadelphia. We tested it at Idaho National Labs. This is the circuit diagram. It's fail safe, fail safe, fail safe, I can't say it enough, because we've got so many safety precautions in here. Somebody asked a question just the other day about what does it do to your neighbors? We modeled here Wisconsin and showed that as you increase neutral blockers along the bottom, you don't really change the amount of current going to your neighbors. It stays the same.

And the whack-a-mole effect, if you use one or two, you got to be careful, but you need to put in sufficient number to really get the advantage. This one's in Wisconsin, it was in Wisconsin for over a year until the engineers got their head around it and got used to it. Now, they love it. Engineers say we need more. It went into operation in March. In June, we had a storm for two days, the 22nd and 23rd. Went into operation 14 times during the two days, and that's the evidence of it working in a small storm.

Chuck Manto: We'll have a chance to, on maybe our best practices section to—one minute, and then we're going to go on the best practices section. I'll be standing right with you.

FRED FAXVOG: Now, on warnings, on St. Patrick's Day, there was a storm. The warning from NOAA came 49 minutes after the storm hit the Earth. I don't think procedures are going to work. And then the storm that we were operating under, we got the message four minutes before the storm hit. Of course, ours is automatic, we don't need a notice.

Here's some things that we've learned. I think I've said most of these. It's a very effective solution, and there we are. Thank you.

Chuck Manto: Thank you very much. We will have a chance to hear a little bit more on the Best Practices panel.

END

The National Space Weather Strategy and Action Plan Impact on the Whole of Community and Preparedness

(Also accepts presentation of the EMP SIG Triple Threat Power Grid Exercise Program)

Corresponding Video:

https://www.youtube.com/watch?v=SzdaJ31ytAY&feature=em-share_video_user

Panelist:

- **Hon. Caitlin Durkovich**, Assistant Secretary, DHS; SWORM Cochair

HON. CAITLIN DURKOVICH: Jerry, thank you very much for that kind introduction. Good afternoon, everybody. It's wonderful to be here. And I would like to just tell Jerry and InfraGard that you are actually a real friend to us, as well. We have a long, constant partnership with the Department of Homeland Security, and especially within the Office of Infrastructure Protection, with InfraGard. And over the course of the year, engage in a number of Joint activities around planning, around sharing tools and information. But most importantly, in some ways, out in the field working across the 16 sectors in the name of advancing physical and cyber security protection, ensuring that we are working closely together to help owners and operators and industry understand the very dynamic risk environment that we're in. So, thank you. I think over the course of the next 12 months, you'll see even more exciting things happen between our two organizations.

I am delighted to be here to talk to you about our efforts on the Space Weather front. I know my colleague, Director Fugate, our Administrator Fugate, was here earlier giving his perspectives on the Space Weather Strategy and Action Plan. And I'm going to focus more on the prevention and mitigation side and the work that we have done around that. I will tell you that this has really been a very rewarding subject for me to work on.

And when I think back three years ago when I stepped into this role and what I thought encompassed the range of threats and hazards that could impact critical infrastructure, Space Weather was not necessarily one of those. But in part because of my good friend, Bill Murtagh, and certainly the advocacy of many of you in this room, I learned a lot both about this hazard, but about the ways that it can impact critical infrastructure. And this is, I think, really exemplary of how we approach the whole all-hazards landscape when we think about critical infrastructure, security and resilience.

We recognize that there are a range of threats and hazards that are out there. Some are higher probability, lower consequence. Some are lower probability, higher consequence. And it's really important for us to work with owners and operators to understand this risk landscape, recognizing that we can't protect against everything. But how do we look at the consequences and build a program that really mitigates the range of these consequences most effectively? And that's, at the end of the day, what we do in the Office of Infrastructure Protection.

To that end, it has been an honor for the Department and certainly for me to serve as the Co-Chair of the National Space Weather Strategy effort alongside my colleagues in the White House, Bill in particular, but certainly across the Interagency and with the National Weather Service. In addressing Space Weather at a national strategic level, the focus of my Office has been understanding the potentially devastating impacts of this hazard within the context of all hazard critical infrastructure security and resilience.

I think one of the most important things that we've done is ensuring the process of developing this National Strategy was collaborative across whole of community. And that really does include owners and operators and making sure that the private sector, those who own and operate our nation's infrastructure, had a voice, had input into this Strategy. Because, at the end of the day, the security and resilience of our nation's infrastructure is not primarily a governmental responsibility, but rather, it requires a national partnership. And to that end, I do want to thank my private sector colleagues both for their enduring efforts to ensure the Security and resilience of our Infrastructure, but certainly for their Partnership on this front.

The other interesting thing about the work that we've done on Space Weather is the recognition that this is a global hazard. And in general, I think when you think about the Security and resilience of infrastructure, we know really, at the end of the day, that it has no boundaries. It's not limited by geographic boundaries because of the nature of these systems and networks. But certainly, when you look at the hazard of Space Weather, it is a global hazard, and so it has enabled us to really think about how we are coordinating internationally on this effort.

So we've worked to integrate an improved understanding of the impacts of Space Weather, whether a geomagnetic phenomenon into how we do our business into sector—the sector coordination structure that we have established under the National Infrastructure Protection Plan, and the National Preparedness System. And that's really what I want to focus on, is how preparing for Space Weather events requires joint action from the public and private stakeholders, really bringing our shared expertise and responsibilities that are embedded in our infrastructure goals.

And the preparation, the work, is really focused on reducing vulnerability, minimizing our risks, addressing the potential cascading impacts, and at the end of the day, enhancing resilience to this particular hazard. So, I hope all of you have had an opportunity to look at the Space Weather Strategy. I think it's been two months, Bill, since we—about then, about rolled it out? But there are six overall goals which are structured to be mutually supportive. The First one is establishing benchmarks for Space Weather events. The Second is enhancing response and recovery capabilities. The Third goal is improving protection and mitigation efforts to Space Weather.

The Fourth is improving assessment, modeling and prediction of impacts on critical infrastructure. The Fifth is improving Space Weather through advanced understanding and forecasting. And then the Sixth, and I touched a little bit on this, is how we increase international collaboration, on this front.

My Office, the Office of Infrastructure Protection, is really focused on the Third goal, which is improving mitigation and protection. And, at the end of the day, risks like Space Weather, but also cyber attacks,

EMP, illustrates the highly interdependent nature of critical infrastructure and demonstrate why we need to take the time to plan, to make investments in the protection and mitigation across different hazards that will have co-benefits.

So Goal Three, which again is improving protection and mitigation efforts, was written in the context of the strategic imperatives that align under Presidential Policy Directive 21, which is critical infrastructure, security and resilience. And it really focuses on the whole of community efforts to protect our nation's critical infrastructure. And then Presidential Policy Directive 8, which focuses on National Preparedness. PPD-8 and the National Preparedness System already contain a robust architecture for national protection, mitigation, response and recovery. And then out of PPD-21 comes the National Infrastructure Protection Plan—and the way that we work with sectors, both to address these threats and hazards specific to their own sectors, but think about them from a cross sector perspective, as well and offer us a framework for how we deal with some of these high profile risks, such as Space Weather.

In addition to the oversight and input from the entire Interagency Space Weather Taskforce, we had a number of active participants in the development of the Strategy and Action Plan. That included the Department of Energy, FEMA again, who you heard from earlier, the Department of Defense, FAA and DHS's own Science and Technology Directorate.

So let's turn now specifically to the core elements of Goal Three, and there are several. And again, I'm going to keep foot stomping this. But, at the end of the day, the way that we've articulated and laid out this goal is the recognition that the Federal Government doesn't own and operate most of the nation's critical infrastructure. And that really, at the end of the day, if we are going to effectively prevent—I shouldn't say prevent—but mitigate the impacts of this hazard, we have to work in close partnership with owners and operators of critical infrastructure.

So to that end, the first element is, how do we look at the relevant legal mechanisms, authorities and incentives that can be used to protect our critical infrastructure and the systems and networks associated with them? So, we knew already that a number of statutory and regulatory Authorities are already in place to protect critical infrastructure, and that there are also incentives that encourage actions by owners and operators. And so part of this is continuing to look at what is in place, but identifying authorities, gaps and issues related to these mechanisms.

I think part of what is unique about what we do in the Office of Infrastructure Protection is, at the end of the day, most of our work is in support of owners and operators, and 95 percent of it is all voluntary. How do we encourage owners and operators, really, given the dynamic environment they're living in, to adopt a risk management approach and to insure that they are as secure and resilient as they can be? But we are learning over time that given, again, this complex environment that we're living in, faced by all of these threats and hazards, that this can become an expensive proposition. And so really, how do we think about how to incentivize owners and operators to make those needed investments? And how can we encourage them to adopt leading practices to address gaps where they may not have thought about a particular hazard. And so that's really what the first element is focused on.

The second, and in some ways is the most important thing that we do within the Department of Homeland Security working across the Interagency, and with owners and operators, but is to develop the plan, right? You got to start somewhere. You need a plan to address the range of threats and hazards that we're dealing with. And so that will be one of the main thrusts of what we work on, is how do we develop hazard mitigation plans that help owners and operators, reduce vulnerabilities, manage the risks and, at the end of the day, if needed, assist with response to the impacts associated with Space Weather.

And again, this is where this Whole of Community Approach comes into place. And it is both looking at how we share information, how do we improve the mechanisms by which we're doing that, including leveraging information sharing and analysis organizations. But thinking about innovative approaches to insure our partners have the information that they need. It is also incumbent upon us, I think, to come together once you have a plan, knowing that no plan survives contact, but really to test and exercise those plans, and then to take the after action that comes out of those exercises and use it to improve the plans and the things that we are working towards.

I think a critical part of this, and this is certainly where we are often most challenged and where you in this room come into play, but how do we achieve long-term vulnerability reduction to Space Weather events by implementing appropriate measures at critical locations that are most susceptible to Space Weather? And I know from going out around this country, and in particular working with some of our partners in the electricity sector, there is investment around this country going on to ensure that we can mitigate some of these vulnerabilities. But really, frankly, more needs to be done. And I think continuing to look for innovative solutions, standards, best practices, operational procedures that will improve the protection and resilience is vital to this whole of community effort.

It is important to point out that the Space Weather benchmark events described in the first strategic goal, which again is established benchmarks for Space Weather events, are going to be used and leveraged—to support the adoption of design standards for enhanced resilience to evaluate strategies for, and the priorities for, and the feasibility of, protecting those critical assets and fostering the mechanisms for sharing best practices that promote the mitigation of, and damage from—of systems and networks impacted by Space Weather.

Another core element of this, and again we've been at this for ten years, but the range of threats and hazards that we focus on from the Public/Private Partnership has evolved. And, so how do we continue to build this Public/Private Partnership to support private action, to reduce overall public vulnerability to Space Weather, given, again, this very kind of complex all-hazards environment that we are living in right now?

And so, at the end of the day, it's how we work with those private sector entities to ensure their resilience to this range of events, including Space Weather. So, one of the efforts that, again, I think is critical to those of you sitting in this room, is, how do we incorporate resilience measures into the systems that span Infrastructure in this country—and continue to look for mechanisms that we can share information?

And as I mentioned earlier, identifying incentives and disincentives for investing in resilience measures. And really help make the business case for making these investments on the front end as opposed to getting to a place where we're in a response-and-recovery phase—and the resources that need to be applied to respond are, at the end of the day, probably more than had we made those investments in the front-end.

I think, at the end of the day, when you think about what are the critical considerations for success, the risk posed by Space Weather highlight the independent nature, or the interdependent nature, of our critical infrastructure systems, and demonstrate how investments and protection and mitigation across multiple hazards can have co-benefits. And again, this is whether we're talking about Space Weather, we're talking about cyber security, whether we're talking about EMP. And this is, frankly, our biggest challenge.

I stand up here as the Assistant Secretary for Infrastructure and Protection, and as much as we're focused on weather, whether here on ground or up in space, we're also very worried about an asymmetric threat about a very changing Counterterrorism environment. And when you're sitting there as an owner and operator trying to figure out where to apply resources given this very complex environment, it can be challenging. And so how do you make sure a potential hazard like Space Weather is considered, given the range of other things that a company is looking at?

I do just want to speak a little bit to Grid resilience. It is something that has been in the news an awful lot lately. This is a topic that is near and dear to my heart, and it applies in the context of whether we're talking about Space Weather, about cyber, about hurricanes, you name it. I think, at the end of the day, we all know that we cannot protect every asset, network and system from every threat and every hazard. And there is certainly a lot of room for improvement when it comes to thinking about how we secure our nation's electric grid, how we ensure it as more resilient. But I will tell you, at the end of the day, I think the Grid is an incredibly resilient system, a very resilient infrastructure.

We have worked over the course of the last 12 years, both in a regulatory framework, but equally important in a voluntary framework, to support the efforts of owners and operators to ensure that we have a resilient Grid. And I think we're buoyed by the fact that, at the end of the day, we've created this very complex system and, in itself, it makes the Grid very resilient. Our former NSA and CIA Director, Michael Hayden, has said that because of the biodiversity of the Grid, it is inherently resilient. And so, then you layer on top of that, the work that we have done around focusing on kind of a defense in-depth approach to how you secure the Grid, and that's deterrence, detection, prevention and prioritization.

And I think, at the end of the day, it is important that we continue to think about how we evolve the resilience of the Grid. But I think that we are overall—can be assured that the Grid is a pretty resilient system. We continue, as we look at how we modernize it, to make sure we are doing what we can as a Federal Government, in support of owners and operators, to maintain the security and the resilience of the Grid. It involves kind of re-thinking the partnership that we have with owners and operators. So as much as we're communicating on a daily basis with Directors of Security, with Chief Security Officers, with Chief Information Security Officers about, again, the range of threats across cyber and physical, we have also recognized that bringing Chief Executive Officers, and others from the C-Suite, to the table to focus on this issue is very important.

Because, at the end of the day, they are managing enterprise for us. They're concerned about regulatory risk, operational risk, brand risk, and they need to be focused on security and resilience. And I will tell you, at the end of the day, this is paying off. It has enormous dividends.

Over the course of the last three years, we've been meeting, we the Federal Government, on a regular basis with the CEOs of Investor-owns, of the Rural Cooperatives, of the Public Power Associations, to talk about these incidents of national significance; again, whether driven by our adversaries or by some sort of weather event. And how we ensure we understand what the roles and responsibilities are of Government, of industry, and where we have some gaps.

And it's led to really some productive efforts and deliverables, ranging from a Playbook, for how we're going to respond - to a national Incident—to really, a renewed look at our National Transformer Strategy. And so again, bringing those executives to the table who have more of a strategic view has proven very beneficial.

We continue to think about how we can continue to share information with our owners and operators, both at the security level, but also at the C-Suite—and I think, really, have evolved both in terms of the sharing of classified information—but on the cyber side, doing this in a more machine-speed fashion where that sharing of information around vulnerability, cyber vulnerability, cyber threats, is relatively instantaneous. And we're evolving to a point where we can share indicators on a real-time basis and do it bi-directionally. So I think that is critical.

And then again, how do we think about some of these strategic issues? I've talked a little bit about the National Transformer Strategy, but I will tell you that EMP, that Space Weather is also on the agenda, as we think about what are these lower-frequency but higher-impact events and the need for industry and Government to come together.

I want to end, and I do want to leave some time for questions and try and keep you all on schedule. But by foot stomping and putting an exclamation point on the fact that we have to think about all of these threats and hazards from a risk management standpoint. And I think you've heard that theme run through my speech today. We can't prevent, we can't protect against every threat and hazard. And so it requires us to look at this from a risk management standpoint. I continue to make the argument that I think this country has gotten better at responding to those higher probability, what I'm now calling kind of medium consequence events, whether it's a hurricane, whether it's a tornado. We really have the ability to quickly—and it's part of the resilient nature of this country, but the fact that owners and operators, I think, do a very good job learning from the last event and improving it so they're in a better place at the next event.

And if you look at kind of what's came out of Hurricane Sandy, even just within the electric sector and how they've re-thought their mutual-aid agreements, how they've re-thought how they're going to need to respond to another Super Storm Sandy, and all of the compacts that they have in place, it really is, I think, a sign of the resilient nature of that particular sector.

But where we're at right now is, I think we've got to pause and look at these lower-frequency, high-impact events and think about, again, how we mitigate the range-of-consequences that are caused by several of what I would call these lower-frequency events.

So that's our goal, is to work with Owners and operators to think about this from a risk management perspective, to make the case that investing up front, whether it's actually the preventive measures that you put in place, your strategies for Planning, Testing and Exercise, all of these investments on the front end are going to save you dollars on the back end.

The other piece of this, and this is certainly something that I think is important we think about—and there's been a lot of conversation about Infrastructure in the last few weeks, and you're starting to see it talked more about on the Campaign Trail. But as we begin to reinvest and we begin to modernize our nation's infrastructure, ensuring that we are baking in security and resilience up front is critical here. And that is something that we have started within the Office of Infrastructure Protection. We've got an Infrastructure Design and Recovery effort that we have stood up that is really aimed at this. As we begin the Design-and-Build phase for new infrastructures, insuring that we understand what the world today looks like, but what the world a hundred years from now is going to look like. And how do we put in place security and resilience measures that are both attuned to today, but a hundred years from now and allow us through the Operations and Maintenance phase to make adjustments, as needed.

I think the last piece that I will leave you with in thinking about this, and I've talked a lot about the electricity sub-sector, but the need really to focus on the interconnections, the interdependencies. And I spent a lot of time going from sector-to-sector meeting, and the electricity sector will maintain they're the most important. The water sector will maintain they're most important and the comm. sector will maintain they're the most important. But at the end of the day, all of these things intersect. They're interconnected. They're interdependent. A major disruption in one is going to have significant consequences in these other lifeline and strategic Infrastructures. And we have to continue to work together to look at these intersections and to think about in the event of a really bad day here—what are we doing to do to make sure that we can get all of those sectors to bounce back quickly? And that includes understanding how to prioritize it, how to make tradeoffs and ensuring that, really, the folks within Government, at Craig's level, at my level, at my Secretary's level, at the President's level, understand those priorities, as well.

So, I will end by just saying this has been an incredibly gratifying project for me to work on, both because of the friendship that I have developed with Bill, who you are going to hear from next, because of the opportunity to work across multiple organizations who are thinking about this. But again, because I think, at the end of the day, we've taken a hazard that maybe in the past we would have just looked at through the Federal Government's eyes, and we've made sure that we've brought industry, the Owners and operators, to the table, really, to help drive those other core goals.

So I think we're off to a good start. We've got a long way to go, but we've got a great strategy that will help guide us.

So with that, I'm a few minutes over 1:30. Gerry, want me to take a question or two?

GERRY BOWMAN: From the microphone.

CAITLIN DURKOVICH: Microphone? Yeah, absolutely.

TOM POPIK: Hello. Tom Popik, Foundation for Resilient Societies. First of all, thank you very much for your engagement on these very important issues, and also for being willing to come and be a speaker at this event. We really do appreciate that. And as you can imagine, we listened very, very carefully to what you have to say. And one of the points you made was that 90 percent of critical infrastructure is owned by the private sector. That leaves still 10 percent that is owned by Government entities. There are two very large transmission systems here in the United States, Bonneville Power and TVA that are Government-owned.

One of the issues that we see is a reluctance among industry to be the first in installing certain protective measures, such as blocking devices. And I just wonder, is there any opportunity for these large Federally-owned entities to be platforms for proof of concept for some of these protective technologies?

CAITLIN DURKOVICH: So, the short answer to that is yes, I've actually been out to Bonneville, which experienced a substation attack some years ago. And they're doing proof of concept around substation security. And so, absolutely. But I will tell you, I've also, on the other side of that, been to investor-owns, and this region that are doing proof of concept work, as well. And not only are they doing it, but they're very willing to share the lessons-learned and the best practices with the rest of industry.

TOM POPIK: Thank you.

CAITLIN DURKOVICH: You're welcome. All right? Very good.

[Applause]

CHUCK MANTO: Our Chairman would like to make a presentation to you. On behalf of the Whole-of-Community that you have called upon to respond to this, a number of them have pulled together top resources in government and industry to come up with a Triple Threat Power Grid Exercise which looks at a one-, three- or 12-month long nationwide disruption of power, either due to Space Weather, EMP or cyber, and we would be glad to share this not only with DHS, but all the affiliated organizations that you work with—and actually offer to be taught and collaborate with you and FEMA and others to make this an ongoing and continually-improved product. So, thank you very much, and maybe you might want to say a word because you know her. Plug in the mic. Thank you.

CAITLIN DURKOVICH: Okay, Chuck, thank you very much. Thank you, absolutely.

GARY GARDNER: Yes. I want to thank Caitlin for coming over and sharing with us her perspective and expertise and interest. And she has been a true supporter as we've gone forward in this effort. And I think that the alliances that we're building are going to further enable us to see the results of all your hard work and better protect this nation against those types of catastrophes.

CAITLIN DURKOVICH: Thank you very much, Gary. That's nice. Thank you.

[Applause]

END

Key Operational Space Weather Facts and the Action Plan

Corresponding Video:

https://www.youtube.com/watch?v=PbBnt4G-pTI&feature=em-share_video_user

Panelist:

- **Mr. William (Bill) Murtagh,** Assistant Director for Space Weather at the White House Office of Science and Technology Policy

WILLIAM MURTAGH: I had a high school teacher, he used to say to me, "Murtagh," he says, "Nothing ever sinks in with you until you hear it twice." And I'm saying that because I'm seeing a lot of familiar faces out there from the presentation yesterday, so there's going to be a lot of similar material here.

However, a slightly different emphasis because Caitlin obviously went before me, and her focus was on Goal Three, and I'm going to emphasize a little bit more on Goal Four, but more specifically on goal Five just because I—largely, even though my position at the White House is all-encompassing on space, weather and preparedness, but the science piece is my background. So I'm going to touch on that a little bit more.

Again, I want to provide an overview of the National Strategy and the mechanisms behind it, the motivation behind it, the process that drove the whole—the development of both the Action Plan and the Strategy, and, of course, the critical Next Steps. Now that we have an implementation plan, what are we going to do with it?

As I've mentioned in many presentations before, I really loathe when I see great efforts going forth and the development of national strategies, and I see that strategy ending up on a shelf and a day, a week, a month, a year or two years later you realize not much has come from that great effort. We do not want to see that happen. So I'll talk about how we're going to make sure that it doesn't happen.

Now, on the motivation, I'm going to jump into a few different things here. One of the—I'm going to state the obvious, but it must be stated, and that is, of course, our reliance on advanced technology for everything we do today, and the vulnerability of this technology to Space Weather was one of the huge motivations behind the White House and the development of a National Strategy. Loss of any of these systems, whether it's GPS, its satellite systems, other satellite systems, communication satellite systems, or of course the electric power, for any loss, may it be for hours, days, weeks, and especially long-term. That's one of the big distinctions on Space Weather, as many of you know, is that we could possibly be looking at a long-term outage of some of these technologies, and are we prepared for that? That's one of the big questions we have to answer.

So needless to say, as we become more and more reliant on this technology, we've got to make sure we're in a place where we know how to manage life without it, should we be without it, or what we need to do so we don't get into a position where we are without it.

Busy chart, it's supposed to be busy. In the back of the room, you probably can't see much. I'm just trying to show here is that one of the distinctions, one of the unique pieces of Space Weather, it is a global threat. It's a distinction we made as we embarked on this effort of the White House. It's a little bit different in hurricane. We know the consequence of Category 5 making landfall in a major U.S. city. We know what it looks like if we see an intense earthquake. It's largely a regional or localized effect. It can be very large region, some of the particular impacts. But in Space Weather, when those big eruptions occur on the sun and they come sweeping past the Earth, the Earth is a little dot of sand as this massive eruption from the sun sweeps past. The whole Earth gets impacted. In an extreme event, Carrington-like event, we're not sure exactly what the impacts of it would be, but in a much lesser event, back on Halloween of 2003, we got a little taste of what it might look like.

The event back in that period-of-time was actually a series of major eruptions that occurred over a two-week period, impacted technology around the world. And we recognized satellite systems of various countries were impacted, power grids. And now the power grid in most of Sweden went down. That was lots of transformers damaged in South Africa, and many, many other examples of impacts on technology around the world. But again, just that distinction that it is a global threat, not a local or regional threat.

One thing I wanted to point out, too, and I just to do this just a little bit more often because if there was a little bit of a criticism on the actual National Strategy, it does focus largely, of course, on an extreme event, and what an extreme event might do to the country. But, it should be pointed out all the time, Space Weather, like regular weather, occurs all the time. And the consequences of Space Weather have been felt quite regularly. So I'm just going to, on more occasion now, throw a few slides up there showing that yes, no big extreme events in the last decade or so, last decent-sized storms we had were those Halloween storms.

But even in more moderate storms, just to show how this stuff can affect technology, March of 2012, you'll see some of the headlines here, "Solar Flare Knocks Out LightSquared Satellite." It was actually the biggest communication satellite in geosynchronous orbit got a whack from one of these growing coronal mass ejections. We thought it would be out for a couple of hours. Meanwhile, three weeks later, it was still down. Others, you'll see in the *Stars and Stripes*, "Recent Solar Storm Interferes with Air Force Satellites," so national security systems will be impacted.

And there, even on the bottom, people often ask about the impact on aviation. This was one particularly interesting one. You'll see INCERFA. The FAA has three stages, or Three Phases of Alarm, alerting system when there's an aircraft potentially in distress. The first one is the INCERFA. And the INCERFA is issued when there's no communication with an aircraft for 30 minutes or more. There's Federal Aviation regulations that require continuous communication when they're without communication 30 minutes or more, it is a problem.

Is it a big problem? Well, that'll be the next two levels. But back in that period, again, March 2012, these INCERFAs were issued because they were not—we were not able to communicate with some of the aircrafts in the higher latitudes. Why? Space Weather. It's very important for situational awareness when you are not communicating with an aircraft, or you are unable to communicate with an aircraft, why is that the case. So, if you're able to turn to the alerts and warnings you might have got from the Space Weather Center, you'll realize, "Okay, very, very likely may be Space Weather that's impacting this."

And even more recently, a couple of stories behind this, and this was just last month. Again, just to demonstrate that Space Weather occurs all the time, we had an event last month. Didn't make—I don't know if anyone in the room even heard about this particular event. But, I can tell you, it did make its way into the Government circles and into the Department of Homeland Security and ended up on my desk. And the question was asked, "Bill, we just had a solar eruption and it shut down the airspace over Sweden." I was just waiting, I'm not quite sure why, and the question was asked, "Bill, can that happen in the United States?"

Because you'll see the Weather Channel headline, "Massive solar storm halts air traffic in Sweden." It wasn't massive, it was a relatively small event. I can't control the media. But this wasn't a big event and it shut down the air traffic. The United Kingdom have embarked four-or-five years ago, 2012—excuse me, back in 2008/'09 timeframe, they had very little interest in Space Weather. Then what happened? Not a Space Weather event, but they had their airspace shut down because of a volcano erupting a thousand miles away. And there was a whole lot of elected officials in the United Kingdom scratching their heads saying, "Oh, I didn't know this could happen. What else out there could happen that we don't know about?" People got up there and they said, "Well, actually sir, ma'am, there's this thing called Space Weather. Should we get one extreme event like that, we would have similar effects to what we saw with the volcanic ash situation by some accounts."

That changed the whole dynamic in the United Kingdom and within a couple of years, actually it was just last year, they opened up a 24/7 Space Weather Operation Center and even in the last discussions between the Prime Minister and the President of the United States, Space Weather was on the agenda. So they're very keenly aware of it in the United Kingdom. So lots of messages in the event that occurred just last month.

But getting back just to touch on one, when folks in the White House and elsewhere were asked, "Bill, can an event like this happen in the United States and close our airspace down," what was the answer? The answer is, we're not so sure. We don't know. And it's, again, part of the motivation behind this National Strategy, to get to a place where we do understand exactly what might happen as best we can, given the limitations in our understanding of the science and, indeed, the impacts.

Another huge motivation, of course, behind the development of a National Strategy was our awareness, our keen awareness, our growing awareness of extreme Space Weather. And you've all heard about this Carrington Event back in 1859, 1-2 September. It was a magnificent event. We measured the intensities of these storms often with a proxy measurement of the Southern extent of the Northern Lights, the aurora borealis, and this particular event, what must it have been like in Cuba and Central America for those folks to look up in the sky and see the Northern Lights, a reflection of the intensity of the geomagnetic storm back then?

But our awareness that a storm like this, should it occur today, what might it do to the nation? And one of the reports that came out, of course, was from Lloyd's of London working with a science company called AER, and they had all sorts of dire suggestions of what might happen, including trillion-dollar-impact outages that would extend past a year. And then one of the big things was highlighting the fact that the vulnerability perhaps great, is not only just in this nation, but across the world as that place between Washington, D. C. and New York City—because of our latitude, because of our geology, because of our proximity to a large conductive body of water—the Atlantic Ocean saltwater—and because of the very interconnected Grid that we have open, this Northeast, all creates that kind of vulnerability.

Another big piece that motivated the development, the necessity for a National Strategy was the tremendous work that we saw going across Government circles over the last couple years. On the Hill, for those who heard yesterday, Congressman Franks gave us an update on some very, very positive steps in the House and, indeed, there's efforts in the Senate, as well, to introduce a Grid. And we heard yesterday, the Shield Act, being folded into the Energy bill.

The U. S. Regulatory action, the Federal Energy Regulatory Commission have stepped up and we're seeing the development of standards now for the—which did not exist years ago—where the operators of the Grid must have operational procedures and do vulnerability assessments. As Caitlin said, Space Weather is now in the Strategic National Risk Assessment. FEMA are doing lots of good things, and then the international scene, especially in the United Nations, all sorts of efforts under way to address the issue of Space Weather, may it be in the aviation side of the House, satellites and the Committee on the Peaceful Uses of Outer Space, and many committees within the United Kingdom.

But all these good things happen, we recognized one thing we really did need; and that was a cohesive, all of Government strategy to address this issue. We're past the point in suggesting that this—we're at that point where we believe this is obviously a real threat, we need to do something about it. So we put together a taskforce, the White House ordered the development of a taskforce, in the summer of last year, and by November of 2014, the SWORM, the Space Weather Operations Research and Mitigation Taskforce, was established and tasked to develop a strategy and an action plan.

And Caitlin, of course, emphasizes, it was—a tremendously valuable piece of this whole effort is we brought the Homeland Security side, the DOD-side, FEMA was in there with the National Security Council. And we assured that when we were going to deal with the private sector, owners and operators, we would be able to work through DHS and all the great connections they have to make sure they were included.

But you could see on the taskforce across so many different Departments, within the Office of the President, it wasn't just the Office of Science and Technology and Policy, but we also had the National Security Council involved. And, of course, key there was the Office of Management and Budget. We need to know what we can and cannot do, where the limitations are during the budget cycle, so we need them involved, and then across all the various Agencies. So it was very much a multi-Agency initiative.

Caitlin emphasized this, that it was not just then, of course, inside the White House and inside Government circles. We had the Public Comment Period where we headed out for public comment for a 30-day period. Got lots of good feedback that we incorporated into the Strategy. And Caitlin also talked about the Six Strategic Goals. But now she emphasized, she focused largely on three. I want to make a few comments here on Goals Number Four and Number Five.

Goal Number Four, in particular, has this critical action, and it is the Vulnerability Assessment, comprehensive vulnerability assessment we must do across all sectors. It's what's missing right now. There's bits and pieces here. There's some great work done here-and-there and everywhere, but we want some comprehensive effort. The U. K. have done one, and I think it's very good, where we can point to a specific document that gives us some sense of what are the vulnerabilities to aviation, what are the vulnerabilities to GPS systems and so on. So that's a key action called out in Goal Four. The deliverable will simply be the report itself and the timeline, which is very important in these action plans is we've identified the Agency responsible, what the deliverable is, and the timeline to provide that deliverable.

In Goal Five, a couple of key actions. So Goal Five largely focused on the science and research piece. So it was NASA, it was NOAA, it was the National Science Foundation, the Air Force Research Lab, Naval Research Lab, other folks involved in the science and research and the operations, issuing alerts and warnings.

But just to give you some examples that I'm very confident we'll see happen because of this Strategy in the very near future, and this is one. DOI, Department of the Interior, specifically the U. S. Geological Survey, USGS, in coordination with Commerce, with NOAA, will sustain existing ground-based magnetometer network and enhance the network through the installation of new observatories, new magnetometers. We understand, it's clearly identified in many documents, that the distribution of magnetometers around the country right now is lacking. These things don't cost many millions of dollars. They're low cost, why are we not doing this? We will do it. We called it out in the national strategy, we will budget for it, a deliverable will achieve a sustained measurement and data-continuity and complete installation of new observatories. The timeline of sustained measurements, enhancements completed within five years. That doesn't mean we're going to start in five years. It means we're going to start in this coming months and years, and be complete within five years.

I'm going to stick with the same theme because there's reason, continuity for it. The Department of Interior, again the GS, will identify and fill gaps in the MT Surveys beginning with the Northeastern United States. One of the things that are lacking right now in our ability to observe and forecast, is understanding truly the geology of the ground beneath us, the conductivity and the resistivity plays big into how electrical field is formed, how induced current is created, and how it impacts the Grid.

Again, not a huge effort here cost-wise. But it will be tremendous benefit should we put a little bit of effort and money into getting this done. And again, we called it out as an action. The deliverable is simply to complete the improvements so we have across the entire continent assessment of the conductivity within one year of publication of this plan. I can tell you already efforts are underway to make this happen.

But these are the kind of things. Over the course working with many of you in this room and many others in the science community, we sat around tables for the last couple of years saying if only we did this, we need to do that. Why haven't we done that? These are all those gaps that we tried to fill here as much as we can in the Action Plan. We've called them out, we've called out the responsible Agency. We put the timeline, we'll bring OMB and other budget people into the room and we'll make this happen.

So, to finish up with The Way Forward, how do we make it happen? What we're going to do is form an

Oversight body in the White House. We've had this taskforce managed from the White House; we'll continue to manage it and the implementation phase from the White House. It'll be quite similar, I think, to the taskforce in that we'll put together that body. It will be lead by Office of Science and Technology Policy working, again, with the OMB, working with the National Security Council. On the Agency side, I want somebody representing, like we had Caitlin, maybe it's somebody from FEMA or from her Office, the DHS. The Science-side, maybe NOAA or NASA again, co-chairing this whole initiative, picture us meeting twice- or four-times-a-year in the White House; this big spreadsheet's been formed. We have a meeting coming up on Tuesday where it'll be laid out. Here's the 90, whatever, actions that are out there. Here's the columns that say who's responsible; here's the point of context. Working Groups will be formed to make sure we pull off these actions. The accountability is there, the reporting will be done, again, twice a year.

And one key piece here, we'll house it on the DNSTC, the National Science and Technology Council in the White House to ensure the perpetuity through—this cannot end with the change of an Administration. So past this Administration, we'll have it, we'll have the Agencies engaged, we'll have it on the DNSTC, so regardless of who the President is, regardless of who the Chief Scientist is, and so on, this thing will live in perpetuity. And we will get it done.

One last thing I just wanted to show, just to give you a sense of how things are working already. This was back in July of this year, the Memorandum comes out from Shaun Donovan, who's the Director of the Office of Management and Budget, along with John Holdren, the Director of the Office of Science and Technology Policy, and this Memorandum is the multi-Agency science and technology priorities for FY2017. So this is the document that gets sent out to NOAA, to NASA, Commerce, everybody gets this document. Here's the Administration's priorities, here's where we need to—where we start focus on our Budget. This is going to help guide you, this is what we want to see in your budget.

And you'll see down at the bottom, Space Weather observations on research and development are essential to address the growing societal needs for accurate timely Space Weather information. The Agency should prioritize investments in Space Weather, according to the National Strategy and Action Plan. Very straightforward.

That's it. So again, thanks very much. I appreciate the opportunity. I certainly appreciate the great work that InfraGard has done and the Triple Threat Power Grid Exercise document. Chuck is very, very good. I shared it with my colleagues in the White House. I think we are going to see a lot of people using it.

One interesting piece, too, the Strategy and the Action Plan is the great enabler. Regardless of what's in there, the fact that it's written to this level at the White House, I've got lots of different interactions over the last month, and two in the last week, on different States across the nation that are trying to figure out how to incorporate Space Weather into their preparedness plans and are planning—one's a workshop, another State's looking at doing a one-day exercise, all based on the fact that now this has been released. It has amplified the issue, is the key point here, and because of that, we'll see more and more effort across the nation to build against this threat.

So, thank you very much for everything you do.

[Applause]

AUDIENCE: Do you have questions? Wait for the microphone, please.

AUDIENCE: One of the key things we mention here is the importance of telecommunications. We're building FirstNet to be there for such emergencies. I've heard from people who are involved with FirstNet who are familiar with the top leadership developing the RFP that FirstNet is developing to build this, that this leadership is not concerned about EMP or GMD and has no intention of making FirstNet resilient to this concern.

First question is—What can you and SWORM do to help assure that FirstNet is resilient, and we don't build something that will not be there at the very time we need it most?

And the second-related question that was asked to Caitlin—We have five or ten percent of the Energy produced in the U. S. by the Federal Government, by the Tennessee Authority along the Colorado, to have a proof of concept effort to show how to make a nuclear—a hydroelectric power plant resilient, the transmission lines. Caitlin responded by saying Bonneville is working on some efforts on the one or two substations. That's nice, that's something. But what about having a proof of concept for the full path from the generation of the hydroelectric power all the way down, as well as making sure that FirstNet is indeed resilient, which might mean having its own alternative power Sources?

WILLIAM MURTAGH: So on the issue of an engineering solution, I think you're touching on there, and the last point I made, I think, when I was talking about that Strategy and Action Plan, is how it is an enabling document. And some people look at me sometimes and say, "That's Bill Murtagh, the Space Weather Forecaster Guy, may not be interested in the engineering solution." I assure you, I believe that that is the key element here, is to build in—so I'm very keen on when I'm seeing the actions on the Hill right now, and Congressman *(Trent)* Franks and the Shield Act and the other efforts, the Legislative efforts on the Hill focus largely on an engineering solution. And I support that fully.

Now that we've got the Strategy and Action Plan out on the street, and obviously it just went out and our first meeting is on Tuesday, now working with the Departments, with Energy in particular, and Homeland Security, and the Federal Energy Regulatory Commission, we are now going to embark on whatever efforts we can to enable those activities; the pilot projects, whatever is necessary, we'll be reaching out to some folks in this room on how best to do that. We've certainly—there's lots of good work done and literature out there that already helps us get to where we need to be. So we'll start moving on it now.

DAVID FICHTENBERG: And First Net?

WILLIAM MURTAGH: What is FirstNet?

DAVID FICHTENBERG: (off microphone) It's the new Emergency Communications Network for First Responders.

WILLIAM MURTAGH: Yeah, I'd have to get with you offline and discuss that separately.

DAVID FICHTENBERG: (off microphone) That's not only one of the most important public communications systems that should be built with protection when new so it's less expensive. So if you do anything built-in, but it's also one of the most important to be there in case of a problem. So it's one of the most important things that has to be done and needs to be done fairly soon before the RFPs are committed to.

DAVID FICHTENBERG: They have to be done by 2018.

WILLIAM MURTAGH: 2018? Okay.

TERRANCE HILL: Bill, I've got a question. Given the emphasis up on the Hill right now being put in both the Energy bills into microgrids, and now this new awareness about Space Weather, is anywhere in the White House, any thought being given to looking at a clean-slate approach to the Grid? Take a white piece of paper and design it as—given what we know today, as opposed to what was happening 120 years ago?

And the other thing is your colleague mentioned in her earlier presentation about trying to build into the future systems, mitigation things. Has anybody come up with some sort of a possibility, a "What-if' on how to do that?

WILLIAM MURTAGH: On system mitigation, like hardening devices, blocking devices? Is that the question?

TERRANCE HILL: Yeah, like Ultra-Wide Band? That's coming. We got to build these things into the Wide Band? Follow the Wide Band?

WILLIAM MURTAGH: I have no idea. I'm the Space Weather person. I can't speak for the engineering folks and the Department of Energy on that specific issue. Like I said, for the technical solutions to engineer around it, it's certainly something we want to do, but it would be—some of the existing proposals out there, they're blocking devices and such. Thanks.

[Applause]

Community Action in Light of High-impact threats Reviewing the Major Culture Shift of Emergency Management Engaging Space Weather, EMP and Cyber Threats

Corresponding Video:

https://www.youtube.com/watch?v=V73HmkJDQmU&feature=em-share_video_user

Panelists:

- **Hon. Andrea Boland**
- **Mr. William Harris**
- **Mr. Thomas Popik**

HON. ANDREA BOLAND: Thank you once again. I guess you're hearing from me a little quickly after the last time. But, the presentation that I wanted to prepare for you is, as you can see, related to the National Space Weather Strategy Plan and how it relates to the experience we had in Maine. It's the whole community effect of that call to action, includes the States, localities, regions and what not. So, I'm just going to review with you just briefly what I said before about the legislation we passed in Maine.

Okay, well, this came up. Okay, this is great. Again, this is what it called for, the Maine Public Utilities Commission, was to examine GMP and DMP, and report back on the most vulnerable components of the Maine system, potential mitigations, estimates of cost (low, medium and high cost), the timeframe for adoption of the mitigation measures and policy implications. That was in 2013. In 2014, Maine LD 1363 to implement mitigations that were recognized in both studies failed by one vote, which I had mentioned to you before.

So, the regulatory resistance was the problem. The PUC dragged its feet, assigned only one person, I think half-time, to the study. It had been offered help from the Federal Regulatory Commission; it did not accept it. It refused to investigate EMP, which actually made it out of compliance with the law we passed. And ultimately had completion of the study shifted to industry control so that Central Maine Power, the predominant electric utility, was made project director of a study taskforce. So, the result was two studies, as I said to you before.

These, and you've heard the National Space Weather Strategy goals from Bill Murtagh and others, and this is just to tell you how well it dovetails with the needs that we discovered in doing our legislation in Maine. Benchmarks were a huge issue for Space Weather events because they were to be pivotal to understanding a modeling base and what we really needed. And the benchmarks for the NERC, North American Electric Reliability Corporation, were found lower than even what had already been experienced in this country. To enhance response and recovery capabilities was a key goal of our legislation. Obviously, to improve protection was a main goal of our legislation, and to improve assessment modeling and prediction of impacts on critical infrastructure. I'm really just reading this to you now.

The heart of the differences between the two studies that emerged were the modeling and the assessments derived from it. The modeling's credibility rested on the benchmarks guiding the modeling. So, going back to the goals of the Strategic Plan, Space Weather Plan, these are big items, the benchmarks and the modeling. And in fact, what happened with the study that was directed by Central Maine Power is they refused to even use their own data, as I had mentioned to you before, even though they were asked repeatedly, particularly by Bill Harris and Tom Popik of the Resilient Societies. They just sort of didn't answer.

And instead, the benchmark they used was from Scandinavia instead of using their own studies. In this country, there were multiple places where they could have derived data, also pertaining to this country. They didn't use that. That's for the Geomagnetically Induced Currents, the GICs. They turned elsewhere and they used a theoretical benchmark rather than something that could have relied on their own data or other American data. Obviously improving the services through advancing understanding and forecasting and increasing international cooperation, it was important.

Though the thing about the National Space Weather Strategy Plan that causes me a little angst, and of course I know it's evolving and it's a wonderful thing that we have it, it's a huge step forward—is that it doesn't seem to involve input from the States. And multiple States have been working on this and they're closer to the population. And so I think that would be valuable to encourage them to do that.

And also in saying to work with industry on coming up with what is called realistic planning in the strategies, is somewhat worrisome also because the industry played many, many tricks on us to try to say what was reasonable and what wasn't reasonable. So that's something we also have to hope we have enough input that they can discern where the truth path will be. And I would encourage support for their calling on outside independent experts also to be involved.

And one of the things that, in talking about my concern about too much accepting industry input as the most valid input, is just an example that we heard yesterday when the NERC Director of the GMD Taskforce spoke at yesterday's event about recovering from an event that for a two-month event, so two- or three-month, anyway, it was—it was two months, I think—that they could, if transformers were destroyed, you could order, receive them and have them up within two months. And I don't think that the man who spoke that way could really believe that, but he's in a position where he sort of has to say what his office wants him to say, I'm sure, because nationwide, the amount of time is a minimum of 18 months to just get them to this country. I don't know about how long it takes to put them up, 18 months-to-two-years in normal times. If we had a wide-ranging event, how many people would be in line to get them? They would be shipped from South Korea or Germany and those places might also have a problem.

This is just to illustrate the difference in standards. This shows the force of the 1921 storm at the top, the red marks that you see. That is experience in 1921. It isn't even considered a huge event. It was kind of a big one, but it wasn't anything as big as we're worried about. The lower line in blue is the NERC benchmark. So, that's what they think is as high as we should meet, and I think that might be kind of a waste of time.

So the value of the Space Weather Action Plan, the Maine PUC and Central Maine Power were not the only parties in the industry regulatory and legislative environment who compromised the rigor of the study. There was plenty of pushback in different places and lobbyists were working hard in one way or another. There were sins of commission and omission by various entities and including Federal Agencies and Departments who should already have been helping us. The Space Weather Action Plan calls all levels of Government, business and the public agencies to account. It assigns duties—this is just great—it assigns duties to specific agencies and it sets time limits, which is really great. And it calls for the electric power industry to engage as a positive force for real, credible benchmark-based strategies for mitigation and protection.

We have to be concerned about keeping everybody honest and making sure that we get the best that we can out of this whole process. And it really supports, at the Federal level, the path that the Maine legislation LD 131, frankly, blazed at the State level. The things that we learned are the things that they're addressing, and that's part of the reason why I suggested they'd be well to reach down to the States for input. It needs more teeth, in my opinion, but it takes a big bite out of this very bizarrely political and sensitive apple.

And thank you all for your attention. Hopefully, we can move forward and make a big difference in this country and make sure that we still have a country when we're finished. Thank you very much.

[Applause]

BILL HARRIS: So I would like to address whether we have a strong need for whole of community involvement or not. So, we start with the Space Weather Action Program that was presented. It appears to recognize the magnitude of a solar storm could be so vast that it's beyond the capability of the federal government, which doesn't control most of the infrastructure resources anyway, to protect us and to mitigate. So we start with the presumption that whole of government is necessary.

But I'd like to then compare what we had—you had earlier today an Assistant Secretary of Homeland Security tell you that the electric Grid was incredibly resilient. So let's take a benchmark from where we were in April 2008. We had the EMP Commission get declassified a major report, because in 2004 they issued a report and there was not much action. So they were able to get, and they had to fight to do it, but they were able to get a major report that took several major critical infrastructures, the energy sector, telecommunications, et cetera, and it identified the electric sector as having about 2,200 large power transformers that had extremely long replacement lead times. And they urged that as high a priority as any that required protection.

At the time, there were no commercial neutral ground blockers available to purchase, but they had been tested, the concept had been developed by the 1980s. So now let's go seven years and eight months later to today. We have one of those 2,200 high voltage transformers with a neutral ground blocker on it. It works fine, but we have one out of 2,200. Let's look at cyber. We had in the Energy Policy Act a requirement to protect communication networks from cyber offense. But the industry was concerned about liability. So in 2006, when we had the Electric Reliability Organization with authority to protect cyber, they went another nine years without requiring protection; and we have the Director of the National Security Agency and other public statements by the Federal Government indicating we have malware embedded in our Grid by Russia and China and, in the water and sewer systems, that's hard to get out.

So, how much progress have we made? We do have from FERC a notice of proposed rule making and they may, in fact, be planning to close this major loophole. There's going to be a technical conference next January. So that's where we are on cyber.

For solar storms, we didn't have much understanding of the frequency of solar storms when the EMP commission reported in 2008. They mainly focused on EMP. Well, you've heard earlier that in the last several years, we have a consensus of scientists that a severe solar storm with the magnitude of the Carrington Event, New York railroad storm of 1921, is about 12 percent in 10 years. So it's not a low frequency event. And so how can we get from the hope that we have an "incredibly resilient Grid" to the reality of an incredibly resilient Grid?

And I suggest partly by using the benchmark of where we were in April 2008 with the EMP Commission Report and where we are now by just objective standards, we need the whole of community. If we don't have work at the state and local levels, we're not going to make much more progress than we made the last seven years and eight months.

So what are some of the opportunities? The first is the Energy Policy Act of 2005 has a savings clause to protect the rights of States to set their own reliability standards. The States have the right to protect the transmission and distribution of electricity in their states for safety, adequacy and reliability. Now, there's also a protection for the bulk power system. It says that the States cannot engage in some regulatory process that makes reliability worse in other States. But the studies that you've seen this year and before show you can protect a high-voltage transformer without spreading much trouble, the GIC currents extensively, if you protect the key gateways at just 20 percent, you reduce the DC current in your transmission system—so even transformers that aren't protected can survive at a higher level of solar storm hazard.

So it could be shown scientifically that you can meet the standard to allow the State to have higher reliability and you're not going to hurt other States. So a lot can be done at the State level through the Public Service Commission or Public Utility Commission. And, better yet, at the State level through legislation, which Andrea was successful in doing.

So, let me just say a little more about Maine, what I think Maine did even though it has not implemented requirements by the Maine Public Utilities Commission. First, in the process of getting the 2013 Legislation, Central Maine Power was asked for the historic data from Chester, Maine. Chester, Maine, is the only site in the country. It has a static VAR compensator. It produces reactive power to cope with voltage collapse in northern Maine. And they have data going back to at least 1990. And they're the only place in the nation which has data going back that far.

The Sunburst system from EPRI had data, but apparently it didn't have a computer backup. It lost about a decade's worth of data, as best we know, and they never released the more recent data, the last 15 years of data. So, we really don't get much for public researchers out of the EPRI system. But we did from Chester, Maine, and we did because when Andrea was in the Maine Legislature, Central Maine Power committed to provide that information publicly. That allowed an expert in Minnesota, John Kappenman and others, to model the different storms and compare the levels of geomagnetic current, impacts on transformers, et cetera.

So Maine has helped the rest of the country by improving modeling just there. And more recently, our foundation had a failure, but sometimes we win through failure. So, we petitioned the Maine Public Utility Commission to use an existing statute that said that if a public utility did not adequately protect against weather that the Public Utility Commission had the right to reduce their rate of return.

And we said that Space Weather is weather, and the statute already exists, we don't need to pass a new one. And we wanted them to open a hearing to determine whether they should reduce the rate of return for Central Maine Power from 11.7 percent, way over the market rates for bonds and similar financial instruments, and they told us that that wasn't the purpose of the statute. And we didn't go to court over it, but we copied every Executive of Central Maine Power, and the holding company, Iberdrola USA [now Avangrid Networks] which is headquartered in Canada,which is ultimately held by a company in Madrid. So all their Executives knew that there were people who wanted to dock them financially for failing to address this issue.

So we lost. But lo and behold, when we next showed up with Andrea this past September, Central Maine Power was already planning to put in two new series capacitors which saved money and protect long transmission lines, the high-voltage DC line from New Brunswick, Canada. But they also protect against the solar storm GIC. And they were prepared to put in a reactive power device that cost four-or-five times what it would have cost to put neutral ground blockers in the Maine state high-voltage transformer system.

But that served other purposes, but it also provides reactive power. The irony is if they had put in neutral blocking devices for 16 transformers, that would have taken out even more reactive power, probably. So we don't fully know why they don't want neutral ground blockers now. They may be waiting for a signal from FERC, which may come when they put out a rule making on the solar storm standard requirement to assess high-voltage transformers. And hopefully, they will signal that one can get cost recovery for doing more than the minimum required.

So we don't know yet, but Maine is already making progress just by getting the attention that it doesn't really want at the State level. Central Maine Power is taking initiatives on their own, and they identified some of the switching equipment. They've looked at other low cost ways to improve the reliability of the Grid in a solar storm.

When they didn't look at EMP, I would have to admit, it's not entirely their fault. The Federal Government has much more knowledge and including significant classified information about how to mitigate EMP. And for whatever reason, the head of the section of FERC was not allowed to go help Maine. So it's not as simple as the utility not wanting to do it. They didn't have the technical knowledge.

So I think we're going to get a major breakthrough with this new EMP Commission because they have a mandate to look at EMP and solar storms together. So I'd like to suggest those in other states that have Public Utility Commissions that haven't assessed the reliability of their Grids, there's a great opportunity to appear before those Commissions. And there's an opportunity to ask your State Legislatures to use the power they have to improve reliability. New Hampshire, for example, passed a statute last August and it requires all the transmission companies in New Hampshire to assess the reliability of their companies and report to the State Public Utility Commission.

Lastly, I'd like to go to the lower level, below the State. I think there are at least two local opportunities for whole of community initiatives. One's the education system. Our children and grandchildren are going to grow up in a society that has so much complexity, so much fragility, so much interdependence, we need to train them to understand that this is a challenge for which they can contribute. And as a result, hopefully some of them will enter careers that will help solve some of these problems. And in a blackout, we hope many of them will join in helping recover rather than help looting. So we have that element.

And another element, we'll have a panel on following, has to do with microgrids. The Federal Government played a major role in initiating the Internet through ARPA, and the Federal Government has a program done jointly by the Department of Energy and Homeland Security and DOD to develop microgrids on military bases. They started at Hickam Field in Pearl Harbor, there are several of them now. But they're not just designed to help individual homeowners. The whole microgrid is managed together. So in a crisis, they can ask for emergency conservation by the homeowners and maintain base operations for critical-mission performance.

So they are serving as a demonstration project. And I think as we get microgrids that are developed that are Net Zero Energy or at least low energy, there's an opportunity for microgrids developed in the private sector, in communities, and they can, I believe, provide protection against Solar Weather because they don't have long line connectivity. And Dr. Baker has told me, because I don't know much myself, that one can protect the inverters for the solar panels. That the cost of protecting inverters has come down, and the cost of the panels has come down.

So if you buy a standard solar system, it's going to have a vulnerable inverter if there's an EMP attack. But one could design methods to shield that are relatively low-cost, George said $100 per inverter. I bet that cost is going to come down.

So to sum up, I take the different view of the state of progress of our "incredibly resilient Grid." I don't think it's that resilient, and I don't think we've made much progress. But the way to get progress is not to just count on the Federal Government, but to work at the State and Local level, in the Public Sector and to encourage and facilitate private initiatives for microgrids and other new adopter systems.

Last aspect—PJM Interconnection. On its own, it's the largest of the regional transmission organizations—13 states and the District of Columbia. On their own initiative, they decided that with the new technologies, wind and solar, they were coming such a high share, potentially, of their transmission market, that they asked them to work with them developing a standard so that when they connect to the transmission system, they have the capability—if the overall transmission system is under voltage to come in at higher voltage and help stabilize the voltage at the right level.

If they come in—if the system is under frequency or over frequency, they can ask the renewables to come in on the opposite side to balance the system. This is not required by the Federal Government, this is not required by State Government. This is a nonprofit Private Sector organization that has a lot of knowledge about Grid reliability that acted on its own to harness the early adopters to make the Grid more reliable while incorporating more renewables.

So I'm just suggesting, don't just think of the Federal Government to fix the Grid. Thank you.

CHUCK MANTO: Tom, a new or no question, and then we have to switch to the next panel.

TOM POPIK: It's almost a yes or no question. Bill, because you talked about Maine and Andrea talked about Maine, under the current NERC standard for GMD protection, how many transformers in Maine would need to be protected?

BILL HARRIS: The sad number is zero. But basically—and the sad thing is they were planning to do better before the Federal proposed standard came in. And the same with American Transmission in Wisconsin. They had one neutral ground blocker, they were planning to do five. So sometimes, this is where we have a dispute with Commissioner LeFleur, who's going to write the Order for FERC. She hopes that the floor of her Order will not be close to the ceiling. We fear that the floor—we've seen the floor being proposed by FERC, and private initiatives to protect more transformers shriveled up and disappeared.

So it's what it is. That's another reason the States could require, at least for their transmission systems, they could require protection, too.

HON. ANDREA BOLAND: Could I just add just one piece to that, and I'm sorry to take any more time, Chuck But, if you know something—if you see something in your State, a piece of legislation or an effort going on, if you'd be willing to call who's ever running it, if you know something that you can contribute, it would be great at a meeting, at a hearing for you to show up. And even if you don't know much, if you can show up and say you care, you're interested, or you can push State and Federal Legislators to address this issue, it would be great.

I think I'm going to get a rubber stamp that I can just put on—when I send my bill payment back, and I still send it in the mail, something to the effect of "Protect the Grid from EMP and GMD". And just sort of push the fact to the utilities that you know there's a problem and you want it fixed. That would be just great. Thank you.

[Applause]

END

Role of Protected Microgrids for Community Protection

Corresponding Video:

https://www.youtube.com/watch?v=2I6sl_Qt6fI&feature=em-share_video_user

Panelists:

- **Hon. R. James Woolsey,** Chairman of the Foundation for Defense of Democracies and a Venture Partner with Lux Capital, former CIA Director
- **Mr. David Geary,** DC Fusion, Direct Current Microgrids
- **Mr. Terrance (Terry) Hill,** Passive House Institute and Community Microgrids
- **Congressman Roscoe Bartlett**
- **Thomas (Tom) Popik**
- **Charles (Chuck) Manto**

JERRY BOWMAN: The current panel that is coming up now is going to talk to us about the role of protected microgrids for community protection. It's a very distinguished panel. Leading that is the Honorable R. James Woolsey who you know best probably as former CIA Director, who is currently Chairman of the Foundation for Defense of Democracies and a venture partner with Lux Capital. One of our panelists is late, I believe, still, or having trouble getting here. We do have Mr. David Geary from dcFUSION, Mr. Terry Hill from the Passive House Institute and Community Microgrids. So without any further delay--oh, and we've got Chuck.

CHUCK MANTO: I'm just here as a foil if something is needed at the last minute, because I know we have to end this one on time, or a little bit early because there's an unexpected requirement for our guest to be on a television program.

JERRY BOWMAN: So let's not get in the way. Go ahead, folks.

AMBASSADOR JAMES WOOLSEY: Thank you. Well, thanks Chuck, and I'm honored at the opportunity to talk to a very knowledgeable and distinguished group such as this. I want to start with a difference of distinction that one of my colleagues at Stanford, when I was teaching out there a few years ago, came up with. She said, "We have to pay attention to the fact that we're dealing with both malignant and malevolent problems when we are dealing with EMP and the Grid." And the solution is not always the same, although sometimes it can overlap. And sometimes instrumentality is the same. For example, small pox is a malignant problem, but it's been used aggressively by giving Indian tribes blankets with small pox in them in the early days of the Western civilization in North America. One can have the same instrumentality be used malevolently, but occur naturally and be malignant in, at least, a general sense, not precisely, in the same way germs are malignant.

Well, I think that there—or rather, cancer is—I think the thing that is of substantial importance is that it is not going to be enough to fix the Grid. I'm delighted to see the world stirring on this issue that several of us have been working on now for several years-- and to see attention to resilience for the Grid, to seeing the stirring of activity on behalf of State legislatures and the like. And let's face it, there are a number of people who are willing, indeed eager, to work on malignant problems, but they just as soon not talk about the malevolent ones—either because they don't want to have to go to the Defense Department for funding and/or because of some other bureaucratic or political reason. There are just a number of people in government and otherwise who really don't want to talk or think about something like a nuclear detonation in low Earth orbit that would with both long- and short wavelength EMP take out not only the Grid, but also your individual devices. And we have to pay attention to that, and we have to find a way to deal with both malignant and malevolent problems, hopefully one can come up with solutions that overlap.

What is the nature of the problems? Well, first of all, let's look at malevolence. This group is probably familiar, largely, with the events in Metcalf, Colorado some now nearly three years ago in which a small group of highly disciplined young men with AK47s having trained exactly how to cut and take out the emergency alerts, carrying the right kind of tools to lift a very heavy manhole covers amongst a group of the transformers, systematically stood and took out, with rifle fire, a large number of the transformers that supply and deal with Silicon Valley. Virtually, everyone connected with looking into that, except for the woman who was a local sheriff, wanted to call it hooliganism, which is a way of avoiding saying terrorism, sort of like saying workplace violence for something that looks an awful lot like terrorism.

But those sorts of euphemisms—don't let me have to think about an intentional group planning to take out large numbers of transformers with rifle fire—I don't want to think about that. I'd rather think about Space Weather. We got to deal with Space Weather. It's a very serious matter and it could take out a huge amount of our electric Grid, but it's not the only problem.

And so not only for rifle fire, so-called kinetic threats, but also for radiation weapons, driving around Wall Street with a panel truck with the right kind of radiation-generating equipment in the panel truck could do huge degrees of disaster to our financial system. And, of course, most devastatingly of all, a nuclear detonation in Low Earth Orbit, and by Low Earth, I mean everything from 20 up to several hundred miles.

And having that detonation be able to generate through the gamma rays both the long wavelength and the short wavelength, EMP is a realistic possibility. We know that the North Koreans, in launching their, I think, three satellites and having tested three or four nuclear weapons, have taken steps to do it in such a way as to make sure when they are launching to the south, to avoid having something come up over the United States from the North, where all our sensors and radars are, you launch to the south in what the Russians called a Fractional Orbital Bombardment System, and the satellite containing the nuclear weapon comes up from the South, southern trajectory so you don't know that it's coming.

The first thing you may know is that the lights go out and you may not know whether it is Space Weather or whether it is an action by the North Koreans launching a Fractional Orbital Bombardment System.

As a result of the different ways in which we have to envision and deal with malevolent problems, as well as with malignant ones, we need to deal with this, I think, doubly. Yes, we have to take steps to build resilience into the Grid. We're going to need the Grid for a very long time, even if we have other emphasis. But there is, I think, also, a real need, a major need, to let Edison have a second crack at his war with Tesla. And to make it possible for Direct Current distributed generation without inverters, all of that, relatively simple, roof of your house, roof of the school, good storage, which requires, I know, some further development. But being able to have those—a simple system like that be resilient against electromagnetic pulse, both the chips and any interconnections, can make it possible in the event of Space Weather of some terrible sort, or any other way of taking out electricity can make it possible to defend against that.

And I would say that steps to have moved toward even a partial independence and resilience by distributed generation, direct current, solar probably, although probably not exclusively, is a sort of thing that encouraging it would be good. Whether it's through tax system, whether it's through pressing utilities to be more tolerant of microgrids than most of them are, whatever steps are necessary for us to be able to have, at least, the basics if something happens to the Grid is essential—and could be the difference between maintaining our civilization and not being able to maintain it.

If the Grid goes and we're not doing anything else, then the Key One of the 18 critical infrastructures is down and the other 17, food, water, transportation, finance, all depend on the 18th, electricity. So if the Grid goes, you are not back in the 1980s pre-Web, you are back in the 1880s, pre-Grid, and very few of us have enough plough horses and seeds to live a 19th century existence successfully.

Is it unpleasant to have to think about that? Yes. Do we feel sorry for you if you have to think about it? No. Think about it. Help us all figure out how distributed generation with simplicity as its principle criterion, and resilience can help farmers, businesses, schools, homes, individual buildings with let's say solar on the roof and the batteries in the basement, take steps to get by even if it's not perfect, even if it's not everything the utility would supply to you on a good day. Even then, to be able to get by is the difference between the civilization being destroyed and its not being destroyed.

And we need to keep that set of values and that set of issues in mind as we look at some of the possibilities and how one can take useful steps in moving toward a world in which Edison and his Direct Current get a second chance. Thank you.

[Applause]

CHUCK MANTO: David Geary, would you like to tell us a little bit more about Direct Current microgrids?

DAVID GEARY: What a great lead in, thank you. I was asked to actually provide an industry update on the DC power and microgrids. And a lot of things—I don't know if everybody has heard about some of the activity, but DC power is sort of regenerating in history. The war of the currents, to some extent, is reoccurring, but in a different way this time, because we know that the Grid's going to be there, and we know that there's new technologies that exist today that didn't exist back then.

Today, we have power electronics, and it's the difference between today and yesterday. It's the difference why we have AC power, for the most part, around the world. But now power electronics is offering some great opportunities to take advantage of a lot of the things that are happening around us and a lot of developments that are occurring, as we speak.

A great history book on how we got to where we are today, some current event ideas and things that are happening around the world that sort of emphasize a lot of the things that Jim just mentioned. This is Intel. Intel's very interested in the whole DC power concept. They see the next generation of what was the Internet in prior years now becoming the ENERNET, the energy net. Their focus, obviously, is that they see a great opportunity of a lot of things being connected to the Internet, the Internet of Things, and all those things require chips.

They also were looking at the other aspect of this, which is the fact that there's still, it's hard to believe, 1.5 billion people without power in the world. So they're looking at this as an enabling technology to actually help develop some of those undeveloped parts of the world as we speak.

So where, actually, all this started was actually in the data center. The data center, as you've probably heard, and all the things that are happening with the Internet, in a few years—in fact, I think they're projecting 2020—will be consuming as much power as the third largest country in the world. Going over Japan and Russia and behind U. S. and China—the data center industry is exploding as we speak. And the base load in the data center is DC. It's a computer. The first thing a computer does is take the AC power that is given, converts it to DC.

So this is where the whole idea started because the Intel's, the Google's, the Apple's, all of whom we've met with over these years, have said, "What's a better way of doing things?" So they've come up with the idea of going to a DC power infrastructure. 380 volts is becoming the worldwide standard of DC power because it's a sort of sweet spot and a lot of other things happening around the world of alternative energy and many other things.

The IT manufacturers (IBM, Lenovo) are all coming on board and developing products and actually writing White Papers that espouse the benefits of DC power. The history actually started early 2000. 2004 was when I did my first demonstration project out in Sun Micro System in the middle of sort of the U.S. timeframe. And you can see that the world has sort of followed pace with that to where we are today. And at this point in time, we're seeing deployments around the world start to increase over time. I have to almost update this slide on a weekly basis these days.

The problem, though, right now, is that China and Japan have accepted this whole concept and are moving forward quite rapidly with it. The United States still has a lot of the things I think we've talked about here today as sort of roadblocks of allowing us to move at a faster pace along with them.

Worldwide Standards are in process and being written. Some are already existing and are edited, some are new, but standards around the world are being developed for this. This is actually a White Paper we just finished with the Society of Cable and Telecommunications Engineers. This is the Cable Industry. One thing that Cable Industry really took a notice of was Hurricane Sandy and all the things that happened to their infrastructure. They don't want to see that happen again and are looking for ways of implementing DC microgrids and other things that allow them to go island mode and survive some of these situations.

So there's actually some White Papers that have been written comparing AC to DC, so all these things are in process on the data center side to show what the business case is. It has to be a business case to show why we would convert from an AC infrastructure to a DC infrastructure, and here are some of the reasons why: More efficiency is how all this started. But then the AC world started to become very efficient as well. But there's a lot of other benefits that we'll see as we go through some of these other slides. It has to be a true business case. It's got to be safe, it's got to be reliable, it's got to be flexible and it has to be efficient. There has to be a business case to make this change.

So again, this started at the data center. These are a few pictures of what a DC data center looks like, some drawings and Proof of Concepts and comparisons. So again, started in the data center. From the data center, groups like EMerge Alliance, which is an industry group that was developed to help promote the idea of DC power started to get members, you can see, of many of the companies that are now part of the EMerge Alliance that want to take this idea outside the data center and now are implementing this idea in office buildings with some of the benefits that occur when you go to DC in office buildings, as well as in microgrids and so forth. And there's been a few projects around the country that are showing this and are in process as we speak.

So the whole idea is to try to go to a Zero Net Energy Building-type of concept. And some of the things that, again, DC power offers is that there's always going to be AC sources and DC sources. There's going to be AC loads and DC loads. In fact, one recent report I saw is that probably over 80 percent of the loads that we have today are basically inherently DC. We still have a few incandescent light bulbs around, but all the fluorescent lighting, the LED lighting, projectors and even the appliances in your home are being developed to operate on a DC power infrastructure.

So a lot of benefits. The only thing we would need to do is need a DC collector bus matching voltages and bring sources together to feed loads. With that, the integration of alternative energy sources, PV, energy storage, all sort of inherently to DC-power systems.

So we're starting to see some of these projects being implemented as we speak. This is actually a data center in Ashburn, Virginia that's testing out this whole concept of doing an integrated DC microgrid within a data center environment. We're looking at 50 different ways of doing DC grids. Again, the DC-grid is a much simpler concept than trying to put together alternating energy or alternating voltage sources when you have to synchronize and worry about phase shifting and so forth.

We're seeing a lot of industry, cross-industry collaboration. We got the alternative energy industry, a lot of DC sources. We have the electric vehicle market. That's inherently about a 400 volt DC system that can plug right into a DC microgrid. You have Department of Defense; the next-generation ship that was just launched is all DC power. Again, there's benefits to that topology.

In fact, the Electric Ship Program has been going on for probably over ten years before this launch, and they've gone through a lot of the research that's already had to take place to basically integrate a DC microgrid at a megawatt-level size. DOD, obviously, they're interested in microgrids. They don't like delivering fuel to the front line. So anything they can do to come up with ideas and configurations that allow them to go to an off-grid islanding-type of mode for DC microgrids is something that they're very interested and are looking at, at all the benefits that come with that.

We, at dcFUSION, basically started up this past year just to try to bring industry together. Because as you would expect, there's a lot of new companies, small companies out there, developing products and systems for this whole topology. They're small companies, but they're doing real projects. Pika Energy out of Maine actually does DC microgrids for high-end homes. Again, a place where the first proof of concepts are basically done because you have people interested in taking care of themselves and have the means to do that and want to make a difference in the world. So they're doing residences, but now they're starting to scale up to commercial.

Yes, these Canadian companies are also doing their versions of a DC microgrid. And all these different other folks that are playing in this arena. Not to mention Tesla and the power wall. I haven't seen the announcement that Solar City is now offering Microgrid-as-a-Service. The whole idea with that is taking the power wall from Tesla, integrating that into a DC microgrid on the load side of the meter.

So the other thing that sort of has happened with this journey is again looking at the benefits of DC power. There's energy savings. There's cost savings. And different universities and folks around the world—this is my friend, Dr. Raj from Clemson showing, trying to show the world, again, that there's over 1.5 billion people without access to electricity, as we speak. This is enabling opportunity to change that, as well as all the other things happening. And he sees PV- generated DC as the enabling technology to help get there.

So we're starting to see this happen. These are Third World countries that don't have electricity, as we speak, but they're implementing this idea and making things work. If we can see folks like this do this, and develop and take advantage of what can be done in a very simple, very cost- effective way, there's got to be lessons that we can learn from them to look at how we can maybe take care of some of these high impact threats that we have to deal with.

So we're looking at going from a little village and hut ideas to neighborhoods, all integrating DC power. We're seeing some designs hitting the drawing board for sustainable communities, where you can basically have maybe a Grid interface at the beginning of a neighborhood—but the whole neighborhood becomes self-sustaining because we're sharing alternative energy resources on a DC microgrid. So all these ideas are starting to percolate.

We had our first conference, IEEE conference, in June in Atlanta on the DC microgrids. We had folks from all over the world coming to give their presentations on what they're doing. Again, quite a bit of activity around the world. We'd like to see more of this happen in the United States.

So again, some of the benefits of DC are the reliability. Less components equals higher reliability, and there's been studies that have proven that. Also, our friends in Japan, one of the reasons why they're big DC proponents of DC microgrid, is that they were doing their own version of DC microgrid research back when the tidal wave hit. And the only things that really stayed up and running and were recoverable were the DC microgrids that they were in the process of testing, and they'd made a full-fledged investment in developing more DC microgrid infrastructure.

The other thing we're doing is now working with some software companies. As an engineer, I'd love to be able to design something, but I'd also like to model it before somebody builds it and proves that it works. So we're starting to see there's tools hit the market as well. Power analytics that has software that allows you to model, analyze, optimize, build, monitor, control and then transact. We're seeing the infrastructure and software being developed to allow us to build these systems, study these systems, model these systems on paper and in the computer. Then we can turn those systems into the monitoring and control system.

We then start gathering real time data. We can then start optimizing. We can then start doing simulations. We can start looking at, okay, what does an EMP event look like on this system that I just modeled in the software that basically is the utility system for this area of the country? And, if you can model that on the computer, and then as things happen and more testing and data gathering is done, the model becomes that much more valuable and the data becomes that much more reliable.

So we're seeing these things hit the market where you can model the system, you can turn it into your control system, and it becomes your real time data monitoring control, real-time arc flash, other things that you can do now with the tools that we have available. And this can happen with both AC and DC, integrated, separate, islanding, all the different modes of operation now are available because they can turn this control system into a Microgrid Power Management System, as well.

So again, it seems like the perfect storm from my perspective. I'm an engineer, so instead of drawing single lines electrically on the drawing board, like I've grown up doing, this to me is sort of like being in a candy shop, allowing me to build it in software, bottle it, play with different scenarios. And where this all is going is the EnergyNet. As time goes on and these microgrids develop, they're going to be integrated with the real Grid. There's going to be the Internet of Power out there, and everything that we're going to do, we do in the Internet right now today, we will have the Power Apps to add to that as we develop the EnergyNet. So that's it.

[Applause]

CHUCK MANTO: Now, Terry Hill, I think you have a couple of slides, too. Terry? Is that the case?

TERRY HILL: Yeah.

CHUCK MANTO: And you could maybe mention what you're doing with—how would you best describe it for us that would differentiate it from what he just did, because you have a very interesting approach—two things that are being done in the residential area.

TERRY HILL: Thanks, Chuck. I'm Terry Hill. I'm on the Board of Directors of the Passive House Institute, and I came here in 2013 to my first one of these. And when I listened to the presentations, it became immediately apparent to me that what I was promoting, the idea of energy-efficient buildings, was a core requirement to fit into all this. So, I've been following Chuck around ever since.

So, last year, Virginia had to revamp its energy plan so they had a listening session around the State. One of them I went to, and I had to have a context of what I wanted to say so I suggested we have a resilient stretch building code. Now, stretch codes have been—Massachusetts has one, actually, where the State office, individual cities and towns have an opportunity to reach beyond the existing building code, and there's some sort of benefit for that. So that's the context that I went through this, the Stretch Code.

You can all read it. I don't have to read it. A little bit stronger than the existing code. I threw the word resilience in—I threw this into the three-minute presentation because 70 percent of the Grid, the energy in the Grid, is basically wasted and we're paying for it. Nobody really talks about that. They talk about 10 percent, 15 percent, maybe. But according to David and our friends, Raj, down there at Clemson, he maintains it's 70 percent. I saw him argue that with a large, very important man from Duke Energy. So

I'm using 70 percent.

So, resilience. I had to throw that in for the people on the Virginia Energy Commission. I like the idea. It goes beyond just what a passive house is. I'd like to see the Passive House Institute expand its idea of what it's promoting into the resilience. After Sandy happened, I'd never really been in a real bad weather event before, so after I drove up to that, and I'd just driven out to Oklahoma and looked at the result of the tornadoes, all of the idea of how do we increase the Passive House Building Code to encompass all these things and make them better, and resilience is the direction I'd like to see it go.

Now, the insurance industry has come up a couple of times here. They're already sick and tired of paying out too much money, so they have their own building code now. And the Department of Homeland Security, I think, is picking up on that, and I think they're calling it Fortified Star. And DOE is also referencing this in its presentations. They're looking at wind events, earthquakes, et cetera, and this fortified section. Their initial effort is to at least build a code—they really want to make it do a good job, go to the Fortified Standard.

Now, in a resilient building, you also want it to be functioning. You want to have good indoor air quality. And the reason I stuck this in here is because Passive House is the only standard that requires an energy recovery ventilator which gives 24/7 fresh air. So this is a Passive House, or a depiction of it. We've got six-sided insulation. This red line here represents a part of the wall you designate as an air barrier. You don't want any air movement through the walls. Triple-pane windows, and therefore you've got a very tight envelope so you've got to take care of the indoor air; and this section over here is the Energy Recovery Ventilator. So, in the—they want 92-, 93-percent efficiency in these things, so you're losing very little of the energy that you've used to condition the air. And the result is you've got a very high-quality indoor air after it's all done.

Now, this is a new home, or a depiction of a new home. But there's 78 million homes in this country that are existing, right? So before I get to that, where does the Passive House fit? This is DOE. Made in 2015. There's the little depiction that DOE's put out—where does the Passive House stand and fit, relative to the other building standards? There's the current standard, or maybe it's got the 2012 now Energy Star, and then there's DOE's Zero Energy House and Passive House, as far as the energy savings, position is right at the top.

There's an existing house in California, a retrofitted house. There are the results. Now, when you build a retrofit to this State, you're going to save 80 percent, or rather, 90 percent of energy for heating and cooling. And this can be retrofitted. So you dramatically reduce the load in the house. What are you left with are plug loads; lighting, appliances, et cetera. David's already told you the impact that Direct Current can have on that. So another 30-percent savings from the plug loads if you had Direct Current inside the envelope.

The PV on the roof produces Direct Current. The batteries use Direct Current. So, homogenous environment with Direct Current inside there has dramatically reduced the load on the Grid for that particular house.

Now, if you expanded that to a city block and tied it all together with a Direct Current microgrid, now you're getting the sense of an island-able microgrid with a dramatically reduced load, is beginning to shift the nature of the Grid. And if we could scale that up, it changes the nature of things, at least from my point of view. Now, other people may not think that, but the possibilities—there are the plug loads, there's a little depiction of the house, the same picture you've got, Dave.

Now, the interesting thing to me is inverters. You know, little microgrids. You only need one point of contact with the big Grid. The other houses wouldn't have to have an inverter. And right now, if you're connected to the Grid with PV, if the Grid goes down, you're down. Except there may be some changes in the later inverters that allow that to island.

What else? There's another depiction of that, we've already done that, resiliency, customer engagement, Internet-of-Things. Who knows what's going to come out when Apple releases its Home Kit on this thing? And, I found, the last two days that these are sort of not affected by EMP and solar flares. I never knew that. It's pretty interesting.

But the final thing I want to mention is DOE's been working on this for years. I don't know whether anybody's come across transactive energy? But it's their effort to model and control distributed generation and to create a retail market for electricity. And they're not—even the last six months, we began to speak about this in the open.

But here's something that we haven't seen too much about—the privacy. Right now, the utility can tell when you switch the light on in your bathroom if they want to. And they can do it from the pole outside. And they really don't need Smart Meters, from what I can understand. They can do it all with solid-state transformers that are sitting on the pole. They can give you Direct Current out of that side, or Alternating Current out of that side, doesn't matter. But with a microgrid, you can aggregate and you can hide behind the point of contact and privacy is an issue that's really not being spoken about much.

So with that, thank you very much.

CHUCK MANTO: Thank you. Since Ambassador Woolsey has to leave for an engagement that is a little bit unexpected, you, and then Dr. Baker, is going to comment on his perspective on all of this. And I wanted to mention that so Dr. Baker doesn't think Ambassador Woolsey doesn't like him or something. But would you like to have a closing word before you go as you walk off aside from goodbye?

AMBASSADOR JAMES WOOLSEY: This is all very interesting and important. I'm learning a lot and I appreciate being invited.

CHUCK MANTO: Thank you very much.

[Applause]

Dr. Baker, you probably have lots of interesting thoughts.

DR. GEORGE BAKER: Well, I'll keep my comments very brief. I was originally supposed to be on the next panel, but this morning Chuck asked if I could—he thought this panel might be a little sparse, asked if I could help with this panel, so I'll make my remarks here.

My main observations are just some thoughts on making the Smart Grids-- they offer an awful lot of promise, but we need to be thinking about their ability to resist EMP and GMD effects. And we're just at the Ground Floor. I mean, for instance, the IEEE, they just had their first international conference on this technology this year. So, we're in a position where now, if we act fast, we can design the protection for EMP and GMD into the systems from the get-go.

And the DOD experience is, if you design in the protection, you'll save a factor of ten. If you retrofit the systems once they're deployed and once they're out—you have them out in the field, it's going to cost you ten-times as much to protect them. So that's one important point.

These systems do offer a tremendous amount of promise; especially, I'm very just fascinated by the data center applications. Because most of data center equipment rides on DC. And you walk into a data center, and I've been in some very large ones that have acres of racks and use megawatts of energy. The equipment in those racks is DC-powered and rides on DC. So that's an application that is really good.

The other thought that occurs to me, my daughter went to St. Olaf College in Northfield, Minnesota. And that campus is powered by wind. They have wind generators. Of course, with their setup, the wind doesn't always blow so they have to be able to—and they're selling power back to the Grid so they're connected into the Grid, as well. So they have to be synchronized, the wind with the 60 cycle on it. And so ideally, if you can run your microgrid stand-alone, that's the best from the EMP and GMD standpoint. The minute you connect to the Grid, all bets are off and the inverters and the electronic synchronization that's required, those would need to be protected.

You also have battery charge monitor and controls that need to be protected. The auto transfer switch, you want to be able to—if there is any kind of major transing on the Grid, you want to be able to automatically disconnect from the Grid to protect the microgrid. And so you need to design in protected auto Transfer switches to enable the necessary protection, and also load balancing. Of course, there's a wind turbine or the sun goes behind the clouds, you need to be able to balance the load, bring it between the Grid, your storage devices and the renewable energy source. And, of course, the photovoltaic cells themselves will need to be—since their P-N Junctions, their Solid-State devices, need to be protected.

So I think there is just a lot of possibilities here. If these things are designed correctly, you can make these systems for an EMP and GMD standpoint hard as a rock.

One other point that I should make, just as I close here, one of the good things about transformers—when you have a system sitting behind a transformer, the transformers block DC, they block quasi-DC, they will block the GMD and the E3, and that's one of the great things about—when I go out and look at big data centers, you know the presence to that transformer is—really helps. If you have long—we need to be careful when we go to microgrids that have long conductors, when they get to be out and if there are situations where you have DC conductors that are kilometers in length, the GMD, quasi-DC currents induced by the sun, the solar effect, will be much larger and that needs to be factored into the design. That's all I have, yeah.

CHUCK MANTO: Thank you. Is there one question, or two? If there is, you have to go to the microphone. I see Congressman Bartlett coming to the microphone.

CONGRESSMAN BARLETT: Thank you very much for your presentations. I've been using solar systems for about 30 years now. And I use almost all of the power—almost all of the energy is DC energy. I had to go to China to get 24 volt LED bulbs. They would sell them to me in any wavelength I wanted, dimmable bulbs. We don't manufacture them yet in our country. Candelabra-based, standard-based bulbs.

I use the energy as DC energy for one simple reason. I have inverters, of course, where I need to have AC power. But they're big, complex, computer-filled devices. They're sure as heck going to fail in an EMP or solar storm. And I would like to be comfortable after they fail. My first array was put on a property on 153 acres, and I put them on the roof. Of course, that's where you put solar panels, is on the roof. And shortly after they were installed, I noticed that 1/8 of my solar panels weren't producing any energy. It was obviously something open up there.

But because I can make do with 7/8 of the energy, and because it was really difficult to get up on the roof, I never did fix them. And then it occurred to me that the sun is 93 million miles away, and I'm not materially closer to the sun on the roof than I am on the ground. So I took them off the roof and I put them on the ground. So now I can more easily control—fix my solar panels and my roof with them on the ground. I don't see any reason for putting solar panels on a roof. The only reason for putting them on a roof is you don't have anywhere else to put them.

And I notice that even when they're showing them on a pedestal, they show that pedestal way the heck up in the air. The sun is 93 million miles away. You ain't get any closer to it up there. Thanks.

[Applause]

CHUCK MANTO: Any comments? The panelists may have a question of each other. We've got 30 seconds.

AUDIENCE: I have a question. What about these advances that the White House is putting so much money into, Ultra-Wide Band semiconductors? That intrigues me, and how will they be hardened in the development process, as you mentioned? How can we foster that?

DR. GEORGE BAKER: Oh, you can actually build protection into the chips themselves, and that's being done by the semiconductor industry. And it turns out that the discharge, when you walk across the room and touch your radiator, you get a spark. That spark has an amplitude, current amplitude and pulse characteristics very similar to what you would get from an EMP. And so that kind of protection is built into many of the semiconductor chips. But you can do the same with the Ultra-Wide Band electronics, as well.

DAVID GEARY: Yeah, that's a very good comment. One thing I wanted to mention is that on all these early Proof of Concepts for doing DC power, we're sort of being forced to make it look a lot like AC power because that's what people are used to using, and they're used to using circuit breakers and running things and installing things in panel boards and so on and so forth.

One of the things, or many of the things, that are going to come out in favor of, I think, DC are the advances and the advantages that DC does provide and there's things that haven't even been invented yet. But things are starting to be invented, solid-state circuit breakers will replace typical rocker switch circuit breakers in the future. They'll be much faster. They'll have things built into them safety-wise, and so forth.

Safety and grounding, a whole new issue that different options in DC power are offering. There's another company that just announced a new product called Voltz Server, if you ever want to Google Voltz Server. But it's basically taking the communications concept of doing a Packetized Internet Communications and turning it into power. You got a power signature now that goes out to every load. If something happens to that signature along the way, it knows exactly within microseconds that something's happened and they're developing protection schemes with that.

So a lot of these different other technologies and other enabling advantages of DC power, quite frankly, coupled with some of the things that we're looking at for EMP protection and other things; if we bring the group together to look at these things specifically, I think there's quite a bit of opportunity there that's going to be very exciting.

CHUCK MANTO: So this goes back and speaks to the value of all of us, for example, in the technology end of—come with our technology societies, all the IEEE organizations, INCOSE and look at this from the big picture perspective. How do we keep everything viable by being more resilient, in the first place, and create a culture from the inventor down to the people who deploy and consume that think of this? And the whole distributed side of things, which basically says I'm not a helpless victim anymore. I can be more resilient myself and with my friends and my local community and not only take responsibility for it, but enjoy the comfort of being able to do something for myself and my neighbors.

Thank you very much today.

[Applause]

DTRA EMP Program and SBIR for EMP Protected Defense Critical Infrastructure

How does DCI include communities for geographically broad long-term events? What is required to protect them? What is underway now?

Corresponding Video:

https://www.youtube.com/watch?v=ODuGbckaTss&feature=em-share_video_user

Panelists:

- **Maj Gen Robert Newman,** USAF Ret and former Adjutant General of Virginia, Cochair EMP SIG Civilian Military Liaison Panel
- **Mr. Kevin Briggs,** National Cybersecurity & Communications Integration Center (NCCIC), DHS staff performing EMP analysis.
- **Dr. George Baker,** Previous DTRA staff, current contractor)
- **Amb. Henry (Hank) F. Cooper,** Chairman,High Frontier; Cochair EMP SIG Civilian Military Liaison Panel
- **Charles (Chuck) Manto,** CEO, Instant Access Networks, LLC

CHUCK MANTO: In the spirit of making certain we stay on track and facilitate slight changes as we move through the day, I'm just going to introduce the basic concept behind this panel. And that's very simply something that happened this year, very similar to the Space Weather Strategy. So earlier today, you heard a lot of people talk about how extreme space weather may shift our view of how emergency management works. We might not be rescued on day four; it might be 44 or 404, so we all have to take a stronger role in thinking about, caring about and doing something about these threats.

On the heels of that announcement, the Defense Threat Reduction Agency, came out with a request for proposals, not to the standard Defense industry, but to every inventor in a garage saying that the threat of EMP is so significant, we might not only lose the grid for weeks or months, but the affected parts of the grid may not be recoverable. And it's not only important for us on the military base in that event, to have protected key elements on the base, but these bases will not be viable if the civilian critical infrastructure on which we're depending fails us. How long will a military base operate without water? Without sewer? Without food? Without the hospital down the road? And that's the purpose of that.

And so given that, we'd like to talk about the significance of these other parts of the civilian infrastructure, the critical infrastructure that the bases depend on and what's required to protect them and what are we thinking about? Earlier, for example, Kevin will be interested in this, someone asked a presenter, "So we're about to launch this new FirstNet network that everybody's going to need to be more capably connected with each other in an emergency. And when we need that network in the most severe emergency, will it be there for us? Will it be there for us when EMP happens, when GMP happens, and so on?"

All of these come back to that basic question—is, how dependent are we on the civilian critical infrastructure that even the military bases need to survive? And to start, because we have—Mr. Kevin Briggs has been very involved in that from the Department of Homeland Security, and he has some slides, and the General wants to be able to comment or criticize afterwards, and he doesn't want to go first.

CHUCK MANTO: That's right. But he knows that he's got to leave some words to the Senior Statesman, Ambassador Cooper. We'll probably start with Mr. Kevin Briggs, but they all outrank me, so they could change their order and do it any way they want. Here's Kevin Briggs. And I just broke your clicker. No, I didn't break it, but I dropped here. There you go.

KEVIN BRIGGS: Thank you. So, thank you very much. Appreciate the opportunity to—you actually get to keep that one if you would. I'll keep one up here in case the lights go out for some reason. I can think of a lot of ways here.

As Chuck said, I'm with the Department of Homeland Security. I'm at the cyber end of the spectrum, but we also do physical risks. So I'm with the NCCIC, acronyms are the National Cybersecurity & Communications Integration Center. And I lead the team that does the modeling analysis of things like EMP.

Okay, I think we have the technology. Just a very quick level setting of what the Department views as EMP. It is a multi-dimensional threat. We do not lean just toward the nuclear, but also look at solar and radio frequency weapons. And the focus of the briefing, though, will be mainly on source-region EMP and on high-altitude EMP.

There are a lot of misconceptions out there with regard to source-region EMP. In some documents, you'll see that they attribute it to about a two-to-five kilometer destruction zone. And what we're finding when we model this is that even with a low-yield pin kiloton-type of warhead, that you could actually have infrastructure outages that go out 13 or more miles. So what you can see here is that the Zone of Destruction is actually 13 miles, but the Zone of Upset is more like 50 miles.

Now, with cordless telephones, you'd think those would be fine. But what we're finding there is that the AC/DC wall boards are actually the problem. And you can have an upset that goes over a hundred miles, and you can have destruction of your AC/DC wall boards out to like 73 miles.

Now, we have a DHS EMP Protection Guidelines in the works. This is actually Version 7, so we hope to get that out this month. The contents, or that it goes through the background on EMP. It describes four different levels of protection. We do actually have two hypothetical scenarios in here for your planning purposes. One is the canonical one-burst over the center of the United States. And it causes quite a lot of disruption to the grid and all the other infrastructures. Then we have a Multi-Burst scenario that's also included in an annex here.

New things that we're adding here include EMP protection device vendors, as well as service providers.

And we've also added an annex for things like priority services, so we have things like the SHARES program, PSP, WPS, and the like. If you have questions on that, we don't have time right now, but just see me afterward and I'll give you a brief on that.

Now, the four levels are outlined here. Unless you've got very good eyesight, I don't expect that you're going to be able to read through all these.

MAJOR GENERAL ROBERT NEWMAN: I can see it fine.

KEVIN BRIGGS: So we're going to have the General read it for you. But essentially, what we're looking at is—Level One is lowest level and that is where you have no-cost, low-cost ways of improving your EMP protection at a facility. And it's everything from unplugging equipment that's not necessary to putting Faraday cages around, or bags, around equipment.

Level Two doesn't require shielding so much, but it does use various protective devices on your Ethernet, your power and other data cables. Level Three isn't at the Military Standard, but it is at the best commercial standards out there with IEC or ITU. And then Level Four is what you would call a normal nuclear-hardening criteria, the military.

We've gone through these at many of the previous presentations, I think. So, essentially what I'm trying here is one of the scenarios would imply that there's not going to be a whole lot of damage, even with a one megaton because the peak here is only about eight kilo holes. But we're learning through the modeling that you really do have to do the coupling calculations.

So in the protection guidelines, we actually include information on what would actually happen in our expectation with different infrastructures. Like this one is with a north-south oriented CAT 5 cable. And you can see that like with a source region, you had a Destruction Zone of maybe 13 miles. Here, with 100 foot long Ethernet cable inside of a building, you've got destruction, and you can't perhaps see it all there, but you've got destruction that goes from the upper portions of the United States, all the way down to Texas, and covers most of the United States.

Again, in comparison with what we said earlier about the AD/DC volt adapters for your cordless phones, or whatever device it's supporting, that most of the United States with one burst would be taken out potentially as far as your AC/DC adaptors. That's one burst. Now, your backups, because if you lose your phone, you lose your Internet, what do you have? You have things like HF radios, which many of us know and love. But HF is in-band, and if you don't protect that, then it's likely to get blown out, as well. And so, what we've done is we've included in these Protection Guidelines just guidance on how do you protect HF radios? But here, with an unprotected HF radio, you can see one burst would cause either disruption or destruction of your components as it couples in the energy.

Now, there is a misperception that even low-yield weapons are really not that big of a concern. And so, what we have here is a burst over the Gulf Coast area. It's a 25 kiloton, it's not that high of a yield device. But what we'll see is, again, that with—like that 100-foot Ethernet cable, most of the East Coast and a lot of the Central part of the U.S. would be in the Danger Zone, or the Damage Zone.

Now, Chuck asked me to put in one slide on RF weapons, and so I wanted to say that we are very concerned about these, as an emerging threat. The guidelines that we're using right now do cover frequencies up-to-18 gigahertz, rather than just stopping at 1 gigahertz. And it's a major concern for us as a Department, because these are threats that are in the thousands-of-dollars range.

So, in conclusion, the risk of not protecting the Grid is rather profound. One burst on the ground, whether it be a 10 kiloton or another small device, is still causing destruction of electronics out to perhaps a hundred miles. Now, that doesn't mean that every device in that area's going to get fried, but a significant portion could be.

Likewise, the HEMP, or the High Altitude EMP, one burst, if it's well placed, could take out a lot of the infrastructure across the continental United States. And so with Dr. Graham and others, we would agree that it's near certain that the grid would go down given the control systems and everything else that would be at disruption. RF weapons are the new area of concern in that all the critical infrastructures are at risk to RF weapons.

EMP protection can be low cost. I keep hearing that it's a high-cost item. But actually as we studied it, and you can see some of the things are almost no cost. But even on the higher end, it's at the beginning of a system. It's like three-to-five percent, and if you are retrofitting, it can be more. But it's still, compared to the threat you're mitigating, it's not that expensive.

The last thing here is that we do need—cost effective EMP solutions through more than just the Grid. We have projects working on microgrids, as well. I can't, because of some of the proprietary nature go into some of that. But we do endorse the need for that, and we're glad that DOD is working on those things. So with that, I'll turn it back over. Thank you.

[Applause]

MAJOR GENERAL ROBERT NEWMAN: Thank you, Kevin. I'll pick it up here and speak a little bit about the Department of Defense and what they're doing, and how some efforts are being made to try to partner with DOD to make the civilian Grid a more resilient, as well.

Kevin's been very clear that the EMP is threatening us, and it can have damaging consequences to our society. We know this from the Department of Defense, as well. For years, DOD has taken steps to harden the National Command Authority and other essential Command elements against an EMP. We've most recently seen that at United States Northern Command in Colorado Springs, where the commander out there testified before Congress that they're putting upwards of $400 million, I believe is the sum—

AMBASSADOR HENRY COOPER: Seven hundred.

MAJOR GENERAL ROBERT NEWMAN: The Ambassador just told me to inflate that here, so $700 million to retrofit Cheyenne Mountain, the old Cold War installation there that was originally designed to provide resiliency against a nuclear threat, or nuclear detonation. Now we're redesigning that to make it resilient against an EMP.

So, prior history, with everything we know from what our Federal Government has done to ensure that we have a continuation of Government, the steps they've taken, some public, mostly private, but we're aware that it's been significant to ensure that EMP does not take down our Government, that's one side.

Now we're hearing that they're concerned about the capabilities of DOD installations throughout CONUS, the continental United States, to continue to operate. And it's obvious from all that we know, and from what we've heard from comments today, is that the Department of Defense is reliant upon the civilian infrastructure to provide for the power, for the water, for everything they need to do, not only day-to-day operations, but power projection capabilities in the event that we're on a wartime footing.

So we're seeing steps taken, some pretty inventive, others not so much. But the steps are being taken in varying degrees for DOD installations to get themselves off the power Grid. The microgrids are being developed, some with solar power, some with other, more conventional power. But to take them off to ensure that should the civilian power Grid go down from either a natural event, such as a solar storm, or from an intentional hostile act from an enemy, that we still have the ability to ensure that our DOD facilities will be able to defend the country.

We've been working on several projects around the country where we're trying to parent with DOD to have, what I would say, a kind of tangential positive effect for local communities. I'll share one specifically with you, and then I yield the floor to Ambassador Cooper. In the Richmond area, we have a little Metropolitan area that includes a smaller city to the South of us, about 20 miles, called Petersburg. Now, Petersburg for you Civil War buffs might be known as the Battle of the Crater, but more recently it has seen a real growth in a DOD installation called Fort Lee.

CASCOM, which is the Headquarters for the Combat Support arena there for training of the soldiers there for the Combat Support and Combat Service Support for Quartermaster Transportation and Ordinance, is headed up there at Fort Lee. Fort Lee is interested in just taking some steps to develop a microgrid for their self-reliance. Now, adjacent to Fort Lee and to Petersburg is a small town called Hopewell. Hopewell sits right on the James River and is the home to many—to a couple very important chemical plants there.

So what we have come up with, and we've talked with the Commonwealth of Virginia's Government, is to talk about assisting Fort Lee in a partnership, or developing one, where we have the economic power of the—it's called the Crater Economic Development Partnership, I believe. But it's the Dinwiddie County Petersburg Hopewell area that surrounds Fort Lee, to get them involved in this, along with Fort Lee, to develop a microgrid that, not only will ensure that Fort Lee has the capability to continue to operate in the event that the grid is down, but also to have power that can be moved over to provide coop capabilities for Virginia in Petersburg, again 20 miles just to the South, hardly a large trek to get there.

Plus, we would have the tangential benefit of ensuring that the very toxic chemicals that are manufactured in Hopewell can have, at least, some type of capability and power there to ensure we don't have a spill or something that could be environmentally damaging, as well.

So one small project, I can't tell you the hurdles we're encountering trying to make this go forward, but I think Kevin highlights the point that DHS is concerned about this. We certainly are from the Virginia point-of-view, and we know DOD is there. And I think that if we work together, we can all benefit from the steps that DOD is taking, perhaps as the lead Agency, and benefit our communities, as well.

So you all know Ambassador Hank Cooper, Chairman of High Frontiers. That is an organization that does a lot to integrate Defense capabilities with local communities, so Ambassador Cooper, I'll ask you to comment on this, and then we'll open the floor for questions.

AMBASSADOR HENRY COOPER: Yeah, what I'd like to take my few minutes to discuss is the role of the National Guard in dealing with this kind of an issue. And in particular, how in the future it might prepare to help restore our nuclear reactors, which I consider to be an island from which we can deal with a black start condition in a constructive way. Although we're not prepared to do it today, and the Guard is not prepared to do it, in our discussions yesterday, the issue came up and we were talking about the circumstances and whether or not the Guardsmen would be around. And the suggestion was made, maybe there would be 15 percent of them, locally, if we had a disaster like this to be helpful.

And my reaction to that was, "Well, I have a lot of confidence in the guys who wear the uniform, whether they're Guardsman or otherwise. And if their leaders know what they're doing and they're trained to do it, I think you can count on them being there." And I think that states the problem. I think our leaders don't know what to do and our Guardsmen are not trained to do the job that would be part of the black-start condition.

So, and I want to say a word or two about the reactors. I assume most in this room recognize that nuclear reactors shut down in the case of a Grid collapse. They do this to protect themselves. They'd be—free-wheeling, their generators, the turbines, if they kept operating. And then the problem becomes whether or not you can restart them and how soon you can restart them. And they have diesel generators, all of them do, sometimes several of them. The problem becomes gasoline or diesel fuel, after a while, to support them. And those are there for the cooling water, to protect the rods that are in neighboring pools, and also ultimately, to assure that the Grid itself, the reactor itself, doesn't melt down.

So they're a potential hazard in the long run for those of us who live around them. And let me just say that it's something on the order of three-quarters of the reactors in this country are in the Eastern Interconnect, which extends all the way out above Texas, as I recall. I don't have a map in front of me. And that's the bulk of the population of the country, as well.

So on the one hand, if we don't deal with these issues and we have the worst case develop, then we have a hazard, we have Fukushima on our hands in the Eastern Seaboard, all the way out to Texas, potentially.

On the other hand, if we're smart about how we protect the reactors, prepare to restart them in the case of a major shutdown of this sort, then they become a resource for reconstructing the Grid. That's because each of these reactors has probably a year's worth of fuel there. So if we can figure out how to set them back up so that they're operationally functional, and so on, then they become a basis for restarting the Grid for the civil population, but also for the military bases, as well.

Now to do this, you have to think about several problems, and it's been a long time since my technical skills were sharply honed. But I remember enough about the past to realize that you have to replace the load. If you don't replace the load on those reactors when you try to restart them, if you force them, then they'll destroy themselves. That's why they shut down in the first place. So we have to have a load that goes on. We have to have the resources to restart them, too. The diesel generators don't have enough power. George Baker and I have done a little bit of speculating on this, and we think you need on the order of—what is it—50 megawatts or so?—of power to restart a reactor.

So where do you get that power? Well, most reactors are drawing water from a lake or from a river, whatever, and many of these have dams on them. And many of those dams are hydroelectric power sources, and they will be there if they're hardened. So you can imagine an architecture that says I will make sure that the hydroelectric plant is hard. I'll make sure that the interconnections between that and the reactor plant is hardened. And I'll also worry about making sure that I have, at least, some buffer zone of the Grid outside the reactor hardened so that I have a load. And I think I need an invention that some of you creative folks might figure out how to do. I need a variable load that I can set up when I restart the reactor at one level. And then as I bring up the Grid, I have a Vernier of sorts, a rheostat, I don't know what the right term is, so that I can back off on my surrogate load as I bring up the real load.

And so now I can figure on how, if I proceed down this path, how I have islands of support from which I can restore the rest of the Grid, whatever they're doing. And I'm not arguing you shouldn't be doing other things, as well.

And by the way, coal-powered plants are more resilient than the other power plants around, and if we don't shut them all down, that's an alternative to the hydroelectric plant that I gave you as a model.

So the thought I'd like to leave the group with, since I haven't heard any discussion at this point, is that we really should be giving serious consideration to the potential solution set that can be built up around our reactor plants. We have on the order of a hundred, I think it's a couple less than that at this point, operating around the nation. It's a resource, and that's the power to keep you going for two years. You'll be down from your main line with all these transformers being shut down; but trust me, if you got 20 percent of what you have, that's better than none of what you have. And in case of South Carolina, where I'm worrying this most carefully, 60 percent of our electricity comes from nuclear reactors, which is why I put it very high on my list of things I'm working on. I think that's the main thought I want to leave you with. Be happy to take questions.

Oh, one more thought. I meant to say, but because it deals with—I forget how Jim rephrased it—but our bureaucratic problems, our regulatory issues that maybe this group doesn't want to talk about very much—I have more confidence in the Nuclear Regulatory Commission, frankly, than I do in the NERC-FERC arrangement that dominates all the rest of the considerations on the Grid. The Nuclear Regulatory Commission still has a heavy heritage that was—that came from our nuclear Navy, Hyman Rickover's world. And there are a lot of competent engineers that have been associated with that institution over the years. And the current Commissioner, I believe, has been associated with the operations there for some 30 years.

So I'm hopeful, at least, that if we can figure out how to deal with this problem, and I don't know what the regulatory problems might be, but we might have a more friendly audience in working it with the NRC than we have with NERC and FERC.

CHUCK MANTO: Al?

AL SMALL: Thank you. Al Small. This question for Ambassador Cooper. If you're going to be using hydroelectric dams to be the start power for the black-start for the nuclear reactors, rather than dummy loads like we use in ham radio, why not pump the water back up into the reservoir and adjust how much you're pumping based on how much load you need? Then you're saving that energy, you're storing it for further use.

AMBASSADOR HENRY COOPER: I think that could—if I understand what you're suggesting, one of the alternatives for the surrogate load that I mentioned that you have to have as a Vernier, is to select—when I studied thermodynamics, somebody showed me studying the Second Law, or something like this, as paddle in the water, one of the ways to have that Vernier operation is to drive some kind of a turbine inside the water from the reactor and scale back the power requirements, as appropriate. So I think I got the gist of your idea. I think that makes sense.

MAJOR GENERAL ROBERT NEWMAN: Al, I'd patent that idea as soon as you leave the room.

[Laughter]

JOE WEISS: Hi, Joe Weiss. What he just talked about is a thing called pump storage. Pump storage exists. The concern, especially when you start talking about things like EMP, when you don't have pump storage, you've got a chance for maybe having older controls. But if you're going to have pump storage, you're going to have to have much newer controls.

AMBASSADOR HENRY COOPER: Much what?

JOE WEISS: Newer control systems, and all of a sudden now, you're back in the middle of EMP.

AMBASSADOR HENRY COOPER: Can't you resist that urge some of the way? Maybe that's the challenge for the inventor, too?

JOE WEISS: No, because you're adding an awful lot of complexity to a system. And just the other technical thing, House Loads in a nuclear plant. In other words, those are the internal pumps and everything else, is probably on the order of maybe 30-to-50 megawatts. So, one of the reasons nukes are so big is they've got an awful lot of internal overhead they've got to meet. And that's why, like Oconee, if I'm not mistaken, their black start is hydro. You'll find each of the nukes have a designated black start-type unit.

But, the other thing is, again, just being a bit technical, boiling water reactors are designed so that you can essentially move load up and down. It's not near as easy with a pressurized water reactor. You know, these things were designed to be base load. So, the idea of trying to use those to play isn't quite as simple as it may sound. That's all I wanted to get across.

CHUCK MANTO: Okay, thank you.

MAJOR GENERAL ROBERT NEWMAN: Well, my only reaction to this is I'm not trying to invent the solution in technical detail here. I'm just saying, on the one hand, you have a problem if you don't deal with it. On the other hand, if you're smart about dealing with the problem, maybe you've got a pathway to a solution.

Industry Best Practices for EMP, Space Weather and other High-impact Events

Corresponding Video:

https://www.youtube.com/watch?v=3bClRjg19fE&feature=em-share_video_user

Panelists:

- **Mr. Charles (Chuck) Manto,** EMP SIG Chair; CEO, Instant Access Networks, LLC
- **Mr. Michael deLamare**, INCOSE and Bechtel
- **Mr. Gale Nordling,** CEO, EMPrimus, MW Transformer protection users
- **Mr. Jack Pressman,** Cyber Innovation Labs, NE Insurance Industry Data Center
- **Mr. William (Bill) Harris,** Foundation for Resilient Societies

CHUCK MANTO: Thank you. This industry panel is on Industry Best Practices for EMP, Space Weather, and other high impact events. The whole idea of best practices is really significant because, while it is important to have regulatory structure in place, and in some cases it's absolutely necessary, it's also useful to have best practices. Because whatever you decide to do as an essential requirement that you may regulate, you always want to improve your game, right? You want to always improve by doing whatever you can, either on the technology side or the business process side or any of the other social systems that might be involved.

And so we all want to continue to improve and enjoy the opportunity to show off or develop best practices of different kinds. This panel today is going to talk about that. The order I'd like to do it in is to have Mr. Gale Nordling talk a little bit about his perspective on best practices, one because he's got some customers who have been adopting some of technology they have been developing. And he can speak sort of, generically, about how customers think about the technology side. But I think he has some insights on processes as well that he'd like to talk about. Michael deLamare, from INCOSE and Bechtel, will also talk about the process side. And then, Mr. Jack Pressman will talk about some of the best practices that the insurance industry has been using, not only from the standpoint of putting people's feet to the fire saying, "We're not going to insure you anymore if you don't do something about this." But some folks in the industry have actually taken measures to do something to protect themselves.

And so that's the sequencing that we're going to begin with. And if Dr. Paul Stockton arrives midway he will just sort of join us. Otherwise, we will start by having Mr. Gale Nordling from Emprimus talk about best practices from your perspective. And you can either do it from the mic there, or you can come here, either one.

GALE NORDLING: I can do it from the mic unless you prefer that I—

CHUCK MANTO: I have no preference.

GALE NORDLING: Okay. First of all, thank you for letting me speak, and thanks to each of you for listening to me. It's getting late in the day. There are many aspects to best practices that could be talked about, whether it's talking about technical issues like Black Start, stockpiling fuel, security, etcetera, they're all important. But I'm going to take a little different approach and talk about some different concepts.

I want to focus on three concepts. Number One, Transparency. Number Two, Transparency. And Number Three, Transparency. Why are they important? Long ago are the days when loss of power was an irritant or an inconvenience. It is now a critical issue to businesses. As an example, a manufacturing firm with simply a blip in the power on a computerized facility can lose a million dollars. And it happens because, if the computer goes down, how do you restart all the manufacturing where it was at, at the location that it was at, and all the different products? You got to throw them away and start over.

We learned this morning that Swiss Re has said they're no longer going to insure for a Carrington kind of event. So don't you think it's time for a full-transparency disclosure with customers? What is the risk that the customers are assuming that they don't know that they're assuming in this day and age?

So, I worked for 20 years with the utility. But I took what the utility best practices and different things are, and let's superimpose transparency on those best practices. One of the first best practices that I looked at was—prepare for the worst event. NEI Electric Power Engineering has an article on that very subject.

Well, the worst case basis and its derivation should be discussed with the customer, including the fact whether a Carrington-level storm is being protected against. I doubt, in this day and age, that the customers out there have any clue as to what's being protected against and to what level, and to what exposure and liability that they are assuming. I don't—we've been going to lots of meetings with utilities over the last eight years. There has been no discussion of that, whatsoever, none.

Another best practice that I found was that utilities should give and publish global estimated restoration times to businesses for major events and outages. How can they protect and restart their business and know what they're faced with if they don't know what they're faced with? So whether it's something simple like a restaurant restarting its business or a manufacturing firm or something a lot bigger than that, it would certainly seem like they need to understand what the restoration process is.

Number Three—I think utilities ought to publish, for very severe GMD and other type of events, the criteria and when other supporting infrastructure will be served and put back into service. You not only need electricity, but you probably need water, you need communications, etcetera. But the utility is behind that as well. So what is their plan for a really significant event to get those things back into service?

For utilities I would suggest, and more power to them, that would suggest that procedures are going to protect the customers. There should be a candid conversation and transparency to say, "Have you tested your plan? How have you analyzed it? To what level do you believe it's going to protect you?" I'm saying that a little bit tongue-in-cheek because the modeling studies we have done would suggest that procedures that are being suggested to date are going to be largely ineffective.

So why should customers believe that there is a solution at hand if it's really not at hand? What is being done to protect against harmonics? In the last benchmark standard and other things that are being talked about, harmonics is not really addressed. And I think that a best practice would absolutely address harmonics. The studies that I now—the studies from Ray Walling and others, would suggest very low GIC currents and transformers, 10-to-30 amps, causes huge harmonics that are beyond the IEEE standards. There needs to be full transparency. We're either not going to protect against that, and that's your problem, customer, or how do you intend to address that?

Today's business is so different, like I said, than when it was more of an irritant or inconvenience of losing power. It's a much more serious consequence today. And finally, some of the old rules of restoring customers, such as one of the rules that always used to exist that still exists today—you restore the most customers first, as far as that. Well, in a very significant—let's say a GMD event like a Carrington, I don't know as that rule works. Don't you have to look at all the different other types of infrastructure and whether it's a community support, for whether it's hospitals or law enforcement, fuel or water or whatever? Maybe those kind of rules don't exactly apply anymore, and there should be a frank discussion of it. So I worked at various industries, and the latest buzzword is "transparency." I think it's long overdue in the electric utility industry. And that is transparency to customers.

MICHAEL dELAMARE: So my field of expertise is actually in systems engineering. So I wanted to talk a little bit about it from the standpoint of a development of systems or modification of systems, and particularly in complex systems. And I wanted to introduce some of the concepts in complex systems.

Now, one thing I wanted to say, up front, is that complex systems is not the same thing as complicated systems. And that's the terminology in the industry has been, where complicated systems might be something, system, like the ICBM system that's sprawled all over the place, very extensive control systems, very extensive higher authority command, all that stuff, and very complicated in terms of the technology deployed, especially in the timeframe in which it was done.

But complex systems are very different. They have—they tend to be system-of-systems. They are things that have emergent behavior, which are things—behavior that emerges out of the integration of things that were not necessarily designed into the system. Now a really simple example of a complex system might be a society. Society is a very complex system. Political system is a very complex system. But these things are not just in social systems, but they're also in technological systems, where you have these systems of systems.

Now one of the things that has been, in the past in systems engineering, it started off with complicated systems, where there was more of a traditional approach of reductionism. You had an idea of, "I have this big black box that I need to create to accomplish a mission, and I'm going to understand that functionally. I'm going to decompose those functions and break it down into, from a very abstract idea to something that's very concrete, that can be designed and built." And that's a reductionist kind of approach.

But the problem with a reductionist approach is that it does not give you all of the picture that you need to see and understand to see how is this system really going to behave under all scenarios? Under all hazards? Nor is it going to be able to tell you what the interactions are outside of it in a way where, how is it going to affect and bring about unintended consequences?

So what you need is to understand the system is a member of an ecosystem before you really need to understand or before you can really understand what it needs to do and how it needs to do it, so how you would, in essence, bring about a response and design that actually addresses all of the scenarios that you're going to be facing in the ecosystem in which it must exist.

So the complex system is an emerging field in systems engineering that would look at those sorts of things. And one of the things that has been kind of a buzzword a lot is the system-of-systems approach, where each system that gets integrated together was this big black box that people built by reductionism. And each one had their own mission that was independent of the others. But now they are trying to bring them together and fuse them to multiple systems together, so that they can accomplish a single mission. But, when you start to do that, now the interactions of the systems, they were never designed to work together necessarily. And when they were not designed to work together, you get the behaviors that, when they do start to work together, you begin to see that overall, the behaviors are not quite what you expected. There's a lot of unintended things that you need to go and now deal with. And you may not understand what seemed to be perfectly okay with the system by itself. And now you start bringing in something else, and now this system has got side effects acting on this other system that may cause the other system to fail.

So some of these things that you might look at in this context of the Grid is that the Grid is a collection of systems, multiple systems that, whether it be generation distribution transmission in multiple municipalities or multiple providers, that each one has their own particular systems that they function together. And in some cases, they are integrated, in some cases they're loosely integrated. But when one fails, you could cause cascading effects to the other.

And let's see. But we—So part of this is, it's not—When you start analyzing a system from the beginning, whether it be to reverse-engineer the system or look at modifying a system or starting a new one, it's more than just the business case analysis that you have to look at. It's more than just what people might think of also as a supply chain analysis, whether you're going vertical integration or horizontal integration across. All these interactions that now you have to start looking at and the emergent behavior.

So one example, another example might be, if you look at Twitter, and look at that as, how did Twitter, which is an independent system, people interact with that, how did that change other systems? How did it change the political system? How did it affect Egypt in the time? You know, it had a lot of things that were unintended consequences. I'm just trying to give you an example of something that might happen in a social-technical system.

Now one of the things about the emergent behaviors is that they're often nonlinear. And they can be over long periods of time. They can appear to be linear, and then all of a sudden you hit a tripping point. And when you hit the tripping point, things just break. They can either collapse on themselves or they can go with uncontrolled growth. And they'll destroy—can destroy each other using those sorts of things. The only way to know what kind of behavior you're going to find is to model them.

And so the—Let me give you just a few characteristics when we're talking about complex systems in the context of system-to-system characteristics. Now each system has its own purpose. It's independent of the other specific systems. Independent authorities. So you think about what we're dealing with in the Grid, you have each system is owned by an independent authority. And decisions about an interface that you have with one of these other systems, you may have no say in what happens on that other side of the interface. It could be completely out of your control what happens, or how they design, or how they neglect to, say, take into account security or take into account resilience on their side, how that's going to affect you. You have no control over that part, so you have to be able to account for that. There's leadership challenges in there, and that there's lack of structured control between these systems.

Capabilities and interactions of interfacing systems aren't necessarily known. So you don't really know all the capabilities of the system on the other side of the interface. You don't necessarily know how it's going to respond when you give a signal or when you apply a certain force upon it. It may cause it to resonate in a way that you did not anticipate. And it's not because you didn't have it designed for the nominal case, it's typically where it's at the edges that these things happen. So when you are working on off-normal conditions, or you're working on design basis accident conditions, or things which could even be beyond design bases, beyond design bases for you might not be beyond design bases for them. And that could result in something that would have an adverse effect upon you, because they're designed to a different level.

So those things need to be understood when you're dealing with systems-of-systems. So the other thing is that they become difficult to test the interactions. So one thing you might seem to think about is that a test settles everything, right? So when you go and you say, "I have these possibilities out here. Will my system be capable of doing this? Can my system withstand that?" When you're dealing with systems on a large scale, and that you don't control everything, you have a difficulty in test because the test and scale, the scale of it makes it very expensive to do. And the interactions that you are trying to induce may not be doable. Maybe the only way you can do it is by analytical methods. And analytical methods needs a means of validation.

Just because somebody builds a computer model doesn't mean that computer model is any good. You got to understand that the assumptions behind which it is based, and that they have to be based on scientific fact and ability, and that maybe stretching beyond our ability to do that. And so verification validation of a model becomes very important in trying to answer that question.

So some of the things that these things might say lie in practicality, when you see these kinds of situations, is the risk is going to be a huge issue that needs to be managed. So best practices and developing complex systems requires a robust risk management problem or risk management approach, and it's not just probability-times-consequence. I know that that's the classic way that risk management is taught, is probability and consequence is the key thing that you look at. When you're doing a failure modes and effects analysis, one of the other things that has to be taken into account is the detectability of the failure mode or detectability when am I getting into a condition where I'm going to fail? Can I predict that? And do I know what my precursors are? Or do I know when the failures happen?

So, if you take a look at some of these incidences or some that have been described to us in cyber, things have been placed in, and nobody has any idea that it's even happened. So the failure is already in place. And you don't have—there's not any kind of detectability to be able to say that it's going to happen. And so what's the worst kind of risk that you have? It's not just the ones which are probable, it's the ones that you don't even know it happened to you when it happened. So that's a key element of risk analysis.

So some of the other best practices then that come into place, that I believe that INCOSE can help bring into this discussion here, is some of the modeling. There has been a tremendous amount of effort in model-based systems engineering. It's been an initiative in INCOSE over the last ten years to bring in new modeling languages and approaches, plus—and that's in combination with modeling methods like system dynamics that have been around for decades, or physical models that have been around also for decades. But it brings a more of a—the new approaches are trying to bring more of a system to systems ability to this.

So I guess I spent enough time on that. So let me hand it off to Jack.

JACK PRESSMAN: Thank you. Appreciate it. Thank you. Always got to make sure the PowerPoint works, so there's a little color for you. I want to talk just a few minutes on an actual implementation. I want to thank Chuck. I want to thank the entire Infragard Board. Been involved with Infragard now for a couple of years.

We've been designing and building mission-critical facilities for about 20 years. And this is a very quick story about a very large multinational insurance company that approached us about four years ago. And we were working with a company called Iron Mountain, which you're probably familiar with. And they approached us and said, "We not only want a data center, but we want a data center to ski and be protected." And we said, "Okay, we can do it."

So we actually began to work very closely with some of the top vendors, many of whom—Mike Caruso, many of you, if you were here yesterday, had a chance to meet. So Mike and I actually began to work on this very closely, and began to look at EMP. And we'll talk about the case study in about a minute and a half. But I thought I'd give a little background.

As we built this, and looked at the cost, and looked at how we engineered it, we began to look at reaching out to a lot of the global risk managers that we work with, a lot of the CTOs and CIOs over the last couple of years. And people began to think about, "Yes, this EMP risk is starting to become significant. But we don't want to pay a lot for it right now, our Board doesn't." We said, "Okay, we understand." And, as we deployed this first major fortune 500 facility, we began to hone not only the model, but actually the whole concept of-- how do you actually build a facility that can protect for EMP? And then we extended that whole concept, because we began to look at geomagnetic storm, and we realized and knew that we had to use different types of methodology to protect that.

So we began to—When I began to present to a lot of global risk managers, and they would start this whole discussion about, "Oh, how much more is it going to cost?" And whether it's green field, new-build, or whether it's retrofit, I said, "Well, think about EMP protection like airbags". And they would say, "What are you talking about?" I said, well 20 years ago, people basically responded, "I'm not going to pay $1,400 dollars more for a car with airbags." It was a very clear cost. People said, "We don't need it, I've got seatbelts." I remember driving with my folks to Florida not wearing a seatbelt. So there was some evolution.

Now you can't buy a car without airbags, and you don't even look at the cost of airbags. It's now a—and you wouldn't buy a car without airbags. I believe that is exactly where EMP and geomagnetic storm protection is going. It is at an adoption scale. And I really like this concept of comparing it to airbags, because I think it's the same emotional concept.

What we did, when we sat down to design this data center for this fortune 500 insurance company, is we began to look at not only just EMP, but how EMP became part of the design of a world-class facility, a world-class data center, a world-class control center.

I'm going to move quickly through this, because it could really get boring for you. But what we did was, we basically looked at parimeterization of the data center, from a conceptual standpoint and a methodology. So we began to look at, what do you have to do? Well, you have to build, basically, a Faraday cage, for all intent and purpose. And I'll have a few snapshots here, a big steel box.

From there, we really start looking at, how do we protect and filter everything that's coming into the facility you're trying to protect? You could build your box. But then, how do you get electricity? And how do you get air? And how do you get low voltage in? So you have to put in filters. And that's a big part of the design, a big part of the cost.

We expanded upon not only that, but taking a look at low voltage. We have to protect our low voltage installations, our PoPs. Big question that is always raised is, okay, you can build a facility that can survive an EMP impact, and yes, we have a methodology to keep it running for 45-to-60 days. But the next big question, and I actually had this conversation with Kevin Briggs from DHS early this year is, how do we communicate? That's a real big question that I think we are going to have to start addressing in a more serious mode.

Because what we're going to do is build these islands of survivability, microgrids, survivable, but they're not necessarily going to be able to communicate with each other. From a corporate fortune 500 perspective, what do these companies want? Or, from a mid-market, low-market, doesn't make a difference. They want to be able to have continuity of their business operations. They need their operations. If the Midwest has an impact, they need to be able to mirror data, from Midwest to East Coast, West Coast, maybe overseas.

We began to also look at geomagnetic storm and realized the same type of filtering that you're putting in for EMP is not necessarily at all going to work for a change in harmonics caused by geomagnetic storm. So we began to work with a number of folks—Bill Radasky, specifically, that many of you know. I know a peer of Dr. Baker's and a peer of Mike Caruso's. And I said, "Well, why don't we just treat a geomagnetic storm event where you've got changes in harmonics that could happen, not only in nanoseconds, but minutes, hours, days? And why don't we just treat it as bad power and just cut it—so it's as if the power source, the Grid went down?"

And we designed a concept called a harmonic kill box, for lack of a better phrase. It's basically an emergency power-off that sees impact of a change in harmonics, kills it, and goes to the internal protective power—so if you will, a true microgrid that's built into these data centers. And DC, by the way, is a methodology that we are adopting, because I do believe in everything that was discussed earlier about DC power.

How did we do this? We began basically to look at the microgrid concept. And not just for power, but also for cooling. Because in a critical data center application, you can energize it, but, if you can't get rid of the heat and get some cooling into it, you're going to just fry everything from the standpoint of just cooking everything.

So what we designed was not only an A-side that had your standard UPS power, your standard cooling systems, but a complete B-side that we also shielded. So the B-side itself, the B-side MEP platform is also in its own shielded facility adjacent to the data center facility. And with that included a sized generator platform and enough diesel fuel adjacent that we could operate this facility for 45-plus days.

But then you have to go down and look at, how are you protecting your pumps? How are you protecting? How are you running through your own exercise when the Grid goes tapioca? How do you get that generator system running?

Again, building a Grid of not just power but also the HVAC that could run the facility for 30-plus days, it impacts how you size it. We began to also look at protecting from a SCADA perspective, because I think you can't just look at EMP. And when you talk about triple threat, it's not just EMP that we're looking at or geomagnetic storm, but we're also looking at cyber. And SCADA is, as I think, the biggest open gap that exists today in the public/private marketplace. And whether it's pumping stations, Grids, any major data center, you can have data centers or Fortune 100 companies that are spending tens of millions of dollars on network protection, but they don't even know who's running their air conditioning systems. And they still have the passwords from the factory from 1986. [laughter]

We started looking at, from an EMP perspective, and from a microgrid perspective, again, telecommunications—how do you actually deploy, rapid deploy the capability of being able to communicate? You've protected a facility, whether it's a hospital, whether it's a data center, whether it's command center, first responder. But you got to be able to communicate. That's a big issue that I think we're all going to have to address with DHS and the States.

Obviously, having the type of DCIM systems, data center monitoring systems and critical operation centers. I mean when an event happens, we're probably not going to get a heads up. So a lot of this is building systems that can immediately fail over instantaneously. Part of the issue with many critical facilities is that they're just not prepared from an operations standpoint to be able to operate a facility instantaneously with a major failure, when you run your own table top from the standpoint of the Grid.

And I come from an area in the country that is the—it is the industry leader in rolling brown-outs. I mean Commonwealth Edison, there's no one who comes close to the success of Commonwealth Edison to kill your power on probably a two-to-three-quarter basis. And they'll just tell you, that's an old Grid, and they're regulated. And they don't have enough money to repair it. And if you want brand new transformers that are grounded, you just got to pay them.

So that concept of how to work with utilities, we learned a lot of this just the hard way, when the power would just go off, and go off for two days, three days in Chicago. Doesn't take—You can pick your natural disaster, and it will trigger it.

So, now I've got to add Kevin's four levels. But basically, we created with our vendors, when we built this data center after about two years, our ad hoc five levels—maybe I have to call it something different—for building this type of resiliency for both EMP and geomagnetic storm. Level 1 meets the MIL-SPEC 188-125, which incorporates, I believe, Kevin Brigg's Level 1-through-4 of the DHA standards.

Level 2 is applying that to critical mechanical. So you have a mechanical infrastructure that can operate instantaneously. Level 3 is building the type of sensor platform that can kill the distortions that are coming in from a natural impact, which an EMP, EMP-filtering and EMP- shielding will not protect. You're going to get all sorts of stuff coming through into your critical IT infrastructure. It's going to fry in a matter of 15-to 20-seconds. Building the type of, we'll call it, microgrid platform, that can self-sustain 30-to-60 days. And again, Level 5 is moving into also protecting all of this from a cyber standpoint.

We strongly believe, after doing this process over the last four or five years, that to just look at one of these threats, isolated, and not looking at an optimized solution for both protecting physical impact and cyber impact, you're wasting, I think—you, whether you're the buyer or you're the vender or you're the facilitator, you're going to waste a lot of money, because the SCADA weakness is, I think, even more dangerous today than EMP weakness, even though I think EMP is a clear threat.

Again, setting these kind of build-specifications. So where were we at the end of the day? Well, we built this really cool steel box. And when there was nothing in it, you know, we'd get a really cool echo chamber. But this basically shows you exactly what an EMP shield and state-of-the art MIL SPEC standard, or exceeding MIL SPEC standard EMP protective facility looks like. It's 360'-degree protected. These doors are specially-milled doors. So when the doors lock, that room tests out.

You're putting in the type of filter systems from the power perspective, from the cooling perspective. Testing is always a big component of it. A lot of people are looking at cheaper ways of building this. You've got to test it. You've got to create a standard. When we were working with our client ,which was a global insurance carrier, they not only were the client, but they also wanted to see what methodology we used to not only build this, but what methodology we used to test it, to certify it, and warranty it. And that testing procedure has to be continuous. Because in every facility, what happens? New things get installed. People are drilling stuff into your—you know, we go back. Two days after we built this, I had somebody in there drilling holes into it. Like, you know, I'm going to give you an emoji, or give you an abbreviation that starts with a W and has a T and an F. [laughter]

So that's a big part of it. It's not just whether you're protecting a small facility or a very, very large command center for a utility producer. A big part of this is then looking at how that facility continues to maintain that testing criteria. What good does it do? Client spent x-amount of money, or the partners x-amount of money. And then eight months later, it fails its test.

With this, we developed a modular system. This is actually for a customer that's about 40 kilometers from the North Korean border. So this is a—I had this concept a long time ago, and modular has been around in data center. But I call it EMP CB. Basically, it's all basically pre-designed, pre-built here. How about that? Building something here, and then shipping it overseas and letting them pay for it. But it's an interesting concept. And it takes all of those components.

Now this is not cheap. Let me talk about, just ten seconds. Building that first level of EMP protection, shielding and filtering at design, is going to add about six-to-eight percent of the project cost. So if you've got a $10 million dollar project, and you want to meet that Level 1, which is DHS is Level 4, which is the MIL SPEC, you can put six-to-eight percent on top of that. You start expanding that, that goes up.

Retrofit doesn't make any sense. I mean it does make sense if you have no other choice. But there's a project, there was a utility that we were working with. They wanted to retrofit two command centers. I said, "I don't—Why not build a new one? For what you're going to pay to try to retrofit a live command center, you can build a brand new one, bring it up, test it, certify it, and there's no down time." So cost is a big part of it.

That's really what we're talking about. I'm part, along with Steve Pappas, John Jackson, a number of others, we've launched, under Chuck's direction and the National Board, the Midwest EMP SIG. And a big part of what we're doing in the Midwest EMP SIG, with a lot of the private sector clients, again, it comes down to money. Everybody wants to be protected. No one doesn't want to be—no one doesn't want the airbag. I mean I don't know, maybe some people don't want airbags. But it's just the cost.

And a lot of this is design. What I will—so this is basically the overview of the client itself. Here was the challenge. This was going to be their brand new next-generation Cloud platform. And they were requiring it to meet MIL SPEC. We built the chamber. You saw the pictures. It's supported and was commissioned. And has been operating now for almost two years, two medium-to-high density data cabinets—60, pardon me, 60 high-density cabinets. And it provides them with their primary, now, global data center.

Value derived, they're meeting a standard that no one else in their industry is meeting. It allows them to now, as they're doing acquisitions, as they're going out to clients, to be able to not only showcase what they've done as an insurer, but to be able to take a lot of what we worked and start talking about how they would offer some type of an insurance product that would meet risk that's starting to be evaluated from the corporate Board.

[Applause]

CHUCK MANTO: So I'll rejoin you. Bill Harris, you had a chance to listen to all these. We've got a minute or two before we get ready to do the next panel. You might have some comments or thoughts about everything you heard. And synthesize it and provide great pearls of wisdom.

BILL HARRIS: Well, I wanted to just give a few pointers for best practices in the public sector. Number One, there is a very important rule making by the Nuclear Regulatory Commission called Beyond Design Basis Rulemaking. They're rethinking what you need to do to protect their licensed nuclear power plants. It's open for comment until roughly the first half of February. And they are not considering how to protect against storms, really. They do ask that the backup power be able to operate indefinitely, but they don't go into detail. They don't require testing of their emergency diesel generators offsite by third party. So they don't consider EMP. So here is a major opportunity to get those nuclear power plants safer.

The Second, the Federal Energy Regulatory Commission now does not require—nor does NRC require geomagnetic disturbance monitors, these GIC monitors. If we had them on the nuclear power plants in most solar storms, they could avoid shutting down if they reported to an operations center, the NRC Operations Center. But they don't have to have them, they don't have to report. So that's a safety factor to have a needless scram of a nuclear power plant.

So we need GIC monitors retrofitted for the nuclear power plants. They cost $15,000 apiece. This is peanuts. At FERC, if we don't have the GIC monitors mandatory and reporting to the Department of Energy Operations Center, under the new authority being granted to the Secretary of Energy for Emergencies, the Secretary won't have visibility of which transformers need protection. If they wanted to switch to the DoE Operations Center into the White House Situation Room, they won't have data—the President can't make a sensible decision.

So we need more transparency, as Gale pointed out. We need it going from the operators in the private sector to the operation centers for FERC and for the Department of Energy has an operations center and NRC has an operations center. None of these can be switched into the White House Situation Room with useful information if they don't have useful information.

Last, the EMP Commission, it has an opportunity to assess jointly what you protect for EMP versus GMD. And you saw Jack talking about some of our assumptions that you can do the same protection may not hold true. So this is a major innovation in modeling and testing. And we need the EMP Commission to get into the business of testing other people's equipment. I think they're going to have to ask the other parties, the manufacturers, to pay for the testing, because the EMP Commission will only have $2 million dollars. We need them to test concurrently for EMP and GMD. It's in their charter to assess both EMP and GMD. And that could make a big difference. And we also need them to test like emergency diesel generators.

If there's a FLEX Program that goes, that sends generators out to the nuclear power plants to avoid a Fukushima, but if these generators—so far, the FLEX facilities that have them, one in Memphis, one near Phoenix, there's going to be a third one, they are not EMP-protected facilities, and the generators aren't EMP-certified. So I think we'll have an opportunity. I hope Peter Pry, when he speaks, will maybe consider some of these issues for the new EMP Commission. Thank you.

CHUCK MANTO: Thank you very much, panel. Thank you.

[Applause]

END

Key EMP/GMD Next Steps

Corresponding Video Link:

https://www.youtube.com/watch?v=XEONsRAK6xI&feature=em-share_video_user

Panelists:

- **Dr. Peter Vincent Pry**, EMP Task Force on National and Homeland Security.
- **Dr. George Baker**, Professor Emeritus, James Madison University, EMP SIG EMP Advisory Panel Chair

DR. PETER VINCENT PRY: Well we've already talked about solar storms enough. I think the next step is actually the first step, the first step. When it comes to thinking about Grid security, we really haven't taken the first step, which is to understand what the real threat is. I think people have heard enough today about the natural threat, but let's talk about the threat from man.

We've also talked about nuclear EMP, this is familiar to everybody at this point, Radio Frequency weapons, which really exist. And, in fact, you can even buy them, you know. You don't need the license to buy them. This is a firm's advertisement. It's not intended to be used as a weapon, it's called the EMP Suitcase. And it's intended to be used as a diagnostic device in industry. In fact, I think Emprimis has actually purchased one and shown it around. You know, anybody who has the money to buy this basically can buy a Weapon of Mass Destruction.

The bad guys know about nuclear EMP, Radio Frequency weapons, physical sabotage, cyber attacks. And if you take away nothing else from my briefing, it is the first step or the next step that we must take, is to understand the first step, which is, the threat is from all of these things. Right now, the way we think about—take our cyber security doctrine. When we think of cyber warfare, it's narrowly—it's defined narrowly as computer viruses and hacking. And most cyber security experts don't talk to the guys who know about nuclear EMP. And the nuclear EMP guys tend not to talk to the radio frequency weapons people. And neither of them—none of them talk to people who are into the Green Beret sabotage commando physical kinetic-type sabotage attacks.

The bad guys plan to use all of these together, all of them in conjunction. So we've got to stop arguing with each other about which of these threats is the highest priority. They're all the highest priority, because we faced all of these threats simultaneously. North Korea has actually practiced the nuclear EMP attack against us. And in fact, it's practiced—it's practiced what I call a blackout war. They have different names for it. They've written about it in their military doctrines—the Russians, the Chinese, Slipchenko's *No-Contact Wars*, Wei-cheng Wang's *Total Information Warfare*. The Iranians have written a booked called *Passive Defense*, which ironically is about the offensive use of all of these elements in combination, physical sabotage, cyber attack, non-nuclear EMP weapons, and nuclear EMP attack against the United States specifically mentioned in a military textbook, you know, as their version of cyber warfare. That's their version of cyber warfare.

They're all patterned after Slipchenko's *No-Contact Wars*, you know, which describes all of these in combination. In fact, the first page of the Iranian military textbook pays tribute to Slipchenko. He's their guru. He basically taught them, you know, how to think about this future revolution in military affairs that involves the attack on electric grids and other critical infrastructures, by all these means.

This is the track of North Korea's KSM [sic] 3 satellite that was launched—that was orbited over us in the midst of the worst-ever North Korean—you won't have read about this in the newspapers unless you've read some articles written by me and Ambassador Cooper, you know. But in the midst of the worst-ever nuclear crisis we ever had with North Korea, and that was taken during of Kim Jong Un, the despot of North Korea in his Command Post in the midst of this crisis. It was the aftermath of their third illegal nuclear test in February of 2013. And the United States was going to impose additional international sanctions to punish them, and they started making nuclear missiles to threaten, and to make a missile strike, a nuclear missile strike against the United States and our lives.

The map on the back wall shows trajectories where they would make this attack. And Kim Jong Un is there with his general staff. And the Obama Administration took this threat so seriously that we were flying B-2 bombers across the Demilitarized Zone doing our own exercises to deter him. But he was not so deterred that he did not orbit a satellite. And this was the satellite, the KSM 3, that in the midst of that worst-ever nuclear crisis, did what Ambassador Woolsey described earlier. It was launched in the South like a fractional bombardment system, so it could evade our Ballistic Missile Early Warning Radars and Interceptors, which are all oriented to the North. We're blind and defenseless to the South. Orbited over the South Pole, came up from the South. On the 10th of April, it passed over the Optimum Altitude and Trajectory to put one of those HEMP fields, an EMP field over all 48 contiguous United States.

On the 16th of April, it was in the Optimum Trajectory to put a peak EMP field over the Washington, D.C./ New York City corridor, which would collapse the Eastern Grid. You know, the United States can't survive without the Eastern Grid. That's 75 percent of our electrical generating capacity there. So even that attack is an existential threat to us. On this very day, the 16th of April, that's when the mysterious attack by the U.S. Navy Seals call it a commando operation against the Metcalf Transformer Substation in California-- on the very day that the KSM 3 was passing over the Washington, D.C./New York City corridor, posing a threat to the Eastern Grid, we had this event that was a threat to the Western Grid happen.

And in the middle of all of this stuff there were cyber attacks, as there are every day, thousands of them being made on our Electric Grid and other critical infrastructure. So you may have had all the elements, or most of the elements, of a Blackout War, or a No-Contact War, or what the U.S. Army War College, some of it call it "Cyber-geddon" which is a combined arms operation using all of these elements and practiced back in 2013 by the North Koreans.

A couple of months after that, one of the nightmare scenarios the EMP Commission had, you know, was the idea that an adversary could launch a Short-Range Ballistic Missile using a nuclear weapon, maybe from the Gulf of Mexico, to put it over the Grid someplace and do an EMP attack that way. It keeps their fingerprints off the attack. You can do it anonymously, because you can use a missile like a Scud and use a freighter that can be owned by anybody, could be done by terrorists.

And indeed, it looks like the North Koreans practiced that too, because in July, the crisis, you know, supposedly was over by that time. But a couple of months after the height of the crisis, you know, we intercepted a North Korean freighter trying to go through the Panama Canal. And, hidden under thousands of bags of sugar were two SA-2 nuclear capable missiles hidden under these bags of sugar on their launchers, on their launchers. And they didn't have nuclear warheads on them, but they are designed to carry a 10 kiloton nuclear—nuclear weapon.

And they had transited the Gulf of Mexico already. They were going back to North Korea, having come in from the Atlantic side. It was almost as if they were trying to see, "Well, we've gotten away with this so far. We've gotten this freighter with nuclear-capable missiles into the Gulf of Mexico. And could we possibly get back through the Panama Canal, and the stupid Americans won't find us?"

Well, the only reason we did find them was because that particular freighter was notorious for providing small arms to terrorists and smuggling drugs. And the people who were inspecting it weren't looking for nuclear capable missiles, they were looking for drugs, but found the SA-2s. And that's actually a picture of one of them that was found.

The bad guys know about—we've been talking theoretically about threats to the Grid. But these threats are actually happening already. In 2013—(There's no clock, Chuck, so tell me when, okay? Okay.) You know, in 2013, you know, the Knights Templars, they're a Mexican drug Cartel terrorist group, you know, they blacked out a whole province in Mexico, put a half a million people into the dark, using primitive weapons, you know, small arms and explosive pipe bombs, so that they could go into the towns and villages and drag village leaders who were opposed to the drug trade into the public square and execute them, you know, cutting the people off from the Federales. Now, if the Knights Templars, you know, these Neanderthals have figured out that the Electric Grid is a major societal vulnerability, what if ISIS and al Qaeda, Iran, North Korea and Russia, what more sophisticated modes of attack may they have in mind for us?

And, while we were fixated on—(Two minutes? Okay.) Anyway, terrorists had actually blacked out—the first time terrorists had blacked out a whole nation was on the 9th of June. Al Qaeda and the Arabian Peninsula blacked out 16 cities, 24 million people living in Yemen. Pakistan, earlier this year in 25th of January, 80 percent of Pakistan was blacked out by a nuclear attack, by an attack on the Grid. And Pakistan is a nuclear weapons State.

Turkey has been blacked out, allegedly by an Iranian cyber attack, according to the reports. If so, it'll be the first time that an entire nation has been blacked out by a cyber attack. Where are we? I think we're like in 1936 to 1938, when basically the Nazis were experimenting with the Blitzkrieg, a new way of warfare, a revolution in military affairs, doing it in a small scale way, before they sprung this surprise on the Western democracies. I think that's where we are.

And I think we need to understand, we need to understand their way of war. And we've got to stop thinking in stovepipe ways about the security of the Grid and the threats as being EMP, nuclear EMP or non-nuclear EMP, or physical threats, or cyber threats. They are thinking of all of this as a combined-arms operation. And they write about it.

And you probably can't read—all right, they write about it. I guess my last suggestion, in terms of next steps, then, in this connection, there are so many other next steps. But the next step has got to be the first step for us to finally understand the threat, really, that's coming at us from the other side, how they think about this, and how they plan about it.

During the Cold War, you know, one of the great benefits—and I think one of the reasons we won the Cold War—is the Department of Defense and the Intelligence Community was smart enough to declassify Soviet military textbooks, things like Sidorenko's *The Offensive* and Sokolovsky's strategy, *Military Strategy.*

There was a U.S. Air Force series that was published that had about 20 of the basic Soviet military textbooks that described how they would fight a future war. And this was available to academics and people who were interested in international security matters and things like that. So we got a whole of nation solution to this, not just a handful of specialists working in DoD and the Intelligence Community, but we were able to harness other brain power in our society, about how do we deter? How would we defeat this kind of a plan that's coming at us? We have no equivalent for that today in this modern war on terrorism, in this modern war of what we call cyber war, what I call a blackout war.

But the textbooks, some of them are translated, but they're For Official Use Only, which is not officially classified, but it keeps it out of the hands of academics. It keeps it out of the hands of business and industry. You know, you get it if you have a specific reason to have it. And, in fact, it does classify it.

So one of the next steps, I would say, is to take the first step by declassifying the textbooks we have in hand now, make them unclassified, not just For Official Use Only—so everybody can read these things and see that this is a real threat, you know, that they're actually practicing it. It isn't theoretical. They're writing about it. They're exercising it. They've actually done it to other nations. And this is their plan. It's spelled out for us. Let us all know about it so that we can apply what wisdom we have to trying to protect our Nation.

So thank you for hearing me out. I've got one step toward that. I just got a book called *Blackout Wars* published on this that has some of what has been unclassified in this area and available to the public to read. Thank you.

[Applause]

DR BAKER: Yeah, thanks Peter, again, very, very good insights. I'll try to keep my remarks brief. I'm trying to capture some of the things I heard in my—Oh, do you have a question?

Q: It's really a comment. I actually just went through the final report of the Turkish blackout. It was not a hack, from everything we can tell. It was control systems cyber, there's no question there. But the report did come out, you know, kind of their version of the final report of the Northeast outage. And there were a lot of issues with it, but it wasn't—And we were actually looking for it, expecting it to have been, but it wasn't.

DR. PRY: Well you know, let me—Can you hear me? You know what? When I first heard in the newspaper account that it was a cyber attack, you know, I was extremely skeptical of that and reluctant to accept that report, because unlike a lot of people, I am one of those who is in the minority view that the ability to collapse an electric grid, just by means of cyber, has been really overblown. That it's much more difficult to do than people suppose.

This goes against what our top experts are telling us, you know, that the Head of Cyber Command has testified to Congress that it is possible to black out the whole national electric grid for 18 months by means of cyber attack.

Q: I believe that.

DR. PRY: So I'm definitely in a minority view, a minority view on this. I know that there was a Turkish report—I've seen that same report myself. But, you know, Turkey is not the United States, sir. Do you think the Turks want to disclose to the Iranians that they really are vulnerable to a cyber attack? I think you still have to be skeptical about the assertions in that report that they are—that it was some other thing.

Q: No. What I'm saying is, when you look at the actual technical data, you look at—

DR. PRY: How do you know you have the technical data? It comes from Turkey.

Q: It comes from Turkey.

DR. PRY: You don't know.

Q: I don't—

DR. PRY: You don't know.

Q: Okay. I will leave it at that.

DR. PRY: I hope you're right, but I doubt it. Turkey is not the United States. And they don't even disclose information to their own people, let alone to foreign actors, especially politically sensitive information like that. Also, one other factor about the Turkish blackout is that it's interesting that the only province in Turkey that didn't go into blackout gets its electricity from Iran.

DR. BAKER: Okay. All right, thanks. So I am dividing my comments—We're talking about next steps and some of the themes that were developed today by the many outstanding presentations that we heard. I'll divide them into some technical observations, and then some programmatic observations.

From the technical side, we know that the protection can be done. I think this presentation we just heard from Jack Pressman was just outstanding. Then we have a private company that has bitten the bullet and protected themselves against a variety of combined effects. And one of the big lessons there is they looked at all these effects together. They didn't stovepipe them. That's a major, major lesson there. So it can be done.

The other theme I heard was complexity, that systems are very complex. And that poses challenges, but this can be done. We can use divide-and-conquer approaches, where we divide the systems up. But we always need to make sure that we have protection end-to-end. So, if you have two data centers talking to each other, you need to protect the data centers and the path in between them. The same goes for the electric grid. You have to do an end-to-end to get to, to overcome the complexity challenges.

And modeling is very important, and that's something that INCOSE, the systems mentioned here, excel at. But I would remind us all that it's very important to test. That was a message that came through from the INCOSI speakers we heard. And he did not just model, but test. And there are always in these EMP installations and GMD and cyber installations, there are always unknown unknowns that you will never identify on paper. You have to do the tests.

A couple of things that need to be tested that haven't been yet, just off the record, you know, just something for your information, that the effects of GMD generators, there have been a couple of papers recently that imply that not only are transformers vulnerable, but generators may be vulnerable as well. And so that needs to be part of our test program. And then large transformers, we have never, to my knowledge, tested a large HV, two-story tall transformer to these effects. And we need to—in fully loaded condition to EMP. And that's something that needs to be done. So that, from a technical perspective.

Then, from a programmatic perspective, we need to come to grips as a nation with these effects and these challenges, and I think this conference. And thanks, Chuck, for organizing them. You know, we are making progress in terms of coming to grips, understanding the effects. We're hearing from all these different sectors and disciplines where people are getting it. They're understanding what we're up against here.

The consequences are preventable. The engineering tools are available to protect. And there are huge cost benefits. When you think about the impact of large portions of the grid going down, we're going to measure those in GNPs. You're talking about trillions of dollars. I mean the insurance industry has done cost studies where the losses are in the trillions. Well, the protection costs are miniscule compared to the effect on the national economy and the effect, in the worst case, on the continued existence of this country.

There are some initiatives that would aid in this endeavor. We need a—The efforts are amorphous. You can tell, just as you see people talk, you know, there's pockets of progress that we've heard about. But they're all very amorphous. Things are being done in bits and pieces. We've got a space-weather strategy that totally ignores EMP and RF weapons and cyber. It's just so piecemeal. We need to have some kind of designated national organization, national executive agency to bring these, coordinate these things.

We need a national trifecta, tri-threat protection plan that includes a set of planning scenarios. And that needs to be coordinated, not just at the federal level, but with state and local stakeholders. And I think my recommendation would be to concentrate on the—try to do this, focusing on the National Power Grid. And we might also—You know, there are also supporting infrastructures that we need to include in the communication, transportation, fuel supply area. But make sure that that grid is on, so that we can have the lights to see how to fix the others.

You know, we found—we've heard this several times. If you don't have electric power, and Katrina and Hurricane Sandy, the first responders are just dead in the water. You have to have that electricity. And that's a real lesson from these huge catastrophes they had, that don't match what we would have with a EMP or GMD scenario.

We also—I also just wanted to mention in passing—important, but we need to address—Congress, actually, will need to address problems inherent in the regulation of the electric grid as conceived in the Energy Policy Act of 2005. The technical solution to any problem is the wrong solution. You have to have the legal and the business end of it covered as well. And so we need to address some of these legal problems, where the industry has become a self-licking ice cream cone. My apologies to industry members. The industry is regulating itself. It's the only industry where that's happening. And we've got to fix that, and it's going to have to go up to Congress to fix that.

And my last comment is the importance of the business end, so technical solution by itself won't work. You have the legal and you have the business end into it. And I think one of the big things in our favor in this country is competition. And I think as the smart grids and these micro grids gain traction, and they expand, and we have these 10-acre data centers that are self-powered and don't need the electric grid, that's going to get their attention. But we need to make use of the competition to get ourselves protected.

The states that have protected their power grid and have protected data center, they're the states that will attract business. And so I think that there's a huge value in competition. So that's my take on it. Thanks.

[Applause]

END

TLP: White

Analysis of the Cyber Attack on the Ukrainian Power Grid

Defense Use Case

March 18, 2016

1325 G Street NW
Suite 600
Washington, DC 20005
404-446-9780 #2 | www.eisac.com

Table of Contents

Preface iii

Summary of Incidents iv

Attacker Tactics Techniques and Procedures Description 1

ICS Cyber Kill Chain Mapping 4

Defense Lessons Learned – Passive and Active Defenses 11

Recommendations 18

Implications and Conclusion 20

Appendix Information Evaluation 22

Preface

Analysis of the Cyber Attack on the Ukrainian Power Grid

This is an analysis by a joint team to provide a lessons learned community resource from the cyber attack on the Ukrainian power grid. The document is being released as Traffic Light Protocol: White (TLP: White) and may be distributed without restriction, subject to copyright controls. This document, the Defense Use Case (DUC), summarizes important learning points and presents several mitigation ideas based on publicly available information on ICS incidents in Ukraine. The E-ISAC and SANS are providing a summary of the available information compiled from multiple publicly available sources as well as analysis performed by the SANS team in relation to this event.[1] This document provides specific mitigation concepts for power system Supervisory Control and Data Acquisition (SCADA) defense, as well as a general learning opportunity for ICS defenders.

Authors, working with the E-ISAC:
Robert M. Lee, SANS
Michael J. Assante, SANS
Tim Conway, SANS

[1] The SANS investigation into this incident should not be confused with the U.S. interagency team investigation or any other organization or company's efforts to include the E-ISAC's past reporting. SANS ICS team has been analyzing the data on their own since December 25, 2015, and has provided its analysis to the wider community. This document is provided to E-ISAC and the North American electricity sector to benefit its members and the larger critical infrastructure community.

Summary of Incidents

On December 23, 2015, the Ukrainian Kyivoblenergo, a regional electricity distribution company, reported service outages to customers. The outages were due to a third party's illegal entry into the company's computer and SCADA systems: Starting at approximately 3:35 p.m. local time, seven 110 kV and 23 35 kV substations were disconnected for three hours. Later statements indicated that the cyber attack impacted additional portions of the distribution grid and forced operators to switch to manual mode.[2, 3] The event was elaborated on by the Ukrainian news media, who conducted interviews and determined that a foreign attacker remotely controlled the SCADA distribution management system.[4] The outages were originally thought to have affected approximately 80,000 customers, based on the Kyivoblenergo's update to customers. However, later it was revealed that three different distribution oblenergos (a term used to describe an energy company) were attacked, resulting in several outages that caused approximately 225,000 customers to lose power across various areas.[5, 6]

Shortly after the attack, Ukrainian government officials claimed the outages were caused by a cyber attack, and that Russian security services were responsible for the incidents.[7] Following these claims, investigators in Ukraine, as well as private companies and the U.S. government, performed analysis and offered assistance to determine the root cause of the outage.[8] Both the E-ISAC and SANS ICS team was involved in various efforts and analyses in relation to this case since December 25, 2015, working with trusted members and organizations in the community.[9]

This joint report consolidates the open source information, clarifying important details surrounding the attack, offering lessons learned, and recommending approaches to help the ICS community repel similar attacks. This report does not focus on attribution of the attack.

---

[2] https://ics.sans.org/blog/2016/01/09/confirmation-of-a-coordinated-attack-on-the-ukrainian-power-grid

[3] http://news.finance.ua/ua/news/-/366136/hakery-atakuvaly-prykarpattyaoblenergo-znestrumyvshy-polovynu-regionu-na-6-godyn

[4] http://ru.tsn.ua/ukrayina/iz-za-hakerskoy-ataki-obestochilo-polovinu-ivano-frankovskoy-oblasti-550406.html

[5] http://www.oe.if.ua/showarticle.php?id=3413

[6] https://ics-cert.us-cert.gov/alerts/IR-ALERT-H-16-056-01

[7] http://www.ukrinform.net/rubric-crime/1937899-russian-hackers-plan-energy-subversion-in-ukraine.html

[8] https://www.rbc.ua/rus/news/pravitelstva-ssha-ukrainy-rassmotryat-otchet-1454113214.html

[10] http://ru.tsn.ua/ukrayina/iz-za-hakerskoy-ataki-obestochilo-polovinu-ivano-frankovskoy-oblasti-550406.html

Summary of Information and Reporting

Background

On December 24, 2015, TSN (a Ukrainian news outlet) released the report "Due to a Hacker Attack Half of the Ivano-Frankivsk Region is De-Energized."[10] Numerous reporting agencies and independent bloggers from the Washington Post, SANS Institute, New York Times, ARS Technica, BBC, Wired, CNN, Fox News, and the E-ISAC Report have followed up on the initial TSN report.[11] These subsequent reports have collectively provided details of a cyber attack that targeted the Ukrainian electric system. The U.S. Department of Homeland Security (DHS) issued a formal report on February 25, 2016, titled IR-ALERT-H-16-056-01.[12] Based on the DHS report, three Ukrainian oblenergos experienced coordinated cyber attacks that were executed within 30 minutes of each other. The attack impacted 225,000 customers and required the oblenergos to move to manual operations in response to the attack.

The oblenergos were reportedly able to restore service quickly after an outage window lasting several hours.[13] The DHS report states that, while electrical service was restored, the impacted oblenergos continue to operate their distribution systems in an operationally constrained mode. Within the Ukrainian electrical system, these attacks were directed at the regional distribution level, as shown in Figure 1.

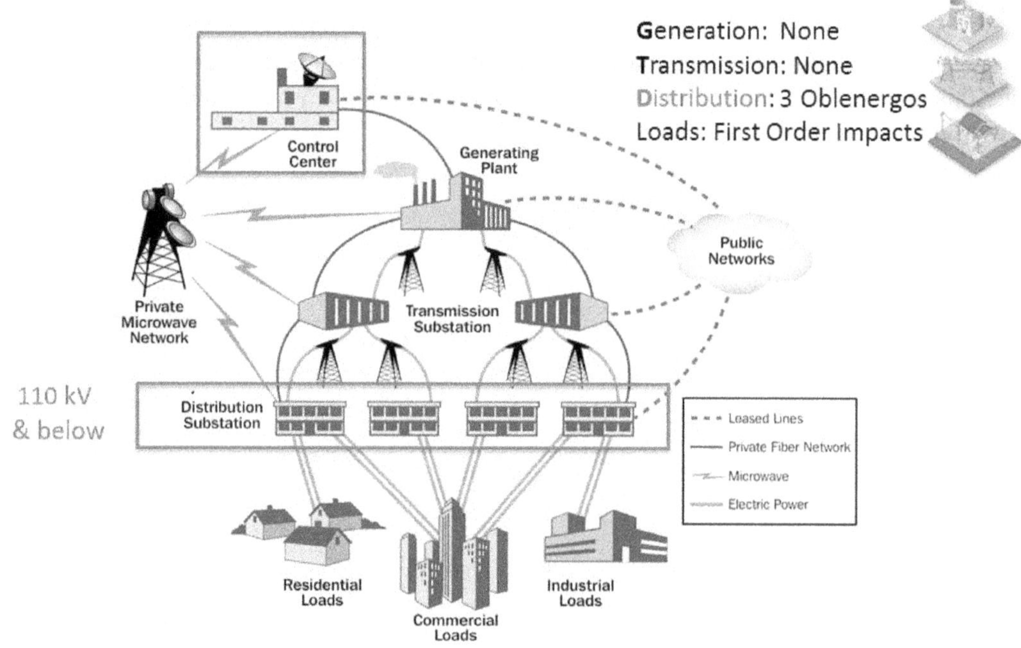

Source: Modification to the DHS Energy Sector-Specific Plan 2010

Figure 1: Electric System Overview

---

[10] http://ru.tsn.ua/ukrayina/iz-za-hakerskoy-ataki-obestochilo-polovinu-ivano-frankovskoy-oblasti-550406.html

[11] E-ISAC: Mitigating Adversarial Manipulation of Industrial Control Systems as Evidenced by Recent International Events, February 9, 2016 (TLP=RED)

12 https://ics-cert.us-cert.gov/alerts/IR-ALERT-H-16-056-01

13 https://www.washingtonpost.com/world/national-security/russian-hackers-suspected-in-attack-that-blacked-out-parts-of-ukraine/2016/01/05/4056a4dc-b3de-11e5-a842-0feb51d1d124_story.html

See the Appendix for an evaluation of the credibility and amount of technical information that is publicly available.

Keeping Perspective

The cyber attacks in Ukraine are the first publicly acknowledged incidents to result in power outages. As future attacks may occur, it is important to scope the impacts of the incident. Power outages should be measured in scale (number of customers and amount of electricity infrastructure involved) and in duration to full restoration. The Ukrainian incidents affected up to 225,000 customers in three different distribution-level service territories and lasted for several hours. These incidents should be rated on a macro scale as low in terms of power system impacts as the outage affected a very small number of overall power consumers in Ukraine and the duration was limited. In contrast, it is likely that the impacted companies rate these incidents as high or critical to the reliability of their systems and business operations.

Attacker Tactics Techniques and Procedures Description

Direct attribution is unnecessary to learn from this attack and to consider mitigation strategies; it is only necessary to use the mental model of how the cyber actor works to understand the capabilities and general profile against which one is defending. The motive and sophistication of this power grid attack is consistent with a highly structured and resourced actor. This actor was co-adaptive and demonstrated varying tactics and techniques to match the defenses and environment of the three impacted targets. The mitigation section of this document provides mitigation concepts related to the attack and how to develop a more lasting mitigation strategy by anticipating future attacks.

Capability

The attackers demonstrated a variety of capabilities, including spear phishing emails, variants of the BlackEnergy 3 malware, and the manipulation of Microsoft Office documents that contained the malware to gain a foothold into the Information Technology (IT) networks of the electricity companies.[14] They demonstrated the capability to gain a foothold and harvest credentials and information to gain access to the ICS network. Additionally, the attackers showed expertise, not only in network connected infrastructure; such as Uninterruptable Power Supplies (UPSs), but also in operating the ICSs through supervisory control system; such as the Human Machine Interface (HMI), as shown in Figure 2.

SCADA Hijacking Techniques

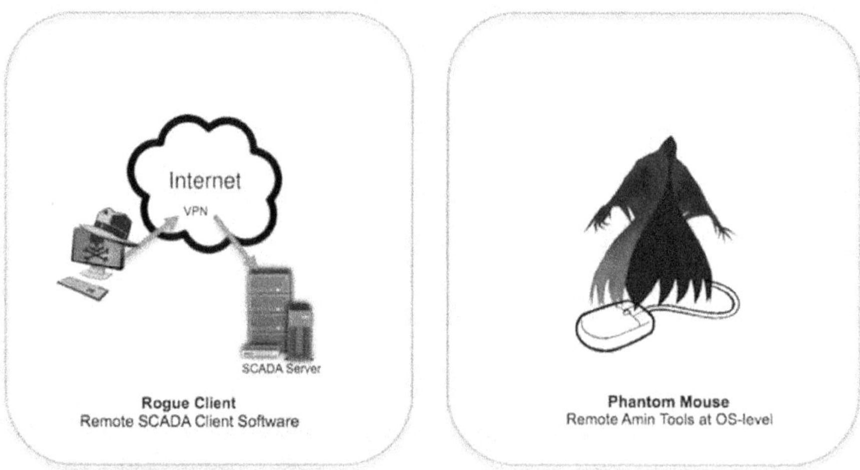

The attackers develop two SCADA Hijack approaches (one custom and one agnostic) and successfully used them across different types of SCADA/DMS implementations at three companies

Figure 2: Control & Operate: SCADA Hijacking Techniques

Finally, the adversaries demonstrated the capability and willingness to target field devices at substations, write custom malicious firmware, and render the devices, such as serial-to-ethernet convertors, inoperable and

[14] For a discussion around the history of the BlackEnergy 3 malware and Sandworm team see the SANS ICS webcast with iSight here: https://www.sans.org/webcasts/analysis-sandworm-team-ukraine-101597

unrecoverable.[15] In one case, the attackers also used telephone systems to generate thousands of calls to the energy company's call center to deny access to customers reporting outages. However, the strongest capability of the attackers was not in their choice of tools or in their expertise, but in their capability to perform long-term reconnaissance operations required to learn the environment and execute a highly synchronized, multistage, multisite attack.

The following is a consolidated list of the technical components used by the attackers, graphically depicted in Figure 3:

- Spear phishing to gain access to the business networks of the oblenergos
- Identification of BlackEnergy 3 at each of the impacted oblenergos
- Theft of credentials from the business networks
- The use of virtual private networks (VPNs) to enter the ICS network
- The use of existing remote access tools within the environment or issuing commands directly from a remote station similar to an operator HMI
- Serial-to-ethernet communications devices impacted at a firmware level[16]
- The use of a modified KillDisk to erase the master boot record of impacted organization systems as well as the targeted deletion of some logs[17]
- Utilizing UPS systems to impact connected load with a scheduled service outage
- Telephone denial-of-service attack on the call center

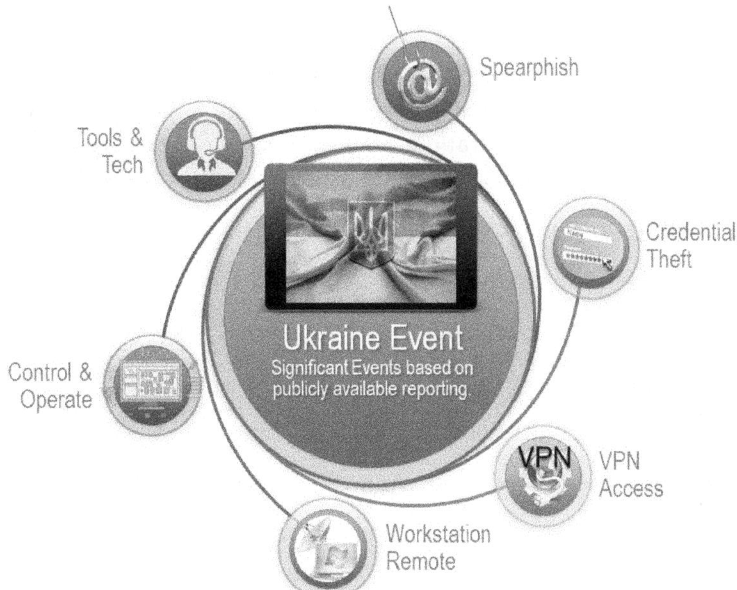

Figure 3: Ukraine Attack Consolidated Technical Components

At various points in the public reporting on the attack, organizations have indicated that BlackEnergy 3 and KillDisk itself could be directly responsible for the outage. One of the items specifically highlighted to support this theory

15 http://mpe.kmu.gov.ua/minugol/control/uk/publish/article;jsessionid=CE1C739AA046FF6BA00FE8E8A4D857F3.app1?art_id=245086886&cat_id=35109

16 To learn about serial to ethernet converters and the types of vulnerabilities that exist to them see DigitalBond's Basecamp report here: http://www.digitalbond.com/blog/2015/10/30/basecamp-for-serial-converters/

17 http://www.symantec.com/connect/blogs/destructive-disakil-malware-linked-ukraine-power-outages-also-used-against-media-organizations

was that KillDisk deleted a process on Windows systems linked to serial-to-ethernet communications.[18] Regardless of the impact of the SCADA network environment, neither BlackEnergy 3 nor KillDisk contained the required components to cause the outage. The outages were caused by the use of the control systems and their software through direct interaction by the adversary. All other tools and technology, such as BlackEnergy 3 and KillDisk, were used to enable the attack or delay restoration efforts.

Opportunities

Multiple opportunities existed for the adversary to execute its attack. External to the oblenergos and prior to the attack, there was a variety of open-source information available; including a detailed list of types of infrastructure such as Remote Terminal Unit (RTU) vendors and versions posted online by ICS vendors.[19] The VPNs into the ICS from the business network appear to lack two-factor authentication. Additionally, the firewall allowed the adversary to remote admin out of the environment by utilizing a remote access capability native to the systems. In addition, based on media reporting, there did not appear to be any resident capability to continually monitor the ICS network and search for abnormalities and threats through active defense measures; like network security monitoring. These vulnerabilities would have provided the adversary the opportunity to persist within the environment for six months or more to conduct reconnaissance on the environment and subsequently execute the attack.[20]

Based on the details provided in the DHS report, the adversary used a consistent attack approach on all three impacted targets. The adversary also used consistent tactics to impact field controllable elements and irreparably damage field devices.

Why these oblenergos were targeted remains an open debate. Based on the public reporting, it is unknown if the targets were selected based on common technologies in use, system architectures, reconnaissance operations, or service territories. Opportunity-based considerations for selecting a specific target may focus on an attacker's confidence and ability to cause an ICS effect. Some example decision factors could include:

- Targets with common systems and configurations
- Multiple systems with common centralized control points
- ICS impact duration estimates (e.g., long term or short term)
- Existing capabilities required to achieve desired results
- Risk level of performing the operation and being discovered
- Achieved access and ability to move and act within the environment

[18] http://www.eset.com/int/about/press/articles/malware/article/eset-finds-connection-between-cyber-espionage-and-electricity-outage-in-ukraine/
[19] http://galcomcomp.com/index.php/ru/nashi-proekty/15-proekt3-material-ru
[20] http://mobile.reuters.com/article/idUSKCN0VL18E

ICS Cyber Kill Chain Mapping

The ICS Cyber Kill Chain was published by SANS in 2015 by Michael Assante and Robert M. Lee as an adaptation of the traditional cyber kill chain developed by Lockheed Martin analysts as it applied to ICSs.[21] The ICS Cyber Kill Chain details the steps an adversary must follow to perform a high-confidence attack on the ICS process and/or cause physical damage to equipment in a predictable and controllable way, as displayed in Figure 4.

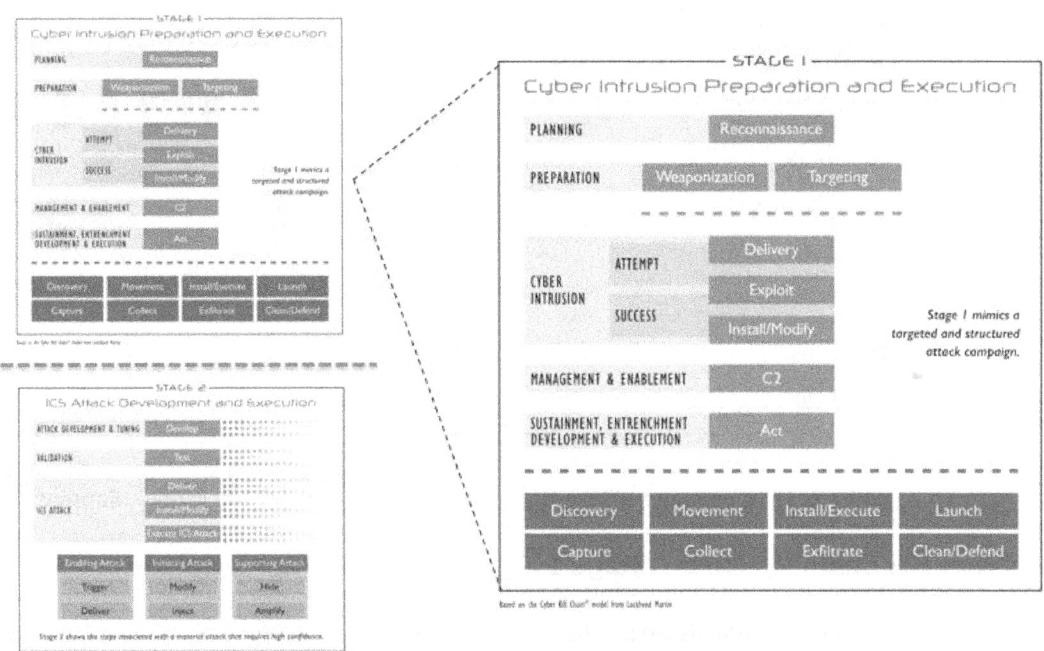

Figure 4: The ICS Cyber Kill Chain with Stage 1 Highlighted

The attack on the Ukrainian power grid followed the ICS Cyber Kill Chain completely throughout Stage 1 and Stage 2. The attack gained access to each level of the ICS, as shown in Figure 5, with the ICS Cyber Kill Chain plotted alongside a segmentation/hierarchy model (e.g., modified Purdue Model). Completing Stage 1 entails a successful cyber intrusion or breach into an ICS system, but is not characterized as an ICS attack. Completion of Stage 2 completed the ICS Kill Chain, resulting in a successful cyber attack that led to an impact on the operations of the ICS. The next section includes a discussion of the two stages using currently available information from the attack.

[21] https://www.sans.org/reading-room/whitepapers/ICS/industrial-control-system-cyber-kill-chain-36297

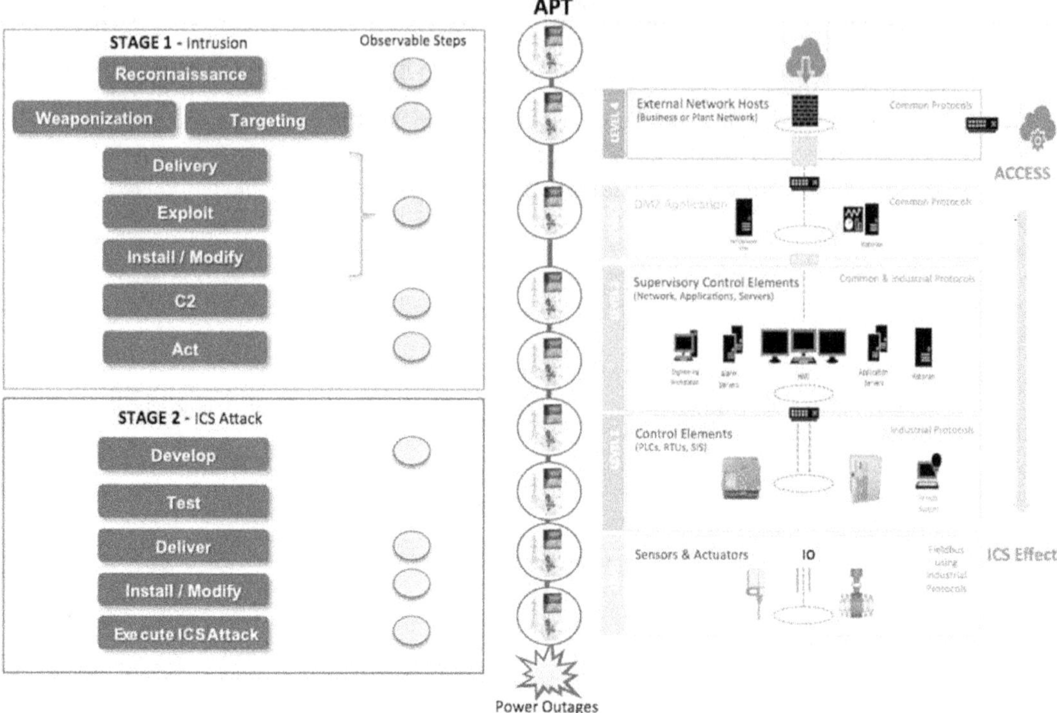

Figure 5: Ukraine Cyber Attack ICS Cyber Kill Chain and Purdue Model Mapping[22]

ICS Cyber Kill Chain Mapping – Stage 1

The first step in Stage 1 is **Reconnaissance**. There were no reports of observed reconnaissance having taken place prior to targeting the energy companies. However, an analysis of the three impacted organizations shows they were particularly interesting targets due to the levels of automation in their distribution system; enabling the remote opening of breakers in a number of substations. Additionally, the targeting and final attack plan for the electricity companies in general were highly coordinated, which indicates that reconnaissance took place at some point. This was very unlikely to have been an opportunistic attack.

The second step is **Weaponization** and/or **Targeting**. Targeting would normally take place when no weaponization is needed; such as directly accessing internet connected devices. In this attack, it does not appear that targeting of specific infrastructure was necessary to gain access. Instead, the adversaries weaponized Microsoft Office documents (Excel and Word) by embedding BlackEnergy 3 within the documents.[23] Samples of Excel and other office documents have been recovered from the broader access campaign that targeted a multitude of organizations in Ukraine; including Office documents used in the specific attack against the three electricity companies.[24, 25]

During the cyber intrusion stage of **Delivery**, **Exploit**, and **Install,** the malicious Office documents were **delivered**

[22] Note, the exact architectures of the impacted utilities are not represented in the figure. The Purdue Model is a standard way of viewing different zones of a well-constructed ICS.

[23] https://securelist.com/blog/research/73440/blackenergy-apt-attacks-in-ukraine-employ-spearphishing-with-word-documents/

[24] https://ics-cert.us-cert.gov/alerts/ICS-ALERT-14-281-01B

[25] Those looking for Indicators of Compromise for the word document, command and control servers, and the malware should look to E-ISAC, ICS-CERT, and iSight private reporting as well as public reporting from Kaspersky Labs, ESET, and CYS Centrum reference: https://cys-centrum.com/ru/news/black_energy_2_3 and https://securelist.com/blog/research/73440/blackenergy-apt-attacks-in-ukraine-employ-spearphishing-with-word-documents/

via email to individuals in the administrative or IT network of the electricity companies. When these documents were opened, a popup was displayed to users to encourage them to enable the macros in the document as shown in Figure 6.[26] Enabling the macros allowed the malware to **Exploit** Office macro functionality to install BlackEnergy 3 on the victim system and was not an exploit of a vulnerability through exploit code. There was no observed exploit code in this incident. The theme of using available functionality in the system was present throughout the adversary's kill chain.

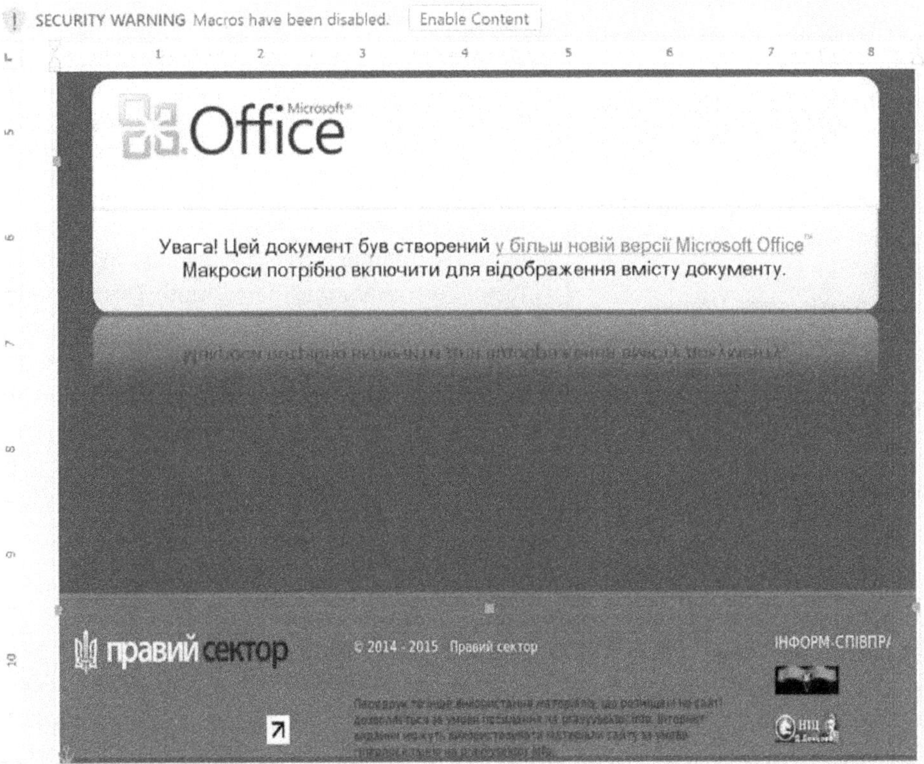

Figure 6: A Sample of a BlackEnergy 3 Infected Microsoft Office Document[27]

Upon the **Install** step, the BlackEnergy 3 malware connected to command and control (C2) IP addresses to enable communication by the adversary with the malware and the infected systems. These pathways allowed the adversary to gather information from the environment and enable access. The attackers appear to have gained access more than six months prior to December 23, 2015, when the power outage occurred.[28] One of their first actions happened when the network was to harvest credentials, escalate privileges, and move laterally throughout the environment (e.g., target directory service infrastructure to directly manipulate and control the authentication and authorization system). At this point, the adversary completed all actions to establish persistent access to the targets. While the initial footholds were used to harvest legitimate credentials for pivoting and systematic takeover of IT systems and remote connections, it is likely that the attackers moved quickly away from their initial footholds and vulnerable C2s in an effort to blend into the target's systems as authorized users. With this information, the attackers would be able to identify VPN connections and avenues from the business network into the ICS network. Using native connections and commands allows the attackers to discover the remainder of the systems and extract data necessary to formulate a plan for Stage 2.

---

[26] For a detailed understanding of the infected Microsoft Office documents and the malicious payload see Kaspersky Lab's write-up here: https://securelist.com/blog/research/73440/blackenergy-apt-attacks-in-ukraine-employ-spearphishing-with-word-documents/

[27] https://securelist.com/blog/research/73440/blackenergy-apt-attacks-in-ukraine-employ-spearphishing-with-word-documents/

[28] http://politicalpistachio.blogspot.com/2016/01/russian-hackers-take-down-power-grid-in.html

Speculation

There was not enough publicly available information to determine how diversified the adversary's attack was to include how many different types of devices were impacted at the firmware level. However, through publicly available information about the Ukrainian networks, as well as knowledge of similar electric distribution systems, it is likely that there was a diverse hardware and software environment.

It is suspected that the administrative and ICS networks contained multiple OS versions such as Windows XP and Windows 7, multiple types of RTUs and gateways, and various industrial switches.

Using the stolen credentials, the adversary was able to pivot into the network segments where SCADA dispatch workstations and servers existed. Upon entry into the network, the actions of the adversaries were consistent in theme but different in technical minutia between the three impacted oblenergoss. In at least one of the oblenergos, the attackers discovered a network connected to a UPS and reconfigured it so that when the attacker caused a power outage, it was followed by an event that would also impact the power in the energy company's buildings or data centers/closets.

There is not sufficient information available to identify if any information was exfiltrated from the environment, but the adversary demonstrated a capability in Stage 2 that indicates internal discovery was performed. This reconnaissance would have needed to include discovering field devices such as the serial-to-ethernet devices used to interpret commands from the SCADA network to the substation control systems. Additionally, the three oblenergos used different distribution management systems (DMSs), and the attackers would have needed to perform some network reconnaissance against these systems and find specific targets to execute their highly coordinated attack.[29]

ICS Cyber Kill Chain Mapping – Stage 2

In most cases, the **Develop** stage occurs in the adversary's networks, thereby limiting any available forensic information, but the attack that follows this stage can reveal a lot about the adversarial process. In the Attack Development and Tuning Stage of Stage 2, the attackers executed the Develop step in at least two ways. First, they learned how to interact with the three distinct DMS environments using the native control present in the system and operator screens. Second, and more importantly, they developed malicious firmware for the serial-to-ethernet devices.[30]

Currently available information indicates that the malicious firmware was consistent amongst devices and uploaded within short periods of each other to multiple sites. Therefore, the malicious uploads of firmware was likely developed prior to the attack for quick and predictable execution.

E-ISAC and the SANS ICS team assess with high confidence that, during the Validation Stage of Stage 2, the adversary did **Test** their capabilities prior to their deployment. It is possible that the adversaries were able to execute this with pure luck, but it is highly unlikely and inconsistent with the professionalism observed throughout the rest of the attack. The adversaries likely had systems in their organization that they were able to evaluate and test their firmware against prior to executing on December 23rd.

---

[29] The three different DMS vendors were discoverable via open-source searching. The names of the vendors are being withheld as it is not important to the discussion of the attack. There were no exploits leveraged against these vendors but they were simply abused with direct access.

[30] http://mpe.kmu.gov.ua/minugol/control/uk/publish/article;jsessionid=CE1C739AA046FF6BA00FE8E8A4D857F3.app1?art_id=245086886&cat_id=35109

During the ICS Attack Stage, the adversaries used native software to **Deliver** themselves into the environment for direct interaction with the ICS components. They achieved this using existing remote administration tools on the operator workstations. The threat actors also continued to use the VPN access into the IT environment.[31]

In final preparation for the attack, the adversaries completed the **Install/Modify** stage by installing malicious software identified as a modified or customized KillDisk across the environment. While it is likely the attackers then ensured their modifications to the UPS were ready for the attack, there was not sufficient forensic evidence available to prove this. The last act of modification was for the adversaries to take control of the operator workstations and thereby lock the operators out of their systems. Figure 7 shows the static analysis of the KillDisk API imports following the event.

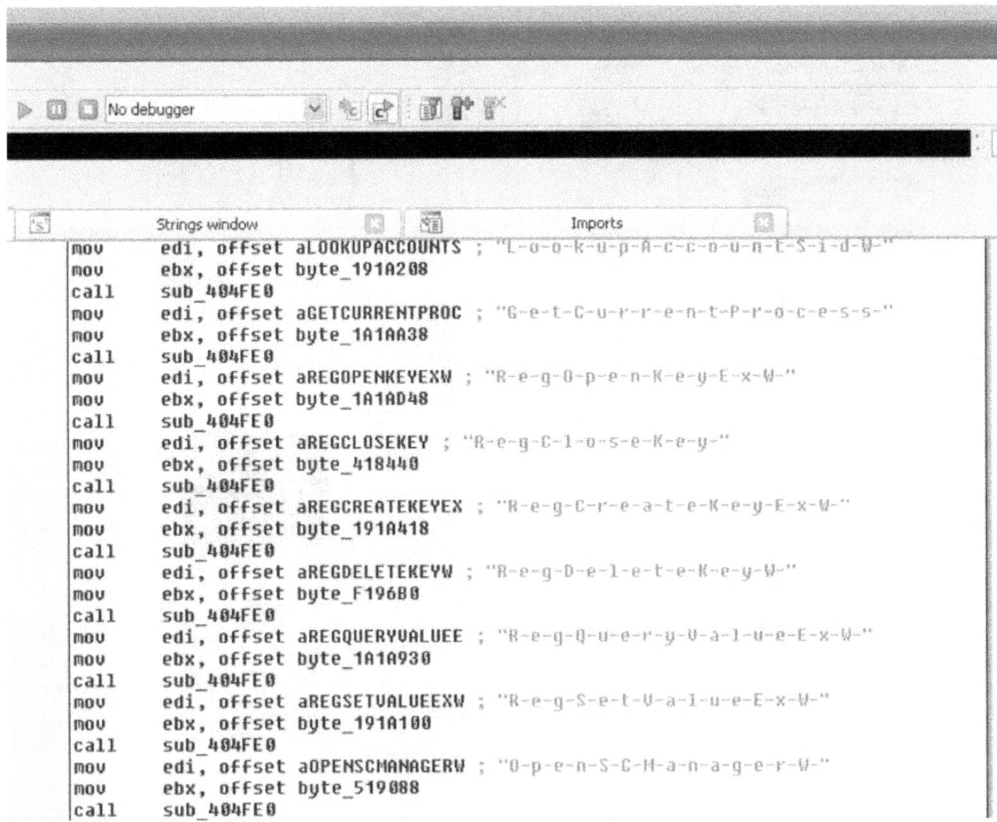

Figure 7: Static Analysis of KillDisk Identifying API Imports[32]

Finally, to complete the ICS Cyber Kill Chain and to **Execute the ICS Attack,** the adversaries used the HMIs in the SCADA environment to open the breakers. At least 27 substations (the total number is probably higher) were taken offline across the three energy companies, impacting roughly 225,000 customers.[33, 34] Simultaneously, the

---

31 http://mpe.kmu.gov.ua/minugol/control/uk/publish/article;jsessionid=CE1C739AA046FF6BA00FE8E8A4D857F3.app1?art_id=245086886&cat_id=35109

32 This image was provided by Jake Williams of Rendition InfoSec. It is included here to note that KillDisk would not run properly in a malware sandbox for analysis. Static analysis was required to fully investigate the malware sample.

33 http://money.cnn.com/2016/01/18/technology/ukraine-hack-russia/

34 In analysis of the impact observed and on the available information on the Ukrainian distribution grid it is assessed with medium confidence that the public number of disconnected substations, 27, is a low number.

attackers uploaded the malicious firmware to the serial-to-ethernet gateway devices. This ensured that even if the operator workstations were recovered, remote commands could not be issued to bring the substations back online (This is characterized this HMIs as "blowing the bridges").

During this same period, the attackers also leveraged a remote telephonic denial of service on the energy company's call center with thousands of calls to ensure that impacted customers could not report outages. Initially, it seemed that this attack was to keep customers from informing the operators of how extensive the outages were; however, in review of the entirety of the evidence, it is more likely that the denial of service was executed to frustrate the customers since they could not contact customer support or gain clarity regarding the outage. The entire attack from March 2015 – December 23, 2015 is graphically depicted below in Figure 8.

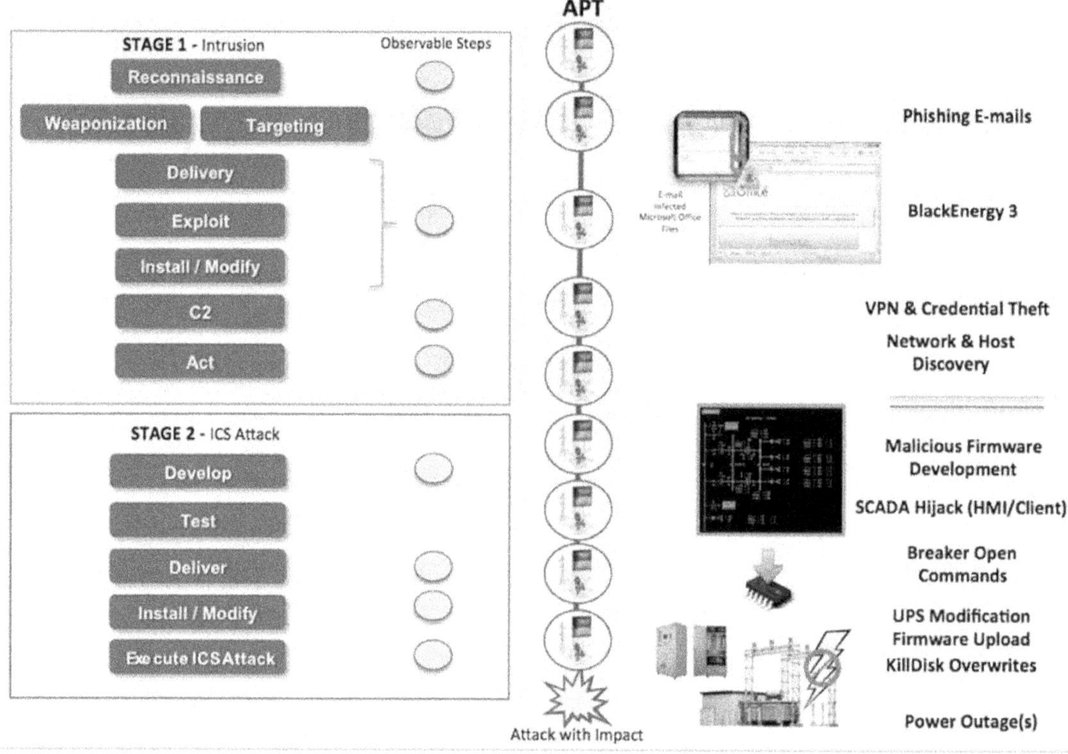

Figure 8: ICS Kill Chain Mapping Chart

It is extremely important to note that neither BlackEnergy 3, unreported backdoors, KillDisk, nor the malicious firmware uploads alone were responsible for the outage. Each was simply a component of the cyber attack for the purposes of access and delay of restoration. For example, on some systems, KillDisk made the Windows systems inoperable by manipulating or deleting the master boot record, but on other systems it just deleted logs and system events.[35, 36] The actual cause of the outage was the manipulation of the ICS itself and the loss of control due to direct interactive operations by the adversary. The loss of view into the system through the wiping of the SCADA network systems simply delayed restoration efforts.

In summary, Stage 2 consisted of the following attack elements:

- **Supporting attacks**:
  - Schedule disconnects for UPS systems
  - Telephonic floods against at least one oblenergos' customer support line
- **Primary attack**: SCADA hijack with malicious operation to open breakers
- **Amplifying attacks**:
  - KillDisk wiping of workstations, servers, and an HMI card inside of an RTU
  - Firmware attacks against Serial-to-Ethernet devices at substations

[35] https://ics-cert.us-cert.gov/alerts/IR-ALERT-H-16-056-01

[36] https://ics.sans.org/blog/2016/01/01/potential-sample-of-malware-from-the-ukrainian-cyber-attack-uncovered

Defense Lessons Learned – Passive and Active Defenses

We reviewed the mitigation strategies provided through the DHS ICS-CERT Alert and considered how an adversary may alter the next attack based on the mitigation taken by a target. We support many of the mitigation recommendations provided to date. However, it is likely that the adversary will modify attack approaches in follow-on campaigns and these mitigation strategies may not be sufficient. In the following section, we discuss mitigations for the attack that took place to extract defense lessons learned. In addition, we discuss future potential attacker methodologies and provide recommendations that could disrupt similar adversary's operations. The mitigations will focus on recommendations for **Architecture, Passive Defense,** and **Active Defense** methodologies along the Sliding Scale of Cyber Security, shown in Figure 9.[37]

ARCHITECTURE
The planning, establishing, and upkeep of systems with security in mind
PASSIVE DEFENSE
Systems added to the Architecture to provide reliable defense or insight against threats without consistent human interaction
ACTIVE DEFENSE
The process of analysts monitoring for, responding to, and learning from adversaries internal to the network
INTELLIGENCE
Collecting data, exploiting it into information, and producing Intelligence
OFFENSE
Legal countermeasures and self-defense actions against an adversary

Figure 9: The Sliding Scale of Cyber Security

Spear Phishing

Ukraine Attack

In the attack, the adversary delivered a targeted email with a malicious attachment that appeared to come from a trusted source to specific individuals within the organizations. Initial mitigation recommendations would point to end-user awareness training and ongoing phishing testing. Efforts to prevent malware have often recommended application whitelisting, which can be effective in ICS environments if the ICS vendor approves of the use. However, based on the details of this attack, application whitelisting would have had a limited role contained to the execution of initial dropper infections in network segments with infected workstations (e.g., users that received and activated infected spear phish emails) where application whitelisting may be more challenging to implement. It is important to note that application whitelisting would not have deterred or prevented the second stage ICS attacks that impacted the Ukrainian oblenergos. In at least one instance, the attacker used a remote rogue client and approved OS-level remote admin features for other components of the attack.

The Next Attack

The adversary may conduct follow-on attacks that pursue alternative forms of social engineering campaigns, like targeting the organization through large-scale phishing campaigns, using water-holing attacks, or conducting direct-call campaigns to users or the help desk. They could also leverage technical exploits not requiring social engineering of personnel.

[37] https://www.sans.org/reading-room/whitepapers/analyst/sliding-scale-cyber-security-36240

Opportunities to Disrupt

The adversary will likely modify attacks to respond to increases or changes in the target's defenses. Defenders need to develop anticipatory responses to attack effects. Since the social engineering components of attacks targeted email and internet accessible cyber assets, these assets and the networks they reside on are untrusted contested territory. Communication with these untrusted areas should be segmented, monitored, and controlled. Operate under the assumption that the environment is accessible by the adversary and ensure appropriate defenses are in place to protect the operations and control environment from the adversary-controlled business cyber assets (while some organizations inherently trust their business systems and networks, additional enforcement and scrutiny of these systems is necessary). Consider using sandboxing technology to evaluate documents and emails coming into the network, using proxy systems to control outbound and inbound communication paths, and limiting workstations to communicate only through the proxy devices by implementing perimeter egress access controls.

Credential Theft

Ukraine Attack

In the attack, the adversary appears to have used BlackEnergy 3 to establish a foothold and utilize keystroke loggers to perform credential theft. As an initial mitigation approach, we recommend that organizations obtain the YARA rules for the latest IOCs. By using the YARA forensic tool, organizations can search for BlackEnergy 3 infections and then utilize antimalware removal tools to eliminate the malware from the infected assets. Defenders should be mindful of the time it takes to detect an infected host as the intruder may have already moved inside the network and secured additional methods to interact and communicate with the infected network. Organizations should change user and shared user passwords (ensure that these steps are approved by operations and the vendor, and tested for impacts to operations and existing security controls).

The Next Attack

Adversaries with persistent access will simply use a different remote access Trojan, an updated version of BlackEnergy 3, or an alternate mode of credential attacks. To detect and mitigate adversary movement throughout an environment and account manipulation, mitigation efforts should be focused on directory (e.g., Active Directory, Domain, eDirectory, and LDAP) segmentation with organizational unit trust models. This approach would allow early detection and prevent some basic attacker approaches.

Opportunities to Disrupt

Monitor user account behavior, network and system communication, and directory-level activity with a focus on identifying abnormalities. Implement alarm capabilities with different priority-level alarms based on the risk of the systems associated with the alarms. It is important to note that YARA is a forensics tool and is not a continuous monitoring solution.

Data Exfiltration

Ukraine Attack

After the attackers achieved the necessary freedom of movement and action in the IT infrastructure, they began exfiltrating the necessary information and discovering the hosts and devices to devise an attack concept to hijack the SCADA DMS to open breakers and cause a power outage. They followed this with destructive attacks against workstations, servers, and embedded devices that provide industrial communications in their distribution substations. The mitigation recommendation here is to understand where this type of information exists inside your business network and ICSs. Minimizing where the information resides and controlling access is a priority for an ICS dependent organization.

The Next Attack

Attackers may look deeper into the ICS configuration and settings or controller and protection/safety logic. Ensure to maintain a vaulted copy of known good project files, control and safety logic, and firmware. Also using file integrity checkers to monitor access or sample loaded files for changes.

Opportunities to Disrupt

Realize that attackers may be able to develop additional attack approaches as they have learned a system and may have stolen information that allows for the development of more powerful future attacks. Defenders should examine their detection and response capabilities. Decision makers should review their restoration plans for attacks with the potential to go deeper into the ICS and could result in damaged equipment. Identify new connections leaving the environment and previously unseen encrypted communications. Network Security Monitoring (NSM) is a great active defense method of detecting exfiltration and ending an adversary's attack path before it disrupts the ICS.

VPN Access

Ukraine Attack

Mitigation guidance based on the attacker approach used in this campaign recommends using two-factor authentication with user tokens to strengthen authentication.

The Next Attack

Attackers may begin looking for existing point-to-point VPN implementations at trusted third party networks or through remote support employee connections where split tunneling is enabled. The immediate mitigation recommendation is to implement trusted jump host or intermediary systems with Network Access Control (NAC) enforcement. Additionally, a VPN configuration approach that disables split tunneling should be enforced.

Opportunities to Disrupt: Defenders are reminded that having remote access through a trusted connection is advantageous for an attacker. Begin by asking why each trusted communication path exists, evaluate the risk, and eliminate each path that does not have an identified need that outweighs the risk of having an attack path. For those communication paths that must remain, consider implementing time of use access for users. Implement the ability to disconnect these paths in an automated way after a defined period of time after access in granted, and a method to disconnect manually if needed. From a passive defense perspective, force choke points in the environment by ensuring that the remote VPNs enter into the environment through a dedicated remote access DMZ. This ensures that traffic and connections can be monitored by active defenders using techniques such as network security monitoring to identify abnormalities in duration of connections, number of connections, and time the connections occur.

Workstation Remote Access

Ukraine Attack

Based on the details provided, the adversaries used the organizations' workstations remotely (while the attacker was physically remote, logically they were local to the host) to conduct Stage 2 of the attack. Mitigation recommendations focus on disabling remote access at the host and at the perimeter firewall.

The Next Attack

Adversaries may modify attack approaches to load additional remote access tools, utilize remote shell capabilities, and tunnel communications over authorized perimeter firewall communications. In response to this modified attack approach, mitigation efforts should focus on host based application aware firewalls, application whitelisting, and configuration management efforts to identify changes in the operation of an asset. Application whitelisting, if installed on the operator HMI to prevent installation of unauthorized remote access software, will not aid in the prevention of authorized software. Also, keep in mind that specific control system vendors may not approve of the whitelisting software.

Opportunities to Disrupt

As a defender prepares for a cyber asset within a trusted environment that may be compromised and remotely controlled, they must consider approaches to quickly move to a conservative operations environment where the ability to issue control signals from untrusted assets is paused. Proper architecture would dictate the ability to segment or disable activities such as remote connections, and unnecessary outbound communications, while conducting active defense mechanisms; such as incident response prior to restoring operational control capabilities to known good assets.

Control and Operate

Ukraine Attack

As the attackers utilized the operator HMI's, they operated numerous sites under the control of the dispatcher. Mitigation approaches for this specific action would focus on application level logic requiring confirmation from the operator, or implement Area of Responsibility (AoR) limitations that only allow an operator to effect certain components of a system. For example: If an entity implemented AoR on one operator workstation that provided east breaker control, and a second operator workstation that provided West breaker control, then an adversary positioned on one workstation would be limited to the AoR allowed on that specific workstation. Some vendor systems allow for Username determined AoR, Workstation determined AoR, and/or an intersection model that combines username and workstation identifier in AoR authorization. There are variations amongst vendor systems in how authentication is handled within the local workstation, directory, or at the application.

The Next Attack

When an attacker identifies a workstation with application security controls in place that limits their capabilities, they may modify their attack to control the system directly by issuing or injecting control commands. Mitigation strategies for this approach would focus on communication path authentication or protocol authentication that would require commands to be issued from an authorized asset. Monitoring communication sessions between hosts can lead to early detection and investigation of suspicious communications.

Opportunities to Disrupt

Preparing for adversarial utilization of cyber assets, or communication paths to control and operate elements of an ICS system, requires system defenders to develop a response approach that eliminates entire sections of cyber asset elements and networks in an effort to inhibit automated control and activate manual operations only. As adversaries learn the environment, they may issue test commands and interact with the SCADA environment without the intention to disrupt it. For mitigation purposes, defenders must talk to operators and ask about abnormal occurrences, and from a passive defense perspective, ensure that logs are collected not only from the host but also from the SCADA applications. Additionally, implement a log aggregation architecture that replicates log files from assets into a log correlation system. Finally, have active defenders routinely review these logs in conjunction with other monitoring activity throughout the ICS to identify abnormalities.

Tools and Technology impacts

Ukraine Attack

The attackers used multiple approaches to impact communication tools, operator technology for restoration efforts, and facility infrastructure essential to many operator activities. Therefore, mitigation recommendations are varied. Items to focus on are:

- Establishing filtering and response capabilities at telecom providers to activate during an ongoing TDoS attack
- Disable remote management of field devices when they are not required.
- Disconnect building control infrastructure systems from the ICS network.
- Consider the number of spares required for embedded systems to regain required communication or control/protection.

The Next Attack

A subsequent attack may progress from resource consumption to a more direct communication path outage that affects communication capabilities. To mitigate this approach, defenders need to establish alternate communications infrastructure for essential service capabilities.

After an attacker identifies increased security requirements for field device management, they may attempt to establish direct access to a field device through a local asset with connectivity or physical presence at the site for direct firmware manipulation. Mitigation strategies for this attack approach focus on electronic and physical access controls and the development of a rapid response capability during an attack or incident.

Opportunities to Disrupt

A determined adversary can impact remote assets either electronically or physically. A defender should develop strong recovery and restoration approaches to replace mission-critical cyber asset components. One option is to rely on inventory and mutual aid assistance from trusted peer organizations and/or suppliers. In cases where specific assets are not immediately recoverable, it is necessary to develop the ability to operate the larger system with operational islands that can be recovered in a timely manner.

Defenders should have access to and visibility of the ICSs to be able to identify abnormal behavior around field device interaction. For example, uploading firmware outside of a scheduled downtime should be quickly observable. Firmware modifications over the network cause spikes in network traffic that active defenders should be consistently looking for. See Figure 10 for an example of a malicious firmware update to an industrial network switch. Even without knowing the baseline of normal activity, which defenders should have, it can be trivial to spot firmware updates in network data.

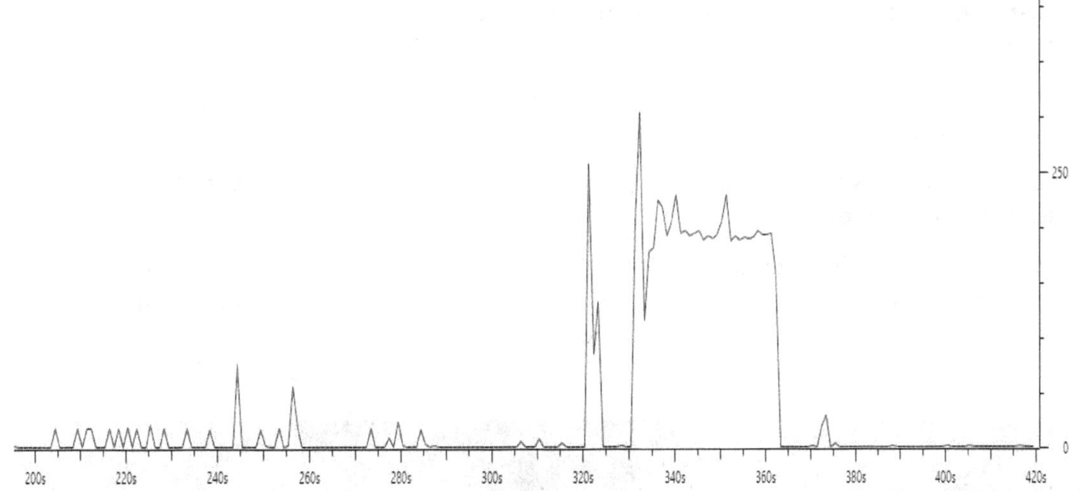

Figure 10: Sample Network I/O Data from a Malicious Firmware Update to an Industrial Ethernet Switch[38]

Respond and Restore

Ukraine Attack

The cyber attacks performed against three Ukrainian oblenergos were well planned and highly coordinated. The attacks consisted of several major elements with both enabling and supporting attack segments. The attackers were remote and interacted with multiple locations within each of their targets to include central and regional facilities. Distribution utilities traditionally have both central business and engineering office(s) and a number of branch facilities used to support line crew, meter reading, bill payment, and distributed supervisory control operations. Certain types of cyber attacks designed to maliciously take over and operate a SCADA DMS may be best performed in a distributed fashion at the lowest or most direct level (from a local dispatch and SCADA server out to the substations that are being monitored and controlled). Preparing for a high-tempo, multifaceted attack is not easy and it requires careful plan review, testing, integrated defense, and operations exercises. Rehearsing steps to more quickly sever or prevent remote access, to safely separate the ICSs from connected networks, or to contain and isolate suspicious hosts is critical.

The Next Attack

The next attack may purposefully differ in its approach to throw off or defeat the defender's plans and expectations. It is critical that defenders exercise and train against different scenarios and be aware that attackers are co-adaptive and creative. It is vital to develop capabilities with flexibility in mind.

Opportunities to Disrupt/Restore

Operations personnel must be involved in planning for restoration from a successful Stage 2 ICS attack. Concepts to consider from an electric operations and engineering perspective include the following and are graphically depicted in Figure 11:

- Cyber contingency analysis: Continuous analysis and preparing the system for the next event.
- Cyber failure planning: Modeling and testing cyber system response to network and asset outages.
- Cyber conservative operations: Intentionally eliminating planned and unplanned changes as well as stopping any potentially impactful processes.

[38] For a good discussion on exploits and malicious firmware updates for industrial ethernet switches see the research by Eireann Leverett, Colin Cassidy, and Robert M. Lee in the DEFCON presentation "Switches Get Stitches" here: https://www.youtube.com/watch?v=yaY3rtA37Uc

- Cyber load shed: Eliminating unnecessary network segments, communications, and cyber assets that are not operationally necessary.
- Cyber Root Cause Analysis (RCA) : RCA forensics to determine how an impactful event occurred and ensure it is contained.
- Cyber Blackstart: Cyber asset base configurations and bare metal build capability to restore the cyber system to a critical service state.
- Cyber mutual aid: Ability to utilize information sharing and analysis centers (ISACs), peer utilities, law enforcement and intelligence agencies, as well as contractors and vendors to respond to large-scale events.

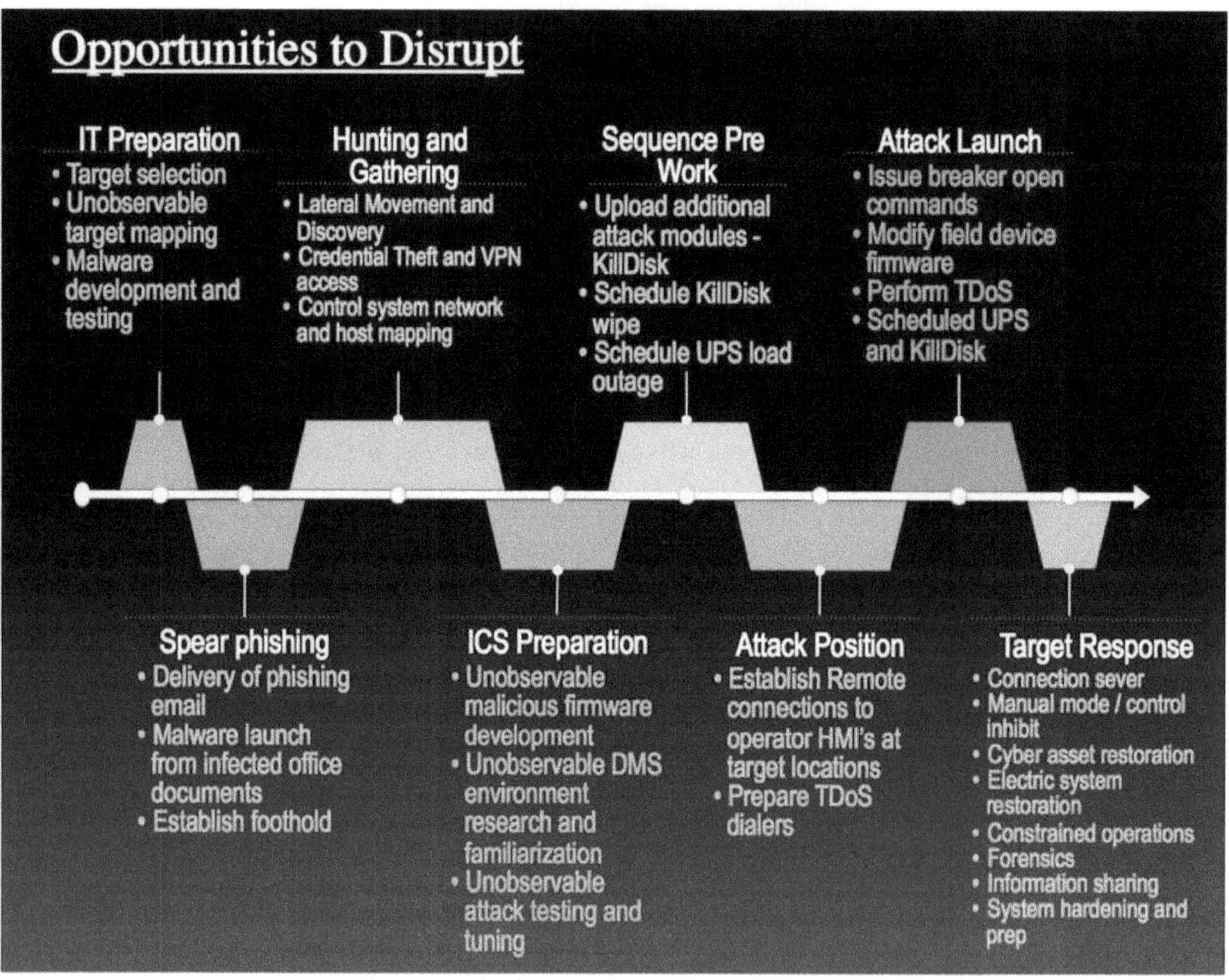
Opportunities to Disrupt
IT Preparation
• Target selection
• Unobservable target mapping
• Malware development and testing
Hunting and Gathering
• Lateral Movement and Discovery
• Credential Theft and VPN access
• Control system network and host mapping
Sequence Pre Work
• Upload additional attack modules - KillDisk
• Schedule KillDisk wipe
• Schedule UPS load outage
Attack Launch
• Issue breaker open commands
• Modify field device firmware
• Perform TDoS
• Scheduled UPS and KillDisk
Spear phishing
• Delivery of phishing email
• Malware launch from infected office documents
• Establish foothold
ICS Preparation
• Unobservable malicious firmware development
• Unobservable DMS environment research and familiarization
• Unobservable attack testing and tuning
Attack Position
• Establish Remote connections to operator HMI's at target locations
• Prepare TDoS dialers
Target Response
• Connection sever
• Manual mode / control inhibit
• Cyber asset restoration
• Electric system restoration
• Constrained operations
• Forensics
• Information sharing
• System hardening and prep

Figure 11: Summary of the opportunities to disrupt the attack

Recommendations

Architecture

Recommendations:

- Properly segment networks from each other.
- Ensure logging is enabled on devices that support it, including both IT and Operational Technology (OT) assets.
- Ensure that network architecture, such as switches, are managed and have the ability to capture data from the environment to support Passive and Active Defense mechanisms.
- Make backups of critical software installers and include an MD5 and SHA256 digital hash of the installers.
- Collect and vault backup project files from the network.
- Test the tools and technologies that passive and active defense mechanisms will need (such as digital imaging software) on the environment to ensure that it will not negatively impact systems.
- Prioritize and patch known vulnerabilities based on the most critical assets in the organization.
- Limit remote connections only to personnel that need them. When personnel need remote access, ensure that if they do not need control that they do not have access to control elements. Use two-form authentication on the remote connections.
- Consider use of a system event monitoring system, configured and monitored specifically for high-value ICS/SCADA systems.

Passive Defense

Recommendations:

- Application whitelisting can help limit adversary initial infection vectors and should be used when not too invasive to the ICSs.
- DMZs and properly tuned firewalls between network segments will give visibility into the environment and allow defenders the time required to identify intrusions.
- Establish a central logging and data aggregation point to allow forensic evidence to be collected and made available to defenders.
- Implement alarm package priorities for abnormal cyber events within the control system.
- Enforce a password reset policy in the event of a compromise especially for VPNs and administrative accounts.
- Utilize up-to-date antivirus or endpoint security technologies to allow for the denial of known malware.
- Configure an intrusion detection system so that rules can be quickly deployed to search for intruders.

Active Defense

Recommendations:

- Train defenders to hunt for odd communications leaving the networked environment such as new IP communications.

- Perform network security monitoring to continuously search through the networked environment for abnormalities.
- Plan and train to incident response plans that incorporate both the IT and OT network personnel.
- Consider active defense models for security operations such as the active cyber defense cycle.
- Ensure that personnel performing analysis have access to technologies such as sandboxes to quickly analyze incoming phishing emails or odd files and extract indicators of compromise (IOCs) to search for infected systems.
- Use backup and recovery tools to take digital images from a few of the systems in the supervisory environment such as HMIs and data historian systems every 6-12 months. This will allow a baseline of activity to be built and make the images available for scanning with new IOCs such as new YARA rules on emerging threats.
- Train defenders on using tools such as YARA to scan digital images and evidence collected from the environment but do not perform the scans in the production environment itself.

Good architecture and passive defense practices build a defensible ICS; active defense processes establish a defended ICS environment. Countering flexible and persistent human adversaries requires properly trained and equipped human defenders.

Implications and Conclusion

Implications for Defenders

The remote cyber attacks directed against Ukraine's electricity infrastructure were bold and successful. The cyber operation was highly synchronized and the adversary was willing to maliciously operate a SCADA system to cause power outages, followed by destructive attacks to disable SCADA and communications to the field. The destructive element is the first time the world has seen this type of attack against OT systems in a nation's critical infrastructure. This is an escalation from past destructive attacks that impacted general-purpose computers and servers (e.g., Saudi Aramco, RasGas, Sands Casino, and Sony Pictures). Several lines were crossed in the conduct of these attacks as the targets can be described as solely civilian infrastructure. Historic attacks, such as Stuxnet, which included destruction of equipment in the OT environment, could be argued as being surgically targeted against a military target.

Infrastructure defenders must be ready to confront highly targeted and directed attacks that include their own ICSs being used against them, combined with amplifying attacks to deny communication infrastructure and future use of their ICSs. The elements analyzed in the attack indicated that there was a specific sequence to the misuse of the ICSs, including preventing further defender use of the ICSs to restore the system. This means that the attacker "burned the bridges" behind them by destroying equipment and wiping devices to prevent automated recovery of the system. The attacks highlight the need to develop active cyber defenses, capable and well-exercised incident response plans, and resilient operations plans to survive a sophisticated attack and restore the system.

Nothing about the attack in Ukraine was inherently specific to Ukrainian infrastructure. The impact of a similar attack may be different in other nations, but the attack methodology, Tactics, Techniques, and Procedures (TTPs) observed are employable in infrastructures around the world.

Conclusion

We have identified five themes for defenders to focus on as they consider what this attack means for their organization:

Theme 1

As defenders of ICSs, consider the sequence of events taken by the adversary in the months leading up to December 23, 2015 when this cyber operation targeting Ukrainian electricity infrastructure was planned and developed. The operation relied upon intrusions that appear to have come from a broader access campaign conducted in the spring of 2015. In a prolonged attack campaign, there are likely numerous opportunities to detect and defend the targeted system. The two-stage ICS cyber kill chain helps note that in an ICS environment, there is an increased window for the detection and identification of the most concerning attack types.

Theme 2

The cyber attacks were conducted within minutes of each other against three oblenergos, resulting in power outages affecting approximately 225,000 customers for a few hours. While the total number of customers across three service territories does not add up to a significant number of customers or load across Ukraine, there may be significance in target selection or specific loads. One critical element of this particular attack was its coordinated nature affecting three target entities and the thoroughness of the adversary sequence of events in achieving their goals. Important opportunities for defenders to disrupt the adversary's sequence of events were identified.

Theme 3

The cyber attacks were mislabeled as solely linked to BlackEnergy 3 and KillDisk. BlackEnergy 3 was simply a tool used in Stage 1 of the attacks and KillDisk was an amplifying tool used in Stage 2 of the attacks. BlackEnergy 3 malware was used to gain initial footholds into a multitude of organizations within Ukraine and not just the three impacted oblenergos. It is unknown if the adversary had planned to use this access campaign to enable their operation or if achieving access was the motivation leading to the development of a concept to attack the power system.

Excessive focus on the specific malware used in this attack places defenders into a mindset in which they are simply waiting for guidance on the specific attack components so they can eliminate them. This attack could have been enabled by a variety of approaches to gain access and utilize existing assets within a target environment. Regardless of the initial attack vector, the ICS tools and environment were ultimately used to achieve the desired effect, not the BlackEnergy 3 malware.

Theme 4

The attack concept had to be able to work across multiple SCADA DMS implementations and target common susceptible elements, such as storage overwrites for Windows-based operating system workstations and servers. The attackers likely developed destructive firmware overwrite techniques after discovering accessible embedded systems. There was likely a significant amount of unobservable adversarial testing performed prior to introducing the attack into the environment. Many capabilities were demonstrated throughout this attack, and they all provide specific lessons learned for defenders to take action on.

Theme 5

Information sharing is key in the identification of a coordinated attack and directing appropriate response actions. Within the Ukraine, an organization with the ability to enable appropriate information sharing and provide incident response guidance should be pursued. In the United States and other countries with established information sharing mechanisms, such as ISACs (Information Sharing and Analysis Centers), the focus should be on maintaining and improving the information provided by asset owners and operators. This increased data sharing will enhance situation awareness within the sector, which will in turn lead to earlier attack detection and facilitate incident response.

In many ways, the Ukrainian oblenergos and their staff, as well as the involved Ukrainian government members deserve congratulations. This attack was a world first in many ways, and the Ukrainian response was impressive with all aspects considered.

As the investigation and analysis of technical data continues and more information regarding this attack surfaces, the authors of this DUC will update this report where appropriate in an effort to maintain the most accurate and beneficial guidance document possible for ICS defenders. The E-ISAC will continue to provide credible reporting and guidance as well.

Appendix Information Evaluation

Credibility: 5[39]

The claims by the Ukrainian government that outages in the service territory of the targeted electricity companies were caused by a series of cyber attacks have been confirmed. The claim was originally met with private skepticism by the SANS ICS team as ICS organizations frequently have reliability issues and incorrectly blame cyber mechanisms such as malware found on the network that is unrelated to the outage. Early reporting on incidents is often rushed and stressful which leads to inaccurate claims. However, in the Ukrainian case, there is a large amount of evidence available; including malware samples, interviews with operators present during the incident, and confirmation by multiple private companies involved in the incident. Lastly, the U.S. government has since also confirmed the attacks due to their own investigation.

The most recent report released from DHS ICS-CERT[40] cites direct interviews with "operations and information technology staff and leadership at six Ukrainian organizations with first-hand experience of the event." Based on the information provided in the report,[41] the U.S. delegation interviewed and considered information from the three impacted organizations as well as others. The format of the interviews, and asset owner and operator discussions, indicated that "the team was not able to independently review technical evidence of the cyber-attack. However, a significant number of independent reports from the team's interviews as well as documentary findings corroborate the events...".[42] However, a large amount of technical information was made available to the larger community including indicators of compromise, malware samples, technical information about the ICS itself and its components, and some samples of logs from the SCADA environment.[43] The majority of sources to date have relied upon initial attempts by Ukrainian power entities to inform customers about the cause of the outage and sources derived from interviews with impacted entities. The DHS report does not attempt to assign attacker attribution and neither will this DUC.

Amount of Technical Information Available: 4[44]

A score of 4 has been assigned for the technical information available due to the fact that malware samples, observable ICS impacts, technical indicators of compromise, and first hand interviews were available. The investigation also included a joint working group between the Ukrainian government, impacted oblenergos, and U.S. government representatives starting on January 18, 2016.[45] This amount of information was sufficient to confirm the attacks.

However, it should be noted that there may be pieces of information missing due to the lack of visibility in various parts of the ICS network. As an example, packet captures from the network during the attack and field device

---

39 Credibility of the information is rated in a scale from [0] Cannot be determined, [1] Improbable, [2] Doubtful, [3] Possibly true, [4] Probably true, [5] Confirmed

40 https://ics-cert.us-cert.gov/alerts/IR-ALERT-H-16-056-01

41 SANS ICS team members have been able to view technical data in both public and non-government private channels to confirm the existence of forensic data and the core components of the analysis based off of the data.

42 https://ics-cert.us-cert.gov/alerts/IR-ALERT-H-16-056-01

43 It should be noted that many in the community would like access to internal forensic logs of the impacted oblenergos. This is an understandable request but it is extremely rare for impacted organizations to make such information publicly available. SANS ICS team members have been able to view technical data in both public and non-government private channels to confirm the existence of forensic data and the core components of the analysis based off of the data.

44 Amount of Technical Information Available is an analyst's evaluation and description of the details available to deconstruct the attack provided with a rating scale from [0] No specifics, [1] high-level summary only, [2] Some details, [3] Many details, [4] Extensive details, [5] Comprehensive details with supporting evidence

45 http://mpe.kmu.gov.ua/minugol/control/uk/publish/article;jsessionid=CE1C739AA046FF6BA00FE8E8A4D857F3.app1?art_id=245086886&cat_id=35109

logging were not available. With this information even more about the technical minutia of the attack would be available. The amount of information available as well as the willingness by the impacted oblenergos and Ukrainian government to share that information publicly was the most seen to date for a confirmed intentional cyber attack that impacted the operations of an ICS.

When considering the technical information provided, the authors of this DUC have considered the larger public reporting of electricity customer outages within Ukraine as a component of the validation and evidence necessary to demonstrate the attacker effects to the electricity system. The official public alert by DHS corroborates prior reporting and is based on interviews and information exchanged with the impacted organizations.

154 FERC ¶ 61,037
UNITED STATES OF AMERICA
FEDERAL ENERGY REGULATORY COMMISSION

18 CFR Part 40

[Docket No. RM15-14-000]

Revised Critical Infrastructure Protection Reliability Standards

(Issued January 21, 2016)

AGENCY: Federal Energy Regulatory Commission.

ACTION: Final rule.

SUMMARY: The Federal Energy Regulatory Commission (Commission) approves seven critical infrastructure protection (CIP) Reliability Standards: CIP-003-6 (Security Management Controls), CIP-004-6 (Personnel and Training), CIP-006-6 (Physical Security of BES Cyber Systems), CIP-007-6 (Systems Security Management), CIP-009-6 (Recovery Plans for BES Cyber Systems), CIP-010-2 (Configuration Change Management and Vulnerability Assessments), and CIP-011-2 (Information Protection). The proposed Reliability Standards address the cyber security of the bulk electric system and improve upon the current Commission-approved CIP Reliability Standards. In addition, the Commission directs NERC to develop certain modifications to improve the CIP Reliability Standards.

DATES: This rule will become effective **[INSERT DATE 65 days after publication in the FEDERAL REGISTER].**

FOR FURTHER INFORMATION CONTACT:

Daniel Phillips (Technical Information)
Office of Electric Reliability
Federal Energy Regulatory Commission
888 First Street, NE
Washington, DC 20426
(202) 502-6387
daniel.phillips@ferc.gov

Simon Slobodnik (Technical Information)
Office of Electric Reliability
Federal Energy Regulatory Commission
888 First Street, NE
Washington, DC 20426
(202) 502-6707
simon.slobodnik@ferc.gov

Kevin Ryan (Legal Information)
Office of the General Counsel
Federal Energy Regulatory Commission
888 First Street, NE
Washington, DC 20426
(202) 502-6840
kevin.ryan@ferc.gov

SUPPLEMENTARY INFORMATION:

154 FERC ¶ 61,037
UNITED STATES OF AMERICA
FEDERAL ENERGY REGULATORY COMMISSION

Before Commissioners: Norman C. Bay, Chairman;
Cheryl A. LaFleur, Tony Clark,
and Colette D. Honorable.

Revised Critical Infrastructure Protection Reliability Standards — Docket No. RM15-14-000

ORDER NO. 822

FINAL RULE

(Issued January 21, 2016)

1. Pursuant to section 215 of the Federal Power Act (FPA),[1] the Commission approves seven critical infrastructure protection (CIP) Reliability Standards: CIP-003-6 (Security Management Controls), CIP-004-6 (Personnel and Training), CIP-006-6 (Physical Security of BES Cyber Systems), CIP-007-6 (Systems Security Management), CIP-009-6 (Recovery Plans for BES Cyber Systems), CIP-010-2 (Configuration Change Management and Vulnerability Assessments), and CIP-011-2 (Information Protection) (proposed CIP Reliability Standards). The North American Electric Reliability Corporation (NERC), the Commission-certified Electric Reliability Organization (ERO), submitted the seven proposed CIP Reliability Standards in response

[1] 16 U.S.C. 824o.

to Order No. 791.[2] The Commission also approves NERC's implementation plan and violation risk factor and violation severity level assignments. In addition, the Commission approves NERC's new or revised definitions for inclusion in the NERC Glossary of Terms Used in Reliability Standards (NERC Glossary), subject to modification. Further, the Commission approves the retirement of Reliability Standards CIP-003-5, CIP-004-5.1, CIP-006-5, CIP-007-5, CIP-009-5, CIP-010-1, and CIP-011-1.

2. The proposed CIP Reliability Standards are designed to mitigate the cybersecurity risks to bulk electric system facilities, systems, and equipment, which, if destroyed, degraded, or otherwise rendered unavailable as a result of a cybersecurity incident, would affect the reliable operation of the Bulk-Power System.[3] As discussed below, the Commission finds that the proposed CIP Reliability Standards are just, reasonable, not unduly discriminatory or preferential, and in the public interest, and address the directives in Order No. 791 by: (1) eliminating the "identify, assess, and correct" language in 17 of the CIP version 5 Standard requirements; (2) providing enhanced security controls for Low Impact assets; (3) providing controls to address the risks posed by transient electronic devices (e.g., thumb drives and laptop computers) used at High and Medium Impact BES Cyber Systems; and (4) addressing in an equally effective and efficient manner the need for a NERC Glossary definition for the term "communication

---

[2] *Version 5 Critical Infrastructure Protection Reliability Standards*, Order No. 791, 78 Fed. Reg. 72,755 (Dec. 3, 2013), 145 FERC ¶ 61,160 (2013), *order on clarification and reh'g*, Order No. 791-A, 146 FERC ¶ 61,188 (2014).

[3] *See* NERC Petition at 3.

networks." Accordingly, the Commission approves the proposed CIP Reliability Standards because they improve the base-line cybersecurity posture of applicable entities compared to the current Commission-approved CIP Reliability Standards.

3. In addition, pursuant to FPA section 215(d)(5), the Commission directs NERC to develop certain modifications to improve the CIP Reliability Standards. First, NERC is directed to develop modifications to address the protection of transient electronic devices used at Low Impact BES Cyber Systems. As discussed below, the modifications developed by NERC should be designed to effectively address, in an appropriately tailored manner, the risks posed by transient electronic devices to Low Impact BES Cyber Systems. Second, the Commission directs NERC to develop modifications to CIP-006-6 to require protections for communication network components and data communicated between all bulk electric system Control Centers according to the risk posed to the bulk electric system. With regard to the questions raised in the Notice of Proposed Rulemaking (NOPR) concerning the potential need for additional remote access controls, NERC must conduct a comprehensive study that identifies the strength of the CIP version 5 remote access controls, the risks posed by remote access-related threats and vulnerabilities, and appropriate mitigating controls.[4] Third, the Commission directs

[4] *Revised Critical Infrastructure Protection Reliability Standards*, Notice of Proposed Rulemaking, 80 Fed. Reg. 43,354 (July 22, 2015), 152 FERC ¶ 61,054, at 60 (2015).

NERC to develop modifications to its definition for Low Impact External Routable Connectivity, as discussed in detail below.

4. The Commission, in the NOPR, also proposed to direct that NERC develop requirements relating to supply chain management for industrial control system hardware, software, and services.[5] After review of comments on this topic, the Commission scheduled a staff-led technical conference for January 28, 2016, in order to facilitate a structured dialogue on supply chain risk management issues identified by the NOPR. Accordingly, this Final Rule does not address supply chain risk management issues. Rather, the Commission will determine the appropriate action on this issue after the scheduled technical conference.

I. Background

A. Section 215 and Mandatory Reliability Standards

5. Section 215 of the FPA requires a Commission-certified ERO to develop mandatory and enforceable Reliability Standards, subject to Commission review and approval. Reliability Standards may be enforced by the ERO, subject to Commission oversight, or by the Commission independently.[6] Pursuant to section 215 of the FPA, the

[5] *Id.* P 66.

[6] 16 U.S.C. 824o(e).

Commission established a process to select and certify an ERO,[7] and subsequently certified NERC.[8]

B. Order No. 791

6. On November 22, 2013, in Order No. 791, the Commission approved the CIP version 5 Standards (Reliability Standards CIP-002-5 through CIP-009-5, and CIP-010-1 and CIP-011-1).[9] The Commission determined that the CIP version 5 Standards improve the CIP Reliability Standards because, *inter alia*, they include a revised BES Cyber Asset categorization methodology that incorporates mandatory protections for all High, Medium, and Low Impact BES Cyber Assets, and because several new security controls should improve the security posture of responsible entities.[10] In addition, pursuant to section 215(d)(5) of the FPA, the Commission directed NERC to: (1) remove the "identify, assess, and correct" language in 17 of the CIP Standard requirements; (2) develop enhanced security controls for Low Impact assets; (3) develop controls to protect transient electronic devices; (4) create a NERC Glossary definition for the term

---

[7] *Rules Concerning Certification of the Electric Reliability Organization; and Procedures for the Establishment, Approval, and Enforcement of Electric Reliability Standards*, Order No. 672, FERC Stats. & Regs. ¶ 31,204, *order on reh'g*, Order No. 672-A, FERC Stats. & Regs. ¶ 31,212 (2006).

[8] *North American Electric Reliability Corp.*, 116 FERC ¶ 61,062, *order on reh'g and compliance*, 117 FERC ¶ 61,126 (2006), *aff'd sub nom. Alcoa, Inc. v. FERC*, 564 F.3d 1342 (D.C. Cir. 2009).

[9] Order No. 791, 145 FERC ¶ 61,160 at P 41.

[10] *Id.*

"communication networks;" and (5) develop new or modified Reliability Standards to protect the nonprogrammable components of communications networks.

7. The Commission also directed NERC to conduct a survey of Cyber Assets that are included or excluded under the new BES Cyber Asset definition and submit an informational filing within one year.[11] On February 3, 2015, NERC submitted an informational filing assessing the results of a survey conducted to identify the scope of assets subject to the definition of the term BES Cyber Asset as it is applied in the CIP version 5 Standards.

8. Finally, Order No. 791 directed Commission staff to convene a technical conference to examine the technical issues concerning communication security, remote access, and the National Institute of Standards and Technology (NIST) Risk Management Framework.[12] On April 29, 2014, a staff-led technical conference was held pursuant to the Commission's directive. The topics discussed at the technical conference included: (1) the adequacy of the approved CIP version 5 Standards' protections for bulk electric system data being transmitted over data networks; (2) whether additional security controls are needed to protect bulk electric system communications networks, including remote systems access; and (3) the functional differences between the respective methods utilized for the identification, categorization, and specification of appropriate levels of

---

[11] *Id.* PP 76, 108, 136, 150.

[12] *Id.* P 225.

protection for cyber assets using the CIP version 5 Standards as compared with those employed within the NIST Cybersecurity Framework.

C. NERC Petition

9. On February 13, 2015, NERC submitted a petition seeking approval of Reliability Standards CIP-003-6, CIP-004-6, CIP-006-6, CIP-007-6, CIP-009-6, CIP-010-2, and CIP-011-2, as well as an implementation plan,[13] associated violation risk factor and violation severity level assignments, proposed new or revised definitions,[14] and retirement of Reliability Standards CIP-003-5, CIP-004-5.1, CIP-006-5, CIP-007-5, CIP-009-5, CIP-010-1, and CIP-011-1.[15] NERC states that the proposed Reliability Standards are just, reasonable, not unduly discriminatory or preferential, and in the public interest because they satisfy the factors set forth in Order No. 672 that the Commission applies when reviewing a proposed Reliability Standard.[16] NERC maintains that the

[13] The proposed implementation plan is designed to match the effective dates of the proposed Reliability Standards with the effective dates of the prior versions of those Reliability Standards under the implementation plan for the CIP version 5 Standards.

[14] The six new or revised definitions proposed for inclusion in the NERC Glossary are: (1) BES Cyber Asset; (2) Protected Cyber Asset; (3) Low Impact Electronic Access Point; (4) Low Impact External Routable Connectivity; (5) Removable Media; and (6) Transient Cyber Asset.

[15] The proposed Reliability Standards are available on the Commission's eLibrary document retrieval system in Docket No. RM15-14-000 and on the NERC website, www.nerc.com.

[16] *See* NERC Petition at 13 and Exhibit C (citing Order No. 672, FERC Stats. & Regs. ¶ 31,204 at PP 323-335).

proposed Reliability Standards "improve the cybersecurity protections required by the CIP Reliability Standards[.]"[17]

10. NERC avers that the proposed CIP Reliability Standards satisfy the Commission directives in Order No. 791. Specifically, NERC states that the proposed Reliability Standards remove the "identify, assess, and correct" language, which represents the Commission's preferred approach to addressing the underlying directive.[18] In addition, NERC states that the proposed Reliability Standards address the Commission's directive regarding a lack of specific controls or objective criteria for Low Impact BES Cyber Systems by requiring responsible entities "to implement cybersecurity plans for assets containing Low Impact BES Cyber Systems to meet specific security objectives relating to: (i) cybersecurity awareness; (ii) physical security controls; (iii) electronic access controls; and (iv) Cyber Security Incident response."[19]

11. With regard to the Commission's directive that NERC develop specific controls to protect transient electronic devices, NERC explains that the proposed Reliability Standards require responsible entities "to implement controls to protect transient devices connected to their high impact and medium impact BES Cyber Systems and associated [Protected Cyber Assets]."[20] In addition, NERC states that the proposed Reliability

[17] NERC Petition at 4.

[18] *Id.* at 4, 15.

[19] *Id.* at 5.

[20] *Id.* at 6.

Standards address the protection of communication networks "by requiring entities to implement security controls for nonprogrammable components of communication networks at Control Centers with high or medium impact BES Cyber Systems."[21] Finally, NERC explains that it has not proposed a definition of the term "communication network" because the term is not used in the CIP Reliability Standards. Additionally, NERC states that "any proposed definition would need to be sufficiently broad to encompass all components in a communication network as they exist now and in the future."[22] NERC concludes that the proposed Reliability Standards "meet the ultimate security objective of protecting communication networks (both programmable and nonprogrammable communication network components)."[23]

12. Accordingly, NERC requests that the Commission approve the proposed Reliability Standards, the proposed implementation plan, the associated violation risk factor and violation severity level assignments, and the proposed new and revised definitions. NERC requests an effective date for the Reliability Standards of the later of April 1, 2016 or the first day of the first calendar quarter that is three months after the effective date of the Commission's order approving the proposed Reliability Standards, although NERC proposes that responsible entities will not have to comply with the

---

[21] *Id.* at 8.

[22] *Id.* at 51-52.

[23] *Id.* at 52.

requirements applicable to Low Impact BES Cyber Systems (CIP-003-6, Requirement R1, Part 1.2 and Requirement R2) until April 1, 2017.

D. Notice of Proposed Rulemaking

13. On July 16, 2015, the Commission issued a NOPR proposing to approve Reliability Standards CIP-003-6, CIP-004-6, CIP-006-6, CIP-007-6, CIP-009-6, CIP-010-2 and CIP-011-2 as just, reasonable, not unduly discriminatory or preferential, and in the public interest.[24] The NOPR stated that the proposed CIP Reliability Standards appear to improve upon the current Commission-approved CIP Reliability Standards and to address the directives in Order No. 791.

14. While proposing to approve the proposed Reliability Standards, the Commission also proposed to direct that NERC modify certain proposed standards or provide additional information supporting its proposal. First, the Commission directed NERC to provide additional information supporting the proposed limitation in Reliability Standard CIP-010-2 to transient electronic devices used at High and Medium Impact BES Cyber Systems. Second, the Commission stated that, while proposed CIP-006-6 would require protections for communication networks among a limited group of bulk electric system Control Centers, the proposed standard does not provide protections for communication network components and data communicated between all bulk electric system Control Centers. Therefore, the Commission proposed to direct that NERC develop

---

[24] NOPR, 152 FERC ¶ 61,054 (2015).

modifications to Reliability Standard CIP-006-6 to require physical or logical protections for communication network components between all bulk electric system Control Centers. Third, while the Commission proposed to approve the new or revised definitions for inclusion in the NERC Glossary, it sought comment on the proposed definition for Low Impact External Routable Connectivity. The Commission noted that, depending on the comments received, it may direct NERC to develop modifications to this definition to eliminate possible ambiguities and ensure that BES Cyber Assets receive adequate protection.

15. In addition, the Commission raised a concern that changes in the bulk electric system cyber threat landscape, identified through recent malware campaigns targeting supply chain vendors, have highlighted a gap in the protections under the CIP Reliability Standards. Therefore, the Commission proposed to direct NERC to develop a new Reliability Standard or modified Reliability Standard to provide security controls for supply chain management for industrial control system hardware, software, and services associated with bulk electric system operations.[25]

16. In response to the NOPR, 41 entities submitted comments. A list of commenters appears in Appendix A. The comments have informed our decision making in this Final Rule.

[25] *Id.* P 18.

II. Discussion

17. Pursuant to section 215(d)(2) of the FPA, we approve Reliability Standards CIP-003-6, CIP-004-6, CIP-006-6, CIP-007-6, CIP-009-6, CIP-010-2 and CIP-011-2 as just, reasonable, not unduly discriminatory or preferential, and in the public interest. We find that the proposed Reliability Standards address the Commission's directives from Order No. 791 and are an improvement over the current Commission-approved CIP Reliability Standards. Specifically, the CIP Reliability Standards improve upon the existing standards by removing the "identify, assess, and correct" language and addressing the protection of Low Impact BES Cyber Systems. With regard to the directive to create a NERC Glossary definition for the term "communication networks," we approve NERC's proposal as an equally effective and efficient method to achieve the reliability goal underlying that directive in Order No. 791. We also approve NERC's proposed implementation plan, and violation risk factor and violation severity level assignments. Finally, we approve NERC's proposed new or revised definitions for inclusion in the NERC Glossary, subject to certain modifications, discussed below.

18. In addition, pursuant to section 215(d)(5) of the FPA, we direct NERC to develop modifications to the CIP Reliability Standards to address our concerns regarding: (1) the need for mandatory protection for transient electronic devices used at Low Impact BES Cyber Systems in a manner that effectively addresses, and is appropriately tailored to address, the risk posed by those assets; and (2) the need for mandatory protection for communication links and data communicated between bulk electric system Control Centers in a manner that reflects the risks posed to bulk electric system reliability.

In addition, we direct NERC to modify the definition of Low Impact External Routable Connectivity in order to eliminate ambiguities in the language. Finally, we direct NERC to complete a study of the remote access protections in the CIP Reliability Standards within one year of the implementation of the CIP version 5 Standards for High and Medium Impact BES Cyber Systems.

19. As noted above, in the NOPR, the Commission proposed to direct that NERC develop requirements on the subject of supply chain management for industrial control system hardware, software, and services. After review of comments on the subject, the Commission scheduled a staff-led technical conference for January 28, 2016. The Commission will determine the appropriate action on this issue after the scheduled technical conference.

20. Below, we discuss the following matters: (A) protection of transient electronic devices; (B) protection of bulk electric system communication networks; (C) proposed definitions; and (D) NERC's implementation plan.

A. Protection of Transient Electronic Devices

NERC Petition

21. In its Petition, NERC states that the revised CIP Reliability Standards satisfy the Commission's directive in Order No. 791 by requiring that applicable entities: (1) develop plans and implement cybersecurity controls to protect Transient Cyber Assets and Removable Media associated with their High Impact and Medium Impact BES Cyber Systems and associated Protected Cyber Assets; and (2) train their personnel on the risks associated with using Transient Cyber Assets and Removable Media. NERC states that

the purpose of the proposed revisions is to prevent unauthorized access to and use of transient electronic devices, mitigate the risk of vulnerabilities associated with unpatched software on transient electronic devices, and mitigate the risk of the introduction of malicious code on transient electronic devices. NERC explains that the standard drafting team determined that the proposed requirements should only apply to transient electronic devices associated with High and Medium Impact BES Cyber Systems, concluding that "the application of the proposed transient devices requirements to transient devices associated with low impact BES Cyber Systems was unnecessary, and likely counterproductive, given the risks low impact BES Cyber Systems present to the Bulk Electric System."[26]

22. NERC further explains that the controls required under Attachment 1 to CIP-010-2, Requirement R4 address the following areas: (1) protections for Transient Cyber Assets managed by responsible entities; (2) protections for Transient Cyber Assets managed by another party; and (3) protections for Removable Media. NERC indicates that these provisions reflect the standard drafting team's recognition that the security controls required for a particular transient electronic device must account for the functionality of that device and whether the responsible entity or a third party manages the device. NERC also states that Transient Cyber Assets and Removable Media have

[26] NERC Petition at 34-35.

different capabilities because they present different levels of risk to the bulk electric system.[27]

NOPR

23. In the NOPR, the Commission stated that proposed Reliability Standard CIP-010-2 appears to provide a satisfactory level of security for transient electronic devices used at High and Medium Impact BES Cyber Systems. The Commission noted that the proposed security controls required under proposed CIP-010-2, Requirement R4, taken together, constitute a reasonable approach to address the reliability objectives outlined by the Commission in Order No. 791. Specifically, the Commission stated that proposed security controls outlined in Attachment 1 should ensure that responsible entities apply multiple security controls to provide defense-in-depth protection to transient electronic devices in the High and Medium Impact BES Cyber System environments.[28]

24. The Commission raised a concern, however, that proposed CIP-010-2 does not provide adequate security controls to address the risks posed by transient electronic devices used at Low Impact BES Cyber Systems, including Low Impact Control Centers, due to the limited applicability of Requirement R4. The Commission stated that this omission may result in a gap in protection for Low Impact BES Cyber Systems where malware inserted at a single Low Impact substation could propagate through a network of many substations without encountering a single security control. The NOPR noted that

---

[27] *Id.* at 38.

[28] NOPR, 152 FERC ¶ 61,054 at P 41.

"Low Impact security controls do not provide for the use of mandatory anti-malware/antivirus protections within the Low Impact facilities, heightening the risk that malware or malicious code could propagate through these systems without being detected."[29]

25. The Commission also indicated that the burden of expanding the applicability of Reliability Standard CIP-010-2 to transient electronic devices at Low Impact BES Cyber Systems is not clear from the information in the record, nor is it clear what information and analysis led NERC to conclude that the application of the transient electronic device requirements to Low Impact BES Cyber Systems "was unnecessary." Therefore, the Commission directed NERC to provide additional information supporting the proposed limitation in Reliability Standard CIP-010-2 to High and Medium Impact BES Cyber Systems, stating that the Commission "may direct NERC to address the potential reliability gap by developing a solution, which could include modifying the applicability section of CIP-010-2, Requirement R4 to include Low Impact BES Cyber Systems, that effectively addresses, and is appropriately tailored to address, the risks posed by transient devices to Low Impact BES Cyber Systems."[30]

Comments

26. While two commenters support the Commission's proposal, most commenters, including NERC, advocate approval of CIP-010-2 without expanding the applicability

[29] *Id.* P 42.

[30] *Id.* P 43.

provision of Requirement R4 to include Low Impact BES Cyber Systems. NERC questions the Commission's assertion that "malware inserted via a USB flash drive at a single Low Impact substation could propagate through a network of many substations without encountering a single security control under NERC's proposal."[31] In particular, NERC and others commenters assert that the proposed security controls in CIP-003-6 adequately address the potential for propagation of malicious code or other unauthorized access by requiring: (1) all routable protocol communications between low impact assets be controlled through a Low Impact Electronic Access Point; (2) mandatory cyber security awareness activities; (3) physical security controls; (4) electronic access controls; and (5) incident response activities.[32] Trade Associations assert that all asset-to-asset routable communications must go through the security control of the Low Impact Electronic Access Point under the proposed controls, other than extremely time sensitive device-to-device coordination.[33] Trade Associations and NIPSCO suggest that the impact on reliability in the event of a successful compromise is inherently low.

27. NERC, Trade Associations, Arkansas, G&T Cooperatives, and ITC argue that any Commission proposal to expand the protections of CIP-010-2, Requirement R4 to transient electronic devices used at Low Impact BES Cyber Systems would contradict the underlying principles of the risk-based approach that was adopted in the Commission-

---

[31] NERC Comments at 26 (quoting NOPR, 152 FERC ¶ 61,054 at P 42).

[32] *Id.* at 27. *See also* Trade Associations Comments at 12; Southern Comments at 5-6; Luminant Comments at 2; G&T Cooperatives Comments at 7.

[33] Trade Associations Comments at 12.

approved CIP version 5 Standards. Likewise, these commenters argue that the resource burden to develop and implement security controls for low impact transient devices would be substantial. NERC, Consumers Energy, and G&T Cooperatives express concern that any requirements for transient electronic devices used at Low Impact BES Cyber Systems may divert resources from the protection of Medium and High Impact BES Cyber Systems.[34]

28. Trade Associations and Southern assert that developing security controls for low impact transient cyber assets would be difficult given that, under CIP-003-6, responsible entities are not required to identify Low Impact BES Cyber Assets. Trade Associations conclude that additional transient cyber asset protections would need to be at the asset level to avoid creating administrative burdens disproportionate to the risk. Arkansas and G&T Cooperatives claim that the Commission's proposal to modify CIP-010-2 could require the implementation of device level controls and assert that the cost for complying with such regulations would be unprecedented because they would be driven by the number of devices and the number of people interacting with those devices.[35]

29. ITC and NIPSCO state that the lack of specificity in CIP-010-2, Requirement R4 raises concerns with how responsible entities will demonstrate compliance, noting that the methods included are general and non-exclusive such that a responsible entity cannot

---

[34] NERC Comments at 24; Consumers Energy Comments at 3-4; G&T Cooperatives Comments at 5.

[35] Arkansas Comments at 2-3; G&T Cooperatives Comments at 5.

be expected to know with reasonable confidence whether its plan will be deemed compliant. ITC states that, if the Commission intends to approve Standards that contain such broad latitude, it must also be prepared to accept a wide variety of plans as compliant.

30. NERC requests that, should the Commission determine that the risk associated with transient electronic devices used at Low Impact BES Cyber Systems requires expanding protections to those devices, it should recognize the varying risk levels presented by Low Impact BES Cyber Systems and the need to focus on higher risk issues. Other commenters, including Arkansas, KCP&L, and G&T Cooperatives, request that the Commission allow the implementation of the low impact controls in CIP-003-6 and the transient device controls in CIP-10-2 before directing further initiatives to expand the scope of the standards. Reclamation suggests that, if the Commission decides to direct NERC to address this potential reliability gap, the transient device and removable media controls for Low Impact BES Cyber Systems should be less stringent than the controls in CIP-010-2 given the facilities with which they are associated. Luminant and Reclamation also request that any new requirements for low impact transient electronic devices be placed in CIP-003-6.

31. APS and SPP RE generally express support for changes to CIP-010-2, Requirement R4 to address mandatory protection for transient devices used at Low Impact BES Cyber Systems. APS states that extending transient device protection to low impact systems would likely afford some additional security benefits, but notes that there may be cases where these controls would be unduly burdensome. SPP RE states that the

burden of extending certain elements of the Attachment 1 requirements to environments containing Low Impact BES Cyber Systems is reasonable, with the benefit far outweighing the cost if the controls are carefully considered with risk and potential burden in mind. SPP RE suggests that the compliance burden could be reduced by allowing Transient Cyber Assets and Removable Media to be readily moved between assets containing only Low Impact BES Cyber Systems without having to re-perform the Attachment 1 requirements between sites. Finally, NIPSCO seeks clarification on how to determine the "manager" of a Transient Cyber Asset under CIP-010-2, Requirement R4, noting that the requirement appears to allow a Transient Cyber Asset to be owned by the responsible entity, but used by a vendor on a day-to-day basis.[36]

Commission Determination

32. After consideration of the comments received on this issue, we conclude that the adoption of controls for transient devices used at Low Impact BES Cyber Systems, including Low Impact Control Centers, will provide an important enhancement to the security posture of the bulk electric system by reinforcing the defense-in-depth nature of the CIP Reliability Standards at *all* impact levels. Accordingly, we direct that NERC, pursuant to section 215(d)(5) of the FPA, develop modifications to the CIP Reliability Standards to provide mandatory protection for transient devices used at Low Impact BES Cyber Systems based on the risk posed to bulk electric system reliability. While NERC

[36] NIPSCO Comments at 9-10.

has flexibility in the manner in which it addresses the Commission's concerns, the proposed modifications should be designed to effectively address the risks posed by transient devices to Low Impact BES Cyber Systems in a manner that is consistent with the risk-based approach reflected in the CIP version 5 Standards.

33. We are not persuaded by NERC and other commenters that the security controls in CIP-003-6 adequately address the potential for propagation of malicious code or other unauthorized access stemming from transient devices used at Low Impact BES Cyber Systems. CIP-003-6 requires responsible entities, for any Low Impact External Routable Connectivity, to implement a Low Impact Electronic Access Point to "permit only necessary inbound and outbound bi-directional routable protocol access." In doing so, however, responsible entities may not foresee and configure their devices to limit all unwanted traffic. Firewalls only accept or drop traffic as dictated by a preprogrammed rule set. In other words, if a piece of malicious code were to leverage permissible traffic or protocol patterns, the firewall could not detect a malicious file signature. In short, under this requirement of CIP-003-6, responsible entities have discretion to determine what access and traffic are necessary, which does not provide enough certainty that the protocols used or ports targeted by future, as-yet-unknown malware would result in the firewall rules dropping the malicious traffic.

34. Second, the firewalls and other security devices installed at Low Impact Electronic Access Points for Low Impact BES Cyber Systems may not be actively monitored. The system security management controls in CIP-007-6 that require logging, alerting, and event review are not mandated for low impact BES Cyber Systems under CIP-003-6. As

a result, even if a security device installed at a Low Impact Electronic Access Point successfully logged suspicious network traffic, there is no assurance that a responsible entity would have processes in place to take swift action to prevent malicious code from spreading to other Low Impact BES Cyber Systems.

35. In addition, we disagree with the assertion raised by some commenters that directing NERC to address the reliability gap created by the limited applicability of CIP-010-2 contradicts the risk-based approach adopted in the CIP version 5 Standards,[37] or will result in an unreasonable resource burden or diversion of resources from the protection of Medium and High Impact BES Cyber Systems. Rather, in the NOPR, the Commission noted that *one means* to address the identified reliability concern would be to modify the applicability section of CIP-010-2, Requirement R4 to include Low Impact BES Cyber Systems. This is not, however, the only means available to address the Commission's concerns. The Commission was clear that any proposal submitted by NERC should be designed to effectively address, in a manner that is "appropriately tailored to address, the risks posed by transient devices to Low Impact BES Cyber Systems."[38] We intend that NERC's proposed modifications will be designed to address the risk posed by the assets being protected in accordance with the risk-based approach reflected in the CIP version 5 Standards, i.e., the modifications to address Low Impact

[37] *See* NERC Comments at 24; G&T Cooperatives Comments at 6.

[38] NOPR, 152 FERC ¶ 61,054 at P 43.

BES Cyber Systems may be less stringent than the provisions that apply to Medium and High Impact Cyber Systems – commensurate with the risk.

36. We agree with the Trade Associations that controls for low impact transient cyber assets could be adopted at the asset level (i.e., facility or site-level) to avoid overly-burdensome administrative tasks that could be associated with identifying discrete Low Impact BES Cyber Assets.[39] While responsible entities are not explicitly required by the CIP standards to maintain a list of discrete Low Impact BES Cyber Assets, entities should be aware of where such assets reside in order to apply the existing protections already reflected in the policies required under CIP-003-6. As noted above, the Commission offered that one possible solution to address the reliability gap could be to modify the applicability section of CIP-010-2, Requirement R4. However, should modifying CIP-010-2 prove overly burdensome as asserted by Arkansas and G&T Cooperatives, NERC may propose an equally effective and efficient solution. For example, we believe it would be reasonable for NERC to consider modifications to CIP-003-6, as suggested by Luminant and Reclamation, since the existing low impact controls reside in that standard.

37. With respect to ITC and NIPSCO's comments regarding potential ambiguity in CIP-010-2, Requirement R4, we reiterate that CIP-010-2, Requirement R4 contains sufficiently clear control objectives to inform responsible entities about the activities that

---

[39] Trade Associations Comments at 13.

must be performed in order for a transient device program to be deemed compliant. We believe that the flexibility reflected in Requirement R4 will help responsible entities to develop secure and cost effective compliance solutions. To the extent that concerns arise in the implementation process, we encourage responsible entities to work with NERC and the Regional Entities to ensure that responsible entities will have reasonable confidence about compliance expectations. Finally, regarding NIPSCO's request for clarification, we clarify our understanding that the phrase "managed by" as it is used in CIP-010-2, Requirement R4, is intended to distinguish between situations where a responsible entity has complete control over a Transient Cyber Asset as opposed to situations where a third party shares some measure of control, as discussed in the Guidelines and Technical Basis section of CIP-010-2.

B. Protection of Bulk Electric System Communication Networks

NERC Petition

38. In its Petition, NERC states that the standard drafting team concluded that it need not create a new definition for communication networks because the term "is generally understood to encompass both programmable and nonprogrammable components (i.e., a communication network includes computer peripherals, terminals, and databases as well as communication mediums such as wires)."[40] According to NERC, the revised CIP Reliability Standards contain reasonable controls to secure the types of equipment and

[40] NERC Petition at 52 (citing *North American Electric Reliability Corp.*, 142 FERC ¶ 61,203, at PP 13-14 (2013)).

components that responsible entities must protect based on the risk they pose to the bulk electric system, as opposed to a specific definition of communication networks. Further, NERC explains that the standard drafting team focused on nonprogrammable communication components at control centers with High or Medium Impact BES Cyber Systems because those locations present a heightened risk to the Bulk-Power System, warranting the increased protections.[41]

39. NERC states that proposed Reliability Standard CIP-006-6 provides flexibility for responsible entities to implement the physical security measures that best suit their needs and to account for configurations where logical measures are necessary because the entity cannot effectively implement physical access restrictions. According to NERC, responsible entities have the discretion as to the type of physical or logical protections to implement pursuant to Part 1.10 of this Standard, provided that the protections are designed to meet the overall security objective.[42]

NOPR

40. In the NOPR, the Commission indicated that NERC's proposed alternative approach to addressing the Commission's Order No. 791 directive regarding the definition of communication networks adequately addresses part of the underlying concerns set forth in Order No. 791.[43] The Commission proposed to accept NERC's

---

[41] *Id.* at 48.

[42] *Id.* at 49-50.

[43] NOPR, 152 FERC ¶ 61,054 at P 53.

explanation that responsible entities must develop controls to secure the nonprogrammable components of communication networks based on the risk they pose to the bulk electric system, rather than develop a specific definition of communication networks to identify assets for protection.

41. However, the Commission also indicated that NERC's proposed solution for the protection of nonprogrammable components of communication networks does not fully meet the intent of the Commission's Order No. 791 directive, because proposed CIP-006-6, Requirement R1, Part 1.10 would only apply to nonprogrammable components of communication networks within the same Electronic Security Perimeter, excluding from protection other programmable and non-programmable communication network components that may exist outside of a discrete Electronic Security Perimeter.[44] Therefore, the Commission proposed to direct that NERC develop a modification to proposed Reliability Standard CIP-006-6 "to require responsible entities to implement controls to protect, at a minimum, all communication links and sensitive bulk electric system data communicated between all bulk electric system Control Centers," including communication between two (or more) Control Centers, but not between a Control Center and non-Control Center facilities such as substations.[45] In addition, the Commission sought comments that address "the value achieved if the CIP Standards were to require the incorporation of additional network segmentation controls, connection

---

[44] *Id.* P 55.

[45] *Id.* P 59.

monitoring, and session termination controls behind responsible entity intermediate systems," including whether these or other steps to improve remote access protection are needed, and whether the adoption of any additional security controls addressing this topic would provide substantial reliability and security benefits.[46]

Comments

42. NERC and a number of commenters generally agree that inter-Control Center communications play a critical role in maintaining bulk electric system reliability and do not oppose further evaluation of the risks described by the Commission in the NOPR.[47] NERC states that timely and accurate communication between Control Centers is important to maintaining situational awareness and reliable bulk electric system operations, and notes that the interception or manipulation of data communicated between Control Centers "could be used to carry out successful cyberattacks against the [bulk electric system]."[48]

43. However, NERC and other commenters also assert that NERC should take steps to ensure that reliability is not adversely impacted with the adoption of any additional controls.[49] SPP RE and EnergySec indicate that latency should not be a concern for protecting Control Center communications. Specifically, SPP RE states that the latency

---

[46] *Id.* P 60.

[47] NERC Comments at 20. *See also* Comments of IRC, IESO and ITC.

[48] NERC Comments at 20.

[49] NERC Comments at 20. *See also* Arkansas Comments at 3-4; APS Comments at 4; EnergySec Comments at 4; IESO Comments at 4.

introduced by encryption is typically not an operational issue for inter-Control Center communications, since regular inter-Control Center communications do not require the same millisecond response time as communications between protective relays in substations. In addition, SPP RE states that protections other than encryption are not as effective in protecting sensitive operational data from alteration or replay.

44. A number of commenters request that the Commission provide flexibility to the extent that it issues a directive on this topic. NERC, EnergySec, APS, and IESO state that the Commission should allow NERC the opportunity to develop an appropriate and risk informed approach to any new Reliability Standard or requirement, while APS and EnergySec also suggest that NERC be granted the flexibility to determine the placement of any new security controls in the body of standards.[50] Trade Associations and Arkansas state that NERC should determine the appropriate controls to implement to meet the Commission's objectives. Luminant, PNM Resources, and Southern suggest that any new standard or requirement should be results-based and not prescriptive, affording some measure of flexibility to responsible entities.

45. Trade Associations, Southern, Wisconsin, and NEI generally agree that protections should be applied to the High and Medium Impact BES Cyber System environment, but oppose extending mandatory protection to the Low Impact Control Center environment without additional study. Trade Associations and PNM also take issue with the blanket

[50] NERC Comments at 20-21; EnergySec Comments at 4; APS Comments at 4; IESO Comments at 4.

application of security controls over all bulk electric system Control Center data and believe that NERC should have the opportunity to determine what data is truly sensitive.

46. A number of commenters oppose the Commission's proposal to require responsible entities to implement controls to protect all communication links and sensitive bulk electric system data communicated between all bulk electric system Control Centers. NIPSCO and G&T Cooperatives argue that the risks posed by such communication networks do not justify the costs of implementing a new standard and, therefore, the standard should, at a minimum, not apply to Low Impact BES Cyber Systems. NIPSCO opines that the Commission's proposal may cause unintentional consequences since data and communications exchanged between Control Centers is often time-sensitive. SCE suggests that the Commission's proposal is premature and that the risks should be studied before taking further actions. Foundation opposes the Commission's proposal because it objects to the exclusion of secure connections to grid facilities other than Control Centers, stating that the Commission should do more to protect the grid.[51]

47. Other commenters request clarification of the Commission's proposal. KCP&L, PNM, UTC, TVA, Idaho Power, and NIPSCO seek clarification whether Control Centers owned by multiple, different registered entities would be included in the Commission's proposal. TVA asks whether the Commission's proposal is focused on protecting the

---

[51] Foundation Comments at 47-48.

data link or the data itself. UTC questions the nature of the reliability gap described in the NOPR given the protections in CIP-005-5 for inbound and outbound communications. In addition, APS and EnergySec seek clarification regarding the term "control center" in the context of adopting controls to protect reliability-related data. APS and EnergySec note that transmission owner SCADA systems do not meet the current definition of control centers despite the fact that these systems contain identical reliability data as the systems operated by reliability coordinators, balancing authorities, and transmission operators. As a result, APS and EnergySec ask that the Commission clarify what constitutes a "control center" for the purposes of communication security.[52] Finally, Idaho Power, KCP&L, and UTC seek clarification whether responsible entities would be held individually accountable for implementing the controls adopted under the CIP Standards when there may be overlapping responsibilities associated with the protection of inter-entity control center communication.[53] For example, Idaho Power opines that two neighboring responsible entities with control centers that communicate with each other should both be equally responsible for implementing the CIP Standards, but states that it is unclear how compliance would be measured.

48. PNM and NIPSCO suggest that, if the NOPR proposal is aimed at protecting intra-control center communications, the Commission should consider modifications to Reliability Standard EOP-008-1. TVA requests that the Commission consider removing

[52] *See* APS Comments at 4; EnergySec Comments at 3.

[53] Idaho Power Comments at 2; UTC Comments at 2; KCP&L Comments at 5.

the requirement for protecting "all communication links" and focus on the "sensitive bulk electric system data" moving between Control Centers. TVA states that physical and logical protections for communications network components between bulk electric system Control Centers should be limited to only essential communications networks.

49. With regard to the Commission's question on the potential need for additional remote access protections, NERC and a number of commenters argue that there are not enough data to conclude that the proposed controls for remote access will be ineffective and suggest that the Commission delay consideration of additional remote access protections until after the CIP version 5 remote access provisions are implemented.[54] NERC and IRC provide a list of the relevant controls applied to remote access systems as evidence that there are substantial controls already in place to address threats associated with remote access. APS and Arkansas assert that the current Standards and industry-developed guidance provide sufficient tools for securing interactive remote access and, thus, additional controls would not provide significant reliability or security benefits. TVA claims that the current requirement language is too prescriptive because it precludes a registered entity's usage of specific technologies due to prejudices against certain "architectures."[55]

---

[54] NERC Comments at 21-23. *See also* Trade Association Comments at 14; KCP&L Comments at 4; Southern Comments at 7; IRC Comments at 6.

[55] TVA Comments at 5.

50. Commenters supporting the development of additional remote access controls for the CIP Standards contend that the current suite of CIP Standards fails to adequately address specific threats and vulnerabilities. SPP RE and CyberArk note the lack of restrictions on what systems remote users can access after successfully logging on to the intermediate system.[56] CyberArk also asserts that there is a lack of protection for remote user credentials after successfully logging onto the intermediate system and a lack of controls to regulate encryption strength and key management. Waterfall states that the proposed controls lack methods to detect and prevent compromised endpoint devices, which, according to Waterfall and SPP RE, presents the opportunity for an attacker to access multiple remote sites from a compromised central site.

51. PNM agrees that some of the controls mentioned by panelists at the April 2014 FERC technical conference may improve reliability and security. However, PNM states that such controls may have only marginal benefits to reliability and security since the increased complexity of these steps would present problems with staff support for such systems.[57] AEP asserts that, while additional controls may enhance a defense-in-depth strategy, prescriptive requirements on intermediate systems may create a need for technical feasibility exceptions for situations where security could impede reliability.

---

[56] SPP RE Comments at 7-8; CyberArk Comments at 1-2.

[57] PNM Comments at 2.

Commission Determination

52. We adopt the NOPR proposal and find that NERC's alternative approach to addressing the Commission's Order No. 791 directive regarding the definition of communication networks adequately addresses part of the underlying concerns set forth in Order No. 791.[58] In accepting this alternative approach, we accept NERC's explanation that responsible entities must develop controls to secure the nonprogrammable components of communication networks at Control Centers with High or Medium Impact BES Cyber Systems.

53. As discussed in detail below, however, the Commission concludes that modifications to CIP-006-6 to provide controls to protect, at a minimum, communication links and data communicated between bulk electric system Control Centers are necessary in light of the critical role Control Center communications play in maintaining bulk electric system reliability. Therefore, we adopt the NOPR proposal and direct that NERC, pursuant to section 215(d)(5) of the FPA, develop modifications to the CIP Reliability Standards to require responsible entities to implement controls to protect, at a minimum, communication links and sensitive bulk electric system data communicated between bulk electric system Control Centers in a manner that is appropriately tailored to address the risks posed to the bulk electric system by the assets being protected (i.e., high, medium, or low impact).

[58] NOPR, 152 FERC ¶ 61,054 at P 53.

54. NERC and other commenters recognize that inter-Control Center communications play a critical role in maintaining bulk electric system reliability by, among other things, helping to maintain situational awareness and reliable bulk electric system operations through timely and accurate communication between Control Centers.[59] We agree with this assessment. In order for certain responsible entities such as reliability coordinators, balancing authorities, and transmission operators to adequately perform their reliability functions, their associated control centers must be capable of receiving and storing a variety of sensitive bulk electric system data from interconnected entities. Accordingly, we find that additional measures to protect both the integrity and availability of sensitive bulk electric system data are warranted.[60] We also understand that the attributes of the data managed by responsible entities could require different information protection controls.[61] For instance, certain types of reliability data will be sensitive to data manipulation type attacks, while other types of reliability data will be sensitive to

---

[59] NERC Comments at 20.

[60] Protecting the integrity of bulk electric system data involves maintaining and ensuring the accuracy and consistency of inter-Control Center communications. Protecting the availability of bulk electric system data involves ensuring that required data is available when needed for bulk electric system operations.

[61] Moreover, in order for certain responsible entities to adequately perform their Reliability Functions, the associated control centers must be capable of receiving and storing a variety of sensitive data as specified by the IRO and TOP Standards. For instance, pursuant to Reliability Standard TOP-003-3, Requirements R1, R3 and R5, a transmission operator must maintain a documented specification for data and distribute its data specification to entities that have data required by the transmission operator's Operational Planning Analyses, Real-time Monitoring and Real-time Assessments. Entities receiving a data specification must satisfy the obligation of the documented specification.

eavesdropping type attacks aimed at collecting operational information (such as line and equipment ratings and impedances). NERC should consider the differing attributes of bulk electric system data as it assesses the development of appropriate controls.

55. With regard to NERC's development of modifications responsive to our directive, we agree with NERC and other commenters that NERC should have flexibility in the manner in which it addresses the Commission's directive. Likewise, we find reasonable the principles outlined by NERC that protections for communication links and sensitive bulk electric system data communicated between bulk electric system Control Centers: (1) should not have an adverse effect on reliability, including the recognition of instances where the introduction of latency could have negative results; (2) should account for the risk levels of assets and information being protected, and require protections that are commensurate with the risks presented; and (3) should be results-based in order to provide flexibility to account for the range of technologies and entities involved in bulk electric system communications.[62]

56. We disagree with the assertion of NIPSCO and G&T Cooperatives that the risk posed by bulk electric system communication networks does not justify the costs of implementing controls. Communications between Control Centers over such networks are fundamental to the operations of the bulk electric system, and the record here does not persuade us that controls for such networks are not available at a reasonable cost (through

---

[62] *See* NERC Comments at 20-21.

encryption or otherwise). Nonetheless, we recognize that not all communication network components and data pose the same risk to bulk electric system reliability and may not require the same level of protection. We expect NERC to develop controls that reflect the risk posed by the asset or data being protected, and that can be implemented in a reasonable manner. It is important to recognize that certain entities are already required to exchange necessary real-time and operational planning data through secured networks using a "mutually agreeable security protocol," regardless of the entity's size or impact level.[63] NERC's response to the directives in this Final Rule should identify the scope of sensitive bulk electric system data that must be protected and specify how the confidentiality, integrity, and availability of each type of bulk electric system data should be protected while it is being transmitted or at rest.

57. With regard to Foundation's argument that the Commission should do more to promote grid security by mandating secure communications between all facilities of the bulk electric system, such as substations, the record in the immediate proceeding does not support such a broad requirement at this time. However, if in the future it becomes evident that such action is warranted, the Commission may revisit this issue.

58. Several commenters sought clarification whether Control Centers owned by multiple registered entities would be included under the Commission's proposal. We clarify that the scope of the directed modifications apply to Control Center

---

[63] *See* Reliability Standards TOP-003-3, Requirement R5 and IRO-010-2, Requirement R3.

communications from facilities at all impact levels, regardless of ownership. The directed modification should encompass communication links and data for intra-Control Center and inter-Control Center communications.

59. Idaho Power, KCP&L, and UTC seek clarification whether entities would be held individually accountable for implementing the Standard when there may be overlapping responsibilities. We clarify that responsible entities may be held individually accountable depending upon the security arrangements with their neighbors and functional partners. Many organizations currently use joint and coordinated functional registration agreements to assign accountability for reliability tasks with joint functional obligations.[64] These mechanisms could be leveraged to address responsibilities under the CIP Standards. For example, if several registered entities have joint responsibility for a cryptographic key management system used between their respective Control Centers, they should have the prerogative to come to a consensus on which organization administers that particular key management system.

60. UTC seeks further explanation regarding the nature of the reliability gap described in the NOPR given the protections in CIP-005-5 for inbound and outbound communications. We clarify that the reliability gap addressed in this Final Rule pertains to the lack of mandatory security controls to address how responsible entities should protect sensitive bulk electric system communications and data. As noted above, while

[64] *See* NERC Compliance Public Bulletin #2010-004, available on the NERC website at www.NERC.com.

responsible entities are required to exchange real-time and operational planning data necessary to operate the bulk electric system using mutually agreeable security protocols, there is no technical specification for how this transfer of information should incorporate mandatory security controls. Although the CIP Standards provide a measure of defense-in-depth for responsible entity information systems, the current security controls primarily focus on boundary protection controls. For instance, CIP-005-5 focuses on access control and malicious code prevention, which requires authentication of the user and ensuring that no malware is included in the communication, but does not provide for security of the actual data while it is being transmitted between Electronic Security Perimeters. Thus, the current CIP Reliability Standards do not adequately address how to protect the transfer of sensitive bulk electric system data between facilities at discrete geographic locations.

61. With respect to APS and EnergySec's request for clarification regarding the meaning of the term "control center" in the context of adopting controls to protect reliability-related data, we clarify that we are using here the NERC Glossary definition of a Control Center.[65] Whether particular facilities meet or do not meet this definition

---

[65] The NERC Glossary defines Control Center as "One or more facilities hosting operating personnel that monitor and control the Bulk Electric System (BES) in real-time to perform the reliability tasks, including their associated data centers, of: 1) a Reliability Coordinator, 2) a Balancing Authority, 3) a Transmission Operator for transmission Facilities at two or more locations, or 4) a Generator Operator for generation Facilities at two or more locations."

should be determined outside of this rulemaking. However, the proposed modification will apply to Control Centers at all impact levels (high, medium, or low).

62. Several commenters addressed encryption and latency. Based on the record in this proceeding, it is reasonable to conclude that any lag in communication speed resulting from implementation of protections should only be measureable on the order of milliseconds and, therefore, will not adversely impact Control Center communications. Several commenters raise possible technical implementation difficulties with integrating encryption technologies into their current communications networks. Such technical issues should be considered by the standard drafting team when developing modifications in response to this directive, and may be resolved, e.g., by making certain aspects of the revised CIP Standards eligible for Technical Feasibility Exceptions.

63. We reject the suggestion of two commenters that any efforts to protect intra-Control Center communications should be considered through modifications in Reliability Standard EOP-008-1. As an initial matter, Reliability Standard EOP-008-1 focuses on backup functionality in the event that primary control center functionality is lost.[66] Reliability Standard EOP-008-1 also does not provide security for communication links or data and, therefore, does not provide for the protection of communication links and sensitive bulk electric system data communicated between bulk electric system Control Centers.

---

[66] *See* http://www.nerc.com/files/eop-008-1.pdf.

64. Finally, with regard to the NOPR discussion regarding the potential need for additional protections related to remote access,[67] we are persuaded by commenters' suggestions that it would be prudent to assess the extent to which the CIP version 5 Standards provide effective controls for remote access before pursuing additional revisions to the CIP Standards.[68] Therefore, we direct NERC to conduct a study that assesses the effectiveness of the CIP version 5 remote access controls, the risks posed by remote access-related threats and vulnerabilities, and appropriate mitigating controls for any identified risks. NERC should consult with Commission staff to determine the general contents of the directed report. We direct NERC to submit a report on the above-outlined study within one year of the implementation of the CIP version 5 Standards for High and Medium Impact BES Cyber Systems.

C. Proposed Definitions

NERC Petition

65. In its Petition, NERC proposes the following definition for Low Impact External Routable Connectivity:

> Direct user-initiated interactive access or a direct device-to-device connection to a low impact BES Cyber System(s) from a Cyber Asset outside the asset containing those low impact BES Cyber System(s) via a bidirectional routable protocol connection. Point-to-point communications between intelligent electronic devices that use routable communication protocols for time-sensitive protection or control functions between

[67] *See* NOPR, 152 FERC ¶ 61,054 at P 60.

[68] *See* NERC Comments at 21-23; Trade Association Comments at 14; KCP&L Comments at 4; Southern Comments at 7; IRC Comments at 6.

> Transmission station or substation assets containing low impact BES Cyber Systems are excluded from this definition (examples of this communication include, but are not limited to, IEC 61850 GOOSE or vendor proprietary protocols).[69]

66. NERC explains that the proposed definition describes the scenarios where responsible entities are required to apply Low Impact access controls under Reliability Standard CIP-003-6, Requirement R2 to their Low Impact assets. Specifically, if Low Impact External Routable Connectivity is used, a responsible entity must implement a Low Impact Electronic Access Point to permit only necessary inbound and outbound bidirectional routable protocol access.[70]

NOPR

67. In the NOPR, the Commission sought comment on the proposed definition for Low Impact External Routable Connectivity. First, the Commission sought comment on the purpose of the meaning of the term "direct" in relation to the phrases "direct user-initiated interactive access" and "direct device-to-device connection" within the proposed definition.[71] In addition, the Commission sought comment on the implementation of the "layer 7 application layer break" contained in certain reference diagrams in the Guidelines and Technical Basis section of proposed Reliability Standard CIP-003-6, noting that the guidance provided in the Guidelines and Technical Basis section of the

---

[69] NERC Petition at 28.

[70] *Id.* at 29.

[71] *See* NOPR, 152 FERC ¶ 61,054 at P 70.

proposed standard may conflict with the plain reading of the term "direct."[72] The Commission noted a concern that a conflict in the reading of the term "direct" could lead to complications in the implementation of the proposed CIP Reliability Standards, hindering the adoption of effective security controls for Low Impact BES Cyber Systems. The Commission indicated that, depending upon the responses received, the final rule may direct NERC to develop a modification to the definition of Low Impact External Routable Connectivity to eliminate ambiguities.

Comments

68. NERC and other commenters do not oppose a modification of the Low Impact External Routable Connectivity definition, so long as it remains consistent with the Guidelines and Technical Basis for section for CIP-003-6.[73] NERC, referencing the Guidelines and Technical Basis section of proposed CIP-003-6, explains that the purpose of the term "direct" is to distinguish between the scenarios where an external user or device could electronically access the Low Impact BES Cyber System without a security break (i.e., direct access) from those situations where an external user or device could

[72] *See* CIP-003-6 Guidelines and Technical Basis Section, Reference Model 6 at p. 39. The layer 7 application layer break concept appears to permit a responsible entity to log into an intermediate application or device to access the Low Impact BES Cyber System or device to avoid implementing Low Impact Electronic Access Point security controls under CIP-003-6, Attachment 1, Section 3.

[73] NERC Comments at 31. *See also* Trade Associations Comments at 15; Southern Comments at 8.

only access the Low Impact BES Cyber System following a security break (i.e., indirect access).

69. NERC explains further that Low Impact External Routable Connectivity would exist and a Low Impact Electronic Access Point would be required if an entity's implementation of a layer 7 application layer break does not provide a sufficient security break (*i.e.*, the layer 7 application does not prevent direct access to the Low Impact BES Cyber System).[74] Southern states that it believes that the Low Impact External Routable Connectivity definition, when combined with the language in the Guidelines and Technical Basis section for CIP-003-6, is sufficiently clear.

70. SPP RE, EnergySec, and APS recommend that the Commission direct NERC to revise the Low Impact External Routable Connectivity definition because the definition, as drafted, would permit transitive connections through out of scope cyber assets at sites containing Low Impact BES Cyber Systems with no required security controls.[75] SPP RE posits that indirect access, through an intervening or intermediate system such as the non-BES Cyber Asset on the same network segment, should also be considered Low Impact External Routable Connectivity because this kind of access would enable "pivot attacks" on low impact networks.

71. SPP RE, EnergySec, TVA, and APS assert that any electronic remote access into a routable network containing BES Cyber Systems should be construed as External

[74] NERC Comments at 30.

[75] SPP RE Comments at 14-18; EnergySec Comments at 2-3; APS Comments at 7.

Routable Connectivity and protected.[76] SPP RE suggests that the layer 7 application layer break language is not well understood by industry, as some responsible entities currently hold the view that a security gateway appliance effectively serves as the layer 7 protocol break eliminating Low Impact External Routable Connectivity. SPP RE asserts that the security gateway appliance acting in this way does not maintain two independent conversations and, as a result, should still be considered as externally routable connected.

72. ITC states that it considers the layer 7 application layer break referenced in Model 6 of the Guidelines and Technical Basis section to be an illustrative example that in no way requires integrity of the data stream down to layer 7 for compliance with CIP-003-6.[77] ITC notes that the illustrative example referenced by the Commission is contained within the non-binding Guidelines and Technical basis section, and does not believe that the controlling language of CIP-003-6 requires such a control.

Commission Determination

73. Based on the comments received in response to the NOPR, the Commission concludes that a modification to the Low Impact External Routable Connectivity definition to reflect the commentary in the Guidelines and Technical Basis section of CIP-003-6 is necessary to provide needed clarity to the definition and eliminate ambiguity surrounding the term "direct" as it is used in the proposed definition.

---

[76] SPP RE Comments at 14-18; EnergySec Comments at 2-3; TVA Comments at 1-2; APS Comments at 7.

[77] ITC Comments at 10-11.

Therefore, pursuant to section 215(d)(5) of the FPA, we direct NERC to develop a modification to provide the needed clarity, within one year of the effective date of this Final Rule. We agree with NERC and other commenters that a suitable means to address our concern is to modify the Low Impact External Routable Connectivity definition consistent with the commentary in the Guidelines and Technical Basis section of CIP-003-6.[78]

74. As discussed above, NERC clarifies that the purpose of the "direct" language in the Low Impact External Routable Connectivity definition is to distinguish between scenarios where an external user or device could electronically access a Low Impact BES Cyber System without a security break (direct access) from those situations where an external user or device could only access a Low Impact BES Cyber System following a security break (indirect access); therefore, in order for there to be no Low Impact External Routable Connectivity, the security break must be "complete" (i.e., it must prevent allowing access to the Low Impact BES Cyber Systems from the external cyber asset). NERC's clarification on this issue resolves many of the concerns raised by EnergySec, APS, and SPP RE regarding the proposed definition, as a complete security break would not appear to permit transitive connections through one or more out of scope cyber assets to go unprotected under the definition, and would appear to require the assets to maintain "separate conversations" as suggested by SPP RE.

---

[78] *E.g.*, NERC Comments at 31; Trade Associations Comments at 15.

75. We decline to adopt the recommendations from EnergySec and APS that the Commission direct NERC to modify the standards to utilize the concept of Electronic Security Perimeters for low impact systems and to leverage existing definitions for Electronic Access Point and External Routable Connectivity. The Commission believes that the electronic security protections developed by the standard drafting team for Low Impact BES Cyber Systems will provide sufficient protection to these systems with the modifications that we are directing to the Low Impact External Routable Connectivity definition. However, we may revisit this decision in the future if we determine that CIP-003-6, Requirement R2 and the Low Impact External Routable Connectivity definition provide insufficient electronic access protection for Low Impact BES Cyber Systems.

D. Implementation Plan

NERC Petition

76. In its Petition, NERC explains that the proposed implementation plan for the revised CIP Reliability Standards is designed to match the effective dates of the proposed Reliability Standards with the effective dates of the prior versions of the related Reliability Standards under the implementation plan of the CIP version 5 Standards. NERC states that the purpose of this approach is to provide regulatory certainty by limiting the time, if any, that the CIP version 5 Standards with the "identify, assess, and correct" language would be effective. Specifically, NERC explains that, pursuant to the CIP version 5 implementation plan, the effective date of each of the CIP version 5 Standards is April 1, 2016, except for the effective date for Requirement R2 of CIP-003-5 (i.e., controls for Low Impact BES Cyber Systems), which is April 1, 2017. NERC

explains further that the proposed implementation plan provides that: (1) each of the proposed reliability Standards shall become effective on the later of April 1, 2016 or the first day of the first calendar quarter that is three months after the effective date of the Commission's order approving the proposed Reliability Standard; and (2) responsible entities will not have to comply with the requirements applicable to Low Impact BES Cyber Systems (CIP-003-6, Requirement R1, Part 1.2 and Requirement R2) until April 1, 2017.[79]

77. NERC also explains that the proposed implementation plan includes effective dates for the new and modified definitions associated with: (1) transient devices (*i.e.*, BES Cyber Asset, Protected Cyber Asset, Removable Media, and Transient Cyber Asset); and (2) Low Impact controls (*i.e.*, Low Impact Electronic Access Point and Low Impact External Routable Connectivity). Specifically, NERC proposes that: (1) the definitions associated with transient device become effective on the compliance date for Reliability Standard CIP-010-2, Requirement R4; and (2) the definitions addressing the Low Impact controls become enforceable on the compliance date for Reliability Standard CIP-003-6, Requirement R2. Lastly, NERC proposes that the retirement of Reliability Standards CIP-003-5, CIP-004-5.1, CIP-006-5, CIP-007-5, CIP-009-5, CIP-010-1 and CIP-011-1 become effective on the effective date of the proposed Reliability Standards.

---

[79] NERC Petition at 53-54.

NOPR

78. In the NOPR, the Commission proposed to approve NERC's implementation plan for the proposed CIP Reliability Standards.[80]

Comments

79. A number of commenters request that the Commission act on the proposed revisions to the CIP Standards in a manner that avoids a different implementation date than the CIP version 5 Standards (i.e., April 1, 2016) in order to avoid confusion and unnecessary burdens.[81] Trade Associations encourage the Commission to take alternative actions to avoid unnecessary burden if a Final Rule facilitating an April 1, 2016 effective date for the revised CIP Standards is not feasible. Reclamation suggests that the Commission update and extend the standards implementation plan for each of the CIP version 5 Standards to April 1, 2017, except for the effective date for Requirement R2 of CIP-003-5, which Reclamation argues should be updated to April 1, 2018. ITC contends that April 1, 2016 is an unreasonably aggressive compliance deadline and urges the Commission to consider extending the deadline by one year to April 1, 2017.

Commission Determination

80. The Commission approves NERC's proposed implementation plan. As a result, the proposed CIP Reliability Standards will be effective the first day of the first calendar

[80] NOPR, 152 FERC ¶ 61,054 at P 73.

[81] Trade Associations Comments at 6; SCE Comments at 4-5; Reclamation Comments at 2-3; Wisconsin Comments at 3; Luminant Comments at 2-3; NextEra Comments at 4.

quarter that is three months after the effective date of the Commission's order approving the proposed Reliability Standard (i.e., July 1, 2016). Responsible entities must comply with the requirements applicable to Low Impact BES Cyber Systems (CIP-003-6, Requirement R1, Part 1.2 and Requirement R2) beginning April 1, 2017, consistent with NERC's proposed implementation plan.

81. We recognize the concerns raised by Trade Associations and other commenters regarding the potential burden of implementing two versions of certain CIP Reliability Standards within a short period of time. The Commission is willing to consider a request to align the implementation dates of certain CIP Reliability Standards or another reasonable alternative approach to addressing potential implementation issues, should NERC or another interested entity submit such a proposal.[82]

III. Information Collection Statement

82. The FERC-725B information collection requirements contained in this Final Rule are subject to review by the Office of Management and Budget (OMB) under section 3507(d) of the Paperwork Reduction Act of 1995.[83] OMB's regulations require approval of certain information collection requirements imposed by agency rules.[84] Upon approval

[82] Given the upcoming April 1, 2016 implementation date for the CIP version 5 Standards, NERC or another interested entity may wish to consider seeking expedited action for any request to address potential implementation issues. The Commission would be cognizant, in considering any request, of the need to provide adequate notice of any changes prior to April 1, 2016.

[83] 44 U.S.C. 3507(d).

[84] 5 CFR 1320.11.

of a collection of information, OMB will assign an OMB control number and expiration date. Respondents subject to the filing requirements of this rule will not be penalized for failing to respond to these collections of information unless the collections of information display a valid OMB control number.

83. The Commission solicited comments on the need for and purpose of the information contained in the proposed CIP Reliability Standards, including whether the information will have practical utility, the accuracy of the burden estimates, ways to enhance the quality, utility, and clarity of the information to be collected or retained, and any suggested methods for minimizing respondents' burden, including the use of automated information techniques. The Commission received no comments regarding the need for the information collection or the burden estimates associated with the proposed CIP Reliability Standards as described in the NOPR.

84. Public Reporting Burden: The Commission based its paperwork burden estimates on the changes in paperwork burden presented by the proposed CIP Reliability Standards as compared to the CIP version 5 Standards. The Commission has already addressed the burden of implementing the CIP version 5 Standards.[85] As discussed above, the immediate rulemaking addresses four areas of modification to the CIP version 5 Standards: (1) removal of the "identify, assess, and correct" language from 17 CIP requirements; (2) development of enhanced security controls for low impact assets;

---

[85] *See* Order No. 791, 145 FERC ¶ 61,160 at PP 226-244.

(3) development of controls to protect transient electronic devices (e.g., thumb drives and laptop computers); and (4) protection of communications networks. We do not anticipate that the removal of the "identify, assess, and correct" language will impact the reporting burden, as the substantive compliance requirements would remain the same, while NERC indicates that the concept behind the deleted language continues to be implemented within NERC's compliance function. The development of controls to protect transient devices and protection of communication networks (as proposed by NERC) have associated reporting burdens that will affect a limited number of entities, i.e., those with Medium and High Impact BES Cyber Systems. The enhanced security controls for Low Impact assets are likely to impose a reporting burden on a much larger group of entities.

85. The NERC Compliance Registry, as of June 2015, identifies approximately 1,435 U.S. entities that are subject to mandatory compliance with Reliability Standards. Of this total, we estimate that 1,363 entities will face an increased paperwork burden under the proposed CIP Reliability Standards, and we estimate that a majority of these entities will have one or more Low Impact assets. In addition, we estimate that approximately 23 percent of the entities have assets that will be subject to Reliability Standards CIP-006-6 and CIP-010-2. Based on these assumptions, we estimate the following reporting burden for entities with Medium and/or High Impact Assets:

Registered Entities	Number of Entities	Total Burden Hours in Year 1	Total Burden Hours in Year 2	Total Burden Hours in Year 3
Entities subject to CIP-006-6 and CIP-	313	75,120	130,208	130,208

010-2 with Medium and/or High Impact Assets				
Totals	313	75,120	130,208	130,208

86. The following shows the annual cost burden for the group with Medium and/or High Impact Assets, based on the burden hours in the table above:

- Year 1: Entities subject to CIP-006-6 and CIP-010-2 with Medium and/or High Impact Assets: 313 entities x 240 hours/entity * $76/hour = $5,709,120.
- Years 2 and 3: 313 entities x 416 hours/entity * $76/hour = $9,895,808 per year.
- The paperwork burden estimate includes costs associated with the initial development of a policy to address requirements relating to transient electronic devices, as well as the ongoing data collection burden. Further, the estimate reflects the assumption that costs incurred in year 1 will pertain to policy development, while costs in years 2 and 3 will reflect the burden associated with maintaining logs and other records to demonstrate ongoing compliance.

Based on the assumptions, we estimate the following reporting burden for entities with Low Impact Assets:

Registered Entities	Number of Entities	Total Burden Hours in Year 1	Total Burden Hours in Year 2	Total Burden Hours in Year 3
Entities subject to CIP-003-6 with low	1,363	163,560	283,504	283,504

impact Assets				
Totals	1,363	163,560	283,504	283,504

87. The following shows the annual cost burden for the group with Low Impact Assets, based on the burden hours in the table above:

- Year 1: Entities subject to CIP-003-6 with Low Impact Assets: 1,363 entities x 120 hours/entity * $76/hour = $12,430,560.
- Years 2 and 3: 1,363 entities x 208 hours/entity * $76/hour = $21,546,304 per year.
- The paperwork burden estimate includes costs associated with the modification of existing policies to address requirements relating to low impact assets, as well as the ongoing data collection burden, as set forth in CIP-003-6, Requirements R1.2 and R2, and Attachment 1. Further, the estimate reflects the assumption that costs incurred in year 1 will pertain to revising existing policies, while costs in years 2 and 3 will reflect the burden associated with maintaining logs and other records to demonstrate ongoing compliance.

88. The estimated hourly rate of $76 is the average (rounded) loaded cost (wage plus benefits) of legal services ($129.68 per hour), technical employees ($58.17 per hour) and

administrative support ($39.12 per hour), based on hourly rates and average benefits data from the Bureau of Labor Statistics.[86]

89. Title: Mandatory Reliability Standards, Revised Critical Infrastructure Protection Standards.

Action: Proposed Collection FERC-725B.

OMB Control No.: 1902-0248.

Respondents: Businesses or other for-profit institutions; not-for-profit institutions.

Frequency of Responses: On Occasion.

Necessity of the Information: This Final Rule approves the requested modifications to Reliability Standards pertaining to critical infrastructure protection. As discussed above, the Commission approves NERC's proposed revised CIP Reliability Standards pursuant to section 215(d)(2) of the FPA because they improve the currently-effective suite of cyber security CIP Reliability Standards.

Internal Review: The Commission has reviewed the proposed Reliability Standards and made a determination that its action is necessary to implement section 215 of the FPA.

90. Interested persons may obtain information on the reporting requirements by contacting the following: Federal Energy Regulatory Commission, 888 First Street, NE, Washington, DC 20426 [Attention: Ellen Brown, Office of the Executive Director, e-mail: DataClearance@ferc.gov, phone: (202) 502-8663, fax: (202) 273-0873].

---

[86] *See* http://bls.gov/oes/current/naics2_22.htm and http://www.bls.gov/news.release/ecec.nr0.htm. Hourly figures as of June 1, 2015.

91. For submitting comments concerning the collection(s) of information and the associated burden estimate(s), please send your comments to the Commission, and to the Office of Management and Budget, Office of Information and Regulatory Affairs, Washington, DC 20503 [Attention: Desk Officer for the Federal Energy Regulatory Commission, phone: (202) 395-0710, fax: (202) 395-7285]. For security reasons, comments to OMB should be submitted by e-mail to: oira_submission@omb.eop.gov. Comments submitted to OMB should include Docket Number RM15-14-000 and OMB Control Number 1902-0248.

IV. Regulatory Flexibility Act Analysis

92. The Regulatory Flexibility Act of 1980 (RFA) generally requires a description and analysis of Proposed Rules that will have significant economic impact on a substantial number of small entities.[87] The Small Business Administration's (SBA) Office of Size Standards develops the numerical definition of a small business.[88] The SBA revised its size standard for electric utilities (effective January 22, 2014) to a standard based on the number of employees, including affiliates (from the prior standard based on megawatt hour sales).[89] Proposed Reliability Standards CIP-003-6, CIP-004-6, CIP-006-6, CIP-007-6, CIP-009-6, CIP-010-2, and CIP-011-2 are expected to impose an additional

[87] 5 U.S.C. 601-12.

[88] 13 CFR 121.101.

[89] SBA Final Rule on "Small Business Size Standards: Utilities," 78 Fed. Reg. 77,343 (Dec. 23, 2013).

burden on 1,363 U.S. entities[90] (reliability coordinators, generator operators, generator owners, interchange coordinators or authorities, transmission operators, balancing authorities, transmission owners, and certain distribution providers).

93. Of the 1,363 affected entities discussed above, we estimate that 444 entities are small entities. We estimate that 399 of these 444 small entities do not own BES Cyber Assets or BES Cyber Systems that are classified as Medium or High Impact and, therefore, will only be affected by the proposed modifications to Reliability Standard CIP-003-6. As discussed above, proposed Reliability Standard CIP-003-6 enhances reliability by providing criteria against which NERC and the Commission can evaluate the sufficiency of an entity's protections for Low Impact BES Cyber Assets. We estimate that each of the 399 small entities to whom the proposed modifications to Reliability Standard CIP-003-6 applies will incur one-time costs of approximately $149,358 per entity to implement this standard, in addition to the ongoing paperwork burden reflected in the Information Collection Statement (a total of $40,736 per entity over Years 1-3), giving a total one-time cost of $190,094 per entity. We do not consider the estimated one-time costs for these 399 small entities a significant economic impact.

94. In addition, we estimate that 14 small entities own Medium Impact substations and that 31 small transmission operators own Medium or High impact control centers. These

---

[90] Public utilities may fall under one of several different categories, each with a size threshold based on the company's number of employees, including affiliates, the parent company, and subsidiaries. For the analysis in this NOPR, we are using a 500 employee threshold for each affected entity to conduct a comprehensive analysis.

45 small entities represent 10.1 percent of the 444 affected small entities. We estimate that each of these 45 small entities may experience an economic impact of $50,000 per entity in the first year of initial implementation to meet proposed Reliability Standard CIP-010-2 and $30,000 in ongoing annual costs.[91] In addition, those 45 small entities will have paperwork burden (reflected in the Information Collection Statement) of $81,472 per entity over Years 1-3. Therefore, we estimate that each of these 45 small entities will incur a total of $191,472 in costs over the first three years. We conclude that 10.1 percent of the total 444 affected small entities does not represent a substantial number in terms of the total number of regulated small entities.

95. Based on the above analysis, the Commission certifies that the proposed Reliability Standards will not have a significant economic impact on a substantial number of small entities. Accordingly, no regulatory flexibility analysis is required.

V. Environmental Analysis

96. The Commission is required to prepare an Environmental Assessment or an Environmental Impact Statement for any action that may have a significant adverse effect on the human environment.[92] The Commission has categorically excluded certain actions from this requirement as not having a significant effect on the human environment. Included in the exclusion are rules that are clarifying, corrective, or procedural or that do

---

[91] Estimated annual cost for year 2 and forward.

[92] *Regulations Implementing the National Environmental Policy Act of 1969*, Order No. 486, FERC Stats. & Regs. ¶ 30,783 (1987).

not substantially change the effect of the regulations being amended.[93] The actions proposed herein fall within this categorical exclusion in the Commission's regulations.

VI. Effective Date and Congressional Notification

97. This Final Rule is effective **[insert date 65 days after publication in the Federal Register]**. The Commission has determined, with the concurrence of the Administrator of the Office of Information and Regulatory Affairs of OMB, that this rule is a "major rule" as defined in section 351 of the Small Business Regulatory Enforcement Fairness Act of 1996. This Final Rule is being submitted to the Senate, House, and Government Accountability Office.

VII. Document Availability

98. In addition to publishing the full text of this document in the Federal Register, the Commission provides all interested persons an opportunity to view and/or print the contents of this document via the Internet through the Commission's Home Page (http://www.ferc.gov) and in the Commission's Public Reference Room during normal business hours (8:30 a.m. to 5:00 p.m. Eastern time) at 888 First Street, NE, Room 2A, Washington, DC 20426.

99. From the Commission's Home Page on the Internet, this information is available on eLibrary. The full text of this document is available on eLibrary in PDF and Microsoft Word format for viewing, printing, and/or downloading. To access this

---

[93] 18 CFR 380.4(a)(2)(ii).

document in eLibrary, type the docket number of this document, excluding the last three digits, in the docket number field.

100. User assistance is available for eLibrary and the Commission's website during normal business hours from the Commission's Online Support at (202) 502-6652 (toll free at 1-866-208-3676) or e-mail at ferconlinesupport@ferc.gov, or the Public Reference Room at (202) 502-8371, TTY (202) 502-8659. E-mail the Public Reference Room at public.referenceroom@ferc.gov.

By the Commission.

(S E A L)

Nathaniel J. Davis, Sr.,
Deputy Secretary.

Note: the following Appendix will not appear in the *Code of Federal Regulations*.

Appendix
Commenters

Abbreviation	**Commenter**
AEP	American Electric Power Service Corporation
ACS	Applied Control Solutions, LLC
APS	Arizona Public Service Company
Arkansas	Arkansas Electric Cooperative
BPA	Bonneville Power Administration
CEA	Canadian Electricity Association
Consumers Energy	Consumers Energy Company
CyberArk	CyberArk
EnergySec	Energy Sector Security Consortium, Inc
Ericsson	Ericsson
Foundation	Foundation for Resilient Societies
G&T Cooperatives	Associated Electric Cooperative, Inc., Basin Electric Power Cooperative, and Tri-State Generation and Transmission Association, Inc.
Gridwise	Gridwise Alliance
Idaho Power	Idaho Power Company
Indegy	Indegy
IESO	Independent Electricity System Operator
IRC	ISO/RTO Council
ISO New England	ISO New England Inc.
ITC	ITC Companies
Isologic	Isologic, LLC
KCP&L	Kansas City Power & Light Company and KCP&L Greater Missouri Operations Company
Luminant	Luminant Generation Company, LLC
NEMA	National Electrical Manufacturers Association
NERC	North American Electric Reliability Corporation
NextEra	NextEra Energy, Inc.
NIPSCO	Northern Indiana Public Service Co.
NWPPA	Northwest Public Power Association
Peak	Peak Reliability
PNM	PNM Resources
Reclamation	Department of Interior Bureau of Reclamation
SIA	Security Industry Association
SCE	Southern California Edison Company
Southern	Southern Company Services

SPP RE	Southwest Power Pool Regional Entity
SWP	California Department of Water Resources State Water Project
TVA	Tennessee Valley Authority
Trade Associations	Edison Electric Institute, American Public Power Association, National Rural Electric Cooperative Association, Electric Power Supply Association, Transmission Access Policy Study Group, and Large Public Power Council
UTC	Utilities Telecom Council
Waterfall	Waterfall Security Solutions, Ltd.
Wisconsin	Wisconsin Electric Power Company
Weis	Joe Weis

156 FERC ¶ 61,215
UNITED STATES OF AMERICA
FEDERAL ENERGY REGULATORY COMMISSION

18 CFR Part 40

[Docket No. RM15-11-000; Order No. 830]

Reliability Standard for Transmission System Planned Performance for Geomagnetic Disturbance Events

(Issued September 22, 2016)

AGENCY: Federal Energy Regulatory Commission.

ACTION: Final rule.

SUMMARY: The Federal Energy Regulatory Commission (Commission) approves Reliability Standard TPL-007-1 (Transmission System Planned Performance for Geomagnetic Disturbance Events). The North American Electric Reliability Corporation (NERC), the Commission-certified Electric Reliability Organization, submitted Reliability Standard TPL-007-1 for Commission approval in response to a Commission directive in Order No. 779. Reliability Standard TPL-007-1 establishes requirements for certain registered entities to assess the vulnerability of their transmission systems to geomagnetic disturbance events (GMDs), which occur when the sun ejects charged particles that interact with and cause changes in the earth's magnetic fields. Applicable entities that do not meet certain performance requirements, based on the results of their vulnerability assessments, must develop a plan to achieve the performance requirements. In addition, the Commission directs NERC to develop modifications to Reliability Standard TPL-007-1: (1) to modify the benchmark GMD event definition set forth in

Attachment 1 of Reliability Standard TPL-007-1, as it pertains to the required GMD Vulnerability Assessments and transformer thermal impact assessments, so that the definition is not based solely on spatially-averaged data; (2) to require the collection of necessary geomagnetically induced current monitoring and magnetometer data and to make such data publicly available; and (3) to include a one-year deadline for the development of corrective action plans and two and four-year deadlines to complete mitigation actions involving non-hardware and hardware mitigation, respectively. The Commission also directs NERC to submit a work plan and, subsequently, one or more informational filings that address specific GMD-related research areas.

EFFECTIVE DATE: This rule will become effective **[INSERT DATE 60 days after publication in the FEDERAL REGISTER]**.

FOR FURTHER INFORMATION CONTACT:

Regis Binder (Technical Information)
Office of Electric Reliability
Federal Energy Regulatory Commission
888 First Street, NE
Washington, DC 20426
Telephone: (301) 665-1601
Regis.Binder@ferc.gov

Matthew Vlissides (Legal Information)
Office of the General Counsel
Federal Energy Regulatory Commission
888 First Street, NE
Washington, DC 20426
Telephone: (202) 502 -8408
Matthew.Vlissides@ferc.gov

SUPPLEMENTARY INFORMATION:

156 FERC ¶ 61,215
UNITED STATES OF AMERICA
FEDERAL ENERGY REGULATORY COMMISSION

Before Commissioners: Norman C. Bay, Chairman;
Cheryl A. LaFleur, Tony Clark,
and Colette D. Honorable.

Reliability Standard for Transmission System Planned Performance for Geomagnetic Disturbance Events — Docket No. RM15-11-000

ORDER NO. 830

FINAL RULE

(Issued September 22, 2016)

1. Pursuant to section 215 of the Federal Power Act (FPA), the Commission approves Reliability Standard TPL-007-1 (Transmission System Planned Performance for Geomagnetic Disturbance Events).[1] The North American Electric Reliability Corporation (NERC), the Commission-certified Electric Reliability Organization (ERO), submitted Reliability Standard TPL-007-1 for Commission approval in response to a Commission directive in Order No. 779.[2] Reliability Standard TPL-007-1 establishes requirements for certain registered entities to assess the vulnerability of their transmission systems to geomagnetic disturbance events (GMDs), which occur when the sun ejects

---

[1] 16 U.S.C. 824o.

[2] *Reliability Standards for Geomagnetic Disturbances*, Order No. 779, 78 FR 30,747 (May 23, 2013), 143 FERC ¶ 61,147, *reh'g denied*, 144 FERC ¶ 61,113 (2013).

charged particles that interact with and cause changes in the earth's magnetic fields. Reliability Standard TPL-007-1 requires applicable entities that do not meet certain performance requirements, based on the results of their vulnerability assessments, to develop a plan to achieve the requirements. Reliability Standard TPL-007-1 addresses the directives in Order No. 779 by requiring applicable Bulk-Power System owners and operators to conduct initial and on-going vulnerability assessments regarding the potential impact of a benchmark GMD event on the Bulk-Power System as a whole and on Bulk-Power System components.[3] In addition, Reliability Standard TPL-007-1 requires applicable entities to develop and implement corrective action plans to mitigate identified vulnerabilities.[4] Potential mitigation strategies identified in the proposed Reliability Standard include, but are not limited to, the installation, modification or removal of transmission and generation facilities and associated equipment.[5] Accordingly, Reliability Standard TPL-007-1 constitutes an important step in addressing the risks posed by GMD events to the Bulk-Power System.

2. In addition, pursuant to section 215(d)(5) of the FPA, the Commission directs NERC to develop modifications to Reliability Standard TPL-007-1: (1) to revise the

---

[3] *See* Reliability Standard TPL-007-1, Requirement R4; *see also* Order No. 779, 143 FERC ¶ 61,147 at PP 67, 71.

[4] *See* Reliability Standard TPL-007-1, Requirement R7; *see also* Order No. 779, 143 FERC ¶ 61,147 at P 79.

[5] *See* Reliability Standard TPL-007-1, Requirement R7.

benchmark GMD event definition set forth in Attachment 1 of Reliability Standard TPL-007-1, as it pertains to the required GMD Vulnerability Assessments and transformer thermal impact assessments, so that the definition is not based solely on spatially-averaged data; (2) to require the collection of necessary geomagnetically induced current (GIC) monitoring and magnetometer data and to make such data publicly available; and (3) to include a one-year deadline for the completion of corrective action plans and two- and four-year deadlines to complete mitigation actions involving non-hardware and hardware mitigation, respectively.[6] The Commission directs NERC to submit these revisions within 18 months of the effective date of this Final Rule. The Commission also directs NERC to submit a work plan (GMD research work plan) within six months of the effective date of this Final Rule and, subsequently, one or more informational filings that address specific GMD-related research areas.

I. Background

A. Section 215 and Mandatory Reliability Standards

3. Section 215 of the FPA requires the Commission to certify an ERO to develop mandatory and enforceable Reliability Standards, subject to Commission review and approval. Once approved, the Reliability Standards may be enforced in the United States by the ERO, subject to Commission oversight, or by the Commission independently.[7]

---

[6] 16 U.S.C. 824o(d)(5).

[7] *Id.* 824o(e).

B. GMD Primer

4. GMD events occur when the sun ejects charged particles that interact with and cause changes in the earth's magnetic fields.[8] Once a solar particle is ejected, it can take between 17 to 96 hours (depending on its energy level) to reach earth.[9] A geoelectric field is the electric potential (measured in volts per kilometer (V/km)) on the earth's surface and is directly related to the rate of change of the magnetic fields.[10] A geoelectric field has an amplitude and direction and acts as a voltage source that can cause GICs to flow on long conductors, such as transmission lines.[11] The magnitude of the geoelectric field amplitude is impacted by local factors such as geomagnetic latitude and local earth conductivity.[12] Geomagnetic latitude is the proximity to earth's magnetic north and south poles, as opposed to earth's geographic poles. Local earth conductivity is the ability of the earth's crust to conduct electricity at a certain location to depths of hundreds of kilometers down to the earth's mantle. Local earth conductivity impacts the magnitude

---

[8] North American Electric Reliability Corp., 2012 Special Reliability Assessment Interim Report: Effects of Geomagnetic Disturbances on the Bulk Power System at i-ii (February 2012), http://www.nerc.com/files/2012GMD.pdf (GMD Interim Report).

[9] *Id.* ii.

[10] *Id.*

[11] *Id.*

[12] NERC Petition, Ex. D (White Paper on GMD Benchmark Event Description) at 4.

(i.e., severity) of the geoelectric fields that are formed during a GMD event by, all else being equal, a lower earth conductivity resulting in higher geoelectric fields.[13]

C. Order No. 779

5. In Order No. 779, the Commission directed NERC, pursuant to section 215(d)(5) of the FPA, to develop and submit for approval proposed Reliability Standards that address the impact of geomagnetic disturbances on the reliable operation of the Bulk-Power System. The Commission based its directive on the potentially severe, wide-spread impact on the reliable operation of the Bulk-Power System that can be caused by GMD events and the absence of existing Reliability Standards to address GMD events.[14]

6. Order No. 779 directed NERC to implement the directive in two stages. In the first stage, the Commission directed NERC to submit, within six months of the effective date of Order No. 779, one or more Reliability Standards (First Stage GMD Reliability Standards) that require owners and operators of the Bulk-Power System to develop and implement operational procedures to mitigate the effects of GMDs consistent with the reliable operation of the Bulk-Power System.[15]

7. In the second stage, the Commission directed NERC to submit, within 18 months of the effective date of Order No. 779, one or more Reliability Standards (Second Stage

---

[13] *Id.*

[14] Order No. 779, 143 FERC ¶ 61,147 at P 3.

[15] *Id.* P 2.

GMD Reliability Standards) that require owners and operators of the Bulk-Power System to conduct initial and on-going assessments of the potential impact of benchmark GMD events on Bulk-Power System equipment and the Bulk-Power System as a whole. The Commission directed that the Second Stage GMD Reliability Standards must identify benchmark GMD events that specify what severity of GMD events a responsible entity must assess for potential impacts on the Bulk-Power System.[16] Order No. 779 explained that if the assessments identified potential impacts from benchmark GMD events, the Reliability Standards should require owners and operators to develop and implement a plan to protect against instability, uncontrolled separation, or cascading failures of the Bulk-Power System, caused by damage to critical or vulnerable Bulk-Power System equipment, or otherwise, as a result of a benchmark GMD event. The Commission directed that the development of this plan could not be limited to considering operational procedures or enhanced training alone but should, subject to the potential impacts of the benchmark GMD events identified in the assessments, contain strategies for protecting against the potential impact of GMDs based on factors such as the age, condition, technical specifications, system configuration or location of specific equipment.[17] Order No. 779 observed that these strategies could, for example, include automatically blocking GICs from entering the Bulk-Power System, instituting specification requirements for

---

[16] *Id.*

[17] *Id.*

new equipment, inventory management, isolating certain equipment that is not cost effective to retrofit or a combination thereof.

D. Order No. 797

8. In Order No. 797, the Commission approved Reliability Standard EOP-010-1 (Geomagnetic Disturbance Operations).[18] NERC submitted Reliability Standard EOP-010-1 for Commission approval in compliance with the Commission's directive in Order No. 779 corresponding to the First Stage GMD Reliability Standards. In Order No. 797-A, the Commission denied the Foundation for Resilient Societies' (Resilient Societies) request for rehearing of Order No. 797. The Commission stated that the rehearing request "addressed a later stage of efforts on geomagnetic disturbances (i.e., NERC's future filing of Second Stage GMD Reliability Standards) and [that Resilient Societies] may seek to present those arguments at an appropriate time in response to that filing."[19] In particular, the Commission stated that GIC monitoring requirements should be addressed in the Second Stage GMD Reliability Standards.[20]

---

[18] *Reliability Standard for Geomagnetic Disturbance Operations*, Order No. 797, 79 FR 35,911 (June 25, 2014), 147 FERC ¶ 61,209, *reh'g denied*, Order No. 797-A, 149 FERC ¶ 61,027 (2014).

[19] Order No. 797-A, 149 FERC ¶ 61,027 at P 2.

[20] *Id.* P 27 (stating that the Commission continues "to encourage NERC to address the collection, dissemination, and use of geomagnetic induced current data, by NERC, industry or others, in the Second Stage GMD Reliability Standards because such efforts could be useful in the development of GMD mitigation methods or to validate GMD models").

E. NERC Petition and Reliability Standard TPL-007-1

9. On January 21, 2015, NERC petitioned the Commission to approve Reliability Standard TPL-007-1 and its associated violation risk factors and violation severity levels, implementation plan, and effective dates.[21] NERC also submitted a proposed definition for the term "Geomagnetic Disturbance Vulnerability Assessment or GMD Vulnerability Assessment" for inclusion in the NERC Glossary of Terms (NERC Glossary). NERC maintains that Reliability Standard TPL-007-1 is just, reasonable, not unduly discriminatory or preferential and in the public interest. NERC further contends that Reliability Standard TPL-007-1 satisfies the directive in Order No. 779 corresponding to the Second Stage GMD Reliability Standards.

10. NERC states that Reliability Standard TPL-007-1 applies to planning coordinators, transmission planners, transmission owners and generation owners who own or whose planning coordinator area or transmission planning area includes a power transformer with a high side, wye-grounded winding connected at 200 kV or higher.[22] NERC explains that the applicability criteria for qualifying transformers in Reliability Standard

---

[21] Reliability Standard TPL-007-1 is not attached to this final rule. Reliability Standard TPL-007-1 is available on the Commission's eLibrary document retrieval system in Docket No. RM15-11-000 and on the NERC website, www.nerc.com. NERC submitted an errata on February 2, 2015 containing a corrected version of Exhibit A (Proposed Reliability Standard TPL-007-1).

[22] A power transformer with a "high side wye-grounded winding" refers to a power transformer with windings on the high voltage side that are connected in a wye configuration and have a grounded neutral connection. NERC Petition at 13 n.32.

TPL-007-1 are the same as that for the First Stage GMD Reliability Standard in Reliability Standard EOP-010-1, which the Commission approved in Order No. 797.

11. Reliability Standard TPL-007-1 contains seven requirements. Requirement R1 requires planning coordinators and transmission planners to determine the individual and joint responsibilities in the planning coordinator's planning area for maintaining models and performing studies needed to complete the GMD Vulnerability Assessment required in Requirement R4.

12. Requirement R2 requires planning coordinators and transmission planners to maintain system models and GIC system models needed to complete the GMD Vulnerability Assessment required in Requirement R4.

13. Requirement R3 requires planning coordinators and transmission planners to have criteria for acceptable system steady state voltage limits for their systems during the benchmark GMD event described in Attachment 1 (Calculating Geoelectric Fields for the Benchmark GMD Event).

14. Requirement R4 requires planning coordinators and transmission planners to conduct a GMD Vulnerability Assessment every 60 months using the benchmark GMD event described in Attachment 1 to Reliability Standard TPL-007-1. The benchmark GMD event is based on a 1-in-100 year frequency of occurrence and is composed of four elements: (1) a reference peak geoelectric field amplitude of 8 V/km derived from statistical analysis of historical magnetometer data; (2) a scaling factor to account for local geomagnetic latitude; (3) a scaling factor to account for local earth conductivity; and (4) a reference geomagnetic field time series or wave shape to facilitate time-domain

analysis of GMD impact on equipment.[23] The product of the first three elements is referred to as the regional geoelectric field peak amplitude.[24]

15. Requirement R5 requires planning coordinators and transmission planners to provide GIC flow information, to be used in the transformer thermal impact assessment required in Requirement R6, to each transmission owner and generator owner that owns an applicable transformer within the applicable planning area.

16. Requirement R6 requires transmission owners and generator owners to conduct thermal impact assessments on solely and jointly owned applicable transformers where the maximum effective GIC value provided in Requirement R5 is 75 amperes per phase (A/phase) or greater.

17. Requirement R7 requires planning coordinators and transmission planners to develop corrective action plans if the GMD Vulnerability Assessment concludes that the system does not meet the performance requirements in Table 1 (Steady State Planning Events).

---

[23] *See* Reliability Standard TPL-007-1, Att. 1; s*ee also* NERC Petition, Ex. D (White Paper on GMD Benchmark Event Description) at 5.

[24] NERC Petition, Ex. D (White Paper on GMD Benchmark Event Description) at 5.

F. Notice of Proposed Rulemaking

18. On May 14, 2015, the Commission issued a notice of proposed rulemaking (NOPR) proposing to approve Reliability Standard TPL-007-1.[25] In addition, the Commission proposed to direct that NERC develop three modifications to Reliability Standard TPL-007-1. First, the Commission proposed to direct NERC to revise the benchmark GMD event definition in Reliability Standard TPL-007-1 so that the definition is not based solely on spatially-averaged data. Second, the Commission proposed to direct NERC to revise Reliability Standard TPL-007-1 to require the installation of GIC monitors and magnetometers where necessary. Third, the Commission proposed to direct NERC to revise Reliability Standard TPL-007-1 to require corrective action plans (Requirement R7) to be developed within one year and, with respect to the mitigation actions called for in the corrective action plans, non-hardware mitigation actions to be completed within two years of finishing development of the corrective action plan and hardware mitigation to be completed within four years. The NOPR also proposed to direct NERC to submit a work plan and, subsequently, one or more informational filings that address specific GMD-related research areas and sought comment on certain issues relating to the transformer thermal impact assessments

---

[25] *Reliability Standard for Transmission System Planned Performance for Geomagnetic Disturbance Events*, Notice of Proposed Rulemaking, 80 FR 29,990 (May 26, 2015), 151 FERC ¶ 61,134 (2015) (NOPR).

(Requirement R6) and the meaning of language in Table 1 of Reliability Standard TPL-007-1.

19. On August 20, 2015 and October 2, 2015, the Commission issued notices setting supplemental comment periods regarding specific documents. On March 1, 2016, Commission staff led a technical conference on Reliability Standard TPL-007-1 and issues raised in the NOPR.[26]

20. On April 28, 2016, NERC made a filing notifying the Commission that "NERC identified new information that may necessitate a minor revision to a figure in one of the supporting technical white papers. This revision would not require a change to any of the Requirements of the proposed Reliability Standard."[27] On June 28, 2016, NERC submitted the revised technical white papers referenced in the April 28, 2016 filing. On June 29, 2016, the Commission issued a notice setting a supplemental comment period regarding the revised technical white papers submitted by NERC on June 28, 2016.

21. In response to the NOPR and subsequent notices, 28 entities filed initial and supplemental comments. We address below the issues raised in the NOPR and comments. The Appendix to this Final Rule lists the entities that filed comments in response to the NOPR and in response to the supplemental comment period notices.

---

[26] Written presentations at the March 1, 2016 Technical Conference and the Technical Conference transcript referenced in this Final Rule are accessible through the Commission's eLibrary document retrieval system in Docket No. RM15-11-000.

[27] NERC April 28, 2016 Filing at 1.

II. Discussion

22. Pursuant to section 215(d) of the FPA, the Commission approves Reliability Standard TPL-007-1 as just, reasonable, not unduly discriminatory or preferential and in the public interest. While we recognize that scientific and operational research regarding GMD is ongoing, we believe that the potential threat to the bulk electric system warrants Commission action at this time, including efforts to conduct critical GMD research and update Reliability Standard TPL-007-1 as appropriate.

23. First, we find that Reliability Standard TPL-007-1 addresses the directives in Order No. 779 corresponding to the development of the Second Stage GMD Reliability Standards. Reliability Standard TPL-007-1 does this by requiring applicable Bulk-Power System owners and operators to conduct, on a recurring five-year cycle,[28] initial and on-going vulnerability assessments regarding the potential impact of a benchmark GMD event on the Bulk-Power System as a whole and on Bulk-Power System components.[29] In addition, Reliability Standard TPL-007-1 requires applicable entities to develop and implement corrective action plans to mitigate vulnerabilities identified through those recurring vulnerability assessments.[30] Potential mitigation strategies identified in the

---

[28] A detailed explanation of the five-year GMD Vulnerability Assessment and mitigation cycle is provided in paragraph 103, *infra*.

[29] *See* Reliability Standard TPL-007-1, Requirement R4; *see also* Order No. 779, 143 FERC ¶ 61,147 at PP 67, 71.

[30] *See* Reliability Standard TPL-007-1, Requirement R7; *see also* Order No. 779, 143 FERC ¶ 61,147 at P 79.

proposed Reliability Standard include, but are not limited to, the installation, modification or removal of transmission and generation facilities and associated equipment.[31] Accordingly, Reliability Standard TPL-007-1 constitutes an important step in addressing the risks posed by GMD events to the Bulk-Power System.

24. The Commission also approves the inclusion of the term "Geomagnetic Disturbance Vulnerability Assessment or GMD Vulnerability Assessment" in the NERC Glossary; Reliability Standard TPL-007-1's associated violation risk factors and violation severity levels; and NERC's proposed implementation plan and effective dates. The Commission also affirms, as raised for comment in the NOPR, that cost recovery for prudent costs associated with or incurred to comply with Reliability Standard TPL-007-1 and future revisions to the Reliability Standard will be available to registered entities.[32]

25. While we conclude that Reliability Standard TPL-007-1 satisfies the directives in Order No. 779, based on the record developed in this proceeding, the Commission determines that Reliability Standard TPL-007-1 should be modified to reflect the new information and analyses discussed below, as proposed in the NOPR. Accordingly, pursuant to section 215(d)(5) of the FPA, the Commission directs NERC to develop and submit modifications to Reliability Standard TPL-007-1 concerning: (1) the calculation of the reference peak geoelectric field amplitude component of the benchmark GMD

---

[31] *See* Reliability Standard TPL-007-1, Requirement R7.

[32] NOPR, 151 FERC ¶ 61,134 at P 49 n.60.

event definition; (2) the collection and public availability of necessary GIC monitoring and magnetometer data; and (3) deadlines for completing corrective action plans and the mitigation measures called for in corrective action plans. The Commission directs NERC to develop and submit these revisions for Commission approval within 18 months of the effective date of this Final Rule.

26. Furthermore, to improve the understanding of GMD events generally, the Commission directs NERC to submit within six months from the effective date of this Final Rule a GMD research work plan.[33] Specifically, we direct NERC to: (1) further analyze the area over which spatial averaging should be calculated for stability studies, including performing sensitivity analyses on squares less than 500 km per side (e.g., 100 km, 200 km); (2) further analyze earth conductivity models by, for example, using metered GIC and magnetometer readings to calculate earth conductivity and using 3-D readings; (3) determine whether new analyses and observations support modifying the use of single station readings around the earth to adjust the spatially averaged benchmark for latitude; (4) research, as discussed below, aspects of the required thermal impact assessments; and (5) in NERC's discretion, conduct any GMD-related research areas generally that may impact the development of new or modified GMD Reliability Standards. We expect that work completed through the GMD research work plan, as

[33] Following submission of the GMD research work plan, the Commission will notice the filing for public comment and issue an order addressing its proposed content and schedule.

well as other analyses facilitated by the increased collection and availability of GIC monitoring and magnetometer data directed herein, will lead to further modifications to Reliability Standard TPL-007-1 as our collective understanding of the threats posed by GMD events improves.

27. Below we discuss the following issues raised in the NOPR and NOPR comments: (1) the benchmark GMD event definition described in Reliability Standard TPL-007-1, Attachment 1 (Calculating Geoelectric Fields for the Benchmark GMD Event); (2) transformer thermal impact assessments in Requirement R6; (3) GMD research work plan; (4) collection and public availability of GIC monitoring and magnetometer data; (5) completion of corrective action plans in Requirement R7; (6) meaning of "minimized" in Table 1 (Steady State Planning Events) of Reliability Standard TPL-007-1; (7) NERC's proposed implementation plan and effective dates; and (8) other issues.

A. Benchmark GMD Event Definition

NERC Petition

28. NERC states that the purpose of the benchmark GMD event is to "provide a defined event for assessing system performance during a low probability, high magnitude GMD event."[34] NERC explains that the benchmark GMD event represents "the most severe GMD event expected in a 100-year period as determined by a statistical analysis

---

[34] NERC Petition at 15.

of recorded geomagnetic data."[35] The benchmark GMD event definition is used in the GMD Vulnerability Assessments and thermal impact assessment requirements of Reliability Standard TPL-007-1 (Requirements R4 and R6).

29. As noted above, NERC states that the benchmark GMD event definition has four elements: (1) a reference peak geoelectric field amplitude of 8 V/km derived from statistical analysis of historical magnetometer data; (2) a scaling factor to account for local geomagnetic latitude; (3) a scaling factor to account for local earth conductivity; and (4) a reference geomagnetic field time series or wave shape to facilitate time-domain analysis of GMD impact on equipment.[36]

30. The standard drafting team determined that a 1-in-100 year GMD event would cause an 8 V/km reference peak geoelectric field amplitude at 60 degree geomagnetic latitude using Québec's earth conductivity.[37] The standard drafting team stated that:

> the reference geoelectric field amplitude was determined through statistical analysis using … field measurements from geomagnetic observatories in northern Europe and the reference (Quebec) earth model …. The Quebec earth model is generally resistive and the geological structure is relatively well understood. The statistical analysis resulted in a conservative peak geoelectric field amplitude of approximately 8 V/km …. The frequency of occurrence of this benchmark GMD event is estimated to be approximately 1 in 100 years.[38]

---

[35] *Id.*

[36] NERC Petition, Ex. D (White Paper on GMD Benchmark Event Description) at 5.

[37] *Id.*

[38] *Id.* (footnotes omitted).

31. The standard drafting team explained that it used field measurements taken from the IMAGE magnetometer chain, which covers Northern Europe, for the period 1993-2013 to calculate the reference peak geoelectric field amplitude used in the benchmark GMD event definition.[39] As described in NERC's petition, the standard drafting team "spatially averaged" four different station groups of IMAGE data, each spanning a square area of approximately 500 km (roughly 310 miles) in width.[40] The standard drafting team justified the use of spatial averaging by stating that Reliability Standard TPL-007-1 is designed to "address wide-area effects caused by a severe GMD event, such as increased var absorption and voltage depressions. Without characterizing GMD on regional scales, statistical estimates could be weighted by local effects and suggest unduly pessimistic conditions when considering cascading failure and voltage collapse."[41]

32. NERC states that the benchmark GMD event includes scaling factors to enable applicable entities to tailor the reference peak geoelectric field to their specific location

[39] *Id.* at 8. The International Monitor for Auroral Geomagnetic Effects (IMAGE) consists of 31 magnetometer stations in northern Europe maintained by 10 institutes from Estonia, Finland, Germany, Norway, Poland, Russia, and Sweden. *See* IMAGE website, http://space.fmi.fi/image/beta/?page=home#.

[40] As applied by the standard drafting team, spatial averaging refers to the averaging of geoelectric field amplitude readings within a given area. NERC Petition, Ex. D (White Paper on GMD Benchmark Event Description) at 9.

[41] NERC Petition, Ex. D (White Paper on GMD Benchmark Event Description) at 9.

for conducting GMD Vulnerability Assessments. NERC explains that the scaling factors in the benchmark GMD event definition are applied to the reference peak geoelectric field amplitude to adjust the 8 V/km value for different geomagnetic latitudes and earth conductivities.[42]

33. The standard drafting team also identified a reference geomagnetic field time series from an Ottawa magnetic observatory during a 1989 GMD event that affected Québec.[43] The standard drafting team used this time series to estimate a geoelectric field, represented as a time series (i.e., 10-second values over a period of days), that is expected to occur at 60 degree geomagnetic latitude during a 1-in-100 year GMD event. NERC explains that this time series is used to facilitate time-domain analysis of GMD impacts on equipment.[44]

34. In the sub-sections below, we discuss two issues concerning the benchmark GMD event definition addressed in the NOPR: (1) reference peak geoelectric field amplitude; and (2) geomagnetic latitude scaling factor.

---

[42] NERC Petition at 18-19.

[43] NERC Petition, Ex. D (White Paper on GMD Benchmark Event Description) at 5-6, 15-16 ("the reference geomagnetic field waveshape was selected after analyzing a number of recorded GMD events… the March 13-14, 1989 GMD event, measured at NRCan's Ottawa geomagnetic observatory, was selected as the reference geomagnetic field waveform because it provides generally conservative results when performing thermal analysis of power transformers").

[44] *Id.* at 5-6.

1. Reference Peak Geoelectric Field Amplitude

NOPR

35. The NOPR proposed to approve the benchmark GMD event definition. The NOPR stated that the "benchmark GMD event definition proposed by NERC complies with the directive in Order No. 779 … [c]onsistent with the guidance provided in Order No. 779, the benchmark GMD event definition proposed by NERC addresses the potential widespread impact of a severe GMD event, while taking into consideration the variables of geomagnetic latitude and local earth conductivity."[45]

36. In addition, the NOPR proposed to direct NERC to develop modifications to Reliability Standard TPL-007-1. Specifically, the NOPR proposed to direct NERC to modify the reference peak geoelectric field amplitude component of the benchmark GMD event definition so that it is not calculated based solely on spatially-averaged data. The NOPR explained that this could be achieved, for example, by requiring applicable entities to conduct GMD Vulnerability Assessments (and, as discussed below, thermal impact assessments) using two different benchmark GMD events: the first benchmark GMD event using the spatially-averaged reference peak geoelectric field value (8 V/km) and the second using the non-spatially averaged peak geoelectric field value cited in the GMD Interim Report (20 V/km). The NOPR stated that the revised Reliability Standard could then require applicable entities to take corrective actions, using engineering judgment,

[45] NOPR, 151 FERC ¶ 61,134 at P 32.

based on the results of both assessments. The NOPR explained that applicable entities would not always be required to mitigate to the level of risk identified by the non-spatially averaged analysis; instead, the selection of mitigation would reflect the range of risks bounded by the two analyses, and be based on engineering judgment within this range, considering all relevant information. The NOPR stated that, alternatively, NERC could propose an equally efficient and effective modification that does not rely exclusively on the spatially-averaged reference peak geoelectric field value.

Comments

37. NERC does not support revising the benchmark GMD event definition. NERC maintains that the spatially-averaged reference peak geoelectric field amplitude value in Reliability Standard TPL-007-1 is "technically-justified, scientifically sound, and has been published in a peer-reviewed research journal covering geomagnetism and other topics."[46] NERC contends that the standard drafting team determined that using the non-spatially averaged 20 V/km figure in the GMD Interim Report would "consistently overestimate the geoelectric field of a 1-in-100 year GMD event."[47] NERC states that, by contrast, spatial averaging "properly associates the relevant spatial scales for the analyzed and applied geoelectric fields and would not distort the complexity of the potential

---

[46] NERC Comments at 6.

[47] *Id.* at 7.

impacts of a GMD event."[48] NERC claims that the 500 km-wide square areas used to determine the areas of spatial averaging are "based on consideration of transmission systems and geomagnetic observation patterns … [and are] an appropriate scale for a system-wide impact in a transmission system."[49] To support this position, NERC cites a June 2015 peer-reviewed publication authored in part by some members of the standard drafting team.[50]

38. Industry commenters, largely represented by the Trade Associations' comments, do not support revising the benchmark GMD event definition.[51] The Trade Associations' reasons largely mirror NERC's. While recognizing that the spatially-averaged reference peak geoelectric field amplitude is lower than the non-spatially averaged figure, the Trade Associations contend that the non-spatially averaged value is inappropriate because: (1) the peak geoelectric field only affects relatively small areas and quickly

[48] *Id.* at 8.

[49] *Id.*

[50] *See* Pulkkinen, A., Bernabeu, E., Eichner, J., Viljanen, A., Ngwira, C., "Regional-Scale High-Latitude Extreme Geoelectric Fields Pertaining to Geomagnetically Induced Currents," Earth, Planets and Space (June 19, 2015) (2015 Pulkkinen Paper).

[51] Trade Associations Comments at 13-18. AEP, APS, ATC, BPA, CEA, Hydro One, ITC, Joint ISOs/RTOs and Exelon indicated that they do not support the NOPR proposal in separate comments and/or by joining the Trade Associations' comments. *See* AEP Comments at 3; APS Comments at 2; ATC Comments at 3; BPA Comments at 3-4; CEA Comments at 8-13; Hydro One Comments 1-2; ITC Comments at 3-5; Joint ISOs/RTOs Comments at 4-5; Exelon Comments at 2.

declines with distance from the peak; (2) Reliability Standard TPL-007-1 is intended to address the wide-scale effects of a GMD event; and (3) the benchmark GMD event definition is designed to provide a realistic estimate of wide-area effects caused by a severe GMD event. The Trade Associations contend that a non-spatially averaged reference peak geoelectric field amplitude "would be weighted by local effects and suggest unrealistic conditions for system analysis … [which] could lead to unnecessary costs for customers, while yielding very little tangible benefit to reliability."[52] Like NERC, the Trade Associations cite to the 2015 Pulkkinen Paper to support the use of 500 km-wide squares in performing the spatial averaging analysis. The Trade Associations note, however, that the selection of 500 km is "only the beginning … [of the] exploration of spatial geoelectric field structures pertaining to extreme GIC."[53]

39. The Trade Associations, while not supportive of the NOPR proposal, recommend that if the Commission remains concerned about relying on NERC's proposed spatially-averaged reference peak geoelectric field amplitude, the Commission should:

> allow NERC to further determine the appropriate localized studies to be performed by moving the "local hot spot" around a planning area. This approach may better ensure that the peak values only impact a local area instead of unrealistically projecting uniform peak values over a broad area. This approach also should better align with the Commission's concerns because this type of study would more accurately reflect the real-world impact of a GMD event on the [Bulk-Power System]. The Trade

---

[52] Trade Associations Comments at 15.

[53] *Id.* at 17 (quoting 2015 Pulkkinen Paper at 6).

Associations understand that existing planning tools may not yet have such capabilities, but the tools can be modified to allow such study.[54]

40. Industry commenters raise other concerns with the NOPR proposal. CEA states that it would be inappropriate to rely on the non-spatially averaged 20 V/km reference peak geoelectric field figure because that figure is found in a single publication. CEA also contends that it is impractical to use "engineering judgment" to weigh the GMD Vulnerability Assessments using the spatially-averaged and non-spatially averaged reference peak geoelectric field amplitudes, as described in the NOPR.[55] ITC states that NERC's proposal is reasonable and that the reference peak geoelectric field amplitude value can be revised periodically based on new information. Joint ISOs/RTOs state that the Commission should afford due weight to NERC's technical expertise.

41. A September 2015 paper prepared by the Los Alamos National Laboratory states that it analyzed the IMAGE data using a different methodology to calculate reference peak geoelectric field amplitude values based on each of eight different magnetometer installations in Northern Europe. However, unlike the standard drafting team, the Los Alamos Paper did not spatially average the IMAGE data. The authors calculated peak geoelectric field amplitudes ranging from 8.4 V/km to 16.6 V/km, with a mean of

---

[54] *Id.* at 16.

[55] *See also* Hydro One Comments at 1-2; Resilient Societies Comments at 24-25.

the eight values equal to 13.2 V/km.[56] The authors used a statistical formula and probability distribution to determine their 1-in-100 year GMD event parameters, as opposed to the 20 V/km non-spatially averaged event from the 2012 paper cited in the GMD Interim Report that visually extrapolated the data.

42. Roodman contends that "NERC's 100-year benchmark GMD event is appropriately conservative in magnitude (except perhaps in the southern-most US) if unrealistic in some other respects."[57] Roodman states that "overall NERC's analytical frame does not strongly clash with the data."[58] However, Roodman contends that actual data support local hot-spots in a larger region of lower magnitude geoelectric fields that are not typically uniform in magnitude or direction.[59] Roodman addresses comments by Kappenman against the benchmark GMD event by stating that the Oak Ridge Report's Meta-R-319 study, authored by Kappenman, modeled a 1-in-100 year GMD event based largely on misunderstandings of historic GMDs, both in magnitude and geographic

---

[56] Rivera, M., Backhaus, S., "Review of the GMD Benchmark Event in TPL-007-1," Los Alamos National Laboratory (September 2015) (Los Alamos Paper).

[57] Roodman Comments at 4. Roodman criticizes the proposed benchmark GMD event definition because it assumes that the induced electrical field resulting from a GMD event is spatially uniform. Roodman also contends that a GMD event that is less than a 1-in-100 year storm could potentially damage transformers. *Id.* at 12-14.

[58] Roodman Comments at 9.

[59] *Id.* at 10, 12-13.

footprint.[60] Roodman recommends that the Commission "require a much larger array of events for simulation" in light of the "deep uncertainty and complexity of the GMD."[61]

43. Commenters opposed to the benchmark GMD event definition proposed by NERC maintain that the standard drafting team significantly underestimated the reference peak geoelectric field amplitude value for a 1-in-100 year GMD event by relying on data from the IMAGE system and by applying spatial averaging to that data set.[62] For example, Resilient Societies states that the standard drafting team should have analyzed "real-world data from within the United States and Canada, including magnetometer readings from the [USGS] and Natural Resources Canada observatories … [h]ad NERC and the Standard Drafting Team collected and analyzed available real-world data, they would have likely found that the severity of GMD in 1-in-100 Year reference storm had been set far below a technically justified level and without a 'strong technical basis.'"[63] Likewise, Kappenman contends that there are multiple examples where the benchmark GMD event and the standard drafting team's model for calculating geoelectric fields under-predict

[60] *Id.* at 5-6 (citing Oak Ridge National Laboratory, Geomagnetic Storms and Their Impacts on the U.S. Power Grid: Meta-R-319 at pages I-1 to I-3 (January 2010), http://www.ornl.gov/sci/ees/etsd/pes/pubs/ferc_Meta-R-319.pdf (Meta-R-319 Study).

[61] *Id.* at 15.

[62] *See*, *e.g.*, JINSA Comments at 2; Emprimus Comments at 1. *See also* Gaunt Comments at 9 (indicating that the proposed benchmark GMD event definition may underestimate the effects of a 1-in-100 GMD event).

[63] Resilient Societies Comments at 20-21.

actual, historical GIC readings.[64] Commenters opposed to NERC's proposal variously argue that the reference peak geoelectric field amplitude should be set at a level commensurate with the 1921 Railroad Storm or 1859 Carrington Event or at the 20 V/km level cited in the GMD Interim Report.[65]

Commission Determination

44. The Commission approves the reference peak geoelectric field amplitude figure proposed by NERC. In addition, the Commission, as proposed in the NOPR, directs NERC to develop revisions to the benchmark GMD event definition so that the reference peak geoelectric field amplitude component is not based solely on spatially-averaged data. The Commission directs NERC to submit this revision within 18 months of the effective date of this Final Rule.

45. NERC and industry comments do not contain new information to support relying solely on spatially-averaged data to calculate the reference peak geoelectric field amplitude in the benchmark GMD event definition. The 2015 Pulkkinen Paper contains the same justifications for spatial averaging as those presented in NERC's petition. In addition, the 2015 Pulkkinen Paper validates the NOPR's concerns with relying solely on

---

[64] Kappenman Comments at 15- 29.

[65] *See, e.g.*, EIS Comments at 2 (advocating use of 20 V/km); Gaunt Comments at 6-9 (contending that NERC's proposed figure results in a "possible underestimation of the effects of GICs" without suggesting an alternative figure); JINSA Comments at 2 (advocating use of 20 V/km); Emprimus Comments at 1 (advocating use of 20 V/km); Briggs Comments at 1 (advocating that the benchmark GMD event should be a "Carrington Class solar superstorm").

spatial averaging generally and with the method used by the standard drafting team to spatially average the IMAGE data specifically. The 2015 Pulkkinen Paper, for example, states that "regional scale geoelectric fields have not been considered earlier from the statistical and extreme analyses standpoint" and "selection of an area of 500 km [for spatial averaging] …[is] subjective."[66] Further, the 2015 Pulkkinen Paper notes that "we emphasize that the work described in this paper is only the beginning in our exploration of spatial geoelectric field structures pertaining to extreme GIC … [and] [w]e will … expand the statistical analyses to include characterization of multiple different spatial scales."[67] On the latter point, NERC "agrees that such research would provide additional modeling insights and supports further collaborative efforts between space weather researchers and electric utilities through the NERC GMD Task Force."[68] These statements support the NOPR's observation that the use of spatial averaging in this context is new, and thus there is a dearth of information or research regarding its application or appropriate scale.

46. While we believe our directive addresses concerns with relying solely on spatially-averaged data, we reiterate the position expressed in the NOPR that a GMD event will have a peak value in one or more location(s) and the amplitude will decline over distance

[66] 2015 Pulkkinen Paper at 2.

[67] *Id.* at 6.

[68] NERC Comments at 8.

from the peak; and, as a result, imputing the highest peak geoelectric field value in a planning area to the entire planning area may incorrectly overestimate GMD impacts.[69] Accordingly, our directive should not be construed to prohibit the use of spatial averaging in some capacity, particularly if more research results in a better understanding of how spatial averaging can be used to reflect actual GMD events.

47. The NOPR proposed to direct NERC to revise Reliability Standard TPL-007-1 so that the reference peak geoelectric field value is not based solely on spatially-averaged data. NERC and industry comments largely focused on the NOPR's discussion of one possible example to address the directive (i.e., by running GMD Vulnerability Assessments using spatially-averaged and non-spatially averaged reference peak geoelectric field amplitudes). However, while the method discussed in the NOPR is one possible option, the NOPR did not propose to direct NERC to develop revisions based on that option or any specific option. The Trade Associations' comments, discussed above, demonstrate that there is another way to address the NOPR directive (i.e., by performing planning models that also assess planning areas for localized "hot spots"). This approach may have merit if, for example, the geographic size of the hot spot is supported by actual data and the hot spot is centered over one or more locations that include an entity's facilities that become critical during a GMD event. Without pre-judging how NERC proposes to address the Commission's directive, NERC's response to this directive

[69] NOPR, 151 FERC ¶ 61,134 at P 35.

should satisfy the NOPR's concern that reliance on spatially-averaged data alone does not address localized peaks that could potentially affect the reliable operation of the Bulk-Power System.

48. We believe our directive should also largely address the comments submitted by entities opposed to NERC's proposed reference peak geoelectric field amplitude. Those commenters endorsed using a higher reference peak geoelectric field amplitude value, such as the 20 V/km cited in the GMD Interim Report. At the outset, we observe that the comments critical of the standard drafting team's use of the IMAGE data only speculate that had the standard drafting team used other sources, the calculated reference peak geoelectric field amplitude value would have been higher.[70] Moreover, among the commenters critical of NERC's proposal, there is disagreement over the magnitude of historical storms which some of these commenters would use as a model.[71] While NERC has discretion on how to propose to address our directive, NERC could revise Reliability Standard TPL-007-1 to apply a higher reference peak geoelectric field amplitude value to assess the impact of localized hot spots on the Bulk-Power System, as suggested by the

---

[70] *See, e.g.,* Resilient Societies Comments at 21 ("Had NERC and the Standard Drafting Team collected and analyzed available real-world data, they *would have likely found* that the severity of GMD in 1-in-100 Year reference storm had been set far below a technically justified level …" (emphasis added)).

[71] *See, e.g.*, Gaunt Comments at 13 (stating that the 1859 Carrington Event is "probably outside the re-occurrence frequency of 1:100 years adopted by NERC for the benchmark event"); Briggs Comments at 1 (advocating using a "'Carrington Class' super storm" as the benchmark GMD event).

Trade Associations. The effects of such hot spots could include increases in GIC levels, volt-ampere reactive power consumption, harmonics on the Bulk-Power System (and associated misoperations) and transformer heating. Moreover, the directive to revise Reliability Standard TPL-007-1 and, as discussed below, the directives to research geomagnetic latitude scaling factors and earth conductivity models as part of the GMD research work plan and to revise Reliability Standard TPL-007-1 to require the collection of necessary GIC monitoring and magnetometer data to validate GMD models should largely address or at least help to focus-in on factors that may be causing any inaccuracies in the standard drafting team's model.

49. Consistent with Order No. 779, the Commission does not specify a particular reference peak geoelectric field amplitude value that should be applied to hot spots given present uncertainties. While 20 V/km would seem to be a possible value, the Los Alamos Paper suggests that the 20 V/km figure may be too high. The Los Alamos Paper analyzed the non-spatially averaged IMAGE data to calculate a reference peak geoelectric field amplitude range (i.e., 8.4 V/km to 16.6 V/km) that is between NERC's proposed spatially-averaged value of 8 V/km and the non-spatially averaged 20 V/km figure cited in the GMD Interim Report.

50. Although the NOPR did not propose to direct NERC to submit revisions to Reliability Standard TPL-007-1 by a certain date with respect to the benchmark GMD event definition, the Commission determines that it is appropriate to impose an 18-month deadline from the effective date of this Final Rule. As discussed below, the Commission approves the five-year implementation period for Reliability Standard TPL-007-1

proposed by NERC. Having NERC submit revisions to the benchmark GMD event definition within 18 months of the effective date of this Final Rule, with the Commission acting promptly on the revised Reliability Standard, should afford enough time to apply the revised benchmark GMD event definition in the first GMD Vulnerability Assessment under the timeline set forth in Reliability Standard TPL-007-1's implementation plan. If circumstances, such as the complexity of the revised benchmark GMD event, require it, NERC may propose and justify a revised implementation plan.

2. Geomagnetic Latitude Scaling Factor

NOPR

51. The NOPR proposed to approve the geomagnetic latitude scaling factor in NERC's proposed benchmark GMD event definition. However, the NOPR sought comment on whether, in light of studies indicating that GMD events could have pronounced effects on lower geomagnetic latitudes, a modification is warranted to reduce the impact of the scaling factors.[72]

[72] NOPR, 151 FERC ¶ 61,134 at P 37 (citing Ngwira, C. M., Pulkkinen, A., Kuznetsova, M. M., Glocer, A., "Modeling extreme 'Carrington-type' space weather events using three-dimensional global MHD simulations," 119 Journal of Geophysical Research: Space Physics 4472 (2014) (finding that in Carrington-type events "the region of large induced ground electric fields is displaced further equatorward … [and] thereby may affect power grids … such as [those in] southern states of [the] continental U.S."); Gaunt, C. T., Coetzee, G., "Transformer Failures in Regions Incorrectly Considered to have Low GIC-Risk," 2007 IEEE Lausanne 807 (July 2007) (stating that twelve transformers were damaged and taken out of service in South Africa (at -40 degrees latitude) during the October 2003 Halloween Storm GMD event)). *See also* Liu, C., Li, Y., Pirjola, R., "Observations and modeling of GIC in the Chinese large-scale high-voltage power networks," Journal Space Weather Space Climate 4 at A03-p6 (2014) (Liu

(continued…)

Comments

52. NERC contends that the geomagnetic latitude scaling factor in Reliability Standard TPL-007-1 "accurately models the reduction of induced geoelectric fields that occurs over the mid-latitude region during a 100-year GMD event scenario … [and] describes the observed drop in geoelectric field that has been exhibited in analysis of major recorded geomagnetic storms."[73] NERC maintains that modifying the scaling factor is not technically justified based on the publications cited in the NOPR. NERC states that the first paper cited in the NOPR is based on models that are not mature and reflect a 1-in-150 year storm. NERC contends that the second paper does not clearly show that the purported transformer damage in South Africa was the result of abnormally high GICs during the October 2003 Halloween Storm. NERC further states that the standard drafting team analyzed the October 2003 Halloween Storm when developing the proposed geomagnetic latitude scaling factor.

53. The Trade Associations support the geomagnetic latitude scaling factor proposed by NERC. Like NERC, the Trade Associations contend that the papers cited in the NOPR do not support modifications because the models in the first paper "remain highly

---

Paper), http://www.swsc-journal.org/articles/swsc/pdf/2014/01/ swsc130009.pdf (finding that GICs of about 25A/phase had been measured in a transformer at a nuclear power plant at 22.6 degrees north latitude (significantly further away from the magnetic pole than Florida)).

[73] NERC Comments at 9 (citing Ngwira, C., Pulkkinen, A., Wilder, F., Crowley, G., "Extended Study of Extreme Geoelectric Field Event Scenarios for Geomagnetically Induced Current Applications," 11 Space Weather 121 (2013) (Ngwira 2013 Paper)).

theoretical and not sufficiently validated" and because the second paper likely involved other causal factors leading to the transformer failure.[74] Joint ISOs/RTOs also support the geomagnetic latitude scaling factor proposed by NERC. ITC states that NERC's proposal is a "reasonable approach given the current state of the science pertaining to GMD … [but] that as the science pertaining to GMD matures and more data becomes available, the scaling factors should be revisited and revised."[75] ITC suggests revisiting the geomagnetic latitude scaling factor every five years to incorporate any new developments in GMD science.

54. Several commenters question or disagree with the geomagnetic latitude scaling factors in Reliability Standard TPL-007-1 based on simulations and reports of damage to transformers in areas expected to be at low risk due to their geomagnetic latitude.[76] EIS contends that the proposed geomagnetic latitude scaling factor's assumption of a storm centered at 60 degrees geomagnetic latitude is inconsistent with a study relied upon by NERC.[77] The Los Alamos Paper's analysis suggests that NERC's proposed geomagnetic latitude scaling factors, while they fit well with weaker historical GMD events from

---

[74] Trade Associations Comments at 18-19.

[75] Joint ISOs/RTOs Comments at 5.

[76] *See, e.g.*, Gaunt Comments at 6; JINSA Comments at 2; Emprimus Comments at 2-3; Roodman Comments at 9; Resilient Societies Comments at 31-31; Kappenman Comments at 41-42.

[77] EIS Comments at 5 (citing Ngwira 2013 Paper).

which they were derived, may not accurately represent the effects of a 1-in-100 year GMD event at lower geomagnetic latitudes. The Los Alamos Paper states that a model of the electrojet is needed to "effectively extrapolate the small to moderate disturbance data currently in the historical record to disturbances as large as the TPL-007-1 Benchmark Event."[78] The Los Alamos Paper uses a larger number of geomagnetic disturbances (122 instead of 12) and a wider range of observatories by using the world-wide SuperMAG magnetometer array data, which includes the INTERMAGNET data used to support NERC's geomagnetic latitude scaling factors. The Los Alamos Paper shows that for more severe storms (Dst < -300, for which there are nine storms in the data set) the NERC scaling factors tend to be low, off by a factor of up to two or three at some latitudes. The Los Alamos Paper also recommends "an additional degree of conservatism in the mid-geomagnetic latitudes" until such time as a model is developed.[79] The Los Alamos Paper authors recommend a factor of 2 as a conservative correction.

Commission Determination

55. The Commission approves the geomagnetic latitude scaling factor in the benchmark GMD event definition. In addition, the Commission directs NERC to conduct further research on geomagnetic latitude scaling factors as part of the GMD research work plan discussed below.

---

[78] Los Alamos Paper at 12.

[79] *Id.*

56. Based on the record, the Commission finds sufficient evidence to conclude that lower geomagnetic latitudes are, to some degree, less susceptible to the effects of GMD events. The issue identified in the NOPR and by some commenters focused on the specific scaling factors in Reliability Standard TPL-007-1 in light of some analyses and anecdotal evidence suggesting that lower geomagnetic latitudes may be impacted by GMDs to a larger degree than reflected in Reliability Standard TPL-007-1.

57. The geomagnetic latitude scaling factor in Reliability Standard TPL-007-1 is supported by some of the available research.[80] In addition, with the exception of the Los Alamos Paper, commenters did not provide new information on the proposed scaling factor nor did commenters suggest alternative scaling factors. However, the Commission finds that there are enough questions regarding the effects of GMDs at lower geomagnetic latitudes to warrant directing NERC to study this issue further as part of the GMD research work plan. The Los Alamos Paper and the sources cited in the NOPR are suggestive that a 1-in-100 year GMD event could have a greater impact on lower

[80] *See* NERC Comments at 9 (citing Ngwira 2013 Paper). We disagree with the contention made by EIS that NERC's proposed geomagnetic latitude scaling factors are inconsistent with the Ngwira 2013 Paper. EIS maintains that the Ngwira 2013 Paper supports the conclusion that the benchmark GMD event should be centered at 50 degrees geomagnetic latitude instead of the 60 degree geomagnetic latitude figure in Reliability Standard TPL-007-1. The Ngwira 2013 Paper contains no such conclusion. Instead, the Ngwira 2013 Paper found that the latitude threshold boundary is a transition region having a definite lower bound of 50 degrees geomagnetic latitude but with an upper range as high as 55 degrees geomagnetic latitude. Ngwira 2013 Paper at 127, 130. The Ngwira 2013 Paper also stated that its findings were "in agreement with earlier observations by [Thomson et al., 2011] and more recently by [Pulkkinen et al., 2012], which estimated the location to be within 50 [degrees]–62 [degrees]." *Id.* at 124.

geomagnetic latitudes than NERC's proposed scaling factor assumes. But, as the Los Alamos Paper recognizes, the current absence of historical data on large GMD events precludes a definitive conclusion based on an empirical analysis of historical observations. Moreover, in prepared comments for the March 1, 2016 Technical Conference, Dr. Backhaus, one of the authors of the Los Alamos Paper, recommended that "the current NERC analysis should be adopted and further analysis performed with additional observational data and severe disturbance modeling efforts with the intent of refining the geomagnetic latitude scaling law in future revisions."[81] The Commission directs NERC to reexamine the geomagnetic latitude scaling factors in Reliability Standard TPL-007-1 as part of the GMD research work plan, including using existing models and developing new models to extrapolate from historical data on small to moderate GMD events the impacts of a large, 1-in-100 year GMD event on lower geomagnetic latitudes.

B. Thermal Impact Assessments

NERC Petition

58. Reliability Standard TPL-007-1, Requirement R6 requires owners of transformers that are subject to the Reliability Standard to conduct thermal analyses to determine if the transformers would be able to withstand the thermal effects associated with a benchmark GMD event. NERC states that transformers are exempt from the thermal impact

---

[81] Statement of Scott Backhaus, March 1, 2016 Technical Conference at 2.

assessment requirement if the maximum effective GIC in the transformer is less than 75 A/phase during the benchmark GMD event as determined by an analysis of the system. NERC explains that "based on available power transformer measurement data, transformers with an effective GIC of less than 75 A/phase during the Benchmark GMD Event are unlikely to exceed known temperature limits established by technical organizations."[82]

59. As provided in Requirements R5 and R6, "the maximum GIC value for the worst case geoelectric field orientation for the benchmark GMD event described in Attachment 1" determines whether a transformer satisfies the 75 A/phase threshold. If the 75 A/phase threshold is satisfied, Requirement R6 states, in relevant part, that a thermal impact assessment should be conducted on the qualifying transformer based on the effective GIC flow information provided in Requirement R5.

60. In its June 28, 2016 filing, NERC states that it identified an error in Figure 1 (Upper Bound of Peak Metallic Hot Spot Temperatures Calculated Using the Benchmark GMD Event) of the White Paper on Screening Criterion for Transformer Thermal Impact Assessment that resulted in incorrect plotting of simulated power transformer peak hot-spot heating from the benchmark GMD event. NERC revised Figure 1 in the White Paper on Screening Criterion for Transformer Thermal Impact Assessment and made corresponding revisions to related text, figures and tables throughout the technical white

[82] NERC Petition at 30.

papers supporting the proposed standard. NERC maintains that even with the revision to Figure 1, "the standard drafting team determined that the 75 A per phase threshold for transformer thermal impact assessment remains a valid criterion … [and] it is not necessary to revise any Requirements of the proposed Reliability Standard."[83]

NOPR

61. The NOPR proposed to approve the transformer thermal impact assessments in Requirement R6. In addition, as with the benchmark GMD event definition, the NOPR proposed to direct NERC to revise Requirement R6 to require registered entities to apply spatially averaged and non-spatially averaged peak geoelectric field values, or some equally efficient and effective alternative, when conducting thermal impact assessments. The NOPR also noted that Requirement R6 does not use the maximum GIC-producing orientation to conduct the thermal assessment for qualifying transformers; instead, the requirement uses the effective GIC time series described in Requirement R5.2 to conduct the thermal assessment on qualifying transformers. The NOPR sought comment from NERC as to why qualifying transformers are not assessed for thermal impacts using the maximum GIC-producing orientation and directed NERC to address whether, by not using the maximum GIC-producing orientation, the required thermal impact assessments could underestimate the impact of a benchmark GMD event on a qualifying transformer.

---

[83] NERC June 28, 2016 Filing at 1.

Comments

62. NERC opposes modifying the thermal impact assessments in Requirement R6 so that the assessments do not rely only on spatially-averaged data. NERC claims that the benchmark GMD event definition will "result in GIC calculations that are appropriately scaled for system-wide assessments."[84] NERC also contends that the "analysis performed by the standard drafting team of the impact of localized enhanced geoelectric fields on the GIC levels in transformers indicates that relatively few transformers in the system are affected."[85] In response to the question in the NOPR of why qualifying transformers are not assessed for thermal impacts using the maximum GIC producing orientation, NERC states that "the orientation of the geomagnetic field varies widely and continuously during a GMD event … [and] would be aligned with the maximum GIC-producing orientation for only a few minutes."[86] NERC concludes that "[i]n the context of transformer hot spot heating with time constants in the order of tens of minutes, alignment with any particular orientation for a few minutes at a particular point in time is not a driving concern."[87] NERC further states that the wave shape used in Reliability

[84] NERC Comments at 17.

[85] *Id.*

[86] *Id.* at 19.

[87] *Id.*

Standard TPL-007-1 provides "generally conservative results when performing thermal analysis of power transformers."[88]

63. The Trade Associations and CEA do not support the proposed NOPR directive because, they state, it focuses too heavily on individual transformers. The Trade Associations maintain that Reliability Standard TPL-007-1 "was never intended to address specific localized areas that might experience peak conditions and affect what we understand to be a very small number of assets that are unlikely to initiate a cascading outage."[89]

64. Certain non-industry commenters contend that the 75 A/phase qualifying threshold for thermal impact assessments is not technically justified. Emprimus contends that "many transformers have GIC ratings less than 75 amps per phase," but Emprimus claims that an Idaho National Lab study showed that "GIC introduced at 10 amps per phase on high voltage transformers exceed harmonic levels allowed under IEEE 519."[90] Emprimus also maintains that a 2013 IEEE paper "suggest[s] that there can be generator rotor damage at GIC levels which exceed 50 amps per phase."[91] Gaunt contends, based on his analysis of historical events, that "degradation is initiated in transformers by

---

[88] *Id.*

[89] Trade Associations Comments at 21.

[90] Emprimus Comments at 4.

[91] *Id.*

currents that are significantly below the 75 amps per phase."[92] Gaunt states that "[u]ntil better records are kept of transformer [dissolved gas in oil analysis] and transformer failure, the proposed level of 75 [A/phase] of GIC needed to initiate assessment of transformer response must be considered excessively high."[93] Gaunt recommends a qualifying threshold of 15 amps per phase. Resilient Societies states that the 75 A/phase threshold is based on a mathematical model for one type of transformer and that several tests referenced in the standard drafting team's White Paper on Transformer Thermal Impact Assessment were carried out under no load or minimal load conditions. In addition, Resilient Societies contends that applying the 75 A/phase threshold and NERC's proposed benchmark GMD event (i.e., using the spatially-averaged reference peak geoelectric field amplitude) results in only "two out of approximately 560 extra high voltage transformers" requiring thermal impact assessments in the PJM region; only one 345 kV transformer requiring thermal impact assessment in Maine; and zero transformers requiring thermal impact assessments in ATC's network.[94] Kappenman contends that the

---

[92] Gaunt Comments at 13.

[93] *Id.* at 14.

[94] Resilient Societies Comments at 5-14. Resilient Societies states that modeling performed by Central Maine Power Co. and Emprimus for the Maine Public Utilities Commission indicates that eight 345 kV transformers (53 percent according to Resilient Societies) would require thermal impact assessments in Maine if the reference peak geoelectric field amplitude were set at 20 V/km. *Id.* at 10. Resilient Societies also contends that this result is consistent with the Oak Ridge Meta-R-319 Study's finding that eight transformers would be "at risk" in Maine under a "'30 Amp At-Risk Threshold scenario.'" *Id.* Central Maine Power Co. calculated that the scaled NERC benchmark

(continued...)

75 A/phase threshold does not consider transformers with tertiary windings or autotransformers which may be impacted at lower GIC levels than 75 A/phase.[95]

Commission Determination

65. Consistent with our determination above regarding the reference peak geoelectric field amplitude value, the Commission directs NERC to revise Requirement R6 to require registered entities to apply spatially averaged and non-spatially averaged peak geoelectric field values, or some equally efficient and effective alternative, when conducting thermal impact assessments.

66. In the NOPR, the Commission requested comment from NERC regarding why Requirement R6 does not use the maximum GIC-producing orientation to conduct the thermal assessment for qualifying transformers. After considering NERC's response, we continue to have concerns with not using the maximum GIC-producing orientation for the thermal assessment of transformers. However, at this time we do not direct NERC to modify Reliability Standard TPL-007-1. Instead, as part of the GMD research work plan discussed below, NERC is directed to study this issue to determine how the geoelectric field time series can be applied to a particular transformer so that the orientation of the

---

GMD event for the northernmost point in Maine would be 4.53 V/km. Resilient Societies' calculations regarding ATC estimate that the scaled benchmark GMD event for Wisconsin would be 2 V/km. *Id.* at 14.

[95] The Commission received two comments following NERC's June 28, 2016 Filing. However, the supplemental comments did not specifically address the revisions submitted in NERC's June 28, 2016 filing.

time series, over time, will maximize GIC flow in the transformer, and to include the results in a filing with the Commission.

67. We are not persuaded by the comments opposed to Requirement R6's application of a 75 A/phase qualifying threshold. The standard drafting team's White Paper on Thermal Screening Criterion, as revised by NERC in the June 28, 2016 Filing, provides an adequate technical basis to approve NERC's proposal. As noted in the revised White Paper on Thermal Screening Criterion, the calculated metallic hot spot temperature corresponding to an effective GIC of 75 A/phase is 172 degrees Celsius; that figure is higher than the original figure of 150 degrees Celsius calculated by the standard drafting team but is still below the 200 degree Celsius limit specified in IEEE Std C57.91-2011.[96] The comments, particularly those of Gaunt, attempt to correlate historical transformer failures to past GMD events (e.g., 2003 Halloween Storm), while arguing that the transformers damaged in those events did not experience GICs of 75 A/phase. The evidence adduced by Gaunt and others is inconclusive.[97] We therefore direct NERC to include further analysis of the thermal impact assessment qualifying threshold in the GMD research work plan.

---

[96] NERC June 28, 2016 Filing, Revised White Paper on Screening Criterion for Transformer Thermal Impact Assessment at 3.

[97] *See, e.g.,* Gaunt Comments at 13 ("Although it has not been possible to assemble an exact model of the power system during the period 29-31 October 2003, and data on the ground conductivity in Southern Africa is not known with great certainty, we are confident that the several calculations of GIC that been carried out are not grossly inaccurate.").

68. In NOPR comments and in comments to the standard drafting team, Kappenman stated that delta winding heating due to harmonics has not been adequately considered by the standard drafting team and that, thermally, this is a bigger concern than metallic hot spot heating.[98] The standard drafting team responded that the vulnerability described for tertiary winding harmonic heating is based on the assumption that delta winding currents can be calculated using the turns ratio between primary and tertiary winding, which is incorrect when a transformer is under saturation.[99] The standard drafting team concluded that Kappenman's concerns regarding delta windings being a problem from a thermal standpoint are unwarranted and that the criteria developed by the standard drafting team use state-of-the-art analysis methods and measurement-supported transformer models. The Commission believes that the heating effects of harmonics on transformers, as discussed at the March 1, 2016 Technical Conference, are of concern and require further research.[100] Accordingly, we direct NERC to address the effects of harmonics, including

[98] Kappenman Comments at 45.

[99] Consideration of Comments Project 2013-03 Geomagnetic Disturbance Mitigation at 39 (December 5, 2014), http://www.nerc.com/pa/Stand/Project201303GeomagneticDisturbanceMitigation/Comment%20Report%20_2013-03_GMD_12052014.pdf.

[100] At the March 1, 2016 Technical Conference, Dr. Horton, a member of the standard drafting team, discussed the potential negative impacts of harmonics generated by GMDs on protection systems, reactive power resources and generators. Slide Presentation of Randy Horton, March 1, 2016 Technical Conference at 2-6.

tertiary winding harmonic heating and any other effects on transformers, as part of the GMD research work plan.[101]

C. GMD Research Work Plan

NOPR

69. The NOPR proposed to address the need for more data and certainty regarding GMD events and their potential effect on the Bulk-Power System by directing NERC to submit informational filings that address GMD-related research areas. The NOPR proposed to direct NERC to submit in the first filing a GMD research work plan indicating how NERC plans to: (1) further analyze the area over which spatial averaging should be calculated for stability studies, including performing sensitivity analyses on squares less than 500 km per side (e.g., 100 km, 200 km); (2) further analyze earth conductivity models by, for example, using metered GIC and magnetometer readings to calculate earth conductivity and using 3-D readings; (3) determine whether new analyses and observations support modifying the use of single station readings around the earth to adjust the spatially averaged benchmark for latitude; and (4) assess how to make GMD data (e.g., GIC monitoring and magnetometer data) available to researchers for study.

70. With respect to GIC monitoring and magnetometer readings, the NOPR sought comment on the barriers, if any, to public dissemination of such readings, including if

---

[101] NERC indicated in its comments that it is already studying the issue of harmonics. NERC Comments at 14 ("NERC is collaborating with researchers to examine more complex GMD vulnerability issues, such as harmonics and mitigation assessment techniques, to enhance the modeling capabilities of the industry").

their dissemination poses a security risk and if any such data should be treated as Critical Energy Infrastructure Information or otherwise restricted to authorized users. The NOPR proposed that NERC submit the GMD research work plan within six months of the effective date of a final rule in this proceeding. The NOPR also proposed that the GMD research work plan submitted by NERC should include a schedule for submitting one or more informational filings that apprise the Commission of the results of the four additional study areas, as well as any other relevant developments in GMD research, and should assess whether Reliability Standard TPL-007-1 remains valid in light of new information or whether revisions are appropriate.

Comments

71. NERC states that continued GMD research is necessary and that the potential impacts of GMDs on reliability are evolving. NERC, however, prefers that the NERC GMD Task Force continue its research without the GMD research work plan proposed in the NOPR. NERC contends that allowing the NERC GMD Task Force to continue its work would "accomplish NERC's and the Commission's shared goals in advancing GMD understanding and knowledge, while providing the flexibility necessary for NERC to work effectively with its international research partners to address risks to the reliability of the North American Bulk-Power System."[102] NERC also claims that, in addition to being unnecessary given the work of the NERC GMD Task Force, the NOPR

[102] NERC Comments at 13.

proposal "poses practical challenges … [because it would] bind[] NERC to a specific and inflexible research plan and report schedule to be determined six months (or even a year) following the effective date of a final rule in this proceeding."[103]

72. The Trade Associations and CEA do not support the GMD research work plan. Instead, they contend that NERC should be allowed to pursue GMD research independently.

73. Several commenters, while not addressing the NOPR proposal specifically, state that additional research is necessary to validate or improve elements of the benchmark GMD event definition.[104]

74. The Trade Associations state that monitoring data should be available for academic research purposes. Resilient Societies contends that monitoring data should be publicly disseminated on a regular basis and that there is no security risk in releasing such data because they relate to naturally occurring phenomena. Emprimus states that it supports making GIC and magnetometer monitoring data available to the public. Bardin supports making GIC and GMD-related information to the public or at least to "legitimate researchers."

---

[103] *Id.* at 16.

[104] *See, e.g.,* USGS Comments at 1 (addressing earth conductivity models), Bardin Comments at 2 (addressing earth conductivity models); Roodman Comments at 3 (addressing reference peak geoelectric field amplitude); Gaunt Comments at 7 (addressing spatial averaging).

75. Hydro One and CEA do not support mandatory data sharing without the use of non-disclosure agreements.

Commission Determination

76. The Commission recognizes, as do commenters both supporting and opposing proposed Reliability Standard TPL-007-1, that our collective understanding of the threats posed by GMD is evolving as additional research and analysis are conducted. These ongoing efforts are critical to the nation's long-term efforts to protect the grid against a major GMD event. While we approve NERC's proposed Reliability Standard TPL-007-1 and direct certain modifications, as described above, the Commission also concludes that facilitating additional research and analysis is necessary to adequately address these threats. As discussed in the next two sections of this final rule, the Commission directs a three-prong approach to further those efforts by directing NERC to: (1) develop, submit, and implement a GMD research work plan; (2) develop revisions to Reliability Standard TPL-007-1 to require responsible entities to collect GIC monitoring and magnetometer data; and (3) collect GIC monitoring and magnetometer data from registered entities for the period beginning May 2013, including both data existing as of the date of this order and new data going forward, and to make that information available.

77. First, the Commission adopts the NOPR proposal and directs NERC to submit a GMD research work plan and, subsequently, informational filings that address the GMD-related research areas identified in the NOPR, additional research tasks identified in this Final Rule (i.e., the research tasks identified in the thermal impact assessment discussion above) and, in NERC's discretion, any GMD-related research areas generally that may

impact the development of new or modified GMD Reliability Standards.[105] The GMD research work plan should be submitted within six months of the effective date of this final rule. The research required by this directive should be informed by ongoing GMD-related research efforts of entities such as USGS, National Atmospheric and Oceanic Administration (NOAA), National Aeronautics and Space Administration, Department of Energy, academia and other publicly available contributors, including work performed for the National Space Weather Action Plan.[106]

78. As part of the second research area identified in the NOPR (i.e., further analyze earth conductivity models by, for example, using metered GIC and magnetometer readings to calculate earth conductivity and using 3-D readings), the GMD research work plan should specifically investigate "coastal effects" on ground conductivity models.

79. In addition, the large variances described by USGS in actual 3-D ground conductivity data raise the question of whether one time series geomagnetic field is sufficient for vulnerability assessments. The characteristics, including frequencies, of the time series interact with the ground conductivity to produce the geoelectric field that

---

[105] The GMD research work plan need not address the fourth research area identified in the NOPR (i.e., assess how to make GIC monitoring and magnetometer data available to researchers for study) given the Commission's directive and discussion below regarding the collection and dissemination of necessary GIC monitoring and magnetometer data.

[106] National Science and Technology Council, National Space Weather Action Plan (October 2015), https://www.whitehouse.gov/sites/default/files/microsites/ostp/final_nationalspaceweatheractionplan_20151028.pdf.

drives the GIC. Therefore, the research should address whether additional realistic time series should be selected to perform assessments in order to capture the time series that produces the most vulnerability for an area.

80. The comments largely agree that additional GMD research should be pursued, particularly with respect to the elements of the benchmark GMD event definition (i.e., the reference peak geoelectric field amplitude value, geomagnetic latitude scaling factor, and earth conductivity scaling factor). There is ample evidence in the record to support the need for additional GMD-related research.[107] For example, USGS submitted comments indicating that USGS's one dimensional ground electrical conductivity models used by the standard drafting team have a "significant limitation" in that they assume that a "[one dimensional] conductivity-with-depth profile can adequately represent a large geographic region," which USGS describes as a "gross simplification."[108] USGS observes that while the "proposed standard attempted to incorporate the best scientific research available … it must be noted that the supporting science is quickly evolving."[109] USGS recommends that "the proposed standard should establish a process for updates and improvements that

---

[107] *See, e.g.*, NERC October 22, 2015 Supplemental Comments at 7-8 (expressing support for additional research regarding geomagnetic latitude scaling factors and earth conductivity models).

[108] USGS Comments at 1.

[109] *Id.*

acknowledges and addresses the quickly evolving nature of relevant science and associated data."[110]

81. Opposition to the proposal centers on the contention that the proposed directive is unnecessary and potentially counterproductive given the continuing work of the NERC GMD Task Force. We do not find these comments persuasive. Our directive requires NERC to submit a work plan for the study of GMD-related issues that are already being examined or that NERC agrees should be studied.[111] Nothing in our directive precludes NERC from continuing to use the NERC GMD Task Force as a vehicle for conducting the directed research or other research. Indeed, we encourage NERC to continue to use

---

[110] *Id.* We note that Reliability Standard TPL-007-1, Att. 1 (Calculating Geoelectric Fields for the Benchmark GMD Event) already provides that a "planner can also use specific earth model(s) with documented justification…" Accordingly, Reliability Standard TPL-007-1 includes a mechanism for incorporating improvements in earth conductivity models when calculating the benchmark GMD event.

[111] *See, e.g.*, NERC Comments at 8 ("NERC agrees that [spatial averaging] research would provide additional modeling insights and supports further collaborative efforts between space weather researchers and electric utilities through the NERC GMD Task Force"), at 10 ("NERC agrees that additional [geomagnetic latitude scaling] research is necessary, and supports the significant research that is occurring throughout the space weather community to develop and validate models and simulation techniques"), at 13 ("Working with EPRI, researchers at USGS, and industry, NERC will work to improve the earth conductivity models that are a vital component to understanding the risks of GMD events in each geographic region"), and at 23 ("efforts are already underway to expand GMD monitoring capabilities … [and] [t]hrough these efforts, NERC and industry should effectively address the concerns noted by the Commission in the NOPR, including ensuring a more complete set of data for operational and planning needs and supporting analytical validation and situational awareness").

the GMD Task Force as a forum for engagement with interested stakeholders. In addition, we do not set specific deadlines for completion of the research; we only require NERC to submit the GMD research work plan within six months of the effective date of a final rule. The GMD research work plan, in turn, should include target dates for the completion of research topics and the reporting of findings to the Commission. The Commission intends to notice and invite comment on the GMD research work plan. An extension of time to submit the GMD research work plan may be available if six months proves to be insufficient. In addition, given the uncertainties commonly associated with complex research projects, the Commission will be flexible regarding changes to the tasks and target dates established in the GMD research work plan.

D. Monitoring Data

NERC Petition

82. Reliability Standard TPL-007-1, Requirement R2 requires responsible entities to "maintain System models and GIC System models of the responsible entity's planning area for performing the study or studies needed to complete GMD Vulnerability Assessment(s)." NERC states that Reliability Standard TPL-007-1 contains "requirements to develop the models, studies, and assessments necessary to build a picture of overall GMD vulnerability and identify where mitigation measures may be necessary."[112] NERC explains that mitigating strategies "may include installation of

[112] NERC Petition at 13.

hardware (e.g., GIC blocking or monitoring devices), equipment upgrades, training, or enhanced Operating Procedures."[113]

NOPR

83. The NOPR proposed to direct NERC to revise Reliability Standard TPL-007-1 to require the installation of monitoring equipment (i.e., GIC monitors and magnetometers) to the extent there are any gaps in existing GIC monitoring and magnetometer networks. Alternatively, the NOPR sought comment on whether NERC should be responsible for installation of any additional, necessary magnetometers while affected entities would be responsible for installation of additional, necessary GIC monitors. The NOPR also proposed that, as part of NERC's work plan, NERC identify the number and location of current GIC monitors and magnetometers in the United States to assess whether there are any gaps. The NOPR sought comment on whether the Commission should adopt a policy specifically allowing recovery of costs associated with or incurred to comply with Reliability Standard TPL-007-1, including for the purchase and installation of monitoring devices.

Comments

84. NERC does not support the NOPR proposal regarding the installation of GIC monitoring devices and magnetometers. NERC contends that the proposed requirement is not necessary because Reliability Standard TPL-007-1 "supports effective GMD

---

[113] *Id.* at 32.

monitoring programs, and additional efforts are planned or underway to ensure adequate data for reliability purposes."[114] NERC also maintains that the proposed directive "poses implementation challenges … [because] GMD monitoring capabilities and technical information have not yet reached a level of maturity to support application in a Reliability Standard, and not all applicable entities have developed the comprehensive understanding of system vulnerabilities that would be needed to deploy GMD monitoring devices for the greatest reliability benefit."[115] NERC also notes that a requirement mandating the installation of monitoring devices for situational awareness purposes would be outside the scope of a planning Reliability Standard.

85. The Trade Associations, CEA, ITC, Hydro One and Tri-State, while agreeing that more data are useful to analytical validation and situational awareness, do not support the NOPR proposal. CEA does not support the proposal because Reliability Standard TPL-007-1 is a planning standard; a one-size-fits-all monitoring approach will not work; the responsibility for monitoring, which in Canada is done by the Canadian government, should not fall to industry or NERC; and the proposal is too costly. Likewise, ITC contends that it would not be prudent or cost effective for entities to have to install

---

[114] NERC Comments at 21. NERC cites as examples the 40 GIC monitoring nodes operated by EPRI's SUNBURST network; the use of GIC monitoring devices by some registered entities (e.g., PJM); and the magnetometer networks operated by USGS and EPRI. *Id.* at 23-25.

[115] *Id.*

monitoring equipment. Hydro One does not support a Reliability Standard that prescribes the number and location of monitoring devices that must be installed. The Trade Associations and ITC, instead, support directing NERC to develop a plan to address this issue. The Trade Associations state that such a plan should involve a partnership between government and industry. Tri-State maintains that NERC, working with USGS and NOAA, should be responsible for determining the need for and installation of any needed magnetometers. If the Commission requires applicable entities to install monitoring devices, the Trade Associations, Tri-State and Exelon agree that there should be cost recovery.

86. BPA supports the NOPR proposal for increased monitoring because BPA believes it will improve situational awareness. As a model, BPA states that the "Canadian government in collaboration with Canadian transmission owners" have developed a "technique that shows real promise of increasing visibility of GIC flows and localized impacts for a regional transmission grid."[116] AEP encourages the Commission to expand the "number and scope of the permanent geomagnetic observatories and install permanent geoelectric observatories in the United States."[117]

87. Resilient Societies supports requiring the installation of GIC monitoring devices and magnetometers, noting that GIC monitors are commercially available and cost as

[116] BPA Comments at 4.

[117] AEP March 29, 2016 Supplemental Comments at 1.

little as $10,000 to $15,000 each. Emprimus supports developing criteria that inform the need for and location of monitoring devices.

Commission Determination

88. We conclude that additional collection and disclosure of GIC monitoring and magnetometer data is necessary to improve our collective understanding of the threats posed by GMD events. The Commission therefore adopts the NOPR proposal in relevant part and directs NERC to develop revisions to Reliability Standard TPL-007-1 to require responsible entities to collect GIC monitoring and magnetometer data as necessary to enable model validation and situational awareness, including from any devices that must be added to meet this need. The NERC standard drafting team should address the criteria for collecting GIC monitoring and magnetometer data discussed below and provide registered entities with sufficient guidance in terms of defining the data that must be collected, and NERC should propose in the GMD research work plan how it will determine and report on the degree to which industry is following that guidance.

89. In addition, the Commission directs NERC, pursuant to Section 1600 of the NERC Rules of Procedure, to collect GIC monitoring and magnetometer data from registered entities for the period beginning May 2013, including both data existing as of the date of this order and new data going forward, and to make that information available.[118] We also provide guidance that, as a general matter, the Commission does not believe that

[118] The Commission's directives to collect and make available GIC monitoring and magnetometer data do not apply to non-U.S. responsible entities or Alaska and Hawaii.

GIC monitoring and magnetometer data should be treated as Confidential Information pursuant to the NERC Rules of Procedure.

Collection of GIC and Magnetometer Data

90. In developing a requirement regarding the collection of magnetometer data, NERC should consider the following criteria discussed at the March 1, 2016 Technical Conference: (1) the data is sampled at a cadence of at least 10-seconds or faster; (2) the data comes from magnetometers that are physically close to GIC monitors; (3) the data comes from magnetometers that are not near sources of magnetic interference (e.g., roads and local distribution networks); and (4) data is collected from magnetometers spread across wide latitudes and longitudes and from diverse physiographic regions.[119]

91. Each responsible entity that is a transmission owner should be required to collect necessary GIC monitoring data. However, a transmission owner should be able to apply for an exemption from the GIC monitoring data collection requirement if it demonstrates that no or little value would be added to planning and operations. In developing a requirement regarding the collection of GIC monitoring data, NERC should consider the following criteria discussed at the March 1, 2016 Technical Conference: (1) the GIC data is from areas found to have high GIC based on system studies; (2) the GIC data comes from sensitive installations and key parts of the transmission grid; and (3) the data comes from GIC monitors that are not situated near transportation systems using direct current

[119] Slide Presentation of Luis Marti (Third Panel), March 1, 2016 Technical Conference at 3, 9.

(e.g., subways or light rail).[120] GIC monitoring and magnetometer locations should also be revisited after GIC system models are run with improved ground conductivity models. NERC may also propose to incorporate the GIC monitoring and magnetometer data collection requirements in a different Reliability Standard (e.g., real-time reliability monitoring and analysis capabilities as part of the TOP Reliability Standards).

92. Our determination differs from the NOPR proposal in that the NOPR proposed to require the installation of GIC monitors and magnetometers. The comments raised legitimate concerns about incorporating such a requirement in Reliability Standard TPL-007-1 because of the complexities of siting and operating monitoring devices to achieve the maximum benefits for model validation and situational awareness. In particular, responsible entities may not have the technical capacity to properly install and operate magnetometers, given complicating issues such as man-made interference, calibration, and data interpretation. Accordingly, the Commission determines that requiring responsible entities to collect necessary GIC monitoring and magnetometer data, rather than install GIC monitors and magnetometers, affords greater flexibility while obtaining significant benefits. For example, responsible entities could collaborate with universities and government entities that operate magnetometers to collect necessary magnetometer data, or responsible entities could choose to install GIC monitors or magnetometers to comply with the data collection requirement. While the Commission's primary concern

---

[120] *Id.* at 8.

is the quality of the data collected, we do not establish a requirement for either approach or promote a particular device for collecting the required data. We also find that cost recovery for prudent costs associated with or incurred to comply with Reliability Standard TPL-007-1 and future revisions to the Reliability Standard, including for the purchase and installation of monitoring devices, will be available to registered entities.[121]

Data Availability

93. We also direct NERC, pursuant to Sections 1500 and 1600 of the NERC Rules of Procedure, to collect and make GIC monitoring and magnetometer data available.[122] We determine that the dissemination of GIC monitoring and magnetometer data will facilitate a greater understanding of GMD events that, over time, will improve Reliability Standard TPL-007-1. The record in this proceeding supports the conclusion that access to GIC monitoring and magnetometer data will help facilitate GMD research, for example, by helping to validate GMD models.[123] To facilitate the prompt dissemination of GIC monitoring and magnetometer data, we address whether GIC monitoring or

---

[121] NOPR, 151 FERC ¶ 61,134 at P 49 n.60.

[122] If GIC monitoring and magnetometer data is already publicly available (e.g., from a government entity or university), NERC need not duplicate those efforts.

[123] *See*, *e.g.*, March 1, 2016 Technical Conference Tr. 58:22-59:13 (Love); 128:5-129:2 (Overbye); ATC Comments at 6-7("as more measuring devices (including magnetometers and GIC monitors) continue to propagate, the body of field data on magnetic fields and the resultant GICs will continue to increase the understanding of this phenomena and result in better models that more closely match real world conditions … [a]bsent this field data, it is difficult to build accurate models that can be used to plan and operate the transmission system").

magnetometer data should qualify as Confidential Information under the NERC Rules of Procedure.[124]

94. Based on the record in this proceeding, we believe that GIC and magnetometer data typically should not be designated as Confidential Information under the NERC Rules of Procedure. We are not persuaded that the dissemination of GIC monitoring or magnetometer data poses a security risk or that the data otherwise qualify as Confidential Information. CEA and Hydro One have objected, without elaboration, to making data available without the use of non-disclosure agreements.[125] At the March 1, 2016 Technical Conference, panelists were questioned on the topic yet could not identify a security-based or other credible reason for not making such information available to requesters. In comments submitted after the March 1, 2016 Technical Conference, the Trade Associations explained that "GIC measurements, while not as sensitive as transmission planning studies, should also be protected … [because a] potentially malicious actor could conceivably combine GIC information with information from other sources to deduce the configuration and operating conditions of the grid or some portion

[124] Providers of GIC and magnetometer data may request that NERC treat their GIC monitoring and magnetometer data as "Confidential Information," as that term is defined in Section 1500 of the NERC Rules of Procedure. Under the NERC Rules of Procedure, disclosure of Confidential Information by NERC to a requester requires a formal request, notice and opportunity for comment, and an executed non-disclosure agreement for requesters not seeking public disclosure of the information. NERC Rules of Procedure, Section 1503 (Requests for Information) (effective Nov. 4, 2015).

[125] CEA Comments at 15; Hydro One Comments at 2.

of it."[126] The Trade Associations' comments, however, do not substantiate the assertion that the release of GIC monitoring (or magnetometer data) alone poses any risk to the Bulk-Power System. The Trade Associations' comment is also vague by not identifying what "information from other sources" could be combined with GIC monitoring "to deduce the configuration and operating conditions of the grid or some portion of it."

95. In conclusion, given both the lack of substantiated concerns regarding the disclosure of GIC and magnetometer data, and the compelling demonstration that access to these data will support ongoing research and analysis of GMD threats, the Commission expects NERC to make GIC and magnetometer data available. Notwithstanding our findings here, to the extent any entity seeks confidential treatment of the data it provides to NERC, the burden rests on that entity to justify the confidential treatment.[127] Exceptions are possible if the providing entity obtains from NERC, at the time it submits data to NERC, a determination that GIC or magnetometer data qualify as Confidential Information.[128] Entities denied access to GIC and magnetometer data by NERC or

---

[126] Trade Associations March 7, 2016 Supplemental Comments at 5.

[127] *See* NERC Rules of Procedure, Section 1502.1. To address any substantiated concerns regarding the need for confidentiality of an entity's GIC or magnetometer data, NERC could develop a policy for disseminating such data only after an appropriate time interval (e.g., six months).

[128] We understand that NERC typically does not determine whether information submitted to it under a claim of confidentiality is Confidential Information when receiving such information. *See North American Electric Reliability Corp.*, 119 FERC ¶ 61,060, at PP 195-196 (2007). We expect that, when a submitter seeks a determination

(continued…)

providers denied Confidential Information treatment of GIC and magnetometer data may appeal NERC's decision to the Commission.

E. Corrective Action Plan Deadlines

NERC Petition

96. Reliability Standard TPL-007-1, Requirement R7 provides that:

> Each responsible entity, as determined in Requirement R1, that concludes, through the GMD Vulnerability Assessment conducted in Requirement R4, that their System does not meet the performance requirements of Table 1 shall develop a Corrective Action Plan addressing how the performance requirements will be met ....

NERC explains that the NERC Glossary defines corrective action plan to mean, "A list of actions and an associated timetable for implementation to remedy a specific problem."[129] Requirement R7.3 states that the corrective action plan shall be provided within "90 calendar days of completion to the responsible entity's Reliability Coordinator, adjacent Planning Coordinator(s), adjacent Transmission Planner(s), functional entities referenced in the Corrective Action Plan, and any functional entity that submits a written request and has a reliability-related need."

NOPR

97. The NOPR proposed to direct NERC to modify Reliability Standard TPL-007-1 to require corrective action plans to be developed within one year of the completion of the

---

by NERC of a claim that GIC or magnetometer data qualify as Confidential Information, NERC will decide promptly.

[129] NERC Petition at 31.

GMD Vulnerability Assessment. The NOPR also proposed to direct NERC to modify Reliability Standard TPL-007-1 to require a deadline for non-equipment mitigation measures that is two years following development of the corrective action plan and a deadline for mitigation measures involving equipment installation that is four years following development of the corrective action plan. Recognizing that there is little experience with installing equipment for GMD mitigation, the NOPR stated that the Commission is open to proposals that may differ from its proposal, particularly from any entities with experience in this area. The NOPR also sought comment on appropriate alternative deadlines and whether there should be a mechanism that would allow NERC to consider, on a case-by-case basis, requests for extensions of required deadlines.

Comments

98. NERC states that it does not oppose a one-year deadline for completing the development of corrective action plans.[130] However, NERC contends that imposing deadlines on the completion of mitigation actions would be problematic because of the uncertainties regarding the amount of time needed to install necessary equipment. NERC maintains that deadlines that are too short may cause entities to take mitigation steps that,

[130] NERC contends that a deadline is unnecessary because "NERC expects that applicable entities would determine necessary corrective actions as part of their GMD Vulnerability Assessments for the initial assessment [due 60 months after a final rule in this proceeding goes into effect] as well as subsequent assessments [due every 60 months thereafter]." NERC Comments at 28.

while quicker, would not be as effective as mitigations that take more time to complete. NERC supports allowing extensions if the Commission adopts the NOPR proposal.

99. AEP states that, even if possible, a one-year deadline for developing corrective action plans is too aggressive and would encourage narrow thinking (i.e., registered entities would address GMD mitigation rather than pursue system improvements generally that would also address GMD mitigation). AEP, instead, proposes a two-year deadline. AEP does not support a Commission-imposed deadline for completing mitigation actions, although it supports requiring a time-table in the corrective action plan. AEP notes that the Commission did not impose a specific deadline for completion of corrective actions in Reliability Standard TPL-001-4 (Transmission System Planning Performance). CEA does not support a deadline for the development of corrective action plans because it is already part of the GMD Vulnerability Assessment process. Like AEP, CEA does not support specific deadlines for the completion of mitigation actions and instead supports including time-tables in the corrective action plan. CEA also contends that an extension process would be impracticable.

100. Trade Associations, BPA and Tri-State support the imposition of corrective action plan deadlines as long as entities can request extensions. Gaunt supports the corrective action plan deadlines proposed in the NOPR. Emprimus supports the imposition of deadlines but contends that non-equipment mitigation actions should be completed in 6 months and that there should be a rolling four-year period for equipment mitigation (i.e., after each year, 25 percent of the total mitigation actions should be completed).

Commission Determination

101. The Commission directs NERC to modify Reliability Standard TPL-007-1 to include a deadline of one year from the completion of the GMD Vulnerability Assessments to complete the development of corrective action plans. NERC's statement that it "expects" corrective action plans to be completed at the same time as GMD Vulnerability Assessments concedes the point made in the NOPR that Reliability Standard TPL-007-1 currently lacks a clear deadline for the development of corrective action plans.

102. The Commission also directs NERC to modify Reliability Standard TPL-007-1 to include a two-year deadline after the development of the corrective action plan to complete the implementation of non-hardware mitigation and four-year deadline to complete hardware mitigation. The comments provide contrasting views on the practicality of imposing mitigation deadlines, with NERC and some industry commenters arguing that such deadlines are not warranted while the Trade Associations and other industry commenters support their imposition. Most of these comments, however, support an extension process if the Commission determines that deadlines are necessary. The Commission agrees that NERC should consider extensions of time on a case-by-case basis. The Commission directs NERC to submit these revisions within 18 months of the effective date of this Final Rule.

103. Following adoption of the mitigation deadlines required in this final rule, Reliability Standard TPL-007-1 will establish a recurring five-year schedule for the identification and mitigation of potential GMD risks on the grid, as follows: (1) the

development of corrective action plans must be completed within one year of a GMD Vulnerability Assessment; (2) non-hardware mitigation must be completed within two years following development of corrective action plans; and (3) hardware mitigation must be completed within four years following development of corrective action plans.

104. As discussed elsewhere in this final rule, the Commission recognizes and expects that our collective understanding of the science regarding GMD threats will improve over time as additional research and analysis is conducted. We believe that the recurring five-year cycle will provide, on a going-forward basis, the opportunity to update Reliability Standard TPL-007-1to reflect new or improved scientific understanding of GMD events.

F. Minimization of Load Loss and Curtailment

NERC Petition

105. Reliability Standard TPL-007-1, Requirement R4 states that each responsible entity "shall complete a GMD Vulnerability Assessment of the Near-Term Transmission Planning Horizon once every 60 calendar months." Requirement R4.2 further states that the "study or studies shall be conducted based on the benchmark GMD event described in Attachment 1 to determine whether the System meets the performance requirements in Table 1."

106. NERC maintains that Table 1 sets forth requirements for system steady state performance. NERC explains that Requirement R4 and Table 1 "address assessments of the effects of GICs on other Bulk-Power System equipment, system operations, and

system stability, including the loss of devices due to GIC impacts."[131] Table 1 provides, in relevant part, that load loss and/or curtailment are permissible elements of the steady state:

> Load loss as a result of manual or automatic Load shedding (e.g. UVLS) and/or curtailment of Firm Transmission Service may be used to meet BES performance requirements during studied GMD conditions. The likelihood and magnitude of Load loss or curtailment of Firm Transmission Service should be minimized.

NOPR

107. The NOPR sought comment on the provision in Table 1 that "Load loss or curtailment of Firm Transmission Service should be minimized." The NOPR stated that because the term "minimized" does not represent an objective value, the provision is potentially subject to interpretation and assertions that the term is vague and may not be enforceable. The NOPR also explained that the modifier "should" might indicate that minimization of load loss or curtailment is only an expectation or a guideline rather than a requirement. The NOPR sought comment on how the provision in Table 1 regarding load loss and curtailment will be enforced, including: (1) whether, by using the term "should," Table 1 requires minimization of load loss or curtailment; or both and (2) what constitutes "minimization" and how it will be assessed.

[131] NERC Petition at 39.

Comments

108. NERC states the language in Table 1 is modeled on Reliability Standard TPL-001-4, which provides in part that "an objective of the planning process should be to minimize the likelihood and magnitude of interruption of Firm transmission Service following Contingency events." NERC explains that Reliability Standard TPL-007-1 "does not include additional load loss performance criteria used in normal contingency planning because such criteria may not be applicable to GMD Vulnerability Assessment of the impact from a 1-in-100 year GMD event."[132] However, NERC points out that the enforcement of Requirement R4 "would include an evaluation of whether the system meets the Steady State performance requirements of Table 1 which are aimed at protecting against instability, controlled separation, and Cascading."[133] NERC further states that "minimized" in the context of Reliability Standard TPL-007-1 means that "planned Load loss or curtailments are not to exceed amounts necessary to prevent voltage collapse."[134]

109. The Trade Associations agree with the NOPR that the lack of objective criteria could create compliance and enforcement challenges and could limit an operator's actions in real-time. The Trade Associations state that the Commission "should consider

[132] NERC Comments at 29.

[133] *Id.*

[134] *Id.*

whether such language in mandatory requirements invites the unintended consequences of raising reliability risks, especially during real-time emergency conditions … [but] [i]n the interim, the Trade Associations envision that NERC will consider further discussions with stakeholders on the issue prior to TPL-007 implementation."[135]

Commission Determination

110. The Commission accepts the explanation in NERC's comments of what is meant by the term "minimized" in Table 1.

G. Violation Risk Factors and Violation Severity Levels

111. Each requirement of Reliability Standard TPL-007-1 includes one violation risk factor and has an associated set of at least one violation severity level. NERC states that the ranges of penalties for violations will be based on the sanctions table and supporting penalty determination process described in the Commission approved NERC Sanction Guidelines. The NOPR proposed to approve the violation risk factors and violation severity levels submitted by NERC, for the requirements in Reliability Standard TPL-007-1, consistent with the Commission's established guidelines.[136] The Commission did not receive any comments regarding this aspect of the NOPR. Accordingly, the Commission approves the violation risk factors and violation severity levels for the requirements in Reliability Standard TPL-007-1.

---

[135] Trade Associations Comments at 28.

[136] *North American Electric Reliability Corp.*, 135 FERC ¶ 61,166 (2011).

H. Implementation Plan and Effective Dates

NERC Petition

112. NERC proposes a phased, five-year implementation period.[137] NERC maintains that the proposed implementation period is necessary: (1) to allow time for entities to develop the required models; (2) for proper sequencing of assessments because thermal impact assessments are dependent on GIC flow calculations that are determined by the responsible planning entity; and (3) to give time for development of viable corrective action plans, which may require applicable entities to "develop, perform, and/or validate new or modified studies, assessments, procedures … [and because] [s]ome mitigation measures may have significant budget, siting, or construction planning requirements."[138]

113. The proposed implementation plan states that Requirement R1 shall become effective on the first day of the first calendar quarter that is six months after Commission approval. For Requirement R2, NERC proposes that the requirement shall become effective on the first day of the first calendar quarter that is 18 months after Commission approval. NERC proposes that Requirement R5 shall become effective on the first day of the first calendar quarter that is 24 months after Commission approval. NERC proposes that Requirement R6 shall become effective on the first day of the first calendar quarter that is 48 months after Commission approval. And for Requirement R3, Requirement

[137] NERC Petition, Ex. B (Implementation Plan for TPL-007-1).

[138] *Id.* at 2.

R4, and Requirement R7, NERC proposes that the requirements shall become effective on the first day of the first calendar quarter that is 60 months after Commission approval.

NOPR

114. The NOPR proposed to approve the implementation plan and effective dates submitted by NERC. However, given the serial nature of the requirements in Reliability Standard TPL-007-1, the Commission expressed concern about the duration of the timeline associated with any mitigation stemming from a corrective action plan and sought comment from NERC and other interested entities as to whether the length of the implementation plan, particularly with respect to Requirements R4, R5, R6, and R7, could be reasonably shortened.

Comments

115. NERC does not support shortening the implementation period. NERC maintains that the proposed implementation period is "appropriate and commensurate with the requirements of the proposed standard" and is based on "industry … projections on the time required for obtaining validated tools, models and data necessary for conducting GMD Vulnerability Assessments through the standard development process."[139] NERC notes that the standard drafting team initially proposed a four-year implementation plan, but received substantial comments expressing concern with only having four years.

---

[139] NERC Comments at 30.

116. The Trade Associations, BPA, CEA, Joint ISOs/RTOs and Tri-State support the proposed implementation plan for largely the same reasons as NERC.

117. Gaunt proposes a shorter implementation period wherein the initial GMD Vulnerability Assessment would be performed 48 months following the effective date of a final rule in this proceeding, as opposed to the proposed implementation plan's 60 months. Subsequent GMD Vulnerability Assessments would be performed every 48 months thereafter. Briggs states that a "3 or 4 year timeline would likely provide industry with enough time to implement corrective measures and should be considered."[140]

Commission Determination

118. The Commission approves the implementation plan submitted by NERC. When registered entities begin complying with Reliability Standard TPL-007-1, it will likely be the first time that many registered entities will have planned for a GMD event, beyond developing the GMD operational procedures required by Reliability Standard EOP-010-1. Registered entities will gain the capacity to conduct GMD Vulnerability Assessments over the course of the five-year implementation plan by complying with, at phased intervals, the foundational requirements in Reliability Standard TPL-007-1 (i.e., establishing responsibilities for planning and developing models and performance criteria). In addition, as discussed above, NERC's implementation plan affords sufficient

[140] Briggs Comments at 7.

time for NERC to submit and for the Commission to consider the directed revisions to Reliability Standard TPL-007-1 before the completion of the first GMD Vulnerability Assessment. As such, the five-year implementation plan will allow for the incorporation of the revised Reliability Standard in the first round of GMD Vulnerability Assessments.

I. Other Issues

119. Several commenters indicated that the Commission should address the threats posed by EMPs or otherwise raised the issue of EMPs.[141] For example, Briggs states that the Commission should "initiate a process to improve the resilience of the U.S. electric grid to the threat of high altitude electromagnetic pulse (HEMP) attacks, which can be more severe than solar superstorms."[142] However, as the Commission stated in Order No. 779 in directing the development of GMD Reliability Standards and in Order No. 797 in approving the First Stage GMD Reliability Standards, EMPs are not within the scope of the GMD rulemaking proceedings.[143]

120. Holdeman contends that the Commission "should modify the current preemption of States preventing them from having more stringent reliability standards for Commission regulated entities than Commission standards."[144] As the Commission

---

[141] *See* Briggs Comments at 7; EIS Comments at 3; JINSA Comments at 2.

[142] Briggs Comments at 7.

[143] Order No. 797, 147 FERC ¶ 61,209 at P 42 (citing Order No. 779, 143 FERC ¶ 61,147 at P 14 n.20).

[144] Holdeman Comments at 2.

indicated in response to similar comments in Order No. 797, section 215(i)(3) of the FPA provides in relevant part that section 215 does not "preempt any authority of any State to take action to ensure the safety, adequacy, and reliability of electric service within that State, as long as such action is not inconsistent with any reliability standard."[145] Moreover, Reliability Standard TPL-007-1 does not preclude users, owners, and operators of the Bulk-Power System from taking additional steps that are designed to mitigate the effects of GMD events, provided those additional steps are not inconsistent with the Commission-approved Reliability Standards.

121. Certain commenters opposed to Reliability Standard TPL-007-1 contend that its approval could absolve industry of any legal liability should a GMD event cause a disruption to the Bulk-Power System. For example, Resilient Societies "ask[s] the Commission to clarify its expectation that the FERC jurisdictional entities will be held to account, and be subject to liability in the event of gross negligence or willful misconduct in planning for and mitigating solar geomagnetic storms."[146] Resilient Societies also contends that the Commission does not have the legal authority "to grant immunity from liability by setting reliability standards."[147]

---

[145] Order No. 797, 147 FERC ¶ 61,209 at P 44 (citing 16 U.S.C. 824o(i)(3)).

[146] Resilient Societies Comments at 62; *see also* CSP Comments at 1 ("It would be far better for FERC to remand Standard TPL-007-1 in its entirety than to approve a reliability standard that would grant liability protection to utilities while blocking the electric grid protection for the public that a 21st century society requires.").

[147] Resilient Societies Comments at 62.

122. The Commission has never stated in the GMD Reliability Standard rulemakings that compliance with Commission-approved Reliability Standards absolves registered entities from legal liability generally, to the extent legal liability exists, should a disruption occur on the Bulk-Power System due to a GMD event. Resilient Societies' comment appears to misconstrue language in Order No. 779 in which the Commission stated, when directing the development of the Second Stage GMD Reliability Standards, that the "Second Stage GMD Reliability Standard should not impose 'strict liability' on responsible entities for failure to ensure the reliability operation of the Bulk-Power System in the face of a GMD event of unforeseen severity."[148] The Commission's statement merely recognized that the Second Stage GMD Reliability Standard should require registered entities to plan against a defined benchmark GMD event, for the purpose of complying with the proposed Reliability Standard, rather than any GMD event generally (i.e., a GMD event that exceeded the severity of the benchmark GMD event). The Commission did not suggest, nor could it suggest, that compliance with a Reliability Standard would absolve registered entities from general legal liability, if any, arising from a disruption to the Bulk-Power System. The only liability the Commission was referring to in Order No. 779 was the potential for penalties or remediation under section 215 of the FPA for failure to comply with a Commission-approved Reliability Standard.

---

[148] Order No. 779, 143 FERC ¶ 61,147 at P 84.

123. Kappenman, Resilient Societies and Bardin filed comments that addressed the NERC "Level 2" Appeal Panel decision.[149] As a threshold issue, we agree with the Appeal Panel that the issues raised by the appellants in that proceeding are not procedural; instead they address the substantive provisions of Reliability Standard TPL-007-1. Section 8 (Process for Appealing an Action or Inaction) of the NERC Standards Process Manual states:

> Any entity that has directly and materially affected interests and that has been or will be adversely affected by any procedural action or inaction related to the development, approval, revision, reaffirmation, retirement or withdrawal of a Reliability Standard, definition, Variance, associated implementation plan, or Interpretation shall have the right to appeal. This appeals process applies only to the NERC Reliability Standards processes as defined in this manual, not to the technical content of the Reliability Standards action.

The appellants, who have the burden of proof under the NERC Rules of Procedure, have not shown that NERC or the standard drafting team failed to comply with any procedural requirements set forth in the NERC Rules of Procedure.[150] Instead, it would appear that the appeal constitutes a collateral attack on the substantive provisions of Reliability Standard TPL-007-1. As the appellants' substantive concerns with Reliability Standard TPL-007-1 have been addressed in this Final Rule, issues surrounding the NERC "Level 2" Appeal Panel decision are, in any case, moot.

---

[149] NERC August 17, 2015 Filing at Appendix 1 (Decision of Level 2 Appeal Panel SPM Section 8 Appeal the Foundation For Resilient Societies, Inc. TPL-007-1).

[150] NERC Rules of Procedure, Appendix 3A (Standard Processes Manual), Section 8 (Process for Appealing an Action or Inaction) (effective June 26, 2013).

III. Information Collection Statement

124. The collection of information contained in this final rule is subject to review by the Office of Management and Budget (OMB) regulations under section 3507(d) of the Paperwork Reduction Act of 1995 (PRA).[151] OMB's regulations require approval of certain informational collection requirements imposed by agency rules.[152]

125. Upon approval of a collection(s) of information, OMB will assign an OMB control number and an expiration date. Respondents subject to the filing requirements of a rule will not be penalized for failing to respond to these collections of information unless the collections of information display a valid OMB control number.

126. The Commission solicited comments on the need for this information, whether the information will have practical utility, the accuracy of the burden estimates, ways to enhance the quality, utility, and clarity of the information to be collected or retained, and any suggested methods for minimizing respondents' burden, including the use of automated information techniques. The Commission asked that any revised burden or cost estimates submitted by commenters be supported by sufficient detail to understand how the estimates are generated. The Commission received comments on specific requirements in Reliability Standard TPL-007-1, which we address in this Final Rule. However, the Commission did not receive any comments on our reporting burden

[151] 44 U.S.C. 3507(d).

[152] 5 CFR 1320.11.

estimates or on the need for and the purpose of the information collection requirements.[153]

Public Reporting Burden: The Commission approves Reliability Standard TPL-007-1 and the associated implementation plan, violation severity levels, and violation risk factors, as discussed above. Reliability Standard TPL-007-1 will impose new requirements for transmission planners, planning coordinators, transmission owners, and generator owners. Reliability Standard TPL-007-1, Requirement R1 requires planning coordinators, in conjunction with the applicable transmission planner, to identify the responsibilities of the planning coordinator and transmission planner in the planning coordinator's planning area for maintaining models and performing the study or studies needed to complete GMD Vulnerability Assessments. Requirements R2, R3, R4, R5, and R7 refer to the "responsible entity, as determined by Requirement R1," when identifying which applicable planning coordinators or transmission planners are responsible for maintaining models and performing the necessary study or studies. Requirement R2 requires that the responsible entities maintain models for performing the studies needed to complete GMD Vulnerability Assessments, as required in Requirement R4. Requirement R3 requires responsible entities to have criteria for acceptable system steady

[153] While noting the uncertainties surrounding the potential costs associated with implementation of Reliability Standard TPL-007-1 and the potential costs that could arise from a revised Reliability Standard, the Trade Associations stated that they "have no specific comments regarding the OMB cost estimate in the NOPR." Trade Associations Comments at 9.

state voltage performance during a benchmark GMD event. Requirement R4 requires responsible entities to complete a GMD Vulnerability Assessment of the near-term transmission planning horizon once every 60 calendar months. Requirement R5 requires responsible entities to provide GIC flow information to transmission owners and generator owners that own an applicable bulk electric system power transformer in the planning area. This information is necessary for applicable transmission owners and generator owners to conduct the thermal impact assessments required by proposed Requirement R6. Requirement R6 requires applicable transmission owners and generator owners to conduct thermal impact assessments where the maximum effective GIC value provided in proposed Requirement R5, Part 5.1 is 75 A/phase or greater. Requirement R7 requires responsible entities to develop a corrective action plan when its GMD Vulnerability Assessment indicates that its system does not meet the performance requirements of Table 1 – Steady State Planning Events. The corrective action plan must address how the performance requirements will be met, must list the specific deficiencies and associated actions that are necessary to achieve performance, and must set forth a timetable for completion. The Commission estimates the annual reporting burden and cost as follows:

FERC-725N, as modified by the Final Rule in Docket No. RM15-11-000 (TPL-007-1 Reliability Standard for Transmission System Planned Performance for Geomagnetic Disturbance Events)[154]						
	Number of Respondents (1)	**Annual Number of Responses per Respondent (2)**	**Total Number of Responses (1)*(2)=(3)**	**Average Burden Hours & Cost Per Response[155] (4)**	**Total Annual Burden Hours & Total Annual Cost (3)*(4)=(5)**	**Cost per Respondent ($) (5)÷(1)**
(One-time) Requirement 1	121 (PC & TP)	1	121	Eng. 5 hrs. ($331.75); RK 4 hrs. ($149.80)	1,089 hrs. (605 Eng., 484 RK); $58,267.55 ($40,141.75 Eng., $18,125.80 RK)	$481.55
(On-going) Requirement 1	121 (PC & TP)	1	121	Eng. 3 hrs. ($199.05); RK 2 hrs. ($74.90)	605 hrs. (363 Eng., 242 RK); $33,147.95 ($24,085.05 Eng., $9,062.90 RK)	$273.95

[154] Eng.=engineer; RK =recordkeeping (record clerk); PC=planning coordinator; TP=transmission planner; TO=transmission owner; and GO=generator owner.

[155] The estimates for cost per response are derived using the following formula: Burden Hours per Response * $/hour = Cost per Response. The $66.35/hour figure for an engineer and the $37.45/hour figure for a record clerk are based on data on the average salary plus benefits from the Bureau of Labor Statistics obtainable at http://www.bls.gov/oes/current/naics3_221000.htm and http://www.bls.gov/news.release/ecec.nr0.htm.

(One-time) Requirement 2	121 (PC & TP)	1	121	Eng. 22 hrs. ($1,459.70); RK 18 hrs. ($674.10)	4840 hrs. (2,662 Eng., 2,178 RK); $258,189.80 ($176,623.70 Eng., $81,566.10 RK)	$2,133.80
(On-going) Requirement 2	121 (PC & TP)	1	121	Eng. 5 hrs. ($331.75); RK 3 hrs. ($112.35)	968 hrs. (605 Eng., 363 RK); $53,736.10 ($40,141.75 Eng., $13,594.35 RK)	$444.10
(One-time) Requirement 3	121 (PC & TP)	1	121	Eng. 5 hrs. ($331.75); RK 3 hrs. ($112.35)	968 hrs. (605 Eng., 363 RK); $53,736.10 ($40,141.75 Eng., $13,594.35 RK)	$444.10
(On-going) Requirement 3	121 (PC & TP)	1	121	Eng. 1 hrs. ($66.35); RK 1 hrs. ($37.45)	242 hrs. (121 Eng., 121 RK); $12,559.80 ($8,028.35 Eng., $4,531.45 RK)	$103.80

(On-going) Requirement 4	121 (PC & TP)	1	121	Eng. 27 hrs. ($1,791.45); RK 21 hrs. ($786.45)	5,808 hrs. (3,267 Eng., 2,541 RK); $311,919.85 ($216,765.45 Eng., $95,154.40 RK)	$2,277.85
(On-going) Requirement 5	121 (PC & TP)	1	121	Eng. 9 hrs. ($597.15); RK 7 hrs. ($262.15)	1936 hrs. (1,089 Eng., 847 RK); $103,975.30 ($72,255.15 Eng., $31,720.15 RK)	$859.30
(One-time) Requirement 6	881 (TO & GO)	1	881	Eng. 22 hrs. ($1,459.70); RK 18 hrs. ($674.19)	35,240 hrs. (19,382 Eng., 15,858 RK); $1,879,957.09 ($1,285,995.70 Eng., $593,961.39 RK)	$2,133.89
(On-going) Requirement 6	881 (TO & GO)	1	881	Eng. 2 hrs. ($132.70); RK 2 hrs. ($74.90)	3,524 hrs. (1,762 Eng., 1762 RK); $182,895.60 ($116,908.70 Eng., $65,986.90 RK)	$207.60

(On-going) Requirement 7	121 (PC & TP)	1	121	Eng. 11 hrs. ($729.85); RK 9 hrs. ($337.05)	2,420 hrs. (1,331 Eng., 1,089 RK); $129,094.90 ($88,311.85 Eng., $40,783.05 RK)	$1,066.90
TOTAL			**2851**		**57,640[156] hrs. (31,792 Eng., 25,848 RK); $3,077,480.04 ($2,109,399.20 Eng., $968,080.84 RK)**	

Title: FERC-725N, Mandatory Reliability Standards: TPL Reliability Standards.

Action: Approved Additional Requirements.

OMB Control No: 1902-0264.

Respondents: Business or other for-profit and not-for-profit institutions.

Frequency of Responses: One time and on-going.

Necessity of the Information: The Commission has reviewed the requirements of Reliability Standard TPL-007-1 and has made a determination that the requirements of

[156] Of the 57,640 total burden hours, 42,137 hours are one-time burden hours, and 15,503 hours are on-going annual burden hours.

this Reliability Standard are necessary to implement section 215 of the FPA. Specifically, these requirements address the threat posed by GMD events to the Bulk-Power System and conform to the Commission's directives regarding development of the Second Stage GMD Reliability Standards, as set forth in Order No. 779.

Internal review: The Commission has assured itself, by means of its internal review, that there is specific, objective support for the burden estimates associated with the information requirements.

127. Interested persons may obtain information on the reporting requirements by contacting the Federal Energy Regulatory Commission, Office of the Executive Director, 888 First Street, NE, Washington, DC 20426 [Attention: Ellen Brown, e-mail: DataClearance@ferc.gov, phone: (202) 502-8663, fax: (202) 273-0873].

128. Comments concerning the information collections proposed in this notice of proposed rulemaking and the associated burden estimates, should be sent to the Commission in this docket and may also be sent to the Office of Management and Budget, Office of Information and Regulatory Affairs [Attention: Desk Officer for the Federal Energy Regulatory Commission]. For security reasons, comments should be sent by e-mail to OMB at the following e-mail address: oira_submission@omb.eop.gov. Please reference FERC-725N and OMB Control No. 1902-0264 in your submission.

IV. Environmental Analysis

129. The Commission is required to prepare an Environmental Assessment or an Environmental Impact Statement for any action that may have a significant adverse effect on the human environment.[157] The Commission has categorically excluded certain actions from this requirement as not having a significant effect on the human environment. Included in the exclusion are rules that are clarifying, corrective, or procedural or that do not substantially change the effect of the regulations being amended.[158] The actions here fall within this categorical exclusion in the Commission's regulations.

V. Regulatory Flexibility Act

130. The Regulatory Flexibility Act of 1980 (RFA)[159] generally requires a description and analysis of final rules that will have significant economic impact on a substantial number of small entities. The Small Business Administration's (SBA) Office of Size Standards develops the numerical definition of a small business.[160] The SBA revised its size standard for electric utilities (effective January 22, 2014) to a standard based on the

---

[157] *Regulations Implementing the National Environmental Policy Act of 1969*, Order No. 486, 52 FR 47897 (Dec. 17, 1987), FERC Stats. & Regs. Preambles 1986-1990 ¶ 30,783 (1987).

[158] 18 CFR 380.4(a)(2)(ii).

[159] 5 U.S.C. 601-12.

[160] 13 CFR 121.101.

number of employees, including affiliates (from a standard based on megawatt hours).[161] Under SBA's new size standards, planning coordinators, transmission planners, transmission owners, and generator owners are likely included in one of the following categories (with the associated size thresholds noted for each):[162]

- Hydroelectric power generation, at 500 employees
- Fossil fuel electric power generation, at 750 employees
- Nuclear electric power generation, at 750 employees
- Other electric power generation (e.g., solar, wind, geothermal, biomass, and other), at 250 employees
- Electric bulk power transmission and control,[163] at 500 employees

131. Based on these categories, the Commission will use a conservative threshold of 750 employees for all entities.[164] Applying this threshold, the Commission estimates that there are 440 small entities that function as planning coordinators, transmission planners, transmission owners, and/or generator owners. However, the Commission estimates that only a subset of such small entities will be subject to the approved Reliability Standard

---

[161] SBA Final Rule on "Small Business Size Standards: Utilities," 78 FR 77,343 (Dec. 23, 2013).

[162] 13 CFR 121.201, Sector 22, Utilities.

[163] This category covers transmission planners and planning coordinators.

[164] By using the highest number threshold for all types of entities, our estimate conservatively treats more entities as "small entities."

given the additional applicability criterion in the approved Reliability Standard (i.e., to be subject to the requirements of the approved Reliability Standard, the applicable entity must own or must have a planning area that contains a large power transformer with a high side, wye-grounded winding with terminal voltage greater than 200 kV).

132. Reliability Standard TPL-007- 1 enhances reliability by establishing requirements that require applicable entities to perform GMD Vulnerability Assessments and to mitigate identified vulnerabilities. The Commission estimates that each of the small entities to whom the approved Reliability Standard applies will incur one-time compliance costs of $5,193.34 and annual ongoing costs of $5,233.50.

133. The Commission does not consider the estimated cost per small entity to impose a significant economic impact on a substantial number of small entities. Accordingly, the Commission certifies that the approved Reliability Standard will not have a significant economic impact on a substantial number of small entities.

VI. Document Availability

134. In addition to publishing the full text of this document in the Federal Register, the Commission provides all interested persons an opportunity to view and/or print the contents of this document via the Internet through FERC's Home Page (http://www.ferc.gov) and in FERC's Public Reference Room during normal business hours (8:30 a.m. to 5:00 p.m. Eastern time) at 888 First Street, N.E., Room 2A, Washington, DC 20426.

135. From FERC's Home Page on the Internet, this information is available on eLibrary. The full text of this document is available on eLibrary in PDF and Microsoft

Word format for viewing, printing, and/or downloading. To access this document in eLibrary, type the docket number excluding the last three digits of this document in the docket number field.

136. User assistance is available for eLibrary and the FERC's website during normal business hours from FERC Online Support at 202-502-6652 (toll free at 1-866-208-3676) or email at ferconlinesupport@ferc.gov, or the Public Reference Room at (202) 502-8371, TTY (202)502-8659. E-mail the Public Reference Room at public.referenceroom@ferc.gov.

VII. Effective Date and Congressional Notification

137. These regulations are effective **[INSERT DATE 60 days after publication in the FEDERAL REGISTER]**. The Commission has determined, with the concurrence of the Administrator of the Office of Information and Regulatory Affairs of OMB, that this rule is not a "major rule" as defined in section 351 of the Small Business Regulatory Enforcement Fairness Act of 1996.

By the Commission.

(S E A L)

Nathaniel J. Davis, Sr.,
Deputy Secretary.

APPENDIX

Commenters

Initial Comments

Abbreviation	Commenter
AEP	American Electric Power Service Corporation
APS	Arizona Public Service Company
ATC	American Transmission Company
Baker	Greta Baker
Bardin	David J. Bardin
BPA	Bonneville Power Administration
Briggs	Kevin Briggs
CEA	Canadian Electricity Association
CSP	Center for Security Policy
EIS	Electric Infrastructure Security Council
Emprimus	Emprimus LLC
Exelon	Exelon Corporation
Gaunt	Charles T. Gaunt
Holdeman	Eric Holdeman
Hydro One	Hydro One Networks Inc.
ITC	International Transmission Company
Lloyd's	Lloyd's America, Inc.
JINSA	Jewish Institute for National Security Affairs
Joint ISOs/RTOs	ISO New England Inc., Midcontinent Independent Transmission System Operator, Inc., Independent Electricity System Operator, New York Independent System Operator, Inc., and PJM Interconnection, L.L.C.
Kappenman	John G. Kappenman and Curtis Birnbach
Morris	Eric S. Morris
NERC	North American Electric Reliability Corporation
Resilient Societies	Foundation for Resilient Societies
Roodman	David Roodman
Trade Associations	American Public Power Association, Edison Electric Institute, Electricity Consumers Resource Council, Electric Power Supply Association, Large Public Power Council, National Rural Electric Cooperative Association
Tri-State	Tri-State Generation and Transmission Association, Inc.
USGS	United States Geological Survey

Supplemental Comments

AEP	American Electric Power Service Corporation
Bardin	David J. Bardin
CSP	Center for Security Policy
Gaunt	Charles T. Gaunt
IEEE	IEEE Power and Energy Society Transformers Committee
Kappenman	John G. Kappenman and Curtis Birnbach
NERC	North American Electric Reliability Corporation
Resilient Societies	Foundation for Resilient Societies
Roodman	David Roodman
Trade Associations	American Public Power Association, Edison Electric Institute, Electricity Consumers Resource Council, Electric Power Supply Association, Large Public Power Council, National Rural Electric Cooperative Association
USGS	United States Geological Survey

156 FERC ¶ 61,050
UNITED STATES OF AMERICA
FEDERAL ENERGY REGULATORY COMMISSION

18 CFR Part 40

[Docket No. RM15-14-002; Order No. 829]

Revised Critical Infrastructure Protection Reliability Standards

(Issued July 21, 2016)

AGENCY: Federal Energy Regulatory Commission.

ACTION: Final rule.

SUMMARY: The Federal Energy Regulatory Commission (Commission) directs the North American Electric Reliability Corporation to develop a new or modified Reliability Standard that addresses supply chain risk management for industrial control system hardware, software, and computing and networking services associated with bulk electric system operations. The new or modified Reliability Standard is intended to mitigate the risk of a cybersecurity incident affecting the reliable operation of the Bulk-Power System.

DATES: This rule will become effective **[INSERT DATE 60 days after publication in the FEDERAL REGISTER].**

FOR FURTHER INFORMATION CONTACT:

Daniel Phillips (Technical Information)
Office of Electric Reliability
Federal Energy Regulatory Commission
888 First Street NE
Washington, DC 20426
(202) 502-6387
daniel.phillips@ferc.gov

Simon Slobodnik (Technical Information)
Office of Electric Reliability
Federal Energy Regulatory Commission
888 First Street NE
Washington, DC 20426
(202) 502-6707
simon.slobodnik@ferc.gov

Kevin Ryan (Legal Information)
Office of the General Counsel
Federal Energy Regulatory Commission
888 First Street NE
Washington, DC 20426
(202) 502-6840
kevin.ryan@ferc.gov

SUPPLEMENTARY INFORMATION:

156 FERC ¶ 61,050
UNITED STATES OF AMERICA
FEDERAL ENERGY REGULATORY COMMISSION

Before Commissioners: Norman C. Bay, Chairman;
Cheryl A. LaFleur, Tony Clark,
and Colette D. Honorable.

Revised Critical Infrastructure Protection Reliability Standards — Docket No. RM15-14-002

ORDER NO. 829

FINAL RULE

(Issued July 21, 2016)

1. Pursuant to section 215(d)(5) of the Federal Power Act (FPA),[1] the Commission directs the North American Electric Reliability Corporation (NERC) to develop a new or modified Reliability Standard that addresses supply chain risk management for industrial control system hardware, software, and computing and networking services associated with bulk electric system operations. The new or modified Reliability Standard is intended to mitigate the risk of a cybersecurity incident affecting the reliable operation of the Bulk-Power System.

2. The record developed in this proceeding supports our determination under FPA section 215(d)(5) that it is appropriate to direct the creation of mandatory requirements

[1] 16 U.S.C. 824o(d)(5).

that protect aspects of the supply chain that are within the control of responsible entities and that fall within the scope of our authority under FPA section 215. Specifically, we direct NERC to develop a forward-looking, objective-based Reliability Standard to require each affected entity to develop and implement a plan that includes security controls for supply chain management for industrial control system hardware, software, and services associated with bulk electric system operations.[2] The new or modified Reliability Standard should address the following security objectives, discussed in detail below: (1) software integrity and authenticity; (2) vendor remote access; (3) information system planning; and (4) vendor risk management and procurement controls. In making this directive, the Commission does not require NERC to impose any specific controls, nor does the Commission require NERC to propose "one-size-fits-all" requirements. The new or modified Reliability Standard should instead require responsible entities to develop a plan to meet the four objectives, or some equally efficient and effective means to meet these objectives, while providing flexibility to responsible entities as to how to meet those objectives.

---

[2] *Revised Critical Infrastructure Protection Reliability Standards*, Notice of Proposed Rulemaking, 80 Fed. Reg. 43,354 (Jul. 22, 2015), 152 FERC ¶ 61,054, at P 66 (2015) (NOPR).

I. Background

A. Section 215 and Mandatory Reliability Standards

3. Section 215 of the FPA requires a Commission-certified Electric Reliability Organization (ERO) to develop mandatory and enforceable Reliability Standards, subject to Commission review and approval. Reliability Standards may be enforced by the ERO, subject to Commission oversight, or by the Commission independently.[3] Pursuant to section 215 of the FPA, the Commission established a process to select and certify an ERO,[4] and subsequently certified NERC.[5]

B. Notice of Proposed Rulemaking

4. The NOPR, *inter alia*, identified as a reliability concern the potential risks to bulk electric system reliability posed by the "supply chain" (i.e., the sequence of processes involved in the production and distribution of, *inter alia*, industrial control system hardware, software, and services). The NOPR explained that changes in the bulk electric system cyber threat landscape, exemplified by recent malware campaigns targeting supply chain vendors, have highlighted a gap in the Critical Infrastructure Protection

---

[3] 16 U.S.C. 824o(e).

[4] *Rules Concerning Certification of the Electric Reliability Organization; and Procedures for the Establishment, Approval, and Enforcement of Electric Reliability Standards*, Order No. 672, FERC Stats. & Regs. ¶ 31,204, *order on reh'g*, Order No. 672-A, FERC Stats. & Regs. ¶ 31,212 (2006).

[5] *North American Electric Reliability Corp.*, 116 FERC ¶ 61,062, *order on reh'g and compliance*, 117 FERC ¶ 61,126 (2006), *aff'd sub nom. Alcoa, Inc. v. FERC*, 564 F.3d 1342 (D.C. Cir. 2009).

(CIP) Reliability Standards.[6] To address this gap, the NOPR proposed to direct that NERC develop a forward-looking, objective-driven Reliability Standard that provides security controls for supply chain management for industrial control system hardware, software, and services associated with bulk electric system operations.[7]

5. Recognizing that developing supply chain management requirements would likely be a significant undertaking and require extensive engagement with stakeholders to define the scope, content, and timing of the Reliability Standard, the Commission sought comment on: (1) the general proposal to direct that NERC develop a Reliability Standard to address supply chain management; (2) the anticipated features of, and requirements that should be included in, such a standard; and (3) a reasonable timeframe for development of a Reliability Standard.[8]

6. In response to the NOPR, thirty-four entities submitted comments on the NOPR proposal regarding supply chain risk management. A list of these commenters appears in Appendix A.

C. January 28, 2016 Technical Conference

7. On January 28, 2016, Commission staff led a Technical Conference to facilitate a dialogue on supply chain risk management issues that were identified by the Commission

---

[6] NOPR, 152 FERC ¶ 61,054 at P 63.

[7] *Id.* P 66.

[8] *Id.*

in the NOPR. The January 28 Technical Conference addressed: (1) the need for a new or modified Reliability Standard; (2) the scope and implementation of a new or modified Reliability Standard; and (3) current supply chain risk management practices and collaborative efforts.

8. Twenty-four entities representing industry, government, vendors, and academia participated in the January 28 Technical Conference through written comments and/or presentations.[9]

9. We address below the comments submitted in response to the NOPR and comments made as part of the January 28 Technical Conference.

II. Discussion

10. Pursuant to section 215(d)(5) of the FPA, the Commission determines that it is appropriate to direct NERC to develop a new or modified Reliability Standard(s) that address supply chain risk management for industrial control system hardware, software, and computing and networking services associated with bulk electric system operations.[10] Based on the comments received in response to the NOPR and at the technical conference, we determine that the record in this proceeding supports the development of

---

[9] Written presentations at the January 28, 2016 Technical Conference and the Technical Conference transcript referenced in this Final Rule are accessible through the Commission's eLibrary document retrieval system in Docket No. RM15-14-000.

[10] 16 U.S.C. 824o(d)(5) ("The Commission . . . may order the [ERO] to submit to the Commission a proposed reliability standard or a modification to a reliability standard that addresses as specific matter if the Commission considers such a new or modified reliability standard appropriate to carry out this section.").

mandatory requirements for the protection of aspects of the supply chain that are within the control of responsible entities and that fall within the scope of our authority under FPA section 215.

11. In its NOPR comments, NERC acknowledges that "supply chains for information and communications technology and industrial control systems present significant risks to [Bulk-Power System] security, providing various opportunities for adversaries to initiate cyberattacks."[11] Several other commenters also recognized the risks posed to the bulk electric system by supply chain security issues and generally support, or at least do not oppose, Commission action to address the reliability gap.[12] For example, in prepared remarks submitted for the January 28 Technical Conference, one panelist noted that attacks targeting the supply chain are on the rise, particularly attacks involving third party service providers.[13] In addition, it was noted that, while many responsible entities are already independently assessing supply chain risks and asking vendors to address the

---

[11] NERC NOPR Comments at 8.

[12] *See* Peak NOPR Comments at 3-6; ITC NOPR Comments at 13-15; CyberArk NOPR Comments at 4; Ericsson NOPR Comments at 2; Isologic and Resilient Societies Joint NOPR Comments at 9-12; ACS NOPR Comments at 4; ISO NE NOPR Comments at 2-3; NEMA NOPR Comments at 1-2.

[13] Olcott Technical Conference Comments at 1-2.

risks, these individual efforts are likely to be less effective than a mandatory Reliability Standard.[14]

12. We recognize, however, that most commenters oppose development of Reliability Standards addressing supply chain management for various reasons. These commenters contend that Commission action on supply chain risk management would, among other things, address or influence activities beyond the scope of the Commission's FPA section 215 jurisdiction.[15] Commenters also assert that the existing CIP Reliability Standards adequately address potential risks to the bulk electric system from supply chain issues.[16] In addition, commenters claim that responsible entities have minimal control over their suppliers and are not able to identify all potential vulnerabilities associated with each of their products or parts; therefore, even if a responsible entity identifies a vulnerability created by a supplier, the responsible entity does not necessarily have any authority,

---

[14] Galloway Technical Conference Comments at 1 ("…ISO-NE supports the Commission's proposal to direct NERC to develop requirements relating to supply chain risk management. We believe that the risks to the reliability of the Bulk Electric System that result from compromised third-party software are real, significant and largely unaddressed by existing reliability standards. While many public utilities are already assessing these risks and asking vendors to address them, these one-off efforts are far less likely to be effective than an industry-wide reliability standard.").

[15] *See* Trade Associations NOPR Comments at 24; Southern NOPR Comments at 14-16; CEA NOPR Comments at 4-5; NIPSCO NOPR Comments at 7.

[16] *See* Trade Associations NOPR Comments at 20-25; Gridwise NOPR Comments at 3; Arkansas NOPR Comments at 6; G&T Cooperatives NOPR Comments at 8-9; NEI NOPR Comments at 3-5; NIPSCO NOPR Comments at 5-6; Luminant NOPR Comments at 4-5; SCE NOPR Comments at 4.

influence or means to require the supplier to apply mitigation.[17] Other commenters argue that the Commission's proposal may unintentionally inhibit innovation.[18] A number of commenters assert that voluntary guidelines would be more effective at addressing the Commission's concerns.[19] Finally, commenters are concerned that the contractual flexibility necessary to effectively address supply chain concerns does not fit well with a mandatory Reliability Standard.[20]

13. As discussed below, we conclude that our directive falls within the Commission's authority under FPA section 215. We also determine that, notwithstanding the concerns raised by commenters opposed to the NOPR proposal, it is appropriate to direct the development of mandatory requirements to protect industrial control system hardware, software, and computing and networking services associated with bulk electric system operations. Many of the commenters' concerns are addressed by the flexibility inherent in our directive to develop a forward-looking, objective-based Reliability Standard that includes specific security objectives that a responsible entity must achieve, but affords flexibility in how to meet these objectives. The Commission does not require NERC to

---

[17] *See* Arkansas NOPR Comments at 5-6; G&T Cooperatives NOPR Comments at 9; Trade Associations NOPR Comments at 25.

[18] *See* Arkansas NOPR Comments at 6; G&T Cooperatives NOPR Comments at 9; NERC NOPR Comments at 13.

[19] *See* Trade Associations NOPR Comments at 23; Southern NOPR Comments at 13; AEP NOPR Comments at 5; NextEra NOPR Comments at 4-5; Luminant NOPR Comments at 5.

[20] *See* Arkansas NOPR Comments at 6; Southern NOPR Comments at 13.

impose any specific controls nor does the Commission require NERC to propose "one-size-fits-all" requirements. The new or modified Reliability Standard should instead require responsible entities to develop a plan to meet the four objectives, or some equally efficient and effective means to meet these objectives, while providing flexibility to responsible entities as to how to meet those objectives. Moreover, our directive comports well with the NOPR comments submitted by NERC, in which NERC explained what it believes would be the features of a workable supply chain management Reliability Standard.[21]

14. We address below the following issues raised in the NOPR, NOPR comments, and January 28 Technical Conference comments: (1) the Commission's authority to direct the ERO to develop supply chain management Reliability Standards under FPA section 215(d)(5); and (2) the need for supply chain management Reliability Standards, including the risks posed by the supply chain, objectives of a supply chain management Reliability Standard, existing CIP Reliability Standards, and responsible entities' ability to affect the supply chain.

---

[21] NERC NOPR Comments at 8-9. The record evidence on which the directive in this Final Rule is based is either comparable or superior to past instances in which the Commission has directed, pursuant to FPA section 215(d)(5), that NERC propose a Reliability Standard to address a gap in existing Reliability Standards. *See, e.g.*, *Reliability Standards for Physical Security Measures*, 146 FERC ¶ 61,166 (2014) (directing, without seeking comment, that NERC develop proposed Reliability Standards to protect against physical security risks related to the Bulk-Power System).

A. Commission Authority to Direct the ERO to Develop Supply Chain Management Reliability Standards Under FPA Section 215(d)(5)

NOPR

15. In the NOPR, the Commission stated that it anticipates that a Reliability Standard addressing supply chain management security would, *inter alia*, respect FPA Section 215 jurisdiction by only addressing the obligations of responsible entities and not directly imposing obligations on suppliers, vendors, or other entities that provide products or services to responsible entities.[22]

Comments

16. Commenters contend that the Commission's proposal to direct NERC to develop mandatory Reliability Standards to address supply chain risks could exceed the Commission's jurisdiction under FPA section 215. The Trade Associations state that the NOPR discussion "appears to suggest a new mandate, over and above Section 215 for energy security, integrity, quality, and supply chain resilience, and the future acquisition of products and services."[23] The Trade Associations assert that the Commission's NOPR proposal does not provide any reasoning that connects energy security and integrity with reliable operations for Bulk-Power System reliability. The Trade Associations seek

[22] NOPR, 152 FERC ¶ 61,054 at P 66.

[23] Trade Associations NOPR Comments at 24.

clarification that the Commission does not intend to define energy security as a new policy mandate.[24]

17. Southern states that it agrees with the Trade Associations that expanding the focus of the NERC Reliability Standards "to include concepts such as security, integrity, and supply chain resilience is beyond the statutory authority granted in Section 215."[25] Southern contends that while these areas "have an impact on the reliable operation of the bulk power system, [...] they are areas that are beyond the scope of [the Commission's] jurisdiction under Section 215."[26] NIPSCO raises a similar argument, stating that the existing CIP Reliability Standards should address the Commission's concerns "without involving processes and industries outside of the Commission's jurisdiction under section 215 of the Federal Power Act."[27]

18. Southern questions how a mandatory Reliability Standard that achieves all of the objectives specified in the NOPR "could effectively address [the Commission's] concerns and still stay within the bounds of [the Commission's] scope and mission under Section 215."[28] Southern asserts that "a reading of Section 215 indicates that [the Commission's]

---

[24] *Id.*

[25] Southern NOPR Comments at 16.

[26] Southern NOPR Comments at 16; *see also* Trade Association NOPR Comments at 24.

[27] NIPSCO NOPR Comments at 7.

[28] Southern NOPR Comments at 14-15.

mission and authority under Section 215 is focused on the *operation* of the bulk power system elements, not on the acquisition of those elements and associated procurement practices."[29] In support of its assertion, Southern points to the definition in FPA section 215 of "reliability standard," noting the use and meaning of the terms "reliable operation" and "operation." Southern contends that "Section 215 standards should ensure that a given BES Cyber System asset is protected from vulnerabilities once connected to the BES, and should not be concerned about how the Responsible Entity works with its vendors and suppliers to ensure such reliability (such as higher financial incentives or greater contractual penalties)."[30]

19. The Trade Associations and Southern also observe that, while the NOPR indicates that the Commission has no direct oversight authority over third-party suppliers or vendors and cannot indirectly assert authority over them through jurisdictional entities, the NOPR proposal appears to assert that authority.[31] The Trade Associations maintain that such an extension of the Commission's authority would be unlawful and, therefore, seek clarification that "the Commission will avoid seeking to extend its authority since such an extension would set a troubling precedent."[32] CEA raises a concern that the

---

[29] *Id.* at 15 (emphasis in original).

[30] *Id.* at 16.

[31] Trade Associations NOPR Comments at 24-25; Southern NOPR Comments at 17; *see also* Trade Associations Post-Technical Conference Comments at 20-21.

[32] Trade Associations NOPR Comments at 24-25.

NOPR proposal "appears to lend itself to the interpretation that authority is indirectly being asserted over non-jurisdictional entities."[33]

20. The Trade Associations also maintain that the Commission's use of the term "industrial control system" in the scope of its proposal suggests that the Commission is seeking to address issues beyond CIP and cybersecurity-related issues. The Trade Associations seek clarification that the Commission does not intend for NERC broadly to address industrial control systems, such as fuel procurement and delivery systems or system protection devices, but intends for its proposal to be limited to CIP and cybersecurity-related issues.[34]

Discussion

21. We are satisfied that FPA section 215 provides the Commission with the authority to direct NERC to address the reliability gap concerning supply chain management risks identified in the NOPR. We reject the contention that our directive could be read to address issues outside of the Commission's FPA section 215 jurisdiction. However, to be clear, we reiterate the statement in the NOPR that any action taken by NERC in response to the Commission's directive to address the supply chain-related reliability gap should respect "section 215 jurisdiction by only addressing the obligations of responsible entities" and "not directly impose obligations on suppliers, vendors or other entities that

---

[33] CEA NOPR Comments at 5.

[34] Trade Associations NOPR Comments at 25.

provide products or services to responsible entities."[35] The Commission expects that NERC will adhere to this instruction as it works with stakeholders to develop a new or modified Reliability Standard to address the Commission's directive. As discussed below, we reject the remaining comments regarding the Commission's authority to direct the development of supply chain management Reliability Standards under FPA section 215(d)(5).

22. Our directive does not suggest, as the Trade Associations contend, a new mandate above and beyond FPA section 215. The Commission's directive to NERC to address supply chain risk management for industrial control system hardware, software, and computing and networking services associated with bulk electric system operations is not intended to "define 'energy security' as a new policy mandate" under the CIP Reliability Standards.[36] Instead, our directive is meant to enhance bulk electric system cybersecurity by addressing the gap in the CIP Reliability Standards identified in the NOPR relating to supply chain risk management for industrial control system hardware, software, and computing and networking services associated with bulk electric system operations. This directive is squarely within the statutory definition of a "reliability standard," which includes requirements for "cybersecurity protection."[37]

---

[35] NOPR, 152 FERC ¶ 61,054 at P 66.

[36] *See* Trade Associations NOPR Comments at 24.

[37] *See* 16 U.S.C. 824o(a)(3) (defining "reliability standard" to mean "a requirement, approved by the Commission under [section 215 of the FPA] to provide for

(continued…)

23. We reject Southern's argument that FPA section 215 limits the scope of the NERC Reliability Standards to "ensur[ing] that a given BES Cyber System asset is protected from vulnerabilities once connected" to the bulk electric system.[38] While Southern's comment implies that the Commission should only be concerned with real-time operations based on the definition of the term "reliable operation," the definition of "reliability standard" in FPA section 215 also includes requirements for "the design of planned additions or modifications" to bulk electric system facilities "necessary to provide for reliable operation of the bulk-power system."[39] Moreover, as noted, FPA section 215 is clear that maintaining reliable operation also includes protecting the bulk electric system from cybersecurity incidents.[40] Indeed, our findings and directives in the Final Rule are intended to better protect the Bulk-Power System from potential cybersecurity incidents that could adversely affect reliable operation of the Bulk-Power System. Accordingly, we would not be carrying out our obligations under FPA section 215 if the Commission determined that cybersecurity incidents resulting from gaps in supply chain risk management were outside the scope of FPA section 215.

---

the reliable operation of the bulk-power system. The term includes requirements for the operation of existing bulk-power system facilities, *including cybersecurity protection*, and *the design of planned additions or modifications to such facilities* to the extent necessary to provide for reliable operation…") (emphasis added).

[38] *See* Southern NOPR Comments at 16.

[39] *See* 16 U.S.C. 824o(a)(4) (defining "reliable operation"); *see also* 16 U.S.C. 824o(a)(3).

[40] *See* 16 U.S.C. 824o(a)(4).

24. With regard to concerns that the NOPR's use of the term "industrial control system" signals the Commission's intent to address issues beyond the CIP Reliability Standards or cybersecurity controls, we clarify that our directive is only intended to address the protection of hardware, software, and computing and networking services associated with bulk electric system operations from supply chain-related cybersecurity threats and vulnerabilities.

B. Need for a New or Modified Reliability Standard

1. Cyber Risks Posed by the Supply Chain

NOPR

25. In the NOPR, the Commission observed that the global supply chain, while providing an opportunity for significant benefits to customers, enables opportunities for adversaries to directly or indirectly affect the operations of companies that may result in risks to the end user. The NOPR identified supply chain risks including the insertion of counterfeits, unauthorized production, tampering, theft, or insertion of malicious software, as well as poor manufacturing and development practices. The NOPR pointed to changes in the bulk electric system cyber threat landscape, evidenced by recent malware campaigns targeting supply chain vendors, which highlighted a gap in the protections under the current CIP Reliability Standards.[41]

26. Specifically, the NOPR identified two focused malware campaigns identified by the Department of Homeland Security's Industry Control System - Computer Emergency

[41] NOPR, 152 FERC ¶ 61,054 at PP 61-62.

Readiness Team (ICS-CERT) in 2014.[42] The NOPR stated that this new type of malware campaign is based on the injection of malware while a product or service remains in the control of the hardware or software vendor, prior to delivery to the customer.[43]

Comments

27. NERC acknowledges the NOPR's concerns regarding the threats posed by supply chain management risks to the Bulk-Power System. NERC states that "the supply chains for information and communications technology and industrial control systems present significant risks to [Bulk-Power System] security, providing various opportunities for adversaries to initiate cyberattacks."[44] NERC further explains that "supply chains risks are … complex, multidimensional, and constantly evolving, and may include, as the Commission states, insertion of counterfeits, unauthorized production, tampering, theft, insertion of malicious software and hardware, as well as poor manufacturing and development practices."[45] NERC states, however, that as to these supply chains, there

---

[42] *Id.* P 63 (citing ICS-CERT, *Alert: ICS Focused Malware (Update A)*, https://ics-cert.us-cert.gov/alerts/ICS-ALERT-14-176-02A; ICS-CERT, *Alert Ongoing Sophisticated Malware Campaign Compromising ICS (Update E)*, https://ics-cert.us-cert.gov/alerts/ICS-ALERT-14-281-01B). ICS-CERT is a division of the Department of Homeland Security that works to reduce risks within and across all critical infrastructure sectors by partnering with law enforcement agencies and the intelligence community.

[43] NOPR, 152 FERC ¶ 61,054 at P 63.

[44] NERC NOPR Comments at 8.

[45] *Id.* at 10.

are "significant challenges to developing a mandatory Reliability Standard consistent with [FPA] Section 215...."[46]

28. IRC, Peak, Idaho Power, CyberArk, NEMA, Resilient Societies and other commenters share the NOPR's concern that supply chain risks pose a threat to bulk electric system reliability. IRC states that it supports the Commission's efforts to address the risks associated with supply chain management.[47] Peak explains that "the security risk of supply chain management is a real threat, and ... a CIP standard for supply chain management may be necessary."[48] Peak notes, for example, that it is possible for a malware campaign to infect industrial control software with malicious code while the product or service is in the control of the hardware and software vendor, and states that, "[w]ithout proper controls, the vendor may deliver this infected product or service, unknowingly passing the risk onto the utility industry customer."[49] Isologic and Resilient Societies comments that supply chain vulnerabilities are one of the most difficult areas of cybersecurity because, among other concerns, entities "are seldom aware of the risks [supply chain vulnerabilities] pose."[50]

---

[46] *Id.* at 2.

[47] IRC NOPR Comments at 1-2.

[48] Peak NOPR Comments at 3.

[49] *Id.* at 3.

[50] Isologic and Resilient Societies Joint NOPR Comments at 9.

29. Idaho Power agrees "that the supply chain could pose an attack vector for certain risks to the bulk electric system."[51] CyberArk states that "infection of vendor web sites is just one of the potential ways a supply chain management attack could be executed" and notes that network communications links between a vendor and its customer could be used as well.[52] NEMA agrees with the NOPR that "keeping the electric sector supply chain free from malware and other cybersecurity risks is essential."[53] NEMA highlights a number of principles it represents as vendor best practices, and encourages the Commission and NERC to reference those principles as the effort to address supply chain risks progresses.[54]

30. Other commenters do not agree that the risks identified in the NOPR support the Commission's NOPR proposal. The Trade Associations, Southern, and NIPSCO contend that the two malware campaigns identified by ICS-CERT and cited in the NOPR do not actually represent a changed threat landscape that defines a reliability gap. Specifically, the Trade Associations state that the two identified malware campaigns "seek to inject malware, while a product is in the control of and in use by the customer and not, as the NOPR suggests, the vendor."[55] In support of this position, the Trade Associations note

---

[51] Idaho Power NOPR Comments at 3.

[52] CyberArk NOPR Comments at 4.

[53] NEMA NOPR Comments at 1.

[54] *Id.* at 2.

[55] Trade Associations NOPR Comments at 20-21.

that the ICS-CERT mitigation measures for the two alerts "focused on the customer and do not address security controls, while the products are under control of the vendors."[56]

31. The Trade Associations and Southern also contend that there is no information from various NERC programs and activities that leads to a reasonable conclusion that supply chain management issues have caused events or disturbances on the bulk electric system.[57] Luminant states that it "does not perceive the same reliability gap that is expressed in the NOPR concerning risks associated with supply chain management" and contends that it is important to understand the potential risks and cost impacts related to any potential mitigation efforts before developing any additional security controls.[58] KCP&L states that it does not share the Commission's view of the supply chain-related reliability gap described in the NOPR and, therefore, does not support the Commission's proposal.[59]

Discussion

32. We find ample support in the record to conclude that supply chain management risks pose a threat to bulk electric system reliability. As NERC commented, "the supply chains for information and communications technology and industrial control systems

---

[56] Trade Associations NOPR Comments at 21; *see also* NIPSCO NOPR Comments at 6.

[57] Trade Associations NOPR Comments at 21; Southern Comments at 11.

[58] Luminant NOPR Comments at 4.

[59] KCP&L NOPR Comments at 7.

present significant risks to [Bulk-Power System] security, providing various opportunities for adversaries to initiate cyberattacks."[60] The malware campaigns analyzed by ICS-CERT and identified in the NOPR are only examples of such risks (i.e., supply chain attacks targeting supply chain vendors). Commenters identified additional supply chain-related threats,[61] including events targeting electric utility vendors.[62]

33. Even among the comments opposed to the NOPR, there is acknowledgment that supply chain reliability risks exist. The Trade Associations state that their "respective

---

[60] NERC NOPR Comments at 8.

[61] Commenters reference tools and information security frameworks, such as ES-C2M2, NIST-SP-800-161 and NIST-SP-800-53, which describe the scope of supply chain risk that could impact bulk electric system operations. *See* Department of Energy, Electricity Subsector Cybersecurity Capability Maturity Model (February 2014), http://energy.gov/sites/prod/files/2014/02/f7/ES-C2M2-v1-1-Feb2014.pdf; NIST Special Publication 800-161, *Supply Chain Risk Management Practices for Federal Information Systems and Organizations* at 51, http://nvlpubs.nist.gov/nistpubs/SpecialPublications/NIST.SP.800-161.pdf; NIST Special Publication 800-53, *Security and Privacy Controls for Federal Information Systems and Organizations*, http://nvlpubs.nist.gov/nistpubs/SpecialPublications/NIST.SP.800-53r4.pdf. These risks include the insertion of counterfeits, unauthorized production and modification of products, tampering, theft, intentional insertion of tracking software, as well as poor manufacturing and development practices. One technical conference participant noted that supply chain attacks can target either (1) the hardware/software components of a system (thereby creating vulnerabilities that can be exploited by a remote attacker) or (2) a third party service provider who has access to sensitive IT infrastructure or holds/maintains sensitive data. *See* Olcott Technical Conference Comments at 1.

[62] Olcott discusses two events targeting electric utility vendors and service providers. Olcott Technical Conference Comments at 2. Specific recent examples of attacks on third party vendors include: (1) unauthorized code found in Juniper Firewalls in 2015; (2) the 2013 Target incident involving stolen vendor credentials; (3) the 2015 Office of Personnel Management incident also involving stolen vendor credentials; and (4) two events targeting electric utility vendors. *See id.* at 1-4.

members have identified security issues associated with potential supply chain disruption or compromise as being a significant threat."[63] Recognizing that such risks exist, we reject the assertion by the Trade Associations and Southern that there is an inadequate basis for the Commission to take action because "[t]he Trade Associations can find nothing within various NERC programs and activities that lead to a reasonable conclusion that supply chain management issues have caused events or disturbances on the bulk power system."[64]

34. We disagree with the Trade Associations' arguments suggesting that the two malware campaigns identified in the NOPR do not represent a change in the threat landscape to the bulk electric system. First, while the Trade Associations are correct that the ICS-CERT alerts referenced in the NOPR describe remediation steps for customers to take in the event of a breach, the vulnerabilities exploited by those campaigns were the direct result of vendor decisions about: (1) how to deliver software patches to their customers and (2) the necessary degree of remote access functionality for their information and communications technology products.[65] Second, the malware campaigns also demonstrate that attackers have expanded their efforts to include the execution of

---

[63] Trade Associations NOPR Comments at 17.

[64] *See* Trade Associations NOPR Comments at 21.

[65] The ICS-CERT alert regarding ICS Focused Malware indicated that "the software installers for … vendors were infected with malware known as the Havex Trojan."

broad access campaigns targeting vendors and software applications, rather than just individual entities. The targeting of vendors and software applications with potentially broad access to BES Cyber Systems[66] marks a turning point in that it is no longer sufficient to focus protection strategies exclusively on post-acquisition activities at individual entities. Instead, we believe that attention should also be focused on minimizing the attack surfaces of information and communications technology products procured to support bulk electric system operations.

2. Objectives of a Supply Chain Management Reliability Standard

NOPR

35. The NOPR stated that the reliability goal of a supply chain risk management Reliability Standard should be a forward-looking, objective-driven Reliability Standard that encompasses activities in the system development life cycle: from research and development, design and manufacturing stages (where applicable), to acquisition, delivery, integration, operations, retirement, and eventual disposal of the responsible

[66] Cyber systems are referred to as "BES Cyber Systems" in the CIP Reliability Standards. The NERC Glossary defines BES Cyber Systems as "One or more BES Cyber Assets logically grouped by a responsible entity to perform one or more reliability tasks for a functional entity." NERC Glossary of Terms Used in Reliability Standards (May 17, 2016) at 15 (NERC Glossary). The NERC Glossary defines "BES Cyber Asset" as "A Cyber Asset that if rendered unavailable, degraded, or misused would, within 15 minutes of its required operation, misoperation, or non-operation, adversely impact one or more Facilities, systems, or equipment, which, if destroyed, degraded, or otherwise rendered unavailable when needed, would affect the reliable operation of the Bulk Electric System. Redundancy of affected Facilities, systems, and equipment shall not be considered when determining adverse impact. Each BES Cyber Asset is included in one or more BES Cyber Systems." *Id.*

entity's information and communications technology and industrial control system supply chain equipment and services. The NOPR explained that the Reliability Standard should support and ensure security, integrity, quality, and resilience of the supply chain and the future acquisition of products and services.[67]

36. The NOPR recognized that, due to the breadth of the topic and the individualized nature of many aspects of supply chain management, a Reliability Standard pertaining to supply chain management security should:

- Respect FPA section 215 jurisdiction by only addressing the obligations of responsible entities. A Reliability Standard should not directly impose obligations on suppliers, vendors or other entities that provide products or services to responsible entities.
- Be forward-looking in the sense that the Reliability Standard should not dictate the abrogation or re-negotiation of currently-effective contracts with vendors, suppliers or other entities.
- Recognize the individualized nature of many aspects of supply chain management by setting goals (the "what"), while allowing flexibility in how a responsible entity subject to the Reliability Standard achieves that goal (the "how").

[67] NOPR, 152 FERC ¶ 61,054 at P 64.

- Given the types of specialty products involved and the diversity of acquisition processes, the Reliability Standard may need to allow exceptions (e.g., to meet safety requirements and fill operational gaps if no secure products are available).
- Provide enough specificity so that compliance obligations are clear and enforceable. In particular, the Commission anticipated that a Reliability Standard that simply requires a responsible entity to "have a plan" addressing supply chain management would not suffice. Rather, to adequately address the concerns identified in the NOPR, the Commission stated a Reliability Standard should identify specific controls.[68]

37. The NOPR recognized that, because security controls for supply chain management likely vary greatly with each responsible entity due to variations in individual business practices, the right set of supply chain management security controls should accommodate, *inter alia*, an entity's: (1) procurement process; (2) vendor relations; (3) system requirements; (4) information technology implementation; and (5) privileged commercial or financial information. As examples of controls that may be instructional in the development of any new Reliability Standard, the NOPR identified the following Supply Chain Risk Management controls from NIST SP 800-161: (1) Access Control Policy and Procedures; (2) Security Assessment Authorization; (3) Configuration Management; (4) Identification and Authentication; (5) System

---

[68] *Id.* P 66.

Maintenance Policy and Procedures; (6) Personnel Security Policy and Procedures; (7) System and Services Acquisition; (8) Supply Chain Protection; and (9) Component Authenticity.[69]

Comments

38. NERC states that a Commission directive requiring the development of a supply chain risk management Reliability Standard: (1) should provide a minimum of two years for Reliability Standard development activities; (2) should clarify that any such Reliability Standard build on existing protections in the CIP Reliability Standards and the practices of responsible entities, and focus primarily on those procedural controls that responsible entities can reasonably be expected to implement during the procurement of products and services associated with bulk electric system operations to manage supply chain risks; and (3) must be flexible to account for differences in the needs and characteristics of responsible entities, the diversity of bulk electric system environments, technologies, risks, and issues related to the limited applicability of mandatory NERC Reliability Standards.[70]

39. While sharing the Commission's concern that supply chain risks pose a threat to bulk electric system reliability, some commenters suggest that the Commission address certain threshold issues before moving forward with the NOPR proposal. IRC notes its

---

[69] NOPR, 152 FERC ¶ 61,054 at P 65 (citing NIST Special Publication 800-161 at 51).

[70] NERC NOPR Comments at 8-9.

concern that the NOPR proposal is overly broad, which IRC states could hamper industry's ability to address the Commission's concerns.[71] Idaho Power expresses a concern "that tightening purchasing controls too tightly could also pose a risk because there are limited vendors" available to industry.[72] Idaho Power states that any supply chain Reliability Standard "should be laid out in terms of requirements built around controls that are developed by the regulated entity rather than prescriptive requirements like many other CIP standards."[73] ISO-NE supports the development of procedural controls "such as requirements that Registered Entities must transact with organizations that meet certain criteria, use specified procurement language in contracts, and review and validate vendors' security practices."[74] Peak notes that "the number of vendors for certain hardware, software and services may be limited" and, therefore, a supply chain-related Reliability Standard should grant responsible entities the flexibility "to show preference for, but not the obligation to use, vendors who demonstrate sound supply chain security practices."[75]

40. NERC, the Trade Associations, Southern, Gridwise, and other commenters request that, should the Commission find it reasonable to direct NERC to develop a new or

---

[71] IRC NOPR Comments at 2.

[72] Idaho Power NOPR Comments at 3.

[73] *Id.* at 3-4.

[74] ISO-NE NOPR Comments at 2 (citing NERC NOPR Comments at 17-18).

[75] Peak NOPR Comments at 4.

modified Reliability Standard for supply chain management, the Commission adopt certain principles for NERC to follow in the standards development process. As an initial matter, NERC and other commenters state that the Commission should identify the risks that it intends NERC to address.[76] In addition, NERC, SPP RE, and AEP state that the Commission should ensure that any new or modified supply chain-related Reliability Standard carefully considers the risk being addressed against the cost of mitigating that risk.[77]

41. NERC states that the focus of any supply chain risk management Reliability Standard "should be a set of requirements outlining those procedural controls that entities should take, as purchasers of products and services, to design more secure products and modify the security practices of suppliers, vendors, and other parties throughout the supply chain."[78] Similarly, SPP RE notes that, while one responsible entity alone may not have adequate leverage to make a vendor or supplier adopt adequate security practices, "the collective application of the procurement language across a broad collection of Responsible Entities may achieve the intended improvement in security

---

[76] NERC NOPR Comments at 9-11; Trade Associations NOPR Comments at 26; Gridwise NOPR Comments at 5; AEP NOPR Comments at 8; SPP RE NOPR Comments at 11; EnergySec NOPR Comments at 4.

[77] NERC NOPR Comments at 11-12; SPP RE NOPR Comments at 11; AEP NOPR Comments at 9.

[78] NERC NOPR Comments at 17.

safeguards."[79] Isologic and Resilient Societies recommends limiting the Reliability Standard requirements to a few that are immediately necessary, such as: (1) preventing the installation of cyber related system or grid components which have been reported by ICS-CERT to be provably vulnerable to a supply chain attack, unless the vulnerability has been corrected; (2) removing from operation any system or component reported by ICS-CERT as containing an exploitable vulnerability; and (3) subjecting hardware and software to penetration testing prior to installation on the grid.[80]

42. In post-technical conference comments, while still opposing the NOPR proposal, APPA suggests certain parameters that should govern the development of any supply chain-related Reliability Standard.[81] Specifically, APPA states that a supply chain-related Reliability Standard should be risk-based and "must embody an approach that enables utilities to perform a risk assessment of the hardware and systems that create potential vulnerabilities," similar to the approach taken in Reliability Standard CIP-014-2, Requirement R1 (Physical Security).[82] In addition, APPA states that a supply chain-related Reliability Standard should not require responsible entities to actively manage third-party vendors or their processes since that would risk involving utilities in areas that

---

[79] SPP RE NOPR Comments at 12.

[80] Isologic and Resilient Societies Joint NOPR Comments at 11.

[81] APPA's post-technical conference comments were submitted jointly with LPPC and TAPS.

[82] APPA Post-Technical Conference Comments at 3-4.

are outside of their core expertise. APPA also argues that "it would be unreasonable for any standard that FERC directs to hold utilities liable for the actions of third-party vendors or suppliers."[83] Finally, APPA states that responsible entities should be able to rely on a credible attestation by a vendor or supplier that it complied with identified supply chain security process. APPA contends that this would be the most efficient way to "establish a standard of care on the suppliers' part."[84]

Discussion

43. We direct that NERC, pursuant to section 215(d)(5) of the FPA, develop a forward-looking, objective-driven new or modified Reliability Standard to require each affected entity to develop and implement a plan that includes security controls for supply chain management for industrial control system hardware, software, and services associated with bulk electric system operations. Our directive is consistent with the NOPR comments advocating flexibility as to what form the Commission's directive should take.

44. We agree with NERC and other commenters that a supply chain risk management Reliability Standard should be flexible and fall within the scope of what is possible using Reliability Standards under FPA section 215. The directive discussed below, we believe, is consistent with both points. In particular, the flexibility inherent in our directive

---

[83] *Id.* at 4-5.

[84] *Id.* at 5.

should account for, among other things, differences in the needs and characteristics of responsible entities and the diversity of BES Cyber System environments, technologies and risks. For example, the new or modified Reliability Standard may allow a responsible entity to meet the security objectives discussed below by having a plan to apply different controls based on the criticality of different assets. And by directing NERC to develop a new or modified Reliability Standard, the Commission affords NERC the option of modifying existing Reliability Standards to satisfy our directive. Finally, we direct NERC to submit the new or modified Reliability Standard within one year of the effective date of this Final Rule.[85]

45. The plan required by the new or modified Reliability Standard developed by NERC should address, at a minimum, the following four specific security objectives in the context of addressing supply chain management risks: (1) software integrity and authenticity; (2) vendor remote access; (3) information system planning; and (4) vendor risk management and procurement controls. Responsible entities should be required to achieve these four objectives but have the flexibility as to how to reach the objective (i.e., the Reliability Standard should set goals (the "what"), while allowing flexibility in how a

[85] We note that the Trade Associations request that the Commission allow "at least one year for discussion, development, and approval by the NERC Board of Trustees." *See* Trade Associations Post-Technical Conference Comments at 22. NERC should submit an informational filing within ninety days of the effective date of this Final Rule with a plan to address the Commission's directive.

responsible entity subject to the Reliability Standard achieves that goal (the "how")).[86] Alternatively, NERC can propose an equally effective and efficient approach to address the issues raised in the objectives identified below. In addition, while in the discussion below we identify four objectives, NERC may address additional supply chain management objectives in the standards development process, as it deems appropriate.

46. The new or modified Reliability Standard should also require a periodic reassessment of the utility's selected controls. Consistent with or similar to the requirement in Reliability Standard CIP-003-6, Requirement R1, the Reliability Standard should require the responsible entity's CIP Senior Manager to review and approve the controls adopted to meet the specific security objectives identified in the Reliability Standard at least every 15 months. This periodic assessment should better ensure that the required plan remains up-to-date, addressing current and emerging supply chain-related concerns and vulnerabilities.

47. Also, consistent with this reliance on an objectives-based approach, and as part of this periodic review and approval, the responsible entity's CIP Senior Manager should consider any guidance issued by NERC, the U.S. Department of Homeland Security (DHS) or other relevant authorities for the planning, procurement, and operation of industrial control systems and supporting information systems equipment since the prior approval, and identify any changes made to address the recent guidance. This periodic

[86] *See* Order No. 672, FERC Stats. & Regs. ¶ 31,204 at P 260.

reconsideration will help ensure an ongoing, affirmative process for reviewing and, when appropriate, incorporating such guidance.

First Objective: Software Integrity and Authenticity

48. The new or modified Reliability Standard must address verification of: (1) the identity of the software publisher for all software and patches that are intended for use on BES Cyber Systems; and (2) the integrity of the software and patches before they are installed in the BES Cyber System environment.

49. This objective is intended to reduce the likelihood that an attacker could exploit legitimate vendor patch management processes to deliver compromised software updates or patches to a BES Cyber System. One of the two focused malware campaigns identified by ICS-CERT in 2014 utilized similar tactics, executing what is commonly referred to as a "Watering Hole" attack[87] to exploit affected information systems. Similar tactics appear to have been used in a recently disclosed attack targeting electric sector infrastructure in Japan.[88] These types of attacks might have been prevented had the

[87] "Watering Hole" attacks exploit poor vendor/client patching and updating processes. Attackers generally compromise a vendor of the intended victim and then use the vendor's information system as a jumping off point for their attack. Attackers will often inject malware or replace legitimate files with corrupted files (usually a patch or update) on the vendor's website as part of the attack. The victim then downloads the files without verifying each file's legitimacy believing that it is included in a legitimate patch or update.

[88] *See* Cylance, Operation DustStorm, https://www.cylance.com/hubfs/2015_cylance_website/assets/operation-dust-storm/Op_Dust_Storm_Report.pdf.

affected entities applied adequate integrity and authenticity controls to their patch management processes.

50. As NERC recognizes in its NOPR comments, NIST SP-800-161 "establish[es] instructional reference points for NERC and its stakeholders to leverage in evaluating the appropriate framework for and security controls to include in any mandatory supply chain management Reliability Standard."[89] NIST SP-800-161 includes a number of security controls which, when taken together, reduce the probability of a successful Watering Hole or similar cyberattack in the industrial control system environment and thus could assist in addressing this objective. For example, in the System and Information Integrity (SI) control family, control SI-7 suggests that the integrity of information systems and components should be tested and verified using controls such as digital signatures and obtaining software directly from the developer. In the Configuration Management (CM) control family, control CM-5(3) requires that the information system prevent the installation of firmware or software without verification that the component has been digitally signed to ensure that hardware and software components are genuine and valid. NIST SP-800-161, while not meant to be definitive, provides examples of controls for addressing the Commission's directive regarding this first objective. Other security controls also could meet this objective.

[89] NERC NOPR Comments at 16-17; *see also* Resilient Societies NOPR Comments at 11.

Second Objective: Vendor Remote Access to BES Cyber Systems

51. The new or modified Reliability Standard must address responsible entities' logging and controlling all third-party (i.e., vendor) initiated remote access sessions. This objective covers both user-initiated and machine-to-machine vendor remote access.

52. This objective addresses the threat that vendor credentials could be stolen and used to access a BES Cyber System without the responsible entity's knowledge, as well as the threat that a compromise at a trusted vendor could traverse over an unmonitored connection into a responsible entity's BES Cyber System. The theft of legitimate user credentials appears to have been a critical aspect to the successful execution of the 2015 cyberattack on Ukraine's power grid.[90] In addition, controls adopted under this objective should give responsible entities the ability to rapidly disable remote access sessions in the event of a system breach.

53. DHS noted the importance of controlling vendor remote access in its alert on the Ukrainian cyberattack: "Remote persistent vendor connections should not be allowed into the control network. Remote access should be operator controlled, time limited, and procedurally similar to "lock out, tag out." The same remote access paths for vendor and employee connections can be used; however, double standards should not be allowed."[91]

---

[90] *See* E-ISAC, *Analysis of the Cyber Attack on the Ukrainian Power Grid* at 3 (Mar. 18, 2016), http://www.nerc.com/pa/CI/ESISAC/Documents/E-ISAC_SANS_Ukraine_DUC_18Mar2016.pdf.

[91] *See* ICS-CERT Alert, *Cyber-Attack Against Ukrainian Critical Infrastructure*, https://ics-cert.us-cert.gov/alerts/IR-ALERT-H-16-056-01.

54. NIST SP-800-53 and NIST SP-800-161 provide several security controls which, when taken together, reduce the probability that an attacker could use legitimate third-party access to compromise responsible entity information systems. In the Systems and Communications (SC) control family, for example, control SC-7 addressing boundary protection requires that an entity implement appropriate monitoring and control mechanisms and processes at the boundary between the entity and its suppliers, and that provisions for boundary protections should be incorporated into agreements with suppliers. These protections are applied regardless of whether the remote access session is user-initiated or interactive in nature.

55. In the Access Control (AC) control family, control AC-17 requires usage restrictions, configuration/connection requirements, and monitoring and control for remote access sessions, including the entity's ability to expeditiously disconnect or disable remote access. In the Identification and Authentication (IA) control family, control IA-5 requires changing default "authenticators" (e.g., passwords) prior to information system installation. In the System and Information Integrity (SI) control family, control SI-4 addresses monitoring of vulnerabilities resulting from past information and communication technology supply chain compromises, such as malicious code implanted during software development and set to activate after deployment. These sources, while not meant to be definitive, provide examples of controls for addressing the Commission's directive regarding objective two. Other security controls also could meet this objective.

Third Objective: Information System Planning and Procurement

56. The new or modified Reliability Standard must address how a responsible entity will include security considerations as part of its information system planning and system development lifecycle processes. As part of this objective, the new or modified Reliability Standard must address a responsible entity's CIP Senior Manager's (or delegate's) identification and documentation of the risks of proposed information system planning and system development actions. This objective is intended to ensure adequate consideration of these risks, as well as the available options for hardening the responsible entity's information system and minimizing the attack surface.

57. This third objective addresses the risk that responsible entities could unintentionally plan to procure and install unsecure equipment or software within their information systems, or could unintentionally fail to anticipate security issues that may arise due to their network architecture or during technology and vendor transitions. For example, the BlackEnergy malware campaign identified by ICS-CERT and referenced in the NOPR resulted from the remote exploitation of previously unidentified vulnerabilities, which allowed attackers to remotely execute malicious code on remotely accessible devices.[92] According to ICS-CERT, this attack might have been mitigated if affected entities had taken steps during system development and planning to: (1) minimize network exposure for all control system devices/subsystems; (2) ensure that

[92] *See* ICS-CERT Alert, *Ongoing Sophisticated Malware Campaign Compromising ICS (Update E)*.

devices were not accessible from the internet; (3) place devices behind firewalls; and (4) utilize secure remote access techniques.[93] The third objective also supports, where appropriate, the need for strategic technology refreshes as recommended by ICS-CERT in response to the 2015 Ukraine cybersecurity incident.[94]

58. NIST SP 800-53 and SP 800-161 provide several controls which, when taken together, reduce the likelihood that an information system will be deployed and/or remain in service with potential vulnerabilities that have not been identified or adequately considered. For example, in the NIST SP 800-53 Systems Acquisition (SA) control family, control SA-3 provides that organizations should: (1) manage information systems using an organizationally-defined system development life cycle that incorporates information security considerations; and (2) integrate the organizational information security risk management process into system development life cycle activities.[95] Similarly, control SA-8 recommends using secure engineering principles during the planning and acquisition phases of future projects such as: (1) developing layered protections; (2) establishing sound security policy, architecture, and controls as the foundation for design; (3) incorporating security requirements into the system

---

[93] *See* ICS-CERT Advisory, *GE Proficy Vulnerabilities*, https://ics-cert.us-cert.gov/advisories/ICSA-14-023-01.

[94] *See* ICS-CERT Alert, *Cyber-Attack Against Ukrainian Critical Infrastructure*.

[95] NIST Special Publication 800-53, Appendix F (Security Control Catalog) at 157.

development life cycle; and (4) reducing risk to acceptable levels, thus enabling informed risk management decisions.[96] Finally, control SA-22 provides controls to address unsupported system components, recommending the replacement of information and communication technology components when support is no longer available, or the justification and approval of an unsupported system component to meet specific business needs. These sources, while not meant to be definitive, provide examples of controls for addressing the Commission's directive regarding objective three. Other security controls also could meet this objective.

Fourth Objective: Vendor Risk Management and Procurement Controls

59. The new or modified Reliability Standard must address the provision and verification of relevant security concepts in future contracts for industrial control system hardware, software, and computing and networking services associated with bulk electric system operations. Specifically, NERC must address controls for the following topics: (1) vendor security event notification processes; (2) vendor personnel termination notification for employees with access to remote and onsite systems; (3) product/services vulnerability disclosures, such as accounts that are able to bypass authentication or the presence of hardcoded passwords; (4) coordinated incident response activities; and (5) other related aspects of procurement. NERC should also consider provisions to help

---

[96] *Id.* at 162.

responsible entities obtain necessary information from their vendors to minimize potential disruptions from vendor-related security events.

60. This fourth objective addresses the risk that responsible entities could enter into contracts with vendors who pose significant risks to their information systems, as well as the risk that products procured by a responsible entity fail to meet minimum security criteria. In addition, this objective addresses the risk that a compromised vendor would not provide adequate notice and related incident response to responsible entities with whom that vendor is connected.

61. The Department of Energy (DOE) Cybersecurity Procurement Language for Energy Delivery Systems document outlines security principles and controls for entities to consider when designing and procuring control system products and services (e.g., software, systems, maintenance, and networks), and provides example language that could be incorporated into procurement specifications. The procurement language encourages buyers to incorporate baseline procurement language that ensures the supplier establishes, documents and implements risk management practices for supply chain delivery of hardware, software, and firmware.[97] In addition, NIST SP 800-161 encourages buyers to use the Information and Communications Technology supply chain risk management (ICT SCRM) plans for their respective systems and missions

---

[97] *See* Energy Sector Control Systems Working Group, *Cybersecurity Procurement Language –Energy Delivery Systems* at 27, http://www.energy.gov/sites/prod/files/2014/04/f15/CybersecProcurementLanguage-EnergyDeliverySystems_040714_fin.pdf.

throughout their acquisition activities.[98] The controls in the ICT SCRM plans can be applied in different life cycle processes.

62. NIST SP 800-161 also provides specific recommendations in control SA-4 pertaining to systems acquisition processes, which are relevant for consideration during the standards development process, including but not limited to: (1) defining requirements that cover regulatory requirements (i.e., telecommunications or IT), technical requirements, chain of custody, transparency and visibility, sharing information on supply chain security incidents throughout the supply chain, rules for disposal or retention of elements such as components, data, or intellectual property, and other relevant requirements; (2) defining requirements for critical elements in the supply chain to demonstrate a capability to remediate emerging vulnerabilities based on open source information and other sources; and (3) defining requirements for the expected life span of the system and ensuring that suppliers can provide insights into their plans for the end-of-life of components. Other relevant provisions can be found in the System and Communications Protection (SC) control family under control SC-18 addressing SCRM guidance for mobile code, which recommends that organizations employ rigorous supply chain protection techniques in the acquisition, development, and use of mobile code to be

---

[98] *See* NIST Special Publication 800-161 at 51.

deployed in information systems.[99] These sources, while not meant to be definitive, provide examples of controls for addressing the Commission's directive regarding objective four. Other security controls also could meet this objective.

3. Existing CIP Reliability Standards

Comments

63. NERC comments that although the CIP Reliability Standards do not explicitly address supply chain procurement practices, existing requirements mitigate the supply chain risks identified in the NOPR. In particular, NERC states that requirements in Reliability Standards CIP-004-6, CIP-005-5, CIP-006-6, CIP-007-6, CIP-008-5, CIP-009-6, CIP-010-2, and CIP-011-2 "include controls that correspond to controls in NIST SP 800-161."[100]

64. For example, NERC explains that responsible entity compliance with Reliability Standard CIP-004-6, addressing the implementation of cybersecurity awareness programs, may include reinforcement of cybersecurity practices to mitigate supply chain risks. NERC also states that requirements in Reliability Standard CIP-004-6 (addressing personnel risk assessment) and requirements in Reliability Standards CIP-004-6, CIP-

---

[99] Mobile code is a software program or parts of a program obtained from remote information systems, transmitted across a network, and executed on a local information system without explicit installation or execution by the recipient. NIST Special Publication 800-53, Appendix B (Glossary) at 14. Mobile code technologies include, for example, Java, JavaScript, ActiveX, Postscript, PDF, Shockwave movies, Flash animations, and VBScript. *Id.*

[100] NERC NOPR Comments at 15-16.

005-5, CIP-006-6, CIP-007-6, and CIP-010-2 (addressing electronic and physical access) apply to any outside vendors or contractors.

65. The Trade Associations, Arkansas, G&T Cooperatives, NIPSCO, Luminant, Southern, NextEra, and SCE contend that the existing CIP Reliability Standards, at least partly, address supply chain risks that are within a responsible entity's control.

66. The Trade Associations state that, while the existing CIP Reliability Standards do not contain explicit provisions addressing supply chain management, "transmission owners and operators already have significant responsibilities to perform under various Commission-approved CIP standards that already address supply chain issues."[101] Specifically, the Trade Associations, NIPSCO, and others state that Reliability Standard CIP-010-2 establishes requirements for cyber asset change management that mandate extensive baseline configuration testing and change monitoring, as well as vulnerability assessments, prior to connecting a new cyber asset to a High Impact BES Cyber Asset.[102]

67. The Trade Associations also contend that the CIP Reliability Standards provide adequate vendor remote access protections by mandating: (1) controls that restrict personnel access (physical and electronic) to protected information systems; (2) controls that prevent direct access to applicable systems for interactive remote access sessions

---

[101] Trade Associations NOPR Comments at 19-20.

[102] Trade Associations NOPR Comments at 20; NIPSCO NOPR Comments at 5; Southern NOPR Comments at 12; Luminant NOPR Comments at 4-5; SCE NOPR Comments at 6.

using routable protocols; (3) the use of encryption for connections extending outside of an electronic security perimeter; (4) the use of two factor authentication when accessing medium and high impact systems; and (5) integration controls which require changing known default accounts and passwords.[103]

68. NIPSCO, Luminant, and G&T Cooperatives point to Reliability Standard CIP-007-6 as an existing Reliability Standard that addresses supply chain risks. Reliability Standard CIP-007-6 requires responsible entities to have processes under which only necessary ports and services should be enabled; security patches should be tracked, evaluated, and installed on applicable BES Cyber Systems; and anti-virus software or other prevention tools should be used to prevent the introduction and propagation of malicious software on all Cyber Assets within an Electronic Security Perimeter.[104]

69. Commenters also identify existing voluntary guidelines that, they contend, augment the existing CIP Reliability Standards to further address any potential risks posed by the supply chain. Southern points to voluntary cybersecurity procurement guidance materials developed by the DHS and the DOE as examples of procurement language that could be used in the course of vendor negotiations. Southern states that the

---

[103] Trade Associations Post-Technical Conference Comments at 6.

[104] NIPSCO NOPR Comments at 5; Luminant NOPR Comments at 4; G&T Cooperatives NOPR Comments at 8-9.

DHS and DOE guidelines recognize the need for flexibility and allow for multiple contractual approaches.[105]

70. Commenters suggest that the Commission direct NERC to develop cybersecurity procurement guidance documents as opposed to a mandatory Reliability Standard. AEP, NextEra, and Southern state that the Commission could direct NERC to develop guidance documents addressing supply chain risk management based, in part, on the DHS and DOE voluntary cybersecurity procurement guidance materials.[106] Luminant asserts that NERC-developed guidance "would effectively communicate key issues while permitting industry the flexibility to effectively protect their BES Cyber Systems in a way most effective for that entity and at the lowest cost."[107]

Discussion

71. While we recognize that existing CIP Reliability Standards include requirements that address aspects of supply chain management, we determine that existing Reliability Standards do not adequately protect against supply chain risks that are within a responsible entity's control. Specifically, we find that existing CIP Reliability Standards do not provide adequate protection for the four aspects of supply chain risk management that underlie the four objectives for a new or modified Reliability Standard discussed

---

[105] Southern NOPR Comments at 13.

[106] AEP NOPR Comments at 7-8; NextEra NOPR Comments at 4-5; Southern NOPR Comments at 12-13.

[107] Luminant NOPR Comments at 5.

above.[108] Moreover, a fundamental premise of cyber security is "defense in depth," and addressing issues in the supply chain (to the extent a utility reasonably can) is an important component of a strong, multi-layered defense.

Software Integrity and Authenticity

72. With regard to software integrity and authenticity, we agree with commenters who state that the existing CIP Reliability Standards contain requirements for responsible entities to implement a patch management process for tracking, evaluating, and installing cybersecurity patches and to implement processes to detect, prevent, and mitigate the threat of malicious code. These provisions, however, do not require responsible entities to verify the identity of the software publisher for all software and patches that are intended for use on their BES Cyber Systems or to verify the integrity of the software and patches before they are installed in the BES Cyber System environment.[109] As discussed above, the CIP Reliability Standards should address compromised software or patches that a responsible entity receives from a vendor, in order to protect the bulk electric system from Watering-Hole or similar cyberattacks. These concerns are not addressed by existing CIP Reliability Standards.

---

[108] Since the directive to NERC to develop a new or modified Reliability Standard is limited to the four objectives discussed above, we limit our analysis of the existing CIP Reliability Standards to requirements that relate to those objectives.

[109] *See* Trade Associations NOPR Comments at 38 (indicating that integrity checking mechanisms used to verify software, firmware, and information integrity found in the NIST SP-800-161 System and Information Integrity (SI) control family are not addressed in the CIP version 5 Reliability Standards).

73. Mandatory controls in the existing CIP Reliability Standards referenced by commenters do not provide sufficient protection against attacks that compromise software and software patch integrity and authenticity. For example, while Reliability Standard CIP-007-6, Requirement R2 requires responsible entities to enforce a patch management process for tracking, evaluating, and installing cyber security patches for applicable systems, including evaluating security patches for applicability, the requirement does not address mechanisms to acquire the patch file from a vendor in a secure manner and methods to validate the integrity of a patch file before installation.

74. With respect to mandatory configuration controls, Reliability Standard CIP-010-2, Requirement R1 requires responsible entities to authorize and document all changes to baseline configurations and, where technically feasible, test patches in a test environment before installing. However, NERC's technical guidance document for CIP-010-2, Requirement R1, Part 1.2 does not require the authorizer to first verify the authenticity of a patch. Similarly, the testing of patches in a test environment under Requirement R1.5 would likely provide insufficient protection as many malware variants are programmed to execute only after the system is rebooted several times. Regarding patch source monitoring, the guidelines and technical basis section for Reliability Standard CIP-007-6 suggests that responsible entities should obtain security patches from original sources, where possible, and indicates that patches should be approved or certified by another

source before being assessed and applied.[110] The Reliability Standard, however, does not require the use of these techniques. Implementing controls that verify integrity and authenticity of software and its publishers may help mitigate security gaps listed above.

75. In sum, the current CIP Reliability Standards do contain certain controls addressing the risks posed by malware, as stated by commenters. Verifying software integrity and authenticity, however, is a reasonable and appropriate complement to these controls, is not required by the current Standards, and is supported by the principle of defense-in-depth. In fact, this verification can be viewed as the first line of defense against malware-infected software.

Vendor Remote Access to BES Cyber Systems

76. On the subject of vendor remote access, which includes vendor user-initiated Interactive Remote Access and vendor machine-to-machine remote access, existing CIP Reliability Standards contain system access requirements, including a requirement for security event monitoring. However, the CIP Reliability Standards do not require remote access session logging for machine-to-machine remote access, nor do they address the ability to monitor or close unsafe remote connections for both vendor Interactive Remote

[110] Reliability Standard CIP-007-6 (Cyber Security – Systems Security Management), Guidelines and Technical Basis at 42-43.

Access and vendor machine-to-machine remote access.[111] The CIP Reliability Standards should address enhanced session logging requirements for vendor remote access in order to improve visibility of activity on BES Cyber Systems and give responsible entities the ability to rapidly disable remote access sessions in the event of a system breach.

77. The existing requirements referenced by NERC, the Trade Associations, and other commenters do not adequately address access restrictions for vendors. For example, while Reliability Standard CIP-004-6, Requirements R4 and R5 provide controls that must be applied to vendors such as restricting access to individuals "based on need," these Requirements do not include post-authorization logging or control of remote access. The existing CIP Reliability Standards do not require a responsible entity to monitor data traffic that traverses remote communication to their BES Cyber Systems. The absence of post-authorization monitoring and logging presents an opportunity for unmonitored malicious or otherwise inappropriate remote communication to or from a BES Cyber System. The inability of a responsible entity to rapidly terminate a connection may allow malicious or otherwise inappropriate communication to propagate, contributing to a degradation of a BES Cyber Asset's function. Enhanced visibility into remote communications and the ability to rapidly terminate a remote communication could mitigate such a vulnerability.

---

[111] *See* Trade Association NOPR Comments at 43 (indicating that mechanisms for monitoring for unauthorized personnel, connections, devices, and software found in the NIST SP-800-161 System and Information Integrity (SI) control family are not addressed in the CIP version 5 Reliability Standards).

78. Reliability Standard CIP-005-5, Requirement R1 provides controls for vendor machine-to-machine and vendor user-initiated Interactive Remote Access sessions by restricting all inbound and outbound communications through an identified Electronic Access Point for bi-directional routable protocol connections. Reliability Standard CIP-005-5, Requirement R2 provides controls for vendor interactive remote access sessions by requiring the use of encryption and requiring multi-factor authentication. However, the provisions of Reliability Standard CIP-005-5, Requirement R2 addressing interactive remote access management do not apply to vendor machine-to-machine remote access. The Reliability Standard CIP-005-5, Requirement R2 controls addressing interactive remote access management only apply to remote connections that are user-initiated (i.e., initiated by a person). Machine-to-machine connections are not user-initiated and, therefore, are not subject to the requirements of Reliability Standard CIP-005-5, Requirement R2. When the interactive remote access management controls of Reliability Standard CIP-005-5, Requirement R2 do not apply, a machine-to-machine remote communication may access a BES Cyber System without any access credentials, over an unencrypted channel, and without going through an Intermediate System.

79. For both Interactive Remote Access and machine-to-machine remote access, Reliability Standard CIP-007-6, Requirement R3 requires monitoring for malicious code and Requirement R4 requires logging of successful and unsuccessful login attempts, as well as logging detected malicious code. However, Reliability Standard CIP-007-6 does not address the risks posed by inappropriate activity that could occur during a remote communication. The lack of a requirement addressing the detection of inappropriate

activity represents a risk because the responsible entity may not be aware if an authorized user is performing inappropriate activity on a BES Cyber Asset via a remote connection. This risk is higher for machine-to-machine communication due to the lack of authentication and encryption requirements in the existing CIP Reliability Standards, lowering the threshold for a malicious actor to execute a man-in-the-middle attack to gain access to a BES Cyber System and conduct inappropriate activity such as reconnaissance or code modification.

80. Therefore, we recognize that the current CIP Reliability Standards do contain certain controls addressing the risks posed by vendor remote access, as noted by commenters. However, the current CIP Reliability Standards do not require monitoring remote access sessions or closing unsafe remote connections for either vendor Interactive Remote Access and vendor machine-to-machine remote access. Accordingly, we determine that vendor remote access is not adequately addressed in the approved CIP Reliability Standards and, therefore, is an objective that must be addressed in the supply chain management plans directed in this final rule.

Information System Planning and Procurement

81. The existing CIP Reliability Standards do not address information system planning. Recent cybersecurity incidents[112] have made it apparent that overall system

[112] *See* E-ISAC, *Analysis of the Cyber Attack on the Ukrainian Power Grid* at 3 (March 18, 2016); *see also* Dell, *Dell Security Annual Threat Report* (2015) at 7, https://software.dell.com/docs/2015-dell-security-annual-threat-report-white-paper-15657.pdf; Olcott Technical Conference Comments at 2.

planning is as important to overall BES Cyber System security and reliability as any other component of security architecture. In general, the CIP Reliability Standards do not provide a framework for maintaining ongoing awareness of information security, vulnerabilities, and threats to support organization risk management decisions;[113] nor do they address the concept of integrating continuous improvement of organizational security posture with supply chain risk management as recommended by NIST SP 800-161.[114] Based on the threats evidenced by recent cybersecurity incidents, the absence of security considerations in system lifecycle processes constitutes a gap in the CIP Reliability Standards that could contribute to pervasive and systemic vulnerabilities that threaten bulk electric system reliability.

82. The existing CIP Reliability Standards also do not provide for procurement controls for industrial control system hardware, software, and computing and networking services. As discussed above, procurement controls are intended to address the threat that responsible entities could enter into contracts with vendors who pose significant risks to their information systems or procure products that fail to meet minimum security criteria, as well as the risk that a compromised vendor would not provide adequate notice and related incident response to responsible entities with whom that vendor is connected.

---

[113] *See* NIST Special Publication 800-137, *Information Security Continuous Monitoring (ISCM) for Federal Information Systems and Organizations* at vi, http://nvlpubs.nist.gov/nistpubs/Legacy/SP/nistspecialpublication800-137.pdf.

[114] NIST Special Publication 800-161 at 46.

83. With regard to commenters' suggestion that the Commission direct NERC to develop cybersecurity procurement guidance documents as opposed to a mandatory Reliability Standard, we agree that the voluntary efforts identified by commenters could provide guidance or otherwise inform NERC's standard development process. We conclude, however, that relying on voluntary guidelines to address the supply chain risks described above is not sufficient to fulfill the Commission's responsibilities under FPA section 215.

4. Vendor Risk Management and Procurement Controls

Comments

84. NERC, G&T Cooperatives, Arkansas and others state that responsible entities have limited influence over vendors and contractors, and, therefore, a limited ability to affect the supply chain for industrial control system hardware, software, and computing and networking services associated with bulk electric system operations.[115] NERC contends that any supply chain management Reliability Standard "must balance the reliability need to implement supply chain management security controls with entities' business need to obtain products and services at a reasonable cost."[116] NERC maintains that responsible entities lack bargaining power to persuade vendors or suppliers to implement cybersecurity controls without significantly increasing the cost of their

[115] NERC NOPR Comments at 11-12; G&T Cooperatives NOPR Comments at 9; Arkansas NOPR Comments at 5.

[116] NERC NOPR Comments at 11-12.

products or services. NERC points to NIST SP 800-161 to highlight that implementing supply chain security management controls "will require financial and human resources, not just from the [acquirer] directly but also potentially from their system integrators, suppliers, and external service providers that would also result in increased cost to the acquirer."[117]

85. G&T Cooperatives contend that they "have minimal control over their suppliers and are not able to identify all potential vulnerabilities associated with each and every supplier and their products/parts."[118] G&T Cooperatives and Arkansas maintain that responsible entities do not have the ability to force a vendor to address all potential vulnerabilities. G&T Cooperatives assert that even if a contract between a responsible entity and a supplier "could include" language requiring the supplier to implement security controls, "it is not feasible for contractual terms … to address all potential vulnerabilities related to supply chain management."[119]

86. NERC, Trade Associations, G&T Cooperatives and Arkansas also raise a concern that the Commission's proposal could place compliance risk on responsible entities for actions beyond their control and, ultimately, incent responsible entities to avoid upgrades

---

[117] *Id.* (citing NIST Special Publication 800-161 at 3).

[118] G&T Cooperatives NOPR Comments at 9.

[119] *Id.* at 9.

that could trigger such compliance risk.[120] NERC states that any supply chain management Reliability Standard should be drafted so that it "creates affirmative obligations to implement supply chain management security controls without holding entities strictly liable for any failure of those controls to eliminate all supply chain threats and vulnerabilities."[121] NERC explains that if a supply chain management Reliability Standard is not reasonably scoped to avoid unreasonable compliance risk, it could create a disincentive for responsible entities to purchase and install new technologies and equipment.

87. G&T Cooperatives state that "placing the compliance risk of vendor and supplier security vulnerability on Responsible Entities could incent Responsible Entities to avoid upgrades to their industrial control system hardware, software, and other services." G&T Cooperatives explain that there are three primary incentives for a responsible entity to avoid upgrades if faced with compliance risks: (1) new regulations would result in additional costs for vendors and suppliers that would be passed on to the end-user; (2) since security patches are not issued by vendors for unsupported hardware and software, there is less security patch management responsibility for the responsible entity; and (3)

---

[120] NERC NOPR Comments at 13; Trade Associations NOPR Comments at 24-25; G&T Cooperatives NOPR Comments at 9-10; Arkansas NOPR Comments at 6.

[121] NERC NOPR Comments at 13.

avoiding new hardware and software reduces the risk of introducing undetected security threats.[122]

Discussion

88. Our directive to NERC to develop a new or modified Reliability Standard that addresses the objectives outlined above balances the supply chain risks facing the bulk electric system against any potential challenges raised by vendor relationships. We believe that the concerns raised in comments with respect to responsible entities' relationships with vendors in relation to supply chain risks are valid. Our directive is informed by this concern and reflects a reasonable balance between the risks facing bulk electric system reliability from the supply chain and concerns over vendor relationships. The directive strikes this balance by addressing supply chain risks that are within responsible entities' control, and we do not expect a new or modified supply chain Reliability Standard to impose obligations directly on vendors. Moreover, entities will not be responsible for vendor errors beyond the scope of the controls implemented to comply with the Reliability Standards.

89. With respect to concerns that the Commission's proposal could place compliance risk on responsible entities for actions beyond their control, which some commenters argue would prompt responsible entities to avoid upgrades that could trigger such compliance risk, we reiterate that the intent of the directive is to address supply chain

---

[122] G&T Cooperatives NOPR Comments at 9.

risks that are within the responsible entities' control. As part of NERC's standard development process, we expect NERC to establish provisions addressing compliance obligations in a manner that avoids shifting liability from a vendor for its mistakes to a responsible entity. Finally, we view the argument that a new or modified Reliability Standard will result in a substantial increase in costs to be speculative because, beyond requiring NERC to address the four objectives discussed above, or some equally effective and efficient alternatives, our directive does not require NERC to develop a Reliability Standard that mandates any particular controls or actions.

III. Information Collection Statement

90. The Paperwork Reduction Act (PRA)[123] requires each federal agency to seek and obtain Office of Management and Budget (OMB) approval before undertaking a collection of information directed to ten or more persons or contained in a rule of general applicability. OMB regulations[124] require approval of certain information collection requirements imposed by agency rules. Upon approval of a collection of information, OMB will assign an OMB control number and an expiration date. Respondents subject to the filing requirements of an agency rule will not be penalized for failing to respond to the collection of information unless the collection of information displays a valid OMB control number.

---

[123] 44 U.S.C. 3507(d).

[124] 5 CFR 1320.

91. The Commission will submit the information collection requirements to OMB for its review and approval. The Commission solicits public comments on its need for this information, whether the information will have practical utility, the accuracy of burden and cost estimates, ways to enhance the quality, utility, and clarity of the information to be collected or retained, and any suggested methods for minimizing respondents' burden, including the use of automated information techniques.

92. The information collection requirements in this Final Rule in Docket No. RM15-14-002 for NERC to develop a new or to modify a Reliability Standard for supply chain risk management, should be part of FERC-725 (Certification of Electric Reliability Organization; Procedures for Electric Reliability Standards (OMB Control No. 1902-0225)). However, there is an unrelated item which is currently pending OMB review under FERC-725, and only one item per OMB Control No. can be pending OMB review at a time. Therefore, the requirements in this Final Rule in RM15-14-002 are being submitted under a new temporary or interim collection number FERC-725(1A) to ensure timely submittal to OMB. In the long-term, Commission staff plans to administratively move the requirements and associated burden of FERC-725(1A) to FERC-725.

93. Burden Estimate and Information Collection Costs: The requirements for the ERO to develop Reliability Standards and to provide data to the Commission are included in the existing FERC-725. FERC-725 includes information used by the Commission to implement the statutory provisions of section 215 of the FPA. FERC-725 includes the burden, reporting and recordkeeping requirements associated with: (a) Self-Assessment and ERO Application, (b) Reliability Assessments, (c) Reliability Standards

Development, (d) Reliability Compliance, (e) Stakeholder Survey, and (f) Other Reporting. In addition, the Final Rule will not result in a substantive increase in burden because this requirement to develop standards is covered under FERC-725. However because FERC is using the temporary information collection number, FERC-725(1A), FERC will use "placeholder" estimates of 1 response and 1 burden hour for the burden calculation.

IV. Regulatory Flexibility Act Analysis

94. The Regulatory Flexibility Act of 1980 (RFA)[125] generally requires a description and analysis of final rules that will have significant economic impact on a substantial number of small entities. The Small Business Administration (SBA) revised its size standard (effective January 22, 2014) for electric utilities from a standard based on megawatt hours to a standard based on the number of employees, including affiliates.[126] The entities subject to the Reliability Standards developed by the North American Electric Reliability Corporation (NERC) include users, owners, and operators of the Bulk-Power System, which serves more than 334 million people. In addition, NERC's current responsibilities include the development of Reliability Standards. Accordingly, the Commission certifies that the requirements in this Final Rule will not have a

[125] 5 U.S.C. 601-612.

[126] SBA Final Rule on "Small Business Size Standards: Utilities," 78 FR 77,343 (Dec. 23, 2013).

significant economic impact on a substantial number of small entities, and no regulatory flexibility analysis is required.

V. Environmental Analysis

95. The Commission is required to prepare an Environmental Assessment or an Environmental Impact Statement for any action that may have a significant adverse effect on the human environment.[127] The Commission has categorically excluded certain actions from this requirement as not having a significant effect on the human environment. Included in the exclusion are rules that are clarifying, corrective, or procedural or that do not substantially change the effect of the regulations being amended.[128] The actions proposed herein fall within this categorical exclusion in the Commission's regulations.

VI. Effective Date and Congressional Notification

96. This Final Rule is effective **[INSERT DATE 60 days after publication in the FEDERAL REGISTER]**. The Commission has determined, with the concurrence of the Administrator of the Office of Information and Regulatory Affairs of OMB, that this rule is not a "major rule" as defined in section 351 of the Small Business Regulatory Enforcement Fairness Act of 1996. This Final Rule is being submitted to the Senate, House, and Government Accountability Office.

---

[127] *Regulations Implementing the National Environmental Policy Act of 1969*, Order No. 486, FERC Stats. & Regs. ¶ 30,783 (1987).

[128] 18 CFR 380.4(a)(2)(ii).

VII. Document Availability

97. In addition to publishing the full text of this document in the Federal Register, the Commission provides all interested persons an opportunity to view and/or print the contents of this document via the Internet through the Commission's Home Page (http://www.ferc.gov) and in the Commission's Public Reference Room during normal business hours (8:30 a.m. to 5:00 p.m. Eastern time) at 888 First Street, NE, Room 2A, Washington, DC 20426.

98. From the Commission's Home Page on the Internet, this information is available on eLibrary. The full text of this document is available on eLibrary in PDF and Microsoft Word format for viewing, printing, and/or downloading. To access this document in eLibrary, type the docket number of this document, excluding the last three digits, in the docket number field.

99. User assistance is available for eLibrary and the Commission's website during normal business hours from the Commission's Online Support at (202) 502-6652 (toll free at 1-866-208-3676) or e-mail at ferconlinesupport@ferc.gov, or the Public Reference

Room at (202) 502-8371, TTY (202) 502-8659. E-mail the Public Reference Room at public.referenceroom@ferc.gov.

By the Commission.

(S E A L)

Nathaniel J. Davis, Sr.,
Deputy Secretary.

Note: the following Appendix will not appear in the *Code of Federal Regulations*.

Appendix
Commenters

Abbreviation	**Commenter**
AEP	American Electric Power Service Corporation
ACS	Applied Control Solutions, LLC
APS	Arizona Public Service Company
Arkansas	Arkansas Electric Cooperative
BPA	Bonneville Power Administration
CEA	Canadian Electricity Association
Consumers Energy	Consumers Energy Company
CyberArk	CyberArk
EnergySec	Energy Sector Security Consortium, Inc.
Ericsson	Ericsson
Resilient Societies	Foundation for Resilient Societies
G&T Cooperatives	Associated Electric Cooperative, Inc., Basin Electric Power Cooperative, and Tri-State Generation and Transmission Association, Inc.
Gridwise	Gridwise Alliance
Idaho Power	Idaho Power Company
Indegy	Indegy
IESO	Independent Electricity System Operator
IRC	ISO/RTO Council
ISO New England	ISO New England Inc.
ITC	ITC Companies
Isologic	Isologic, LLC
KCP&L	Kansas City Power & Light Company and KCP&L Greater Missouri Operations Company
Luminant	Luminant Generation Company, LLC
NEMA	National Electrical Manufacturers Association
NERC	North American Electric Reliability Corporation
NextEra	NextEra Energy, Inc.
NIPSCO	Northern Indiana Public Service Co.
NWPPA	Northwest Public Power Association
Peak	Peak Reliability
PNM	PNM Resources
Reclamation	Department of Interior Bureau of Reclamation
SIA	Security Industry Association
SCE	Southern California Edison Company

Southern	Southern Company Services
SPP RE	Southwest Power Pool Regional Entity
SWP	California Department of Water Resources State Water Project
TVA	Tennessee Valley Authority
Trade Associations	Edison Electric Institute, American Public Power Association, National Rural Electric Cooperative Association, Electric Power Supply Association, Transmission Access Policy Study Group, and Large Public Power Council
UTC	Utilities Telecom Council
Waterfall	Waterfall Security Solutions, Ltd.
Wisconsin	Wisconsin Electric Power Company

UNITED STATES OF AMERICA
FEDERAL ENERGY REGULATORY COMMISSION

Revised Critical Infrastructure Protection Reliability Standards — Docket No. RM15-14-002

(Issued July 21, 2016)

LaFLEUR, Commissioner *dissenting*:

In today's order, the Commission elects to proceed directly to a Final Rule and require the development of a new reliability standard on supply chain risk management for industrial control system hardware, software, and computing and networking services associated with bulk electric system operations. I fully support the Commission's continued attention to the threat of inadequate supply chain risk management procedures, which pose a very real threat to grid reliability.

However, in my view, the importance and complexity of this issue should guide the Commission to proceed cautiously and thoughtfully in directing the development of a reliability standard to address these threats. I am concerned that the Commission has not adequately considered or vetted the Final Rule, which could hamper the development and implementation of an effective, auditable, and enforceable standard. I believe that the more prudent course of action would be to issue today's Final Rule as a Supplemental Notice of Proposed Rulemaking (Supplemental NOPR), which would provide NERC, industry, and stakeholders the opportunity to comment on the Commission's proposed directives. Accordingly, and as discussed below, I dissent from today's order.[1]

I. The Commission's Decision to Proceed Directly to Final Rule is Flawed and Could Delay Protection of the Grid Against Supply Chain Risks

Last July, as part of its NOPR addressing revisions to its cybersecurity critical infrastructure protection (CIP) standards, the Commission raised for the first time the prospect of directing the development of a standard to address risks posed by lack of controls for supply chain management.[2] The Commission indicated that new threats

---

[1] I do agree with one holding in the order: that the Commission has authority under section 215 of the Federal Power Act to promulgate a standard on this issue.

[2] *Revised Critical Infrastructure Protection Reliability Standards*, Notice of Proposed Rulemaking, 80 Fed. Reg. 43,354 (July 22, 2015), 152 FERC ¶ 61,054 (2015). I will refer to the section of that order addressing supply chain issues as the "Supply Chain NOPR," and the remainder of the order as the "CIP NOPR."

might warrant directing NERC to develop a standard to address those risks. While the Commission noted a variety of considerations that might shape the standard, including, among others, jurisdictional limits and the individualized nature of companies' supply chain management procedures, the Commission notably did not propose a specific standard for comment. Instead, the Commission sought comment on (1) the general proposal to require a standard, (2) the anticipated features of, and requirements that should be included in, such a standard, and (3) a reasonable timeframe for development of a standard.[3]

The record developed in comments responding to the Supply Chain NOPR and through the January 28, 2016 technical conference reflects a wide diversity of views regarding the need for, and possible content of, a reliability standard addressing supply chain management. Notwithstanding these diverse views, there was broad consensus on one point: that effectively addressing cybersecurity threats in supply chain management is tremendously complicated, due to a host of jurisdictional, technical, economic, and business relationship issues. Indeed, in the Supply Chain NOPR, the Commission recognized "that developing a supply chain management standard would likely be a significant undertaking and require extensive engagement with stakeholders to define the scope, content, and timing of the standard."[4]

Yet, the Commission is proceeding straight to a Final Rule without in my view engaging in sufficient outreach regarding, or adequately vetting, the contents of the Final Rule. As to those contents, it is worth noting that the four objectives that will define the scope and content of the standard were not identified in the Supply Chain NOPR. Therefore, even though the Final Rule reflects feedback received on the Supply Chain NOPR, and is not obviously inconsistent with the Supply Chain NOPR, no party has yet had an opportunity to comment on those objectives or consider how they could be translated into an effective and enforceable standard.[5] This is a consequence of: (1) the lack of outreach on supply chain threats prior to issuing the Supply Chain NOPR; (2) the lack of detail in the Supply Chain NOPR regarding what a standard might look like; and (3) the decision today to proceed straight to a Final Rule rather than provide additional opportunities for public feedback.

---

[3] *Id.* P 66.

[4] *Id.*

[5] To be clear, I am less concerned about whether the Final Rule satisfies minimal notice requirements than whether the Final Rule represents reasoned decision making by the Commission.

A. The Commission and the Public's Consideration of Supply Chain Risks Would Benefit from Additional Stakeholder Engagement

First, I believe that meaningful stakeholder input on the content of any proposed rule is essential to the Commission's deliberative process. This is especially important in our reliability work, as any standard developed by NERC must be approved by stakeholder consensus before it may be filed at the Commission. I do not believe that the record developed to date establishes that the Final Rule will lead to an appropriate solution to address supply chain risks. I note that much of the feedback we received in response to the Supply Chain NOPR was not focused on the merits of particular approaches to address supply chain threats. Yet, in this order, the Commission directs the development of a standard based on objectives not reflected in the Supply Chain NOPR, depriving the public of the ability to comment, and the Commission of the benefit of that public comment.

In retrospect, given both the preliminary nature of the consideration of the issue and the lack of a concrete idea regarding what a proposed standard would look like, I believe that the Supply Chain NOPR was, in substance, a *de facto* Notice of Inquiry and should have been issued as such, rather than as a subsection of the broader CIP NOPR on changes to the CIP standards. For example, it is instructive to compare the Supply Chain NOPR with two other documents: (1) the Notice of Inquiry being issued today on cybersecurity issues arising from the recent incident in Ukraine,[6] and (2) the NOPR concerning the proposed development of a reliability standard to address geomagnetic disturbances.[7] The level of detail and consideration of the issues presented in the Supply Chain NOPR are much more consistent with that in a Notice of Inquiry than a traditional NOPR. As a result, I am concerned that the Commission, by styling its prior action as a NOPR, has skipped a critical step in the rulemaking process: the opportunity for public comment on its directive to develop a standard and the objectives that will frame the design and development of that standard. As explained below, I believe this procedural decision actually makes it less likely that an effective, auditable, and enforceable standard will be implemented on a reasonable schedule, particularly given the acknowledged complexity of this issue.[8]

---

[6] *Cyber Systems in Control Centers*, Notice of Inquiry, Docket No. RM16-18-000.

[7] *Reliability Standards for Geomagnetic Disturbances*, Notice of Proposed Rulemaking, 77 FR 64,935 (Oct. 24, 2012), 141 FERC 61,045 (2012).

[8] I believe that *Reliability Standards for Physical Security Measures*, 146 FERC ¶ 61,166 (2014) (Physical Security Directive Order), which is cited in the Final Rule as support for today's action, is primarily relevant to demonstrate a different point than the order indicates. The Physical Security Directive Order followed focused outreach with

(continued…)

B. The Lack of Adequate Stakeholder Engagement Will Have Negative Consequences for the Standards Development Process

I am also concerned about the consequences for the standards development process of the Commission's decision to proceed straight to a Final Rule. In particular, I am concerned that the combination of insufficient process and discussion to develop the record and inadequate time for standards development (since the Commission substantially truncated NERC's suggested timeline)[9] will handicap NERC's ability to develop an effective and enforceable proposed standard for the Commission to consider. As noted above, NERC, industry, and other stakeholders will have no meaningful opportunity before initiating their work to provide feedback on the contents of the rule, to seek clarification from the Commission, or to propose revisions to the rule. Yet, this type of feedback is a critical component of the rulemaking process, to ensure that the entities tasked with implementing the Commission's directive have been heard and understand what they are supposed to do. I believe that the Commission is essentially giving the standards development team a homework assignment without adequately explaining what it expects them to hand in.

NERC and other stakeholders to discuss how a physical security standard could be designed and implemented within the parameters of section 215 of the Federal Power Act. As a result of that outreach, the directives in the Physical Security Directive Order were clear, targeted, and reflected shared priorities between the Commission and NERC. Physical Security Directive Order, 146 FERC ¶ 61,166 at PP 6-9. Consequently, NERC was able to develop and file a physical security standard with the Commission in less than three months, and the Commission ultimately approved that standard in November 2014, only roughly eight months after directing its development. *Physical Security Reliability Standard*, 149 FERC ¶ 61,140 (2014). In my view, this example demonstrates how essential outreach is to the timely and effective development of NERC standards.

[9] In its comments responding to the Supply Chain NOPR, NERC requested that, if the Commission decides to direct the development of a standard, the Commission provide *a minimum* of two years for the standards development process. However, the Commission disregards that request and directs NERC to develop a standard in just one year, apparently based solely on the Trade Associations' request that the Commission allow *at least* one year for the standards development process. I believe this timeline is inconsistent with the Commission's own recognition of the complexity of this issue, and, as discussed herein, likely to delay rather than expedite the implementation of an effective, auditable, and enforceable standard.

I do not believe that the Final Rule's flexibility is a justification for proceeding straight to a Final Rule. Indeed, given the inadequate process to date, I fear that the flexibility is in fact a lack of guidance and will therefore be a double-edged sword. The Commission is issuing a general directive in the Final Rule, in the hope that the standards team will do what the Commission clearly could not do: translate general supply chain concerns into a clear, auditable, and enforceable standard within the framework of section 215 of the Federal Power Act. While the Commission need not be prescriptive in its standards directives, the Commission's order assumes that the standards development team will be able to take the "objectives" of the Final Rule and translate them into a standard that the Commission will ultimately find acceptable. I believe that issuing a Supplemental NOPR would benefit the standards development process by enabling additional discussion and feedback regarding the design of a workable standard.

C. By Failing to Engage in Adequate Stakeholder Outreach Before Directing Development of a Standard, the Commission Increases the Likelihood that Implementation of a Standard Will be Delayed

A compressed and possibly compromised standards development process also has real consequences for the Commission's consideration of that proposed standard, whenever it is filed for our review. Unlike our authority under section 206 of the FPA, the Commission lacks authority under section 215 to directly modify a flawed reliability standard. Instead, to correct any flaws, the statute requires that we remand the standard to NERC and the standards development process.[10] Thus, notwithstanding the majority's desire to quickly proceed to Final Rule, the statutory construct constrains our ability to timely address a flawed standard, which could actually delay implementation of the protections the Commission seeks to put in place.

Given the realities of the standards development and approval process, we are likely years away from a supply chain standard being implemented, even under the aggressive schedule contemplated in the order. I believe that the Commission should endeavor to provide as much advance guidance as possible before mandating the development of a standard, to increase the likelihood that NERC develops a standard that will be satisfactory to the Commission and reduce the need for a remand. I worry that the limited process that preceded the Final Rule and the expedited timetable will make it extremely difficult for NERC to file a standard that the Commission can cleanly approve. Had the Commission committed itself to conducting adequate outreach, I believe we could have mitigated the likelihood of that outcome, and more effectively and promptly addressed the supply chain threat in the long term. "Delaying" action for a few months thus would, in the long run, lead to prompter and stronger protection for the grid.

---

[10] 18 U.S.C. § 824o(d)(4).

II. Conclusion

The choice the Commission faces today on supply chain risk management is not between action and inaction. Rather, given the importance of this issue, I believe that more considered action and a more developed Commission order, even if delayed by a few months, is better than a quick decision to "do something." Ultimately, an effective, auditable, and enforceable standard on supply chain management will require thoughtful consideration of the complex challenges of addressing cybersecurity threats posed through the supply chain within the structure of the FERC/NERC reliability process. In my view, the Commission gains very little and does not meaningfully advance the security of the grid by proceeding straight to a Final Rule, rather than taking the time to build a record to support a workable standard.

Accordingly, I respectfully dissent.

____

Cheryl A. LaFleur
Commissioner

Report on the FERC-NERC-Regional Entity Joint Review of Restoration and Recovery Plans

Prepared by the Staffs of the
Federal Energy Regulatory Commission
and the
North American Electric Reliability Corporation and its Regional Entities

January 2016

This report was prepared by the staff of the Federal Energy Regulatory Commission in consultation with staff from the North American Electric Reliability Corporation and its Regional Entities.

This report does not necessarily reflect the views of the Commission.

Table of Contents

I. Executive Summary i

II. Introduction: What are System Restoration and Recovery Plans and Why Are They Important? 1

III. Joint-Staff Review Process 3

IV. Review of System Restoration Plans and Related Standards Assessment 5

- A. Strategies and Priorities for Restoration 6
- B. Roles, Interrelationships and Coordination 12
- C. Situational Awareness Tools for Quick and Orderly Restoration 17
- D. System Restoration Resources 20
- E. Island Development and Synchronization 28
- F. Testing of System Restoration Resources 39
- G. Testing, Verification, and Updating of System Restoration Plans 44
- H. System Restoration Drills and Training Exercises 50
- I. Incorporating Lessons Learned from Prior Outage Events 58

V. Review of Cyber Security Incident Response and Recovery Plans, and Related Standards Assessment 64

- A. Resources, Processes, and Tools for Cyber Incident Response and Recovery 65
- B. External Roles, Interrelationships, and Coordination 70
- C. Monitoring for and Detection of Cyber Incidents and Triggers for Incident Response 73
- D. Initial Event Response Actions 80
- E. Recovery Planning 84

F. Review and Verification of Incident Response and Recovery Plans 87
G. Drills and Training Exercises 93
H. Improving Cyber Security Response and Recovery Plans Based on Actual Events and Other Feedback 97
VI. Appendix 1– Joint Staff Review Team 103
VII. Appendix 2 –Request Letter for Participation in Reliability Assessment 104
VIII. Appendix 3 – Standards and Requirements Assessed 107
IX. Appendix 4 – Glossary of Terms Used in Report 113
X. Appendix 5 - Acronyms Used in Report 118

I. Executive Summary

In September 2014, the Federal Energy Regulatory Commission (FERC or the Commission) initiated a joint staff review, in partnership with the North American Electric Reliability Corporation (NERC) and the Regional Entities,[1] to assess entities' plans for restoration and recovery of the bulk power system following a widespread outage or blackout.[2] The objective of the review was to assess and verify the electric utility industry's bulk power system recovery and restoration planning, and to test the efficacy of related Reliability Standards in maintaining and advancing reliability in that respect. The joint staff review was not a compliance or enforcement initiative. This report presents the results of that joint staff review.

In conducting this review, the joint staff review team gathered information from a representative sample of nine registered entities with significant bulk power grid responsibilities (the participants), including some entities that are registered with NERC in multiple functions.

The review team examined the restoration, response and recovery plans of each participant, along with supporting information. Documents reviewed included, but were not limited to, reliability coordinator-approved restoration plans, procedures for deploying blackstart resources, steady state and dynamic simulations testing the effectiveness of the plans, and cyber security incident response plans and recovery plans for critical cyber assets. The team also met with or conferred with the participants to discuss the above plans, as well as their experiences with recent restoration, response and recovery exercises or drills, and observed a number of restoration training exercises. The team assessed the relative strengths as well as any shortcomings of the plans across the various stages and topics of restoration, cyber security incident response and critical cyber asset recovery. The joint staff review team then reviewed the associated Reliability Standard requirements for clarity and efficacy to determine any reliability gaps, also taking into consideration relevant recommendations from the NERC-convened Independent Experts Review Panel (IERP).

---

[1] Pursuant to section 215(e)(4) of the Federal Power Act, NERC has delegated certain compliance and enforcement authority to eight Regional Entities.

[2] NERC maintains a Compliance Registry that identifies all entities, referred to as "registered entities," which must comply with mandatory Reliability Standards.

This report provides the team's observations on the participants' plans, assesses the related Reliability Standards, and makes recommendations for potential enhancements to the plans, related practices, and the provisions of certain Reliability Standards.[3]

Overall, the joint staff review team found that the participants have system restoration plans that, for the most part, are thorough and highly-detailed. The reviewed plans require identification and testing of blackstart resources, identification of primary and alternate cranking paths, and periodic training and drilling on the restoration process under a variety of outage scenarios.[4] Likewise, the joint staff review team found that participants had extensive cyber security incident response and recovery plans for critical cyber assets covering the majority of the response and recovery stages. In addition, the team observed that each participant has full time personnel dedicated to the roles and responsibilities defined in their respective response and recovery plans.

The joint staff review team identified several opportunities for improving system restoration and cyber incident response and recovery planning and readiness through, among other things, improvements to the clarity of certain Reliability Standard requirements. The joint staff review team accordingly recommends that measures be taken, including (in accordance with NERC's standards development process), considering changes to the current Reliability Standards to address the issues and recommendations as set out below and further discussed in the body of this report. In addition, the joint staff review team recommends that further studies be performed in certain areas, including those in which new Critical Infrastructure Protection (CIP) Reliability Standards have yet to go into effect.[5] Finally, the joint staff review team observed numerous beneficial practices employed by individual participants. The joint

---

[3] Appendix 4 includes a glossary of terms, and Appendix 5 includes a list of acronyms used in this report.

[4] A cranking path is a portion of the electric system that can be isolated, and then energized to deliver electric power from a generation source to enable the startup of one or more other generating units. *See NERC Glossary of Terms.*

[5] The joint staff review team recommends that FERC and NERC staff discuss, following report issuance, responsibility for performing, and prioritization of, the recommended studies along with the associated details to accomplish them.

staff team recommends that other registered entities responsible for system restoration, cyber security incident response, or recovery readiness consider incorporating similar practices into their plans and practices, where and as appropriate.

System Restoration Planning

Recommendations for Changes

1. **Clarify when system changes will trigger a requirement to update restoration plans.** The joint staff review team recommends that measures be taken (including considering changes to the Reliability Standards) to address the need for updating restoration plans for all system modifications that would change the implementation of an entity's restoration plan for an extended period of time, not just permanent or planned system modifications. In considering these measures, the kinds of events that may warrant an update to the system restoration plan should be identified, taking into account the length of time the system is affected, as well as the overall objective of ensuring that restoration plans are generally flexible enough so that system modifications can be addressed without continuous updates. **[Section IV.E]**

2. **Verification/testing of modified restoration plan**. The joint staff review team recommends that measures be taken (including considering changes to the Reliability Standards) to address the need for re-verification of a system restoration plan when a system change precipitates the need to determine whether the plan's restoration processes and procedures, when implemented, will operate reliably, i.e., when needed to ensure that the restoration plan, when implemented, allows for restoration of the system within acceptable operating voltage and frequency limits.[6] In considering such measures, the types of system changes that could impact reliable implementation of the restoration plan should be taken into account (e.g., identification of a new blackstart generator location or on redefinition of a cranking path). **[Section IV.G]**

3. **Operator training: Exercises on transferring control back to the balancing authority.** The joint staff review team recommends that measures be taken

---

[6] The Reliability Standards currently require verification of an applicable entity's system restoration plan every five years. While the review participants currently test and verify the effectiveness of their restoration plans following significant changes that could impact the viability of their plans, they are not obligated to do so under the current Reliability Standards.

(including considering changes to the Reliability Standards) to address system restoration training and drilling for transitioning from transmission operator island control to balancing authority ACE/AGC[7] control. **[Section IV.H.]**

Recommended Studies and Coordination Efforts

4. **Planning for loss of SCADA and loss of other data sources.** Given the possibility that Supervisory Control and Data Acquisition (SCADA) computer systems,[8] Inter-Control Center Communications Protocol (ICCP), or Energy Management System (EMS) functionality may be compromised during a major disturbance (e.g., portions of SCADA may not be available after a significant blackout), the joint staff review team recommends that further study be conducted to (a) assess system restoration plan steps that may be difficult in the absence of SCADA, ICCP data, and/or EMS; and (b) identify viable resources, methods or practices that would enable timely system restoration to occur absent SCADA/EMS functionality, which could then be incorporated into entities' system restoration training. The study should also examine and identify best practices that may be shared across the industry. Pending such study, individual entities should initiate or update consideration of resources, methods and practices they can use in these circumstances. **[Section IV.C]**

5. **Gain further understanding of recent blackstart resource changes.** The joint staff review team recommends study of the availability of blackstart resources, including the identification of strategies for replacing blackstart resources going forward and factors to be considered for such replacement resources (e.g., locational diversity, dual fuel, etc.). **[Section IV.D]**

6. **Gain further understanding of the use of direct current (DC) facilities for restoration**. The joint staff review team recommends that a study be conducted to determine the benefits of including existing or future voltage source converter DC lines in system restoration plans. **[Section IV.D]**

---

[7] Area Control Error (ACE) and Automatic Generation Control (AGC) are mechanisms to assess and adjust the instantaneous difference between a balancing authority's actual and scheduled interchange.

[8] A SCADA system operates with coded signals over communication channels to monitor and provide control of remote equipment (using typically one communication channel per remote station).

7. **Blackstart resource testing under anticipated blackstart conditions.** The joint staff review team recommends a study be performed to identify options for expanding restoration plan testing beyond the currently-required blackstart resource testing, to ensure the blackstart resource can energize equipment needed to restore the system as intended in the restoration plan. Any expanded testing requirements should take into consideration whether such testing is practical while maintaining system reliability, and whether such expanded testing requirements could affect the identification of blackstart resources in the future. **[Section IV.F]**

8. **Obtaining insight from entities that have experienced a widespread outage.** The team recommends that applicable entities that have not recently experienced a blackout or other events which impacted, or could have the potential to impact, the viability of their restoration plans reach out to those who have experienced such events, in an effort to continuously improve their restoration plans. Entities could benefit from the sharing of experiences across different regions of the country to gain insight into events that may not have ever occurred locally, including:

    - Severe flooding and storm impacts on facilities and equipment depended on for system restoration;

    - Effects of extreme temperatures, including severe cold weather impacts on facilities and equipment depended on for system restoration; and

    - Preparedness training for the above impacts. **[Section IV.I]**

Cyber Incident Response and Recovery Plans

Recommendations for Changes

9. **Response and recovery plan ownership.** The joint staff review team recommends that cyber security incident response plans and recovery plans for critical cyber assets specifically designate accountability at the cyber asset level (e.g., EMS servers, remote terminal unit (RTU) concentrators, network routers, etc.). The team recommends that measures be taken (including considering changes to the Reliability Standards) to address this. **[Section V.A]**

10. **Require details on types of cyber security events that should trigger a response and reporting.** The joint staff review team recommends that measures be taken (including considering changes to the Reliability Standards) to address the need for cyber security incident response plans to include details around the types of events that should trigger a response, and what types should be reported. While the team recognizes that CIP version 5 will require responsible entities to have processes to

identify cyber security incidents, consideration should be given as to whether any additional clarification or improvements are needed once some experience is gained with CIP version 5. **[Section V.C]**

11. **Use of technical expertise and advanced tools.** The joint staff review team has concluded that cyber event monitoring and response would be greatly improved by expanding the use of cyber security technical expertise and advanced technical tools, and recommends that measures be taken (including considering changes to the Reliability Standards) to address the use of these tools to improve cyber event monitoring and response. In considering such changes, it may be appropriate to allow for some experience with CIP versions 5 and 6. In addition, the team recommends that such measures clarify that these advanced tools and resources should be employed in a manner that does not negate the benefits by making the cyber security event monitoring process more cumbersome or unnecessarily burdensome. **[Section V.C.]**

12. **Recovery plan inventory assumptions risk.** The joint staff review team recommends that measures be taken (including considering changes to the Reliability Standards) to eliminate, to the extent possible, "inventory assumptions" in cyber asset recovery plans that could significantly affect prompt recovery of critical cyber assets. For example, entities may assume that hardware from external sources or other third-party vendor support needed for recovery of critical cyber assets will be available, without necessarily having measures to ensure availability. Likewise, entities may not consider interdependent or common-mode failure scenarios, which can create the need to recover multiple critical cyber assets concurrently from the same vendors. **[Section V.E]**

Recommended Studies and Practices

13. **Independent review of cyber security response and recovery plans.** The joint staff review team recommends that recovery plans for critical cyber assets and cyber security incident response plans be reviewed by an independent authority or third party for the purpose of supporting thoroughness and technical reliability, using a trusted or qualified third party to ensure a proper security review.**[Section V.F]**

14. **Exercises of response and recovery plans using paper drills.** The joint staff review team observed that participation in full operational exercises and other more complex simulations provides greater insight into the viability of a given cyber response and recovery plan, and believes that participation in such exercises by the industry is valuable for developing robust recovery and response plans. The joint staff review team recommends that applicable entities participate in exercise scenarios and simulations structured to gain insight into the viability of cyber response and recovery

plans (i.e., beyond paper drills and tabletop exercises), including testing for interdependencies and other vulnerabilities. **[Section V.G]**

15. **Gain further understanding of response and recovery plan updating following testing or actual cyber events.** The joint staff review team recommends that a study be conducted to better understand the associated plan improvements made by entities where testing or an actual cyber event reveals the need or opportunity for improvements to a response and recovery plan. This study would support a better understanding of the effectiveness and existence of continuous improvement processes. In addition, the study should examine and identify best practices with regard to the types of plan improvements made from entities' analyses of actual cyber events and/or testing. Such information could reveal the need or opportunity for improvements to other entities' response and recovery plans and be a valuable component of a continuous improvement process. **[Section V.H]**

Observed Practices for Consideration

Throughout its review, the joint staff review team found that the participants have many practices and protocols that serve to enhance their restoration and recovery planning and readiness but go beyond the requirements of the Reliability Standards. The joint staff review team recognizes that these practices may not be appropriate for all entities in all situations, but believes that wider understanding and incorporation of these practices will have significant value to certain entities and to the industry as a whole. Examples of these beneficial practices include the following:

- Some review participants include in their restoration plans illustrations and accompanying steps to assist operators in system restoration. Illustrations and guidelines include electrical (i.e., one-line) diagrams, tables, or charts of reference information to augment the steps of restoration. The inclusion of these additional details can be a valuable aid to operators in the execution of the plan.

- Many participants have extra personnel in place to augment operators and other support staff during system restoration. The additional personnel can perform tasks in support of the restoration effort, including performing off-line power flow studies, so system operators are able to focus on essential system restoration tasks with minimal distractions.

- Some participants perform exercises or drills that involve the actual transfer of control center operations to an alternate site for a period of time, in order to test the functionality of the recovery resources. This practice goes beyond the requirements of the Reliability Standards to provide a more realistic test of response and recovery readiness. The actual evacuation and verification of functionality of recovery resources can reveal unknown issues or problems through use of the alternate site's cyber assets.

Such sound practices, which were voluntarily implemented by review participants, serve to enhance the industry's preparation for a major event, and provide training to recover more quickly and efficiently when an event occurs. A discussion of beneficial practices observed by the joint staff review team can be found in the relevant sections of this report.

II. Introduction: What are System Restoration and Recovery Plans and Why Are They Important?

In the United States, electric customers depend on reliable and continuous service. Unexpected loss of power is inconvenient. Moreover, sustained and widespread outages may lead to more severe circumstances that are potentially catastrophic. Typically, power losses are confined to relatively small areas of the electric system, and the vast majority of outages experienced by customers are the result of the loss of distribution level facilities.[9] Despite the overall reliability of the transmission system as a whole, however, widespread outages do occur, as seen with the August 2003 blackout, the September 2011 outages in Arizona and Southern California, and the outages caused by Hurricane Sandy in 2012.

These major events can cause significant disruption of the bulk power system, and often require the use of blackstart resources[10] and coordinated, multi-entity efforts to restore the system. Because these events are significant, although uncommon, it is critical that all entities potentially involved in the system restoration process be prepared to respond to potential widespread outage scenarios. This report focuses on evaluating the readiness of the electric utility industry to restore the bulk power system following a widespread outage.

While utilities have historically developed their own formal plans and procedures to restore their systems after widespread outages, they were not subject to a mandatory requirement to do so prior to the August 2003 blackout. Following that outage, Congress passed the Energy Policy Act of 2005, which, among other things, required the

---

[9] Generation or transmission line outages often do not impact electric customers. During storms, for example, one or more transmission lines may trip offline due to lightning strikes or other causes. However, customer service may not be interrupted because the transmission systems are typically designed to isolate the affected circuits and prevent a shutdown.

[10] Blackstart resources are generating units that have the ability to be started without support from the rest of the bulk power system, or are designed to remain energized without connection to the remainder of the bulk power system, and can be used to re-start other generating units as part of the process of re-energizing the system.

Commission to certify an independent Electric Reliability Organization (ERO) tasked with developing and enforcing mandatory reliability standards. NERC was certified as the ERO in 2006, and works with industry to develop mandatory reliability standards, the first set of which were approved by the Commission in 2007.[11]

One approved Reliability Standard, EOP-005-2 (System Restoration from Blackstart Resources), requires transmission operators and reliability coordinators to develop and maintain adequate system restoration plans. Specifically, each transmission operator is required to have a system restoration plan to reestablish its electric system in a stable and orderly manner in the event of a partial or total shutdown of its system. These plans are required to include necessary processes and procedures to cover emergency conditions and the loss of vital telecommunications channels. The standard also requires generator operators with blackstart resources to establish procedures related to those units, and to coordinate and communicate with other entities regarding the status of those units.

Although most entities have system restoration plans that cover multiple situations, the scope of the restoration plan required by the Reliability Standards is as follows:[12]

> The restoration plan shall allow for restoring the Transmission Operator's System following a Disturbance in which one or more areas of the Bulk Electric System (BES) shuts down and the use of Blackstart Resources is required to restore the shut down area to service . . . .[13]

The Reliability Standards require that the system restoration plan restore "the shut down area to a state whereby the choice of the next Load to be restored is not driven by the need to control frequency or voltage regardless of whether the Blackstart Resource is located within the Transmission Operator's System."[14]

---

[11] Electric power entities that own or operate infrastructure or systems that comprise the bulk power system are generally required to register with NERC and comply with the mandatory Reliability Standards, including those pertaining to system restoration.

[12] Some entities also have restoration plans or procedures to address restoration of services at the distribution level, which plans and procedures do not fall within the Commission's jurisdiction and are not the subject of this report.

[13] NERC Reliability Standard EOP-005-2 (System Restoration from Blackstart Resources) at Requirement R1.

[14] *Id.*

In addition to the system restoration plan requirements, the approved Reliability Standards also require applicable entities to have a *cyber security incident response plan*, as well as a *recovery plan for critical cyber assets*. In most cases, the computer systems used to remotely monitor and control the electric system are identified as "critical cyber assets," and therefore subject to the Reliability Standard requirements related to cyber security responses and critical cyber asset recovery plans. Having appropriate cyber security and cyber response plans in place is thus a critical part of system restoration.

III. Joint-Staff Review Process

The primary objective of the joint staff review was to assess participants' plans and readiness for system restoration and recovery efforts following a widespread outage, and to evaluate the efficacy and clarity of the associated Reliability Standards to help ensure the adequacy of these plans. The objectives of the review included:

- Gathering information via outreach to a representative sample of selected entities with significant bulk power system responsibilities.

- Gaining an understanding of the overall state of restoration plans by comparing and contrasting their content, scope and interrelationships.

- Assessing the clarity of the Reliability Standards in supporting the adequacy and efficacy of restoration and recovery plans.

- Identifying good industry practices and making recommendations to ensure that effective restoration and recovery plans are in place to support reliability.[15]

The recovery and restoration plan review focused on reviewing the adequacy of three Reliability Standards (as discussed further below in Section IV):

- EOP-005-2 System Restoration Plans from Blackstart Resources
- CIP-008-3 Cyber Security—Incident Reporting and Response Planning
- CIP-009-3 Cyber Security—Recovery Plans for Critical Cyber Assets[16]

---

[15] *See* Appendix 2 – Request Letter for Participation in Reliability Assessment at 2 (sent Sept., 2014).

[16] The cyber-related Reliability Standards reviewed reflect the CIP standards currently in effect, i.e., CIP Version 3. The Commission has approved a revised version of CIP

The joint staff review team adopted a collaborative model for conducting the review. Subject matter experts from the Commission, NERC and the Regional Entities collaborated to form the review team, collectively providing the necessary planning, operations and cyber security expertise.[17]

Once assembled, the joint staff review team identified a representative sample of entities with significant bulk power system responsibilities, to achieve comprehensive review of the wider area restoration capabilities.

The joint staff review team contacted each identified entity to request its participation. All contacted entities agreed to participate in the review, and without exception, were exemplary in their cooperation with the joint staff review team, sharing the detailed technical rationale behind their restoration and recovery plans. The joint staff team commends the participating entities for their open and active contributions.

In order to facilitate a full and open discussion of each participant's methodologies and strategies for restoration, their underlying rationale, and the resulting list of critical assets, the joint review team agreed not to disclose entity-specific information outside each review group. This report accordingly provides the results of the reviews without attribution to individual entities.

The joint staff review team reviewed each participant's restoration and recovery plans and supporting information, and engaged in discussions with the participants to gain additional information and insights regarding individual plans. The reviews were comprehensive and thorough, with some involving on-site visits. The team evaluated the participants' plans and procedures for each stage of restoration, response, and recovery, to ensure completeness and consistency of review from one participant to the next.

---

standards, i.e., CIP Version 5, which will become enforceable for certain assets starting on April 1, 2016. *See Version 5 Critical Infrastructure Protection Reliability Standards,* Order No. 791, 145 FERC ¶ 61,160 (2013). While the joint staff review team focused on the currently-effective CIP Version 3 Reliability Standards in conducting their review, this report also indicates whether the team expects CIP Version 5 to address or otherwise affect an identified area of concern.

[17] Appendix 1 lists the joint staff review team members.

For the various stages and topics of restoration, cyber security incident response and critical cyber asset recovery, the team undertook the following steps:

> Step 1: Gain understanding of participants' bulk power system restoration and recovery reliability activities;
>
> Step 2: Identify strengths and shortcomings of individual plans and procedures;
>
> Step 3: Using the results from steps 1 and 2, perform an assessment of relevant Reliability Standards; and
>
> Step 4: Form recommendations to improve reliability.

IV. Review of System Restoration Plans and Related Standards Assessment

The joint staff review team reviewed the system restoration plans, procedures and resources of the participants to assess their readiness to restore the electric system to a normal condition in the event of a partial or total system shutdown. This report provides a breakdown of the review by various restoration topics. These topics include:

- Strategies and Priorities for Restoration;
- Roles, Interrelationships and Coordination;
- Situational Awareness Tools for Quick and Orderly Restoration;
- System Restoration Resources;
- Island Development and Synchronization;
- Testing of System Restoration Resources;
- Testing, Verification, and Updating of System Restoration Plans;
- System Restoration Drills and Training Exercises; and
- Incorporation of Lessons Learned from Prior Outage Events.

As noted above, included at the close of each topic is analysis of the participants' plans against the relevant Reliability Standards, to see where improvements in clarity or efficacy of the standards may be warranted. In reviewing the Reliability Standards, the team also considered relevant recommendations for improvement to Reliability Standards as made by the IERP.[18]

A. Strategies and Priorities for Restoration

1. Summary

The overall objective of a restoration plan is timely restoration of the transmission operator's system, with priority placed on restoring the interconnection as a whole. To accomplish this, the transmission operator assesses the initial conditions to determine the restoration strategy. In its review of the participants' strategies and priorities for restoration, the joint staff review team examined initial assessments to determine various restoration plan strategies, and priorities for restoring loads and tie-lines.

As described below, the joint staff review team found that the participants' restoration plans address the need to identify restoration strategies and priorities. The team found that the participants' plans require highly-detailed initial status assessments using templates, computer applications, or forms to identify and convey to system operators and reliability coordinators the extent of the outages and affected facilities. These initial status assessments ultimately determine the strategy(ies) to employ for restoration.

The joint staff review team concludes, as a result of its examination of the plans, that the relevant EOP-005-2 Reliability Standard requirements that address system restoration strategies and priorities are clear and effective. The joint staff review team also observed certain practices and approaches that appear to enhance an entity's ability to assess or address a given disturbance, and recommends that applicable entities consider implementing these approaches in their own restoration plans. Observations by the joint staff review team are detailed below.

2. Review of Participants' Restoration Plans

a) Initial Assessment of Conditions

All of the participants' restoration plans require an initial assessment of the status of the system as a critical first step, including assessment of the status of major transmission

---

[18] See NERC, *Standards Independent Experts Review Project: An Independent Review by Industry Experts* (June 2013) http://www.nerc.com/pa/Stand/Standards%20Development%20Plan%20Library/Standards_Independent_Experts_Review_Project_Report.pdf) (IERP Report).

lines, generating units available to be ramped up or started on demand, and electrical islands that may still be operating. This initial assessment allows the participants to determine an appropriate restoration strategy(ies).[19]

All of the participants rely on SCADA as their primary data gathering tool during the assessment phase following a disturbance. Some of the participants have recognized the likelihood that their data reporting systems will be inundated with data during a large disturbance, and are using special algorithms to filter the data, to assist in evaluating events and alarms and other status indicators received. These participants indicated that they developed this approach in response to previous events, and the joint staff review team believes that these kinds of alarm management approaches can enhance an entity's ability to accurately assess system conditions and initiate prompt system restoration.

One reliability coordinator has instituted the use of a status reporting form as part of its restoration plan. All transmission operators within its footprint are familiar with the form, as it is used during the reliability coordinator's regular restoration training drills. The joint staff review team found that use of a status reporting form, including training and drilling based on that form, should improve the speed and accuracy of reporting and appears to be a best practice worthy of consideration by other reliability coordinators. Use of a common form also enables the integration of individual reports from multiple entities to better enable the reliability coordinator to understand the state of the system within its footprint.

b) Restoration Strategies

All of the participants' restoration plans are designed around a worst-case, total blackout scenario baseline, although several participants' plans include a range of scenarios in addition to a full blackout, as discussed further below.

Participants employ an "inside-out" island development strategy, in that the viability of their plans is not dependent on outside sources (i.e., not dependent on tie-line connections with other entities for restoration). The only exception is for pre-arranged external blackstart resources. Since most participants use more than one blackstart generator in their system restoration plans, their plans generally address the simultaneous

[19] In making their initial assessment, the participants analyze a range of factors, including, for example: (1) frequency monitoring locations for restoration; (2) availability and location of blackstart resources; (3) available transmission paths to start up generating plants; and (4) boundaries of energized areas and status of interconnected systems.

development of multiple islands, in which transmission and generation operators work together to develop electrical islands within their respective footprints. Developing multiple islands can limit the impact of an outage during restoration, by preventing problems experienced during restoration in one in-development island from affecting another island. In addition, it allows multiple areas to be restored at the same time.

The joint staff review team compared the island development methods contained in participants' restoration plans and procedures and found that several participants use a "core-island" approach, while others use a "backbone-island" approach or some combination of the two. The core-island approach involves the start-up of a blackstart generator, which is then used to energize a transmission cranking path and provide cranking power for a nearby generator and priority loads. Other loads are then added incrementally, and additional generators are synchronized via additional paths. With this incremental addition of generators and loads, the participant can develop a core electric island, while maintaining reserve generator capacity for island stability. The "backbone-island" approach involves starting up a larger blackstart generator and energizing higher nominal voltage and longer transmission lines (e.g. 230 kV, 345 kV) to develop a cross-system backbone to which core-developed islands can subsequently synchronize.

There are advantages to each island development approach. The core-island approach provides more island stability during the early stages of restoration. It also allows for underfrequency relay-controlled load to be restored sooner, but it may delay station service power to transmission substations, which may result in loss of SCADA for those substations due to back-up power supply (e.g., battery) depletion. The backbone-island approach can be a quicker method to restore auxiliary power to generators and transmission substations, including SCADA functionality. In addition, the larger amount of generation capacity brought online for the backbone method can provide transient stability and dynamic reactive reserve for voltage stability. However, this approach carries a risk of excessive voltages and may require additional voltage control facilities and equipment settings to mitigate these higher voltages.

As noted above, several participants include a range of initial scenarios in their restoration plans, (e.g., no blackout, with area internal to the transmission operator footprint becoming islanded) providing guidance to operators to respond to a wider range of emergency conditions. Also, based on their particular experience, lessons learned, and planning and engineering studies, some participants have identified areas vulnerable to a voltage collapse, and have developed strategies to contain the impact from such a collapse using automatic separation schemes. Thus, in addition to the total blackout scenario, some participants have incorporated these limited outage scenarios into their restoration plans to provide the operators a range of strategies and procedures for restoration.

c) Priority Loads and Tie-lines

All the participants include priorities for restoring loads in their system restoration plans, with the underlying objective of restoring the interconnection as a whole. Those transmission operators responsible for providing primary or back-up service to nuclear power plants prioritize the restoration of off-site power to those plants for safe shut-down. In addition, those participants serving metropolitan high-density loads place priority on restoration of those areas, recognizing the need to protect human safety.

Some participants' plans include criteria for identifying other high priority loads, including the following:

- Start-up power - otherwise referred to as "cranking power" - to non-blackstart generators that are designated to start-up quickly (e.g., in 4 hours or less) as part of the system restoration plan;
- Power to electric-powered pumps for natural gas pipelines that pressurize and provide the large volumes of natural gas deliveries to quick-start generators, such as combustion turbines;
- Auxiliary power needed for steam generator plants that do not have their own auxiliary power resources;
- Power to pumping stations for oil pipelines, nuclear military installations and floodwater or floodwall control installations; and
- Power to pumps that maintain oil pressure on underground electric transmission cables, to prevent failure of these cables during system restoration.[20]

All of the plans reviewed establish priorities for restoring load consistent with the requirements of the Reliability Standards (discussed further below), and most go beyond identifying the highest priorities and objectives to provide the system operator with a clear understanding of restoration priorities as restoration moves forward. For example, the joint staff review team observed that some participants' plans identify priority loads beyond those identified in the standards (i.e. nuclear power plants). These participants

[20] Distribution-level restoration plans also often identify higher-priority loads, including, for example, hospitals, critical water systems, and critical natural gas facilities.

indicated that identification of the next-level priority loads provides system operators with a clear understanding of the priority actions to take during system restoration.

All of the participants' restoration plans also recognize the priority of reconnecting to the rest of the system after a partial shutdown. However, the participants' restoration plans do not specify which external interconnection to restore first, i.e., they do not specify which interconnection neighboring transmission operators should restore first. The participants indicated that this practice was by design, in order to allow for flexibility in their restoration based upon the initial assessment of conditions. Participants emphasized that designing the plans to be flexible in this regard allows them to be adaptable to a range of initial conditions.

3. Related Standards Assessment

Reliability Standard EOP-005-2 has as its stated purpose to ensure plans, facilities, and personnel are prepared to enable system restoration from Blackstart Resources to assure reliability is maintained during restoration and priority is placed on restoring the Interconnection. The standard includes broadly written requirements for transmission operators to have strategies for system restoration based on expected blackout conditions, and procedures for restoring loads and interconnections, including prioritization and provision of off-site power supply for nuclear power plants. Sub-requirements R1.1 – R1.3 and R1.8 require transmission operators to have Reliability Coordinator-approved system restoration plans that include the following:

> **R1.1.** Strategies for system restoration that are coordinated with the Reliability Coordinator's high level strategy for restoring the Interconnection.
>
> **R1.2.** A description of how all Agreements or mutually agreed upon procedures or protocols for off-site power requirements of nuclear power plants, including priority of restoration, will be fulfilled during System restoration.
>
> **R1.3.** Procedures for restoring interconnections with other Transmission Operators under the direction of the Reliability Coordinator.
>
> . . .
>
> **R1.8.** Operating Processes to restore Loads required to restore the System, such as station service for substations, units to be restarted or stabilized, the Load needed to stabilize generation and frequency, and provide voltage control.

Requirements R1.1-R1.3 and R1.8 allow for flexibility in identifying strategies and priorities for restoration, which the team found to be appropriate, since strategies and priorities for each entity would be different based on the entity's size, system topography, etc. Moreover, while certain standard requirements allow entities the flexibility to determine the level of detail to include in their restorations plans, this flexibility is balanced by the fact that Reliability Standard EOP-005-2 also requires simulation testing of the plan (R6) and reliability coordinator review and approval of each entity's plan (R3). This approach is supported by the joint staff review team's observations that, consistent with the requirements of Reliability Standard EOP-005-2 Requirements R1.1, R1.2, R1.3, and R1.8, the participants' plans each include detailed strategies and priorities for restoration under varying circumstances. Thus, the joint staff review team did not identify any issues related to such strategies and priorities that would suggest that modification of these requirements or other additions to the Reliability Standards is needed in this regard. The joint staff review team found that participants' plans include highly-detailed initial assessments, often involving the use of templates or forms to identify and convey the extent of the outages and affected facilities, to ultimately determine the strategy(ies) to employ for restoration. In addition, all participants' plans require restoration strategies to: be coordinated with the reliability coordinator's high-level strategy of restoring the interconnection (R1.1); clearly identify as a top priority re-establishing off-site power supply to nuclear power plants (R1.2), and identify priority loads for restoration (R1.8).

Finally, consistent with Requirement R1.3, all participants' restoration plans include procedures for restoring interconnections with other transmission operators under the direction of the reliability coordinator. While the plans did not place any priority on restoration of the connection with one transmission operator over another, the joint staff review team considers this to be appropriate given that restoration priority should depend on the initial assessment of conditions. The joint staff review team concurs with participants that prioritization of restoration of particular interconnected transmission operators should not be required.

4. Observed Practices for Consideration

In evaluating participants' strategies and priorities for system restoration, the joint staff review team observed the following practices and recommends consideration of them by entities:

- Some participants' plans include steps for addressing a range of scenarios in addition to a total blackout, including:
    - transmission operator area islanded
        - area within the transmission operator footprint becomes islanded, no blacked out area

        - transmission operator area separation occurs from the rest of the interconnection, no blackout area

    - transmission operator area blacked out with external/interconnection assistance available to aid in restoration

    - transmission operator area blacked out without external/interconnection assistance available to aid in restoration

    - transmission operator area becomes split (areas expected to break apart) in some pre-determined manner, requiring use of restoration plan processes to re-establish connection

  Addressing multiple scenarios in the restoration plan provides flexibility and adaptable guidance for the operators to follow, enabling them to better respond to a wider range of emergency conditions.

- Some participants have highly-detailed load restoration priority guidance when developing their restoration plans, such as criteria for identification.

- Some participants employ applications, algorithms or other sorting and filtering mechanisms to analyze the high influx of alarms and other status-related data that may accompany a disturbance.

- Some participants use status reporting forms to expedite and clarify reporting of facility status information, and include the use of such forms in their restoration training and drills.

B. Roles, Interrelationships and Coordination

1. Summary

It is crucial that affected entities understand each other's roles and expected responsibilities in restoring the system to interconnected operations. The joint staff review team accordingly examined how the participants' restoration plans and procedures address or define the roles of the various entities involved in or affected by the participants' system restoration plans. The joint staff review team examined:

- Functional roles and interrelationships, according to NERC registration;
- Contractual roles and interrelationships, such as those covered by agreements or arrangements;
- Actual operational roles and interrelationships, such as those understood to exist based on participant discussions; and

- Coordination and communication that occurs during system restoration.

The joint review team found that participants' restoration plans address the respective roles of each entity involved in the restoration plan and are organized accordingly, with several that include tables of internal tasks or responsibilities, and of tasks expected of and approvals needed from the other entities involved in restoration. The joint staff review team determined that the participants' restoration plans, together with any related arrangements, are generally clear and sufficient in defining the roles and relationships among entities.

2. Review of Restoration Plans and Related Arrangements

Of the nine registered entities that participated in the review, the functional entity categories for the participants reviewed are as follows:

- Three entities were registered as reliability coordinators;
- Seven entities were registered as transmission operators;
- Five entities were registered as transmission owners (three of which were also registered as the transmission operator);
- One entity was registered as a generator operator;
- Two entities were registered as generator owners (one of which was also registered as the generator operator); and
- Five entities were registered as balancing authorities.

Two of the participants that perform transmission operator tasks have local control centers and operators who maintain SCADA-control of transmission facilities, but are not registered as transmission operators, and as such, are not required under the Reliability Standards to have a restoration plan that is approved by the reliability coordinator. However, the joint staff review team found that these two entities have detailed restoration procedures, which were prepared in coordination with, or as an appendix to, their respective transmission operator's restoration plan. The steps covered in these transmission owners' restoration procedures are similar to those covered in the transmission operator's system restoration plan, including, but not limited to: strategies for system restoration, steps for providing nuclear power plant off-site power, procedures for restoring interconnections with external entities and identification of cranking paths and initial switching requirements.

The review team also examined the following arrangements affecting restoration plans and the implementation of those restoration plans:

- Coordinated Functional Registrations: Some transmission operators have coordinated functional registrations with other transmission operators, covering responsibilities that include system restoration. A coordinated functional registration represents an agreement between two or more registered entities sharing and/or splitting compliance responsibility for requirements/sub-requirements within particular Reliability Standard(s).[21]

- Transmission operator-transmission owner member agreements or operating agreements: In some situations, a transmission owner may not also register as the transmission operator with NERC, although it does retain some operational control over transmission facilities. In this arrangement, the transmission owner's system operators (i.e., those that have control room operators with direct operational control of transmission facilities) are given authority to take actions to operate their system with transmission operator oversight, including actions required during system restoration.

- Blackstart generator agreements between transmission operators and generator operators: Entities have established blackstart service agreements that provide performance specifications for the blackstart unit. Some entities have blackstart resource agreements for resources located outside of the transmission operator footprint.

The joint staff review team determined that the participants' restoration plans, together with any related arrangements, are generally clear and sufficient in defining the roles and relationships among entities. For all of the arrangements reviewed, participants who had delegated restoration tasks through contractual arrangements or coordinated functional registrations included tables of tasks or responsibilities defining the roles, tasks and approvals needed by each entity involved in restoration. Such tables and charts provide clarity and reduce confusion as to who is responsible for performing each task, and help to define the associated communication, coordination and approval protocols during system restoration.

---

[21] *See* NERC Rules of Procedure § 508 – Provisions Relating to Coordinated Functional Registration (CFR) Entities, http://www.nerc.com/FilingsOrders/us/RuleOfProcedureDL/NERC_ROP_Effective_20151104.pdf.

3. Related Standards Assessment

Reliability Standard EOP-005-2, Requirement R1, includes several sub-requirements that address the need for transmission operators to coordinate with other entities and define roles through the development of procedures as part of a system restoration plan (sub-requirements R1.1 – R1.3 and R.1.9). These sub-requirements state that the restoration plan must include:

> **R1.1.** Strategies for system restoration that are coordinated with the Reliability Coordinator's high level strategy for restoring the Interconnection.
>
> **R1.2.** A description of how all Agreements or mutually agreed upon procedures or protocols for off-site power requirements of nuclear power plants, including priority of restoration, will be fulfilled during System restoration.
>
> **R1.3.** Procedures for restoring interconnections with other Transmission Operators under the direction of the Reliability Coordinator.
>
> . . .
>
> **R1.9.** Operating Processes for transferring authority back to the Balancing Authority in accordance with the Reliability Coordinator's criteria.

In addition, EOP-005-2 Requirement R13 addresses one aspect of the specification of roles and relationships among entities participating in the restoration process:

> **R13.** Each Transmission Operator and each Generator Operator with a Blackstart Resource shall have written Blackstart Resource Agreements or mutually agreed upon procedures or protocols, specifying the terms and conditions of their arrangement. Such Agreements shall include references to the Blackstart Resource testing requirements.

As noted above, the joint staff review team determined that the participants' restoration plans and related arrangements are generally clear and sufficient in defining the roles and relationships among entities. Though the relevant Requirements provide broad coordination-related topics to be addressed in restoration plans, the restoration plans and related arrangements reviewed by the team provide for appropriate coordination with the reliability coordinator's overall strategy for restoring the interconnection (as required under Requirement R1.1). Likewise, the team observed that the individual plans adequately cover the roles and responsibilities of blackstart resources (as required in R13), establish procedures with nuclear power plants for restoring offsite power (as required in R1.2), and include the necessary reliability coordinator approval steps for restoring interconnections with other transmission operators (as required in R1.3). The joint staff review team determined that EOP-005-2, Requirements R1.1 – R1.3, and R13 are clear and effective in supporting restoration plans. Participants appeared to have a clear understanding of the obligations set forth in these provisions, and no issues of ambiguities were raised in discussions with participants or reflected in their restoration plans.

While Requirement R1.9 requires the transmission operator to have processes for transferring authority back to the balancing authority in accordance with the reliability coordinator's criteria, the joint staff review team found that some restoration plans contain limited information on the triggers, steps involved, or checks necessary for that transfer. However, the team believes that any additional need for clarification and better understanding of the processes for transferring authority back to the balancing authority should be addressed through operator training, as further discussed in section IV.H - System Restoration Drills and Training Exercises.

Reliability Standard EOP-005-2 applies primarily to transmission and generator operators and, as such, does not require transmission owners or generator owners to have an approved system restoration plan. However, the joint staff review team found that some generator owners and transmission owners have active roles and responsibilities in system restoration in maintaining stable operation during island development.[22] In some instances, the team found that the transmission owner has a restoration plan in coordination with, or as an appendix to, their transmission operator's restoration plan (as described above). In each case reviewed where the transmission owner has control center operators with SCADA control of the facilities, the transmission owner has such a plan in place. The team did not assess generator operator blackstart procedures, as the scope of the project was focused on the restoration plans of the transmission operators and the procedures for using blackstart resources. The team noted that each involved or affected entity (including transmission owners and generator operators) appeared to understand its respective roles and relationships for system restoration. Accordingly, while further study of the dependency of transmission operators on transmission owners and generator

---

[22] The IERP recommended that generator owners, distribution providers, transmission owners, and generator operators should be required to have an approved system restoration plan in place (in addition to transmission operators as required in EOP-005-2, and reliability coordinators as required in EOP-006-2). See NERC, *NERC Standards Announcement: Posted - Independent Experts Scoring for Requirements Spreadsheet* (hyperlink: "Independent Experts Scoring for Requirements Spreadsheet," http://www.nerc.com/pa/Stand/Standards%20Development%20Plan%20Library/Ind_Exp_Scoring_Req_Spreadsheet_Announc_082913.pdf) (IERP Scoring Sheet).

owners in the system restoration process may be warranted, the team did not identify any concerns as to the participants' understanding and acceptance of their role in system restoration and does not recommend modification of the Reliability Standards to address this issue at this time. However, there may be similarly situated entities that were not part of this review that warrant future examination of transmission owner and generator owner roles and responsibilities during system restoration.

4. Observed Practices for Consideration

In evaluating participants' roles and responsibilities for system restoration, the joint staff review team observed that most participants include tables of tasks or responsibilities defining the roles, tasks and approvals needed by each entity involved in restoration. Such tables and charts provide clarity and reduce confusion as to who is responsible for performing each task, and help to define the associated communication, coordination and approval protocols during system restoration. The joint staff review team recommends that other registered entities with responsibility for system restoration consider this practice if they do not already maintain tables or similar methods to clearly defines roles, tasks and approvals needed for restoration.

C. Situational Awareness Tools for Quick and Orderly Restoration

1. Summary

Through discussions with participants and review of their restoration plans and other pertinent procedures, the joint staff review team examined and compared the situational awareness tools used to plan and carry out system restoration. The team found that the participants' plans recognize the importance of maintaining access to vital system data during a disturbance, employing redundant, diversely routed communications systems and redundant EMS or SCADA systems. In addition, participants' procedures include strategies for addressing the loss of SCADA, EMS[23] or ICCP[24], calling for the dispatch of personnel to substations in the event of the loss of such systems. Situational awareness tools and their usage are addressed in Reliability Standards other than EOP-005-2. Given

---

[23] Energy Management System (EMS) is a system of computer-aided tools used by bulk-power system operators to monitor, control and optimize system performance.

[24] The Inter-Control Center Protocol (ICCP) allows the exchange of real time and historical power system information between entities, including status and control data, measured values, scheduling data, energy accounting data and operator messages.

the possibility that SCADA, ICCP or EMS functionality may be compromised during a major disturbance, the team recommends that a study be conducted to assess system restoration plan steps that may be difficult in the absence of SCADA, ICCP data and/or EMS, and to identify viable resources, methods or practices that allow timely system restoration to occur absent SCADA/EMS functionality, which could then be incorporated into entities' system restoration training.

2. Review of Restoration Plans and Procedures

All of the participants' restoration plans rely on the extensive use of SCADA systems in assessing system status and in carrying out the restoration process. The joint staff review team found that SCADA systems facilitate a number of restoration processes, including the operator monitoring of frequency and voltages to ensure stability of developing islands. Likewise, the team observed that SCADA facilities can be particularly useful in the synchronization and interconnection of separate islands. SCADA systems can be used to provide much of the system information needed to identify interconnection opportunities and to evaluate whether conditions necessary to initiate synchronization have been met, and can be used along with other tools to allow system operators to interconnect systems without substation operators on site. In addition to SCADA, participants' restoration plans rely on ICCP data and EMS to remotely monitor and control the electric system. Participants noted the difficulty involved in restoring the system in the event SCADA/EMS or ICCP data are not available. One participant commented to the team that, in the absence of SCADA, restoration would be a long, tedious process. Some participants have a primary and back-up EMS at multiple control center locations with the ability to fully use the EMS from each location. These levels of redundancy help to ensure that EMS and SCADA have high levels of availability for reliable operation of the bulk power system, including reliable restoration.

Despite the redundancy in EMS and SCADA systems, all participants plan for the possibility that SCADA and EMS may be partially or totally unavailable at some time during a restoration event. For example, portions of SCADA functionality may not be available after a longer-duration blackout due to back-up power supply (e.g., battery) depletion for unrestored substations. Participants have procedures for loss of EMS and SCADA that broadly apply during normal and emergency grid conditions, including during restoration events. In the event of an EMS or SCADA failure, participant system operators notify EMS, Information Technology (IT), and/or telecommunications staff responsible for resolving SCADA and EMS concerns. In addition, participants' plans call for system operators to dispatch field personnel to transmission stations, so that field personnel are ready to manually perform feasible restoration activities, and provide field equipment status and data, at the direction of the system operator. Furthermore, most participants plan for system operators to work with operations and/or planning engineers using off-line power-flow models to perform system studies if SCADA, EMS and associated power flow applications are not available. Some participants' operators use off-line data tracking applications specifically developed for system restoration monitoring in the case of EMS unavailability.

The joint staff review team found that EMS/SCADA systems play a significant role in operators' decision-making during system restoration, and the unavailability of these systems following a major system disturbance can delay system restoration. Participants indicated that a priority is placed on recovering SCADA, because there is not currently an alternate means of restoring the system as quickly without SCADA. Dispatching personnel to substations is an option, but lacking alternative or supplemental tools and resources to provide timely situational awareness for operators' decision-making could delay or complicate restoration compared to using SCADA or an equivalent approach. While the restoration of SCADA functionality is thus important for restoration efforts, SCADA functionality is equally important for adequate situational awareness during any system condition (normal or emergency). Accordingly, SCADA system protection and recovery has implications beyond system restoration, which the team believes should be addressed in that broader context.

3. Related Standards Assessment

The Reliability Standard requirements that relate to communications and use of situational awareness tools are covered by Reliability Standards other than EOP-005-2, including communications standards (COM-001-1.1), and a number of current and under-development operational standards (TOP and IRO). This report makes no recommendations as to those Reliability Standards.

4. Recommendations

Planning for loss of SCADA and loss of other data sources. Given the possibility that SCADA, ICCP or EMS functionality may be compromised during a major disturbance (e.g., portions of SCADA may not be available after a longer-term blackout), the joint staff review team recommends that a study be conducted to (a) assess system restoration plan steps that may be difficult in the absence of SCADA, ICCP data, and/or EMS; and (b) identify viable resources, methods or practices that would enable timely system restoration to occur absent SCADA/EMS functionality, which could then be incorporated into entities' system restoration training.[25] The study should also examine and identify best practices that may be shared across the industry. Pending such study, individual

---

[25] The joint staff review team recognizes that the study may be accomplished by performing analyses regionally, where there may exist different capabilities from one area to another.

entities should initiate or update consideration of resources, methods and practices they can use in these circumstances.

5. Observed Practices for Consideration

In evaluating each participants' approach to the use of SCADA, EMS and other situational awareness tools during system restoration (through site visits and discussions), the joint staff review team observed the following practices and recommends that other entities consider adoption of these practices :

- Remote monitoring of parameters necessary to synchronize islands and performance of remote synchronizing of islands.
- Provision of tools for system operators and support staff to allow them to efficiently process the numerous alarms received during the assessment phase of restoration. Participants who used such tools indicated that processing the alarm data quickly offers insight into the initial cause of the event, as well as provides information on the status of equipment after the event and during restoration.
- Use of cranking path displays, highlighting the cranking path transmission substations and transmission lines between the blackstart generator-substation and the next unit to be started.
- Use of SCADA displays which provide underfrequency load shedding (UFLS) load-relay substation locations with UFLS-controlled load values totaled, helping to improve island stability management where UFLS is integral to participants' restoration plans.

D. System Restoration Resources

1. Summary

The joint staff review team examined how the participants' restoration plans and procedures addressed the various resources needed for system restoration, focusing on blackstart generators (including their characteristics and procurement), communications, control center and field resources, and direct current transmission resources.

The joint staff review team found that the participants' restoration plans and procedures address the necessary identification and dedication of blackstart and other resources for system restoration, and the team did not identify any clarity or efficacy concerns with the Reliability Standards based on this area. Specifically, the team found that participants' restoration plans identify system restoration resources and their characteristics, including, at a minimum, blackstart resources' name, location, type, and MW and MVAr capacity, and identify personnel and their assigned responsibilities during system restoration.

Many participants indicated, however, concern about the future availability of blackstart resources currently relied on for system restoration. The joint staff review team recommends further study to gain a better understanding of this issue and to identify strategies for identifying other blackstart resources. Additionally, industry experts from the joint staff review team provided information on recent advances in voltage source converter technology that could facilitate system restoration efforts. The team recommends that the benefits of including voltage source converter DC lines in system restoration also be further studied.

2. Review of Restoration Plans

a) Blackstart Generators - Characteristics and Procurement

Blackstart units are selected, sited and adapted to their service areas, and participants' plans accordingly include a wide of range of blackstart resources. The participants' plans take different approaches to the use of power from blackstart resources, with some participants using power from blackstart units to energize priority loads (before providing auxiliary power to other generators), and others using power from blackstart resources to supply auxiliary power to larger units first during system restoration. The blackstart generators included in the participants' plans range in size from small (e.g., 25 MVA) to larger units (e.g. 100-200 MVA), or even banks of units, exceeding 1,000 MVA in capacity. Some participants rely on a single unit while others included multiple units in designated islands.

Participants needing to procure blackstart services generally have strategies and procedures in place for procurement. Based on the review of participants' plans, the joint staff review team observed that the period for procuring blackstart resources ranges from two to five years. The observed strategies and approaches to such procurement include the following:

- Some participants make an initial determination whether an existing facility can be retrofitted to make it blackstart capable, or whether a new facility can be contracted for blackstart service.
- Some participants may agree to provide a contribution towards a feasibility study that will cover the installation, technical capabilities and cost of installing blackstart capability at the site.
- Some participants may deal with multiple providers simultaneously to determine the most economic and efficient option. In general, the process involves a request for proposals from generator operators to provide blackstart service followed by a determination of the most efficient arrangement/allocation of blackstart resources.

- Where the removal or retirement of existing blackstart units from service creates the need for a blackstart resource, some participants issue requests for proposals to replace the retired units.

- One participant instituted a commitment period for blackstart generators of three years and required that a blackstart generator operator provide a two-year notice to cease providing blackstart service. This requirement allows for timely procurement of replacement blackstart resources in the affected zones.

Participants consider various parameters when selecting or procuring a blackstart resource, including geographic area, the capacity and reactive capability of blackstart generators, start-up time,[26] and proximity to priority load and results from computer simulations of cranking paths to determine viable blackstart solutions. In addition, some participants require the provider of new blackstart services to verify that it is capable of providing the contracted service, through an assessment at commissioning.

Participants also consider the proximity of generator fuel supplies and electric transmission facilities. Some participant transmission operators specifically identify areas within their footprint where it would be beneficial to locate blackstart resources. These transmission operators identify cranking paths for supplying blackstart generation from multiple areas to meet priority load requirements, such as supplying offsite power to a nuclear power plant. Some participants also allow blackstart units to be physically located outside their footprint, as long as the blackstart resource and cranking path(s) to receive the blackstart power are appropriately identified. The advantages of these analyses and arrangements can include improved restoration speed and efficiency, meeting priority load restoration timing requirements and eliminating a blackstart resource shortage in an area.

Finally, some participants take into account the value of diversifying the location of their blackstart resources, to mitigate the risk of multiple blackstart units being unavailable due to a single-point loss or failure. The joint staff review team believes that this is a practice that should be considered by other industry participants in appropriate circumstances.

Regarding the amount of blackstart resources, participants identify blackstart units to meet priority load requirements in alignment with their restoration strategies, and most of

---

[26] The start-up time for blackstart units in the reviewed plans ranges from five minutes to several hours. Although not a requirement, a shorter start-up time is desirable to aid in the speedy restoration of the system.

the participants reviewed have multiple blackstart resources.[27] Some participant transmission operators require that the capability of the identified blackstart units be large enough to provide sufficient power (MW) to restore priority loads and have sufficient reactive capability (MVAr) for voltage control. Others require that the total generating capacity of blackstart units be a certain factor above priority loads.[28]

Participants' blackstart resources include a mix of coal and gas-fired steam units, gas combustion turbines, and hydroelectric units. In order to ensure consistent access to fuel, the participants have taken the following measures:

- To ensure that gas supply to blackstart generators is not interrupted during restoration, some participants include gas compressors as priority loads.

- Most participants' restoration plans have blackstart units with dual fuel capabilities (using both oil and gas). One participant reported that about fifty percent of its blackstart generation capacity has dual fuel capability. With the possibility of limited natural gas supply that may occur during a blackout, some participants with dual-fuel blackstart capability have procured onsite oil or gas, which could run the generator for a limited period (e.g., 48 to 72 hours).

- While some participants indicated that their hydroelectric blackstart generator output may be restricted during certain times of the year when water is low,[29] some participants' plans contemplate coordinating with environmental or other associated regulatory agencies in emergencies, e.g., through the issuance of waivers of environmental restrictions for brief periods of time.

---

[27] Some transmission operators plan for a minimum of two blackstart units for defined areas within their footprint to meet priority load and cranking path needs.

[28] For instance, one participant requires that the total capacity of blackstart units be maintained at 110 percent of the total priority load.

[29] Some participants indicated that at certain times during the year, their small hydro facility or pumped storage facility output may be restricted due to the amount of water stored or available in its reservoir. One participant with a pumped storage unit reported that a minimum amount of stored water is required in the reservoir to maintain the ability to blackstart. Other participants relying on hydroelectric blackstart services indicated that the environmental restrictions on their units are relatively minimal, and may only occur during periods when water levels at the hydro station are below what is necessary to sustain fish life or during periods of drought.

In addition to traditional blackstart units, which are able to start without power from the interconnection, some participants include generating units capable of automatic load rejection (ALR) in their restoration plans. These units are typically base load, coal fired units that have the ability to immediately disconnect from the grid during a blackout, but can continue operating as an island and can be used to re-energize the transmission grid during restoration by providing startup power to larger units, to load, and eventually to interconnect with neighboring systems. ALR units can be vital to these participants' restoration processes, as they can reduce system restoration time.

The joint staff review team observed that some participants have elected to start retiring some of these ALR units. In general, participants indicated that the availability of some traditional blackstart resources is being affected by recent changes in environmental emissions regulations and CIP Reliability Standards, and that some of these units are now being withdrawn as blackstart resources.

b) Direct Current Transmission Lines

The joint staff review team explored the role of direct current (DC) transmission lines during system restoration and recovery, and observed that DC transmission lines are not considered in the participants' restoration plans or restoration plan simulations. Most of the existing DC lines use "line commutated" converter technology, and this technology is not typically operable during the early stages of island development and restoration. In the participants' current plans and procedures, restoration of DC transmission lines occurs during the later stages of restoration. Some evidence indicates that it is more advantageous to use DC lines with voltage source converter technology during early stages of island restoration instead of reenergizing a long EHV, AC transmission line, since the latter creates high voltage issues that must be mitigated.[30] The joint staff review team accordingly sees value in studying the potential benefits of using DC lines with voltage source converter technology during early stages of system restoration.

---

[30] Voltage source converter technology uses transistors, specifically insulated gate bipolar transistors, which are semiconductor devices that act as switches in the converter but function differently from thyristors. Commutation during the inversion process (DC to AC conversion at the receiving terminal) will take place under all system conditions at the receiving end. This allows a voltage source converter to be used when the system is very weak or blacked out. In addition, a voltage source converter has the capability to control active and reactive power at the receiving terminal. For further information, see M. Davies, M. Dommaschk, J. Dorn, J. Lang, D. Retzmann, D. Soerangr, *HVDC PLUS – Basics and Principle of Operation* (2011).

c) Communications, Control Center, and Field Resources

The joint staff review team also examined the participants' deployment and use of communications, control center, and field resources (including personnel) in their restoration plans.

Communications Resources. Participants indicated that system operators primarily use dedicated telephone or radio for voice communications, which systems are typically redundant and diversely routed. Participants indicated that additional phone lines, cell phone, satellite phone, and Government Emergency Telephone Service cards are the primary back-up voice communications facilities. Participants also indicated that their back-up facilities are tested on a regular basis to ensure their operability. In addition, some participants have their control center staff perform normal operations occasionally using their various back-up voice communications facilities to ensure that system operators are familiar with the back-up facilities.

Control Center and Field Resources. The review showed that system restoration is controlled and directed by NERC-certified system operators (reliability coordinators, transmission operators, and balancing authorities). The participants' control centers are staffed with system operators 24 hours a day, seven days a week. These control centers plan for the use of additional system operator staff during a restoration event. Participants indicated that their system operators are assigned specific roles during a restoration event and are trained accordingly. In addition, participants indicated that they typically use additional personnel, planning engineers, operations engineers, schedulers, and others to aid and support the system operators during major storms or a restoration event.[31] Furthermore, participants' plans call for dispatch of field personnel to key transmission substations to perform activities at the direction of the system operator.

As discussed further below in the Testing, Verification, and Updating of System Restoration Plans section, the joint staff review team found that station batteries can be vital resources during system restoration, since they provide power to station equipment when system power is lost. Participants indicated that their station batteries are typically

---

[31] Participants indicated that one challenge faced during a restoration event is staff transportation. Since streetlights and public communications infrastructure may be affected, participants have found that their staff may encounter difficulty moving from one location to another. Therefore, their plans include notifying staff of assigned work locations so they know where to report given a restoration event. Also, the participants noted that system operators on shift when a disturbance occurs may have to work extended hours before additional staff arrives.

sized to provide adequate power for eight hours, based on estimated system restoration times. Some participants go beyond this typical approach, as follows:

- Some participants size batteries to provide power for a longer time (e.g. twenty-four hours) for certain substations that are a priority for restoration, for more remote stations, or where the participant anticipates difficulty reaching the station due to damage from a natural disaster (e.g., areas more prone to hurricane weather).

- Some participants use portable batteries and portable generators to supply station power if needed during restoration.

- Some participants install local generation at control centers and other key facilities as back-up power sources and test these generators regularly to ensure operability.

3. Related Standards Assessment

Under Reliability Standard EOP-005-2, Requirement R1.4, all transmission operators are required to identify each blackstart resource and its characteristics as part of the restoration plan. This sub-requirement provides that the restoration plan must include:

> **R1.4.** Identification of each Blackstart Resource and its characteristics including but not limited to the following: the name of the Blackstart Resource, location, megawatt and megavar capacity, and type of unit.

Other system restoration resources, including communications, control center, and field resources, are covered by other Reliability Standards.[32]

The joint staff review team found that the participants' plans include identification of blackstart resources and characteristics, including name of each blackstart unit, MW, MVAr, location, size and fuel type. The team did not identify any clarity or efficacy concerns with Requirement R1.4. However, recognizing that changes may be occurring with the need to procure and identify different blackstart resources going forward, the

[32] For example, telecommunications resources are covered in Reliability Standard COM-001-1.1 Requirement R1, which currently requires that transmission operators provide adequate and reliable telecommunications facilities which are redundant and diversely routed. Reliability Standard EOP-008-1 requires that transmission operators include in their contingency plans monitoring and control of critical transmission facilities and substation devices during emergencies.

joint staff review team recommends further study regarding these changes as discussed below.

4. Recommendations

Gain further understanding of recent blackstart resource changes. The joint staff review team recommends study of the availability of blackstart resources, including the identification of strategies for replacing blackstart resources going forward and factors to be considered for such replacement resources (e.g., location diversity, dual fuel, etc.). A future study may include discussions with a representative sample of generation owners and operators to gain further understanding.

Gain further understanding on the use of DC facilities for restoration. The joint staff review team recommends that a study be conducted to determine the benefits of including existing or future voltage source converter DC lines in system restoration plans.

5. Observed Practices for Consideration

In evaluating the participants' identification and use of resources necessary for system restoration, the joint staff review team observed the following practices and recommends consideration of them by other entities as appropriate:

- Some participants include generating units with load rejection capability in their system restoration plans, to speed up restoration and recovery.
- Some participants coordinate the use of blackstart facilities across multiple transmission service footprints, which can allow a blackstart generator to contribute in supplying an adjacent area's priority load.
- Many participants maximize the use of dual fuel blackstart units, in order to minimize the risk that the blackstart unit will not be available if one fuel is in short supply or otherwise unavailable at the blackstart unit site.
- Many participants have special procedures in place to augment operators and other support staff during system restoration. The extra personnel can perform tasks in support of the restoration effort, including performing off-line power flow studies, among other things, so system operators are able to focus on essential system restoration tasks with minimal distractions.

E. Island Development and Synchronization

1. Summary

The joint staff review team examined the review participants' approach to island development and synchronization. In doing so, the review team examined three key areas:

- Participants' cranking or restoration paths and methods to connect blackstart units to priority loads in preparation for the next units to be started;
- How frequency and voltage is managed during restoration; and
- How synchronization is performed with other islands and systems.

The review team found that island development protocols in the participants' restoration plans thoroughly cover the above areas. The joint staff review team concludes that the related Reliability Standard requirements are clear and effective for most aspects of island development and synchronization. However, as discussed below, the team recommends that measures be taken (including considering changes to the relevant Reliability Standards) to address the need to update restoration plans for any system modification that would change implementation of an entity's restoration plan for an extended period of time.

2. Review of Restoration Plans

a) Restoration Paths and Initial Loads Energized

Restoration Paths. The joint staff review team found that all of the participants' restoration plans include the identification of initial restoration paths originating from blackstart generator(s) to the initial loads to be energized.[33] The team found that the participants also take into account the possibility that transmission facilities may not be available as planned for system restoration, and have adopted one or more approaches to addressing that issue.

First, as noted earlier in the report, some participants that are more prone to severe coastal weather patterns have developed significant storm response plans in conjunction

---

[33] Initial loads include the entities' priority loads as described earlier, including provision of nuclear power plant off-site power, and generating plant auxiliary loads for the next unit(s) to be started.

with their system restoration plans. Storm response plans reviewed by the team typically include targeted repair and restoration efforts for lines comprising a restoration path.[34] Also, many participants include alternate or back-up cranking paths and priority load paths as part of their restoration plan, to be used in the event a primary path is unavailable. Participants indicated that providing path redundancy, alternate paths, or back-up paths in plans for system restoration is of critical importance in situations where loss of the primary restoration path is likely. Participants take different approaches in identifying these alternate paths, however. Most participants explicitly identify the alternate cranking path, or path from their blackstart generator to the priority load(s), including verifying the viability of the alternate path through simulation. Other participants test multiple paths for restoration, but allow the operators discretion to select the path during restoration based on the conditions at the time.

Participants that execute the SCADA steps to energize cranking paths typically include highly-detailed steps for energizing those paths in their restoration plans, along with subsequent restoration steps. These steps specify the breaker-by-breaker steps to energize the entire cranking or restoration path. Those transmission operator participants that have arrangements with transmission owners to execute the actual steps to energize cranking paths have restoration plans that are more principles-based. These plans include an explanation of the electrical characteristics and associated protocols (e.g., voltage monitoring and control during restoration switching steps) to be observed during restoration. In this case, the transmission operator/transmission owner plans are designed to complement each other, with the combination providing a highly-comprehensive restoration plan.

Initial Loads. The joint staff review team found that in managing island stability, the magnitude of the initial loads planned for energization varies based on the capacity of that participant's blackstart generation. As initial loads are energized, participants take into account that cold load pickup and load inrush currents will be multiple times greater than steady-state values when loads are first energized following a sustained blackout condition. Also, for the initial steps of island path development, participants typically avoid energizing UFLS-enabled load, since initial load pickups are expected to cause large deviations in frequency. If the frequency falls below underfrequency relay trip levels, the resulting load shed could result in high frequency on the developing system, and cause generators to automatically come off-line due to over-speed conditions. Some

---

[34] This includes the transportation and mobilization of personnel and use of equipment inventories of transmission line towers and transmission substation elements (e.g. power transformers, breakers, switches) to storm damage locations.

participants avoid energizing UFLS-enabled load in their restoration plan entirely. For those participants that include energizing UFLS load at some point in their restoration plans, the order of restoration is set such that UFLS load with the lowest underfrequency trip settings is restored first.

Most of the participants' restoration plans include guidelines or factors as to the increment of load to be energized, depending on system conditions, in order to maintain sufficient generator reserves to ensure stability. Examples of factors incorporated in various participants' restoration plans include:

- Maximum increment of load pick-up:
    - 5 percent of online generator capacity
    - 5 percent (steam units), 15 percent (hydro units), 25 percent (combustion turbine units), of online generator capacity
    - Lesser of 5 percent or 25 MW of online generator capacity
    - Lesser of 5 percent or 100 MW of online generator capacity
    - 100 percent of total energized UFLS load (for later stages of island development)
- Island generator reserves, to account for cold load pickup:
    - 50 percent of online generator capacity
    - Approximately eight times the increment of large blocks of load added

In addition to managing reserves for cold load pickup, after the initial stage of island development, participants also verify that enough contingency generation reserves exist to withstand the forced outage of the largest online generator.

b) Managing Island Frequency, Voltage and Stability

Frequency. Participants' restoration plans require certain generators to monitor and control the frequency of the island. For example, for islands under isochronous control,[35]

[35] An isochronous (or zero droop) generator governor maintains the same speed regardless of the load, and ensures that the frequency of the electricity generated is

the frequency control and metering is located at the isochronous generator. Also, for frequency monitoring by transmission operators, frequency metering data from multiple transmission substations is displayed via the EMS/SCADA system along with the generator frequency metering data. The joint staff review team found that monitoring system frequency at diverse locations across their footprint via SCADA can aid transmission operators not only in island monitoring and management of load pick-up, but also in detection of an islanded condition.[36] The joint staff review team found the specific frequency limits in the participants' restoration plans typically provide for larger deviations in frequency than are permitted under normal operations.

Participants' restoration plans vary in the level of detail regarding operator coordination between the transmission control center, blackstart generator(s), and other generator control rooms with respect to managing generator operation. Even those plans that are highly detailed as to SCADA switching steps (i.e., plans of participants that execute the actual SCADA steps), appear to lack guidance on how the system operators should work with the generator control room operators to coordinate the output and operation of multiple generators within the developing island. In many cases, based on the interrelationships of the entities sampled, the system operator tasked with coordinating generator control room operators is the transmission operator or local control center (transmission owner) operator, who only performs these tasks during restoration following a blackout or in a restoration drill.[37] Alternatively, a few participants have the generator control room operators manage a considerable amount of coordination of load pick-up, generator loading, and management of reserves.

Voltage. The joint staff review team found that the participants' plans reflect the critical nature of maintaining a stable voltage on the transmission system during restoration in

---

constant or flat. Isochronous control mode is used to control frequency in an island during system restoration.

[36] In some instances, entities monitor system frequency at diverse locations in their footprint via phasor measurement units (PMUs). During Hurricane Gustav in 2008, operators first detected the electrical island that resulted from the large-scale outage when operator-monitored PMUs showed diverging system frequencies.

[37] One participant explained that the role of managing frequency and stability by a transmission owner/transmission operator is infrequent, and that these operators take on a "pseudo" balancing authority role during restoration.

order to successfully reenergize the grid. The review team found the voltage monitoring and management protocols in the plans to be robust and thorough, as they include identification of acceptable voltage limits, and processes and provisions for voltage control.

All of the transmission operator participants' restoration plans include energization guidelines for operators to limit the impact of sustained high voltages and switching transients during restoration, based on the unique characteristics of their systems.[38] Some participants' plans require that their operators connect load to newly-energized paths prior to energizing additional higher nominal voltage lines, and have a minimum loading requirement per mile of transmission line for energizing these higher nominal voltage lines (e.g., EHV transmission lines). Other participants' plans avoid energizing higher nominal voltage transmission lines early in the restoration process due to their excessive reactive requirements.

Participants that plan to energize higher nominal voltage transmission lines early in the restoration process manage overvoltage risk by placing shunt reactors or static VAr compensators in service. These participants also mitigate the risk of overvoltages by initially restoring sufficient generator capacity to provide dynamic reactive reserve. Some of the reviewed restoration plans also call for use of EHV underground transmission cables, which have a greater risk of overvoltages due to large charging currents.[39] These participants also mitigate the risk of overvoltage through the use of shunt reactors to maintain voltages within a specified bandwidth.

All participants recognized that, even when the SCADA system is available for monitoring and performing system restoration switching steps, the state estimators and real time contingency analysis (RTCA) tools typically used to analyze the impact of a transmission contingency will not be functional during system restoration.[40] When these tools are unavailable, operators are dependent on offline studies to evaluate contingencies and to identify preventative actions to ensure island stability. As discussed above in the

---

[38] High voltages are due to the "Ferranti effect," which is the rise in voltage resulting from energizing a transmission line that is lightly loaded. If not mitigated, this voltage increase could result in equipment damage and tripping of transmission lines.

[39] Charging currents associated with underground EHV cables can be many times greater than that of overhead transmission lines of the same nominal voltage.

[40] In the event of a large outage or blackout, these tools lack usable data inputs to function properly.

"System Restoration Resources" section, some participants' plans call for increasing operations and engineering support staffing to assist in performing off-line contingency studies that can help identify transmission contingency concerns and preventative actions when the state estimator and RTCA tools are not available. In addition, for those contingencies discovered as part of the initial assessment (e.g., impaired transmission facilities due to weather damage), participants account for this by identifying alternate or back-up facilities for system restoration, as described above in the restoration planning stages. This includes accounting for contingencies of each voltage-controlling facility on the primary restoration path (e.g. loads, shunt reactors).

c) Synchronization with Other Islands and Systems

With the initially-developed islands not connected, the islands' frequencies, voltages and phase angles must match within tolerance before interconnecting to create a merged system. Along with other factors, such as merged-system generation reserves, participants take these characteristics into account to ensure stability and avoid the risk of collapse of the merged system. To guard against collapse, participants' restoration plans identify threshold conditions that must be met prior to attempting to synchronize and interconnect separate systems. One participant determines reserves by comparing the islands' total generator capacity and island loads prior to interconnecting. In addition, some participants study contingencies (e.g., loss of largest generator) prior to interconnecting to ensure the merged island will be stable.

A number of participants' restoration plans call for synchronizing to connecting systems using switching equipment at generating stations. This approach has several advantages. First, operating personnel at generating stations perform synchronizing on a routine basis and are therefore very familiar with the process. Moreover, adjustment of the frequencies and voltages is facilitated by synchronizing at generating stations.

Another participant synchronizes islands using the highest voltage line available, which allows it to take advantage of the lower impedance and higher relay loadability of the higher voltage lines. However, that participant's plan also recognizes that possible over-voltages or special considerations could prompt the use of lower voltage lines.

d) Voltage, Frequency, and Phase Angle

Prior to connecting two systems or islands, the frequencies and voltages of the two systems should ideally be close with a near-zero phase angle difference. Such exact matching will rarely if ever be feasible, and many participants have established limits for the acceptable differences in the parameters of systems being interconnected.

Participants' restoration plans typically seek to establish stabilized voltages between 90 and 110 percent of nominal voltage before attempting synchronization. As far as the acceptable voltage difference between two islands, one participant reported bringing the two system voltages to within 2 percent of each other prior to actually tying the systems together, with the lower voltage being on the smaller system. Participants' restoration plans also call for system frequencies to be within a certain nominal range (e.g., 59.75 and 61 Hz) before attempting synchronization.

As far as the acceptable phase angle difference between two islands being connected, the restoration plans of some participants identify 30 degrees as the maximum acceptable phase angle difference between systems being connected. Other participants set synchronizing check relays to block closing the synchronizing breaker for phase angle differences in excess of 20 degrees.[41] Transmission operators coordinate with generator control operators to minimize the phase angle difference between the systems, enabling the synchronizing check relays to permit synchronizing.

e) Synchronizing Coordination

When synchronizing islands within a transmission operator's footprint, the participants' restoration plans rely on the transmission operator, who, either directly or through delegation of the tasks, authorizes operators to perform synchronizations and interconnections of internal islands with minimal involvement by the reliability coordinator.[42] The joint staff review team found that the transmission operators, transmission owners, generator operators, and generator owners generally coordinate on the formation and expansion of islands without intervention by the reliability coordinator. The joint staff review team found that these operating entities are in the best position to coordinate formation of islands and synchronization of smaller islands in the process of restoring the interconnection, as they have access to all necessary information and are more frequently engaged in synchronizing operations.

When synchronizing islands between neighboring transmission operators, and in some cases when synchronizing larger islands within a transmission operator footprint, the participants' plans call for coordination by the reliability coordinator. For external transmission operator interconnections, the reliability coordinator typically validates that the conditions necessary for interconnection have been achieved. The reliability coordinator may not be able to monitor all the synchronizing parameters that are available to the system operators and field personnel (where the execution of the steps to

[41] These transmission phase angle settings are based on engineering analysis of the specific neighboring areas to protect against instability upon closing the breaker.

[42] As conditions permit, these islands will be formed and interconnected in accordance with the restoration plan, modified as required by system conditions.

synchronize is performed). However, the reliability coordinator will monitor the evolving restoration effort, using its wide area view capability, and can identify interconnection synchronizing opportunities not necessarily apparent to the individual transmission operators. In these circumstances, the reliability coordinator may instruct a transmission operator to interconnect.

The joint staff review team found that having the reliability coordinator coordinate inter-transmission operator interconnections has several advantages. The reliability coordinator can include in the Reliability Coordinator Area Restoration Plan specific, well thought out procedures to ensure a uniform approach to synchronization throughout the reliability coordinator's area. The reliability coordinator can then ensure that all the steps in the process have been carefully performed prior to any interconnection operation being attempted. This step is critical when interconnecting areas are in different transmission operator footprints. These will typically be large interconnections, and the consequences of a failed interconnection attempt, such as the loss of both islands, are apt to be severe. Placing such interconnections under the authority of the reliability coordinator better ensures that entities have undertaken all preliminary steps, and that the interconnection will be successful.

Some participants include standard forms and procedures in their restoration plans to guide system operators performing the interconnection of islands. Some participants' plans identify specific islands to be formed, specific synchronizing points to be used, and synchronization parameters. Others include narrative guidance for preparing to synchronize, and forms to be used during the synchronizing process. These forms typically identify those system parameters that the system operator is expected to consider before commencing the synchronizing process. All involved system operators are expected to complete these worksheets and to analyze the system parameters to determine if they are within entity limits before proceeding. Participants train on restoration synchronization in their training programs and drills. Some participants also emphasize synchronization training during regional training exercises.

3. Related Standards Assessment

Several of the sub-requirements of Reliability Standard EOP-005-2 specify elements that an entity must include in its restoration plan, many of which relate to the island development and synchronization topics discussed above. The relevant sub-requirements require the restoration plan to include:

> **R1.3.** Procedures for restoring interconnections with other Transmission Operators under the direction of the Reliability Coordinator.
>
> . . .

R1.5. Identification of Cranking Paths and initial switching requirements between each Blackstart Resource and the unit(s) to be started.

R1.6. Identification of acceptable operating voltage and frequency limits during restoration.

R1.7. Operating Processes to reestablish connections within the Transmission Operator's System for areas that have been restored and are prepared for reconnection.

R1.8. Operating Processes to restore Loads required to restore the System, such as station service for substations, units to be restarted or stabilized, the Load needed to stabilize generation and frequency, and provide voltage control.

As a general matter, the joint team found the relevant Requirements to be sufficiently detailed and specific as to the elements that must be included in a restoration plan related to island development and synchronization. Moreover, the overall viability of the plan, including its approach to island development and synchronization, is tested through simulation testing and is reviewed and approved by the reliability coordinator. With respect to identifying cranking paths and initial switching requirements (Requirement R1.5), the joint staff review team found that the participants' plans and procedures have highly-detailed switching steps and a range of resources for reliable restoration, such as back-up cranking paths. As described earlier, the joint team found that participants are well-prepared for the unavailability of primary restoration paths, by, among other things, identifying and planning for the use of back-up paths.

The team also found that the participants' plans include applicable voltage limits during restoration, and otherwise cover the provision of voltage control in great detail. Participants' plans include identification of acceptable voltage and frequency limits and processes for restoring loads needed for restoration (such as station service for substations, load needed to stabilize generation and frequency, generating units to be started or stabilized, and detailed provisions for voltage control). Similarly, the joint staff review team found that the participants' plans have detailed procedures for restoration, reconnection and synchronization that also reflect the impact of contingencies (such as instability and loss of transmission) on voltage and frequency, on availability of reserves, and on synchronization.

As noted above, the team found that the participants' restoration plans incorporate some level of planning for contingencies (*e.g.,* by identifying and planning for the use of back-up cranking paths). However, the joint staff review team recommends tightening the requirements to modify restoration plans to reflect changed circumstances. For example, a given cranking or restoration path may be modified, which could change the implementation of a restoration plan. Reliability Standard EOP-005-2, Requirement R4 requires:

> **R4.** Each Transmission Operator shall update its restoration plan within 90 calendar days after identifying any unplanned permanent System modifications, or prior to implementing a planned BES modification, that would change the implementation of its restoration plan.
>
> **R4.1.** Each Transmission Operator shall submit its revised restoration plan to its Reliability Coordinator for approval within the same 90 calendar day period.

The standard as currently written does not require updates to the restoration plan for non-permanent unplanned system modifications, even when they may be long-term and affect implementation of a given entity's restoration plan.[43]

Given the critical nature of identifying and planning for the use of restoration paths to the success of the restoration plan, the joint staff review team concludes there is a need for updating restoration plans for system modifications that would change implementation of an entity's restoration plan for an extended period of time. While the joint staff review team recognizes that restoration plans necessarily incorporate some degree of flexibility so that they need not be updated with every change in configuration, the Reliability Standards do not currently require for instance, any update for an unplanned, but not permanent, system modification, regardless of whether the restoration plan is sufficiently flexible to address that change in system configuration. For example, if a transmission operator determines that an extended outage of a generator changes the implementation of the restoration plan, then the plan should be updated. Notably, the team found that some participants currently update their plans when, for example, there is a modification to a cranking path that changes their restoration plan.

4. Recommendations

Clarify when system changes will trigger a requirement to update restoration plans. The joint staff review team recommends that measures be taken (including considering changes to the Reliability Standards) to address the need for updating restoration plans for all system modifications that would change the implementation of an entity's restoration plan for an extended period of time, not just permanent or planned system

[43] When the IERP reviewed this requirement, it recommended requiring entities to update their restoration plan for *all* system changes that impact an entity's plan for an extended period of time (not just permanent system changes). IERP Scoring Sheet at cell S594.

modifications. In considering measures, the kinds of events that may warrant an update to the system restoration plan should be identified, taking into account the length of time the system is affected, as well as the overall objective of ensuring that restoration plans are generally flexible enough so that system modifications can be addressed without frequent updates.

5. Observed Practices for Consideration

In evaluating each participant's island development and synchronization related procedures, the joint staff review team observed the following practices and recommends consideration of these by entities:

- Many of the participants' restoration procedures require identification of back-up or alternate cranking paths during a forced or planned outage of the restoration plan-identified cranking path or segment of the path.

- Some participants include in their restoration plans the use of illustrations and accompanying steps to assist operators in system restoration, which the joint staff review team found to be a valuable aid to the operators in execution of the plan. The types of illustrations and guidelines include: electrical (i.e., one-line) diagrams, and tables or chart of reference information to augment the steps of restoration.

- Some participants include in their restoration plans a summary preceding each section of blackstart cranking path switching procedures, which participants indicated was very helpful to operators during island development.

- Some participants include in their restoration plans load pickup curves or data tables which help operators in planning the amount of online generator capacity needed to ensure island stability.

- Some participants account for seasonality when calculating cold load pickup values.

- Some participants include in their restoration plans multiple, diversely located frequency measurement sources to assist operators during system restoration, as well as in detection of an islanded condition.

- Some participants use island data monitoring methods and tools to manage island development (i.e., methods and tools to monitor frequency, voltage, load and reserves) during system restoration. These tools are used to calculate whether there is enough generating capacity online for the next increment of load pickup, and for evaluating the contingency loss of the largest island generator, as well as in preparation to transition to balancing authority control in the later stages of restoration.

- Some participants incorporate transmission line charging current/MVAr tables into their restoration plans (to help operators in their attempts to balance reactive requirements on transmission lines using line charging so that voltages can be maintained within limits).

- Some participants prioritize the restoration of power to pumps that maintain oil pressure on underground cables used during the restoration process, to maintain the dielectric strength of the cables and to prevent failure of the cables during restoration.

- Some participants include in their restoration plans a checklist for transitioning to balancing authority control, which can be used to track the necessary details for the transfer, such as online generator attributes (capacity, control mode, output, and restrictions), current and forecasted load values, reserve positions, reconnected tie-lines with neighboring transmission operator(s), etc.

F. Testing of System Restoration Resources

1. Summary

The joint staff review team examined the participants' testing of system restoration resources, including blackstart resources and communications and control center resources, under their respective restoration plans.

As described in detail below, the joint staff review team found that all of the participants test their system restoration resources in accordance with the current Reliability Standards, and some participants have testing requirements and procedures that exceed the standards, such as energizing a blackstart unit's cranking path and starting the next unit. The review team recommends a study be performed to identify options for expanding the testing of blackstart resources to ensure they can energize equipment needed to restore the system as intended in the restoration plan, including consideration of whether such testing is practical while maintaining system reliability, and whether such expanded testing requirements could affect the identification of blackstart resources in the future.

2. Review of Restoration Plans

a) Actual Tests of Blackstart Resources

All of the participants' plans require periodic testing of blackstart units. Some participants test blackstart units once every three years, consistent with Reliability

Standard EOP-005-2 requirements (discussed further below), while others do so annually, exceeding the standard's requirements.

In addition, all of the participants require blackstart testing that meets the Reliability Standard's requirements with respect to the form of testing, i.e., they require a demonstration that the blackstart unit is able to start when isolated from the bulk power system and that the unit can energize a bus. Actual tests typically involve energizing the unit auxiliaries without outside power, starting the unit with the unit remaining stable (controlling voltage and frequency), then closing the generator breaker to energize a bus.

Some participants' plans include additional criteria that must be demonstrated during blackstart unit testing, including the following:

- Blackstart unit must be available to serve load within three hours;
- Blackstart unit must remain stable and control voltages while operating isolated from the grid for a period of 10-30 minutes.[44]

As mentioned above, the joint staff team observed that all participants' plans contain provisions for blackstart testing. However, the provisions do not necessarily require verification using conditions that anticipate actual blackout conditions, as some rely on simulations or on assumptions that certain equipment will be in service. The blackstart testing requirements in the Reliability Standards are limited to ensuring that a blackstart unit is functional, but they do not explicitly require verification of the ability to energize equipment under the conditions anticipated during an actual blackout situation.

Some participants test both the blackstart generator *and* cranking path energizing (not just energization of a bus), by isolating the system and supplying the cranking power to the next generating unit to be started. Participants who do this kind of testing only perform it in locations where the cranking paths can be isolated without outages to customers or other adverse impact on reliable operations. To perform these tests without loss of load, these participants must coordinate with all affected parties, including the generator operator, the transmission owner, and the transmission operator. In addition, these entities schedule the tests to minimize any associated cost and reliability impact (e.g., the blackstart unit is offline, the next generating unit to be started is offline, and system loads are at a lower level).

[44] While transmission operators are required to include a minimum duration for each test under Reliability Standard EOP-005-2, the participants' plans varied as to the minimum time specified.

The procedure for this testing is typically as follows:

- Start the blackstart unit after isolating the cranking path to the targeted unit;
- Close the blackstart unit generator breaker and energize the cranking path;
- Establish station service and start the motors and equipment needed for operating the targeted unit; and
- Close the targeted unit generator breaker and start energizing system load.

Because this type of testing requires a demonstration that the blackstart unit can establish station service and start the motors and equipment needed for the next targeted generating unit, as would be required under actual blackstart conditions, it demonstrates that the blackstart unit can and will function "as intended" under that entity's restoration plan. In addition, this type of testing can be used to benchmark against computer simulations to determine whether improvements need to be made to system restoration models or to the restoration plan procedures and/or restoration facilities and controls. Finally, this more robust testing provides an opportunity to test the coordination needed between the blackstart control room operator (i.e., the generator owner or operator), the transmission operator, and the control room operator of the next generator to be started.[45]

While the joint staff review team believes this more robust type of testing can be beneficial, as it provides a more realistic demonstration that the blackstart resource can perform as intended under the restoration plan, the team recognizes that such testing requires significant coordination in order to minimize the reliability and customer impact, and may not even be possible in certain locations, where cranking paths cannot be isolated without outages to customers. Accordingly, it may not be advisable to simply adopt a more stringent blackstart testing requirement that mimics the more robust testing currently done by some participants. Instead, the team recommends further study of the issue, including identifying other means of ensuring that the blackstart resource can function as intended, e.g. through verification of the restoration plan as a whole by analysis of actual events, steady state and dynamic simulations, or other means of testing.

---

[45] Close coordination by these entities can be critical to successful restoration, given that some of the priority auxiliary loads at the next generator to be started may consist of very large motors (e.g., 10,000 horsepower motors). The success of starting the next generating unit is greatly enhanced if the large motors are equipped with technologies which reduce the start-up current needed (e.g., variable frequency drives).

b) Tests of Other System Restoration Resources

The joint staff review team also observed that all of the participants' restoration plans include periodic testing and monitoring of vital communications facilities expected to be used during system restoration, in accordance with the Reliability Standards. In addition to testing of dedicated telephones or radios used for voice communications, participants' plans include periodic testing of their back-up facilities to ensure operability. Participants' plans also include testing of critical substation devices such as breakers, transformers, and protective relays. Furthermore, some participants test the functionality of various back-up voice communications facilities by having their control center staff perform normal operations occasionally using the back-up facilities, and as part of their drills.

3. Related Standards Assessment

Reliability Standard EOP-005-2 currently sets out the following requirements for blackstart resource testing, pertaining to both the frequency of testing and what must be demonstrated.

R9. Each Transmission Operator shall have Blackstart Resource testing requirements to verify that each Blackstart Resource is capable of meeting the requirements of its restoration plan. These Blackstart Resource testing requirements shall include:

R9.1. The frequency of testing such that each Blackstart Resource is tested at least once every three calendar years.

R9.2. A list of required tests including:

R9.2.1. The ability to start the unit when isolated with no support from the BES or when designed to remain energized without connection to the remainder of the System.

R9.2.2. The ability to energize a bus. If it is not possible to energize a bus during the test, the testing entity must affirm that the unit has the capability to energize a bus such as verifying that the breaker close coil relay can be energized with the voltage and frequency monitor controls disconnected from the synchronizing circuits.

R9.3. The minimum duration of each of the required tests.

The joint staff review team found that the noted Requirements are detailed and, as described above, that the participants included the necessary testing parameters of these sub-requirements in their restoration plans. In some cases, the joint team observed that participants exceeded the requirements, e.g., by requiring annual blackstart resource testing.[46]

The current blackstart testing requirements in Requirement R9.2 require a demonstration that such a unit can energize a bus. The provision does not go so far as to require a demonstration that the unit will be able to energize the equipment needed to start the next targeted unit.[47] Based on review of participants' actual testing as described above, the joint staff review team believes that more robust testing could be beneficial in many situations, as it would better demonstrate whether or not a blackstart resource is capable of functioning as intended by the restoration plan, i.e., whether it can start the equipment needed to start the next targeted unit in the plan, which would better mimic "real world" conditions.

The testing of other system restoration resources (non-blackstart resources) is covered by Reliability Standards other than EOP-005-2, including testing of telecommunications facilities (the Communications (COM) family of standards), control center functionality (EOP-008), and testing of protective relays (PRC-005).[48]

---

[46] While it may seem that three years could be too long of an interval, team discussions with participants revealed that there is a significant amount of planning and coordination needed to perform these tests (reliability coordinator, transmission owner, balancing authority, transmission operator, generator owner, and generator operator involvement to plan tests during low system load times, and to ensure availability of all necessary equipment, including all blackstart/cranking path resources).

[47] The IERP review of these requirements resulted in a recommendation to modify or redraft Requirement R9.2 so that it would require a demonstration of "the ability to start the unit and energize equipment under the conditions anticipated during an actual blackstart situation."

[48] Reliability Standard COM-001-1.1, R2 requires that each reliability coordinator, transmission operator, and balancing authority shall manage, alarm, test and/or actively monitor vital telecommunications facilities. Special attention shall be given to emergency telecommunications facilities and equipment not used for routine communications. Reliability Standard EOP-008-1 R1 requires each reliability coordinator, balancing authority, and transmission operator to have an operating plan describing how it will meet its functional obligations with regard to the reliable operation of the bulk electric system during loss of its primary control center functionality. In addition, EOP-008-1 R4 requires each balancing authority and transmission operator to have backup functionality that includes "monitoring, control, logging, and alarming

4. Recommendations

Blackstart resource testing under anticipated blackstart conditions. The joint staff review team recommends a study be performed to identify options for expanding restoration plan testing beyond the currently-required blackstart resource testing, to ensure the blackstart resource can energize equipment needed to restore the system as intended in the restoration plan. Any expanded testing requirements should take into consideration whether such testing is practical while maintaining system reliability, and whether such expanded testing requirements could affect the identification of blackstart resources in the future.

5. Observed Practices for Consideration

In evaluating participants' testing of system restoration resources, the joint staff review team found that some participants perform real time tests of the blackstart generator and cranking path energizing by isolating the system and supplying the cranking power to the next generating unit to be started. The joint staff review team recommends consideration of this practice by entities.

G. Testing, Verification, and Updating of System Restoration Plans

1. Summary

The joint staff review team examined the participants' plans and procedures for conducting the required testing, verification, and updating of their system restoration plans. The team examined the participants' modeling considerations and inputs, types of studies performed, study verification, simulation tools used, frequency of verification, and restoration plan completion time.

---

sufficient for maintaining compliance" with Reliability Standards that depend on a balancing authority and transmission operator's primary control center functionality. EOP-008-1 R7.2 requires an applicable entity to conduct an annual test of its operating plan that demonstrates the backup functionality, in the event that its primary control center functionality is lost, for a minimum of two continuous hours. In addition, Reliability Standard PRC-005-2(i) requires maintenance of protection systems affecting the reliability of the bulk electric system, which includes required testing of relays and other protection system components, as specified in Table 1 of the standard.

As described in detail below, the joint staff review team found that participants test their plans for viability at least every five years, in accordance with the Reliability Standards. The review team recommends that measures be taken (including considering changes to the Reliability Standards) so that plan verification through testing and simulation is performed whenever system changes occur that precipitate the need to determine whether the plan's restoration processes and procedures, when implemented, will operate reliably. In considering such measures, the types of system changes that are significant enough to warrant additional testing and verification (e.g., identification of a new blackstart generator location or on redefinition of a cranking path) should be identified, keeping in mind the overall objective of ensuring that restoration plans are flexible enough so that system changes can be addressed without frequent updates.

2. Review of Restoration Plans

a) Modeling Considerations

All of the participants' restoration plan verification methods include performing computer simulations to analyze whether their plans can accomplish the intended function. Participants employ dynamic and steady-state modeling for analyzing restoration cranking paths, as further discussed below. For accurate modeling of the cranking path loads, participants take into account auxiliary load inrush currents and auxiliary motor characteristics needed to start up the next generating unit(s). A few participants also model other load values at their predicted inrush or cold-load pickup levels, such as where large block loads are planned to be restored or used to control system voltage or frequency. Participants' models also include the dynamic characteristics of the generators needed for stability analysis.

Overall, participants' models are designed to allow for analysis of the effects of the operator switching steps to sequentially energize transmission facilities or segments, adding the expected increments of load, switching other devices in-service (such as shunt reactors), and making adjustments as necessary for island stability and operation within steady-state limits.

b) Studies and Simulations

Participants use a range of approaches to their dynamic studies, but all typically test the following:

- Viability of switching steps to energize the primary or preferred transmission cranking paths to supply the priority loads;
- Viability of switching steps to energize alternate or back-up transmission cranking paths to supply the priority loads;

- Viability of the restoration plan assuming one of the blackstart generators is not available;

- With successful blackstart cranking path power delivered and additional generators on-line, viability of switching steps to energize additional restoration paths; and

- Transient stability assuming a fault on the island system (during later stages of restoration).

Participants use dynamic simulations to ensure frequency is kept within the tolerances needed to keep generators from tripping on underfrequency. The simulations verify the capability of blackstart facilities and of other generation resources to meet real and reactive power requirements of cranking paths, and verify their dynamic capability to supply priority loads. These studies identify the location and magnitude of loads required to control voltage and frequency within acceptable operating limits.

Participants use steady state, dynamic, and contingency analysis of the system to ensure it will perform in accordance with the restoration plans. No single study will verify all aspects of a restoration plan. Participants perform their steady state and dynamic analysis using commercial power-flow software. Generally, participants run their analyses using lightly loaded cases. As mentioned above in the Island Development and Synchronization section (V.E), some participants also use seasonal cases accounting for worst-case cold load pick up values.

Steady-state Analyses. The participants use steady-state analyses to verify that steady-state voltages are maintained within limits, and to determine the real and reactive load output and voltage controlling device adjustments necessary to balance generation and load. Participants also monitor any thermal limit exceedances on transmission facilities as part of their steady-state analysis, although thermal limits during the early stages of island development are typically not as much of a concern as they are during later stages when system transfers between loads and generators increase.

Dynamic Analyses. As noted above, participants perform simulations of their plans to ensure that there are no stability issues during switching or when an event occurs. Participants commonly perform voltage analysis for the cranking path switching steps to determine any transient switching over-voltages that could result from energizing transmission lines and cables. Participants that use an EHV transmission system during the restoration process also verify the effectiveness of shunt reactors at both ends of EHV transmission lines, to ensure they can control over-voltages. The joint staff review team notes that these kinds of analyses are becoming more important as an increasing number of entities use higher-voltage cranking paths. The recent industry trend of replacing blackstart coal-fired generators which were not connected to the EHV transmission system, with new blackstart gas-fired combustion turbine generators connected to the EHV transmission system, has increased the usage of higher voltage cranking paths.

Participants also use voltage stability analysis to verify that incremental load pickup takes cold load pickup into account, and that inrush currents do not result in voltage or generator instability. In addition, participants use voltage stability analysis to determine the generator, load, and voltage controlling device adjustments necessary to maintain acceptable voltage levels. Participants monitor frequency and voltage deviations in the analyses to prevent an element from tripping (e.g., underfrequency relay trip settings, generator low frequency limits or trip points). In an iterative manner, participants would then make any necessary adjustments to the plan to ensure acceptable voltage and frequency levels.

Some participants include testing for short circuit fault stability by imposing a simulated fault on the transmission restoration path following the initial stages of restoration to test generator stability. These N-1 simulations are generally focused on later stages of restoration because a fault during the early stage would most likely result in a generator tripping off line. As the islands grow and load is restored in their studies, these participants will run offline cases and simulate N-1 conditions.

c) Frequency of Performing Simulations

All of the participants perform their offline dynamic and steady state studies at least every five years, as required by the Reliability Standards, or more frequently if something significant changes on their system. Several participants indicated that they regularly perform their analysis on a more frequent (e.g., annual) basis.

d) Restoration Plan Completion Time

Participants indicated that estimating the time to complete a restoration plan and fully restore the system is a difficult task. Many factors must be taken into consideration, including the extent of the blackout, damage to the system, state of generating units and the system status of neighboring utilities. Participants' restoration plans are designed on the assumption of a total blackout with no help from their neighbors. For those participants that do incorporate a target restoration time into their restoration plans, the estimates on restoring the transmission operator's system range from as few as seven hours to as many as 16 hours. However, these estimates assume that all goes according to plan and few issues arise during restoration. Most participants have encountered longer restoration times during simulations.

Some participants factor these potentially lengthy restoration times into the sizing of station batteries, which are considered to be vital resources during system restoration since they provide power to station equipment when system power is lost. As discussed earlier in the System Restoration Resources section (Section IV.D), some participants size batteries to provide power for a longer time (e.g., 24 hours) for certain substations that are a priority for restoration, for more remote stations, or where the participant anticipates difficulty reaching the station due to damage from a natural disaster (e.g. areas more prone to hurricane weather). Portable batteries and portable generators are also employed to supply station power if needed during restoration. Some participants also install local generation at key facilities as back-up power sources and test these generators regularly to ensure operability.

3. Related Standards Assessment

Requirement R6 of Reliability Standard EOP-005-2 requires transmission operators to verify that their restoration plan can accomplish its "intended function" through analysis of actual events, simulations, and testing, and sets out specific capabilities that must be confirmed:

R6. Each Transmission Operator shall verify through analysis of actual events, steady state and dynamic simulations, or testing that its restoration plan accomplishes its intended function. This shall be completed every five years at a minimum. Such analysis, simulations or testing shall verify:

R6.1. The capability of Blackstart Resources to meet the Real and Reactive Power requirements of the Cranking Paths and the dynamic capability to supply initial Loads.

R6.2. The location and magnitude of Loads required to control voltages and frequency within acceptable operating limits.

R6.3. The capability of generating resources required to control voltages and frequency within acceptable operating limits.

The joint staff review team found that the noted Requirements are sufficiently detailed, noting that the accuracy of the simulations and models are influenced by the Requirements of Reliability Standard TPL-001-4, assessment modeling, and related system modeling (MOD) standards to ensure their restoration plans are viable in the event of an actual blackout. These requirements, taken collectively, would ensure sufficiency and expose any inadequacies of a bare-bones or inaccurate model. In addition, the joint staff review team found that participants perform extensive steady-state and dynamic simulations, including short-circuit fault stability analyses in some cases, to test whether the blackstart resources can meet the requirements to supply initial loads (R6.1), and to verify the capability of loads and generating resources to ensure voltages and frequency are kept within acceptable limits (EOP-005-2 R6.2 and R6.3). However, the team identified one concern with the scope of Requirement R6 of this standard: transmission operators are required to verify the effectiveness of their restoration plan every five years, but are not required to perform additional simulations or testing if their restoration plan is impacted by a change.[49] For example, given recent changes occurring with blackstart resources (as described earlier in the report), the joint staff review team is concerned that many entities may have to modify their restoration plans going forward, but may not verify that the modified plan can accomplish its intended function until as late as five years after making the change.

While the reliability goal is for transmission operators to have up-to-date and verified restoration plans, the joint staff review team recognizes that the triggers for re-verification of the plan should be clearly set out, and that re-verification of the full plan may not be necessary in all situations where a restoration plan has been or should be updated. For example, the addition of new blackstart generation or redefinition of a cranking path may warrant additional verification, but it may not necessitate computer stability simulations of other areas of the plan. At a minimum, however, re-verification should occur when needed to ensure that the restoration plan can, when implemented, allow for restoration of the system within acceptable operating voltage and frequency limits.[50]

4. Recommendation

Verification/testing of modified restoration plan. The joint staff review team recommends that measures be taken (including considering changes to the Reliability Standards)s to address the need for re-verification of a system restoration plan when a system change precipitates the need to determine whether the plan's restoration processes and procedures, when implemented, will operate reliably; i.e., when needed to ensure that the restoration plan, when implemented, allows for restoration of the system within acceptable operating voltage and frequency limits. In considering and developing such measures, the types of system changes that could impact reliable implementation of the restoration plan should be taken into account (e.g., identification of a new blackstart generator or on redefinition of a cranking path).

[49] The joint staff review team recognizes that the Reliability Standard EOP-005-2, Requirement R4 currently addresses the need to update restoration plans given a planned system modification that would change the implementation of the plan, but does not require a re-verification of the plan's effectiveness following the modification.

[50] *See* Section IV.E Island Development and Synchronization.

5. Observed Practices for Consideration

Due to the potential length of the restoration process, some participants size batteries to provide power for a relatively long time (e.g., 24 hours) for certain substations that are a priority for restoration, for more remote stations, or where the participant anticipates difficulty reaching the station due to damage from a natural disaster (e.g. areas more prone to hurricane weather). Portable batteries and portable generators are also employed to supply station power if needed during restoration. Some participants also install local generation at key facilities as back-up power sources and test these generators regularly to ensure operability. The joint staff review team recommends consideration of these practices by entities.

H. System Restoration Drills and Training Exercises

1. Summary

The joint staff review team examined the participants' plans and procedures for conducting system restoration drills and training exercises, and observed some of the participants' actual system restoration exercises. The joint staff review team observed that the participants' restoration plans address restoration plan training and drilling, including training on coordination with other entities, restoration priorities, building cranking paths and synchronizing to the interconnection.

However, the joint staff review team found that participants' plans are not clear regarding training on the processes for transfer of control from transmission operator to balancing authority during system restoration. This transfer of control is a crucial step in the restoration process and can require coordination between several entities. Therefore, the joint staff review team recommends that measures be taken (including considering changes to the Reliability Standards) to address drills and training on the operating processes for transferring authority from the transmission operator back to the balancing authority. These measures would allow transmission operators, reliability coordinators, and relevant generator operators to gain experience on the coordination needed through all the stages of restoration, including coordination needed in the transfer of control back to the balancing authority.

2. Review of Restoration Plans

The joint staff review team found that participants' system restoration drills facilitate the review of their system restoration plans and emergency operating procedures, and provide coordinated training for operators through simulation exercises. Participants use restoration drills to review and understand their (and other entities') plans, to coordinate with neighboring entities, to identify weaknesses in restoration plans while identifying ways to improve them, to verify results of system studies pertaining to restoration, and to maintain familiarity with established processes. Depending on the sponsor, system restoration drills may include reliability coordinators, transmission operators, generator operators, and in some areas, certain transmission and generator owners. By promoting regional coordination between entities, participants' system restoration drills provide greater regional exposure for operators, promote sharing of knowledge and lessons learned among system operators, and can improve future interaction between entities.

As discussed further below, the joint staff review team found that the system restoration drills allow for coordinated training of operators using simulation exercises, and also test the interaction between operating entities during the execution of their system restoration plans. These drills require interaction between neighboring entities, which mimics the required response to a real-life event and promotes better communications and cooperation among neighboring entities. The drills also facilitate the development of the skills, experience and tools required to effectively manage the system restoration process. In general, the drills are designed to improve system operator skills and prepare them to efficiently respond to rare, catastrophic events such as a partial or total shutdown of the bulk power system. The drills also require interactions between entities that do not routinely work together that may have to cooperate to remedy a blackout.

The joint staff review team also found that execution of organized periodic system restoration drills provides the participants a mechanism by which any weaknesses or defects in the restoration plan may be exposed. The periodic drills, along with associated debriefs and operator feedback, facilitate the evaluation of existing restoration methods and provide an opportunity for continuous improvement. Because Reliability Standard EOP-005-2 Requirement R10 mandates annual system restoration training for all system operators, it is likely that on-duty operators will have been trained prior to having to implement the plan during an actual restoration event.

a) Planning a Restoration Drill

Participants indicated that some of the drill exercises are sponsored and developed at the reliability coordinator level. These drills are developed in collaboration with drill coordinators and operators from the reliability coordinator, transmission operators, generator operators, and certain transmission owners who perform restoration steps in accordance with the transmission operators' restoration plans. The joint staff review team found that a reliability coordinator-sponsored system restoration drill is normally planned several months prior to the actual drill. During annual training, a reliability coordinator typically schedules training sessions to prepare operators for specific topics related to system restoration. Similarly, system restoration drills sponsored at the transmission operator level are planned in advance, and operating personnel are trained on the system restoration procedures prior to the drill.

The joint staff review team observed that generally, prior to finalizing drill scenarios, details of the simulated event are reviewed by the coordinators of participating entities to identify potential opportunities for interacting with neighboring entities, and to resolve issues that may adversely impact the execution of the drill. A final system restoration drill plan is then provided to participating entities, which typically includes the following:

- Scope of the drill – initial condition of the interconnection from which the operators intend to execute their plans, including information regarding whether the drill involves a total or partial shutdown of the interconnection.

- Checklist of information (restoration tracking form) to be completed by the participants and passed on to the drill sponsor – generation and transmission facility information is used to record and track equipment status and to provide updates on restoration progress to the drill sponsor.

- "Injects" or specific unexpected constraints to be included by the drill sponsor to mimic possible failures of equipment or issues that may impact the restoration process - injects can include the loss or unavailability of major transmission lines, other transmission equipment, cranking paths, generation, or substations. Injects can also include factors such as loss or impairment of communication with field personnel or the loss of battery power at a substation.

Review of participants' plans and other procedures showed that the participants typically conduct or otherwise participate in restoration drills during the spring and fall each year (during lower load periods of the year). This affords greater availability of operators to dedicate to the restoration training. For some drills, blackstart generator operators and other generator operators needed for the restoration plans are invited to participate. Some participants indicated that, in their experience, involving these generator operators makes the drill more realistic and provides valuable collaborative experience for all participants.

b) Training Scenarios and Exercise Tools

Some participants undertake training drills more frequently than is required by the Reliability Standards, in order to allow for the use of different outage scenarios, or initial conditions, for each drill. These scenarios and initial conditions include, but are not limited to:

- Entire blackout of a region encompassing multiple entities;

- Partial blackout of a region, with separation into electrical islands within the region; and

- Regional blackout, with a cyber impact condition, resulting in loss of tools such as SCADA for some entities.

The drills conducted by participants also typically cover operator roles and responsibilities, and plans for communications.

Operator Training Simulators. The joint review team found that most participants use operator or dispatcher training simulators (DTS) to provide computer simulations of system restoration drills. In many applications, the DTS allows the participating operators to use the same tools and computer applications used in normal operations for the many steps executed during system restoration. The DTS system also tracks the status of generation and dynamically updates and displays the latest simulated system conditions. Moreover, accuracy of simulations to the actual system is created by building DTS systems from snapshots of actual system conditions accessed from the entity's SCADA system.

Participants also use tabletop exercises during which operating personnel discuss, in an informal setting, the effectiveness of their restoration plans, policies, and procedures under various possible scenarios, and the expected restoration steps to take. Some tabletop exercises also involve drilling on communications and coordination between operators (e.g., operators located at separate training facilities). The joint staff review team found that the structure of a tabletop exercise allows for open discussion among participants. However, tabletop exercises do not provide the full simulation experience possible with an operator training simulator, where the computer simulations for operator-execution of the restoration plan steps provide a more realistic experience.

Communications Tools. The joint staff review team found that during the early stages of restoration drills, participants typically test their primary and back-up communications systems, including telephones and other systems used for messaging. Participants send and receive messages and information via email, telephone, or facsimile (fax) during restoration drills, i.e. using the same media that would be used during a system emergency. The joint staff review team found that the communications tools provide an effective way for the participants to communicate with neighboring entities and to send summaries of the status of their restoration to the reliability coordinator or transmission operator, as well as providing the reliability coordinator or transmission operator a means to provide notification or feedback to participants. The drills may also exercise communications tools normally reserved for emergency situations, thus improving familiarity with these systems.

Restoration Maps and Tracking Forms. In one of the training drills observed by the joint staff review team, the sponsoring entity (in that case, the reliability coordinator) provided participants with a restoration tracking form as described above. As the system restoration drill progressed, the participants provided updated information to the reliability coordinator using the tracking forms, and the reliability coordinator updated its system models with the latest information as provided by participants.[51] In addition to restoration tracking forms, participants use as part of the restoration training and drilling process restoration maps, which geographically illustrate the location of substations, generation and interconnected transmission. Maps also show opportunities for connecting neighboring systems, voltage class, and lines out of service. Some participants use electronic maps, which are updated as information is received from other participants. In these cases, information received is processed and linked to the map on a periodic basis (e.g., within an hour). Other participants overlay restored lines on a geographic map board.

c) Drill Observations

As noted above, the joint staff review team observed system restoration exercises in which some of the participants took part. In the observed drills, the reliability coordinators initially focused on coordinating the restoration of offsite power to nuclear generating units for safe shutdown. This effort included transmission operators identifying blackstart units within their footprints and coordinating with the reliability coordinator to deliver power to nuclear generating units via specified transmission paths. The observed drills also included scenarios that involved coordination of such offsite power supply from outside a participating transmission operator's footprint, using pre-established plans that could involve multiple adjacent transmission operators.

During reliability coordinator-sponsored drills, the team observed several conference calls between staff of the reliability coordinator and transmission operators, which allowed the transmission operators to discuss outstanding issues or obstacles faced in the restoration process with the reliability coordinator. In addition, the team observed that the reliability coordinators make open party communication lines available to all participating transmission operators during the restoration drills.

During the mid-stage of observed drills, some participating transmission operators were able to connect islands within their respective footprints, connect to external islands outside of the reliability coordinator's footprint, or connect to other islands within the reliability coordinator's footprint with approval from the reliability coordinator. The

[51] The tracked information can include information on blackstart resources providing offsite power to nuclear units, the largest contingency in a particular island, amount of load restored, power available in a particular transmission operator's island or footprint, dynamic reserves available in each island, number of islands for each transmission operator, and tie line schedules and locations for synchronization with other transmission operators.

team also observed that the reliability coordinator operators usually recommend which ties to begin connecting between the transmission operators, based on system topography and proximity to stable external power.

During some of the observed restoration drills, the reliability coordinator focused on restoring the high voltage transmission system or "backbone" transmission early in the restoration process, using multiple cranking paths, and then coordinated the delivery of offsite power to nuclear units for safe shutdown. Next, the transmission operators used power from the "backbone" transmission to power islands created within their footprints. These system restoration drills used a "top-down" approach, since participating transmission operators used power from the "backbone" to restore their respective areas.

Generally, for transmission operator-sponsored drills, the transmission operator monitors island frequency, voltage and VArs as drills progress, using computer displays of SCADA data. The team observed transmission operators coordinating generation start up to regulate island frequency and switching of reactive components to maintain voltage and VArs in the islands created. In certain situations, neighboring transmission operators connected adjacent islands, thereby establishing larger islands. After multiple larger islands were created, more opportunities to simulate synchronization with other adjacent islands or external stable interconnections were identified.

Generally, at the conclusion of the observed system restoration drills, participating reliability coordinators and transmission operators requested and received feedback from drill participants. The transmission operators typically report any deficiencies found in their restoration plans during the drill to the reliability coordinator and subsequently seek to correct those deficiencies. Both reliability coordinators and transmission operators typically review the feedback provided by the participating entities to inform future system restoration drills and improve existing restoration plans. For example, as feedback, participating transmission operators in one observed drill reported issues with their blackstart plans, simulator issues, and coordination issues with other transmission operators.

The joint staff review team identified one area for improvement through its observation of these various drills and training scenarios, finding that in some instances, the process for transferring control of generation back to the balancing authority was not focused on as part of the drill.

3. Related Standards Assessment

Reliability Standard EOP-005-2 currently sets out explicit requirements for system restoration training and drills, including what must be demonstrated. Requirement R10 of EOP-005-2 requires each transmission operator to conduct annual system restoration training for its system operators, including training on specific areas, as follows:

R10. Each Transmission Operator shall include within its operations training program, annual System restoration training for its System Operators to assure the proper execution of its restoration plan. This training program shall include training on the following:

R10.1. System restoration plan including coordination with the Reliability Coordinator and Generator Operators included in the restoration plan.

R10.2. Restoration priorities.

R10.3. Building of cranking paths.

R10.4. Synchronizing (re-energized sections of the System).

In addition, Requirements R12 and R18 of Reliability Standard EOP-005-2 require transmission operators and generator operators to take part in their reliability coordinator's restoration training drills and exercises:

R12. Each Transmission Operator shall participate in its Reliability Coordinator's restoration drills, exercises, or simulations as requested by its Reliability Coordinator.

R18. Each Generator Operator shall participate in the Reliability Coordinator's restoration drills, exercises, or simulations as requested by the Reliability Coordinator.

The joint staff review team found that the training-related requirements (R10, including R10.1-R10.4) of the Reliability Standard are clear and effective. Participants' restoration plans, training procedures and scenarios extensively cover the coordination and exercising of the steps of restoration identified in these requirements. However, the team concludes from its review and discussion with participants that training on the criteria and steps for the transfer of control from the transmission operator back to the balancing authority during the late stages of restoration may not be sufficient. Some of the concern regarding the lack of training in this area may be attributable to the lack of guidance in some participants' plans regarding the initiating factors, methods and permissions for this important transfer. The joint team also observed that exercises related to the transfer of control from the transmission operator to the balancing authority are usually planned for the latter part of the training sessions. Therefore, exercises may not be implemented sufficiently to train operators on these topics. Thus, the joint staff review team recommends considering revisions to the Reliability Standards to require training focused on the transfer of control from transmission operators to the balancing authority.

Reliability Standard EOP-005-2, Requirements R12 and R18 require that each transmission operator and generator operator participate in its reliability coordinator's restoration drills, exercises or simulations as requested by its reliability coordinator. The joint staff review team did not identify any clarity or efficacy concerns with these requirements, or any other concerns, except as described above, with generator operators' or transmission operators' actual participation in reliability coordinator-convened restoration drills and training.

4. Recommendations

Operator training: exercises on transferring control back to balancing authority. The joint staff review team recommends that measures be taken (including considering changes to the Reliability Standards) to address system restoration training and drilling for transitioning from transmission operator island control to balancing authority ACE/AGC control. These measures will allow transmission operators, reliability coordinators, and relevant generator operators to gain experience on the coordination needed through all the stages of restoration, including coordination needed in the transfer of control back to the balancing authority.

5. Observed Practices for Consideration

In evaluating participants' system restoration drills and training exercises, the joint staff review team observed the following practices and recommends consideration of these by entities:

- Most participants use operator or dispatcher training simulators (DTS) to provide computer simulations of system restoration drills. In many applications, the DTS allows the participating operators to use the same tools and computer applications used in normal operations for the many steps executed during system restoration. The DTS system also tracks the status of generation and dynamically updates and displays the latest simulated system conditions.
- Some participants use unexpected scenarios and added visualization as part of participants' restoration training and drilling process, which included the following:
    - Constraints (injects) to mimic possible failures of certain equipment or existing system issues that may impact the restoration process.
    - The use of electronic maps that update dynamically and provide the most up-to-date visual display of the restored system to operators.

I. Incorporating Lessons Learned from Prior Outage Events

1. Summary

The joint staff review team examined the extent to which lessons learned from major outage events are incorporated in the restoration plans of the review participants. The review team examined recommendations from:

- 2011 Arizona-Southern California Outages,
- Hurricanes Gustav (2008) and Sandy (2012), and
- 2011 Cold Snap and 2014 Polar Vortex.

The joint staff review team observed that all participants incorporate the analysis of actual outage events through review of recommendations and lessons learned to enhance their restoration plan procedures. The team found that this practice helps to ensure the viability of the participants' restoration plans, and that the relevant Reliability Standards addressing the analysis of actual events are clear and effective. With the understanding that some areas have never experienced a blackout, the team recommends that applicable entities that have not experienced a blackout or other events which impacted, or could have the potential to impact, the viability of their restoration plans reach out to those who have, in order to gain more knowledge on improving their own restoration plans.

2. Reports from Recent Events

a) 2011 Arizona-Southern California Outages

On September 8, 2011, a disturbance occurred in the Southwest, leading to cascading outages and approximately 2.7 million customers without power. The outages affected parts of Arizona, Southern California, and Baja California, Mexico. All of the San Diego area lost power, with nearly one-and-a-half million customers losing power, some for up to 12 hours.[52]

The joint FERC-NERC Staff report analyzing this event made certain recommendations that pertain to the restoration process. The report recommended that the reliability coordinator involved in that event should clarify its role, including the real-time information it can provide in emergency situations like a multi-system restoration. In addition, the report recommended that that reliability coordinator should specifically

---

[52] *See* FERC and NERC Staff Report, *Arizona-Southern California Outages on September 8, 2011*, at 1 (April 2012) (Southwest Outage Report).

address coordination among balancing authorities and transmission operators in its operating area, outlining the areas of responsibility during system restoration and other emergencies.[53]

The joint staff review team found that participants have incorporated tables of tasks or responsibilities defining the roles, tasks, and approvals needed by each entity involved in restoration. Also, participants have identified points of contact during emergency situations dedicated to providing information. This identification of roles and responsibilities allows operators to focus on restoring the system during an emergency (see Roles, Interrelationships and Coordination section above).

b) Hurricanes Gustav (2008) and Sandy (2012)

On September 1, 2008, Hurricane Gustav made landfall in Louisiana as a strong Category 2 hurricane – 1 mph below Category 3 level. Hurricane Gustav, with impacts compared to Hurricane Katrina, resulted in outages to more than 1.3 million customers. The impacts of Gustav were concentrated primarily in Louisiana, Mississippi and Arkansas, and caused severe flooding, which slowed the restoration efforts. Moreover, due to damage to several high-voltage transmission lines, a portion of the transmission system was "islanded" during the event.[54]

On October 29, 2012, Hurricane Sandy, a Category 1 hurricane, made landfall on the New Jersey shore at around 8 p.m. Eastern time, with an unprecedented storm surge. Over the course of the event, over 8,000 MW of generation capacity was forced off line, and seven interconnections to southeastern New York, from Connecticut and New Jersey, were disconnected. By late Monday, October 29, approximately 2.2 million electric

---

[53] *See id.* at 62.

[54] *See* Infrastructure Security and Energy Restoration Office of Electricity Delivery and Energy Reliability, U.S. Department of Energy, *Comparing the Impacts of the 2005 and 2008 Hurricanes on U.S. Energy Infrastructure* at 7, 9 (February, 2009). The electrical island was first detected by operator-monitoring of PMUs installed across Mississippi, Louisiana, Arkansas and East Texas, where operators were alerted to the island's creation by the diverging system frequencies. The electrical island existed for approximately 33 hours, until transmission facilities were restored and re-synchronization with the Eastern Interconnection occurred. *See* Kolluri, S.; Mandal, S.; Galvan, F.; Thomas, M., *The Role of Phasor Data in Emergency Operations*, Transmission T&D World Magazine (Dec. 1, 2008), and Power & Energy Society General Meeting, 2009. PES '09. IEEE, doi: 10.1109/PES.2009.5275340, 1, 5, and 26-30 (July 2009).

customer outages were reported. The storm surge was so extensive that transmission owners reported that low-lying stations were flooded to the degree that staff had to evacuate for safety reasons.[55]

With respect to Gustav, the use of PMUs from multiple transmission substations to monitor system frequency at diverse locations aided in island monitoring and management of load pick-up, and in detection of an islanded condition. Lessons learned from Sandy included the value of tracking the *combined* effects of tides and storm surge, and in increased operator awareness that storm surge projections became accurate only within one day of the storm.

The joint staff review team found through participant discussions, on-site observations and reviews of emergency response and restoration plans regarding these and other similar recent events (e.g., Hurricane Katrina in 2005), that participants more prone to severe coastal weather patterns and associated damage have incorporated several lessons learned from those events. These lessons include maintaining large inventories of transmission line and substation equipment, along with establishing storm response plans for closely monitoring forecasted weather conditions, mobilizing equipment, and for activating operations and field personnel to expedite restoration.[56]

---

[55] System impacts included outages to 28 345 kV transmission lines, one 230 kV transmission line, 42 138 kV transmission lines, and 15 115 kV transmission lines. Generating facilities over a very wide footprint were forced off line. Some generators were rendered unavailable due to the loss of interconnecting transmission. There were also reports of other generators that were forced into preemptive "shut-downs" to protect assets from long-term damage or for human safety reasons. *See* New York Independent System Operator, *Hurricane Sandy: A report from the New York Independent System Operator* (March 2013), http://www.nysrc.org/pdf/MeetingMaterial/RCMSMeetingMaterial/RCMS%20Agenda%20159/Sandy_Report___3_27_133.pdf.

[56] While beyond the scope of this report, the possibility of damage to major equipment has been discussed in reports by various others, such as the National Research Council of the National Academies, "Terrorism and the Electric Power Delivery System," 69-91 (2012) (focusing on transmission towers, mobile generators and transformers and shared inventories of transformers); Department of Energy, "Large Power Transformers and the U.S. Electric Grid Report," (2012, updated in 2014); Center for the Study of the Presidency & Congress, "Securing the U.S. Electrical Grid," (2014) (citing mutual assistance agreements and shared inventories for equipment such as transformers); and

c) 2011 Cold Snap and 2014 Polar Vortex

During the first week of February 2011, the southwest region of the United States experienced unusually cold and windy weather, with lows in the teens for five consecutive mornings and many sustained hours of below freezing temperatures throughout Texas and in New Mexico. Between February 1 and February 4, 2011 individual generating units throughout Texas experienced either an outage, a derate, or a failure to start. These reductions in available generation were severe enough to trigger a controlled load shed of 4,000 MW. In total, 4.4 million customers were affected over the course of the event.[57]

In early January 2014, the Midwest, South Central and East Coast regions of the United States experienced a polar vortex, where some areas were 35° F or more below their average temperatures, resulting in record high electrical demand. One of the largest issues affecting gas-fired generators during the polar vortex was the curtailment or interruption of fuel supply. Extreme cold weather also had a major impact on generator equipment. Of the approximately 19,500 MW of generator capacity lost due to cold weather, over 17,700 MW was due to frozen equipment.[58]

NERC, "Severe Impact Resilience: Considerations and Recommendations," 50-51 (2012).

[57] System operators initiated controlled rolling blackouts during the event. Although emergency conditions existed, entities' restoration plans did not need to be deployed. Had a total blackout occurred in the region, the unavailability of 10 blackstart resources, comprising 687 MW out of a total 1150 MW of blackstart capacity, could have jeopardized the ability to promptly restore the system. *See Report on Outages and Curtailments During the Southwest Cold Weather Event of February 1-5, 2011* (August 2011), http://www.ferc.gov/legal/staff-reports/08-16-11-report.pdf (2011 Cold Snap Report).

[58] Many generator outages, including a number of those in the southeastern United States, were the result of temperatures that fell below the plant's design basis for cold weather. At the height of generation outages (January 7, 2014 at 0800 Eastern time), the southeastern United States accounted for approximately 9,800 MW of the outages attributed to cold weather. While widespread outages occurred, no blackouts occurred and system operators were able to successfully maintain reliability. *See* North American Electric Reliability Corp., *Polar Vortex Review*, iii, 2, 19 (Sept. 2014).

One recommendation from the 2011 Cold Snap Report pertains to system restoration, advising that balancing authorities, transmission operators and generator owners/operators take the steps necessary to ensure that blackstart generators can be utilized during adverse weather and emergency conditions.[59]

The joint staff review team found during its review that participants do not typically test blackstart units during extreme temperatures. Many blackstart resources are needed as peaking generators during times of high demand, and entities commonly do not risk scheduling a test during these periods. However, other lessons learned have been incorporated in participants' restoration planning, including implementing weather–related emergency procedures which include alert levels and triggers. These procedures include steps for requesting additional generator reserves, including the distribution or location of additional reserves (through means such as adjusting reserve requirements), as well as the cancelling of upcoming maintenance or limiting planned outages to enhance more operational flexibility during the storm or possible severe weather periods.

3. Related Standards Assessment

Reliability Standard EOP-005-2, Requirement R6 requires transmission operators to verify the functionality of their restoration plan based on actual events, among other things:

R6. Each Transmission Operator shall verify through analysis of actual events, steady state and dynamic simulations, or testing that its restoration plan accomplishes its intended function . . . .

Based on the joint staff review of participants' analyses of actual events, no related clarity or efficacy concerns were identified. The joint staff review team found that participants routinely review lessons learned from events such as those above, and similar weather events (e.g., Hurricane Katrina), and incorporate them into their emergency response and restoration plans where applicable, recognizing there is no substitute for experience. The incorporation of lessons learned from actual events into these plans is mostly done on an intra-regional basis through restoration working groups. The joint team learned that some participants at the ISO level regularly share experiences and lessons learned from drills with ISOs from adjacent regions, which these participants found beneficial.

---

[59] Recommendation 7 from the 2011 Cold Snap Report.

4. Recommendations

Obtaining insight from entities that have experienced a widespread outage. The joint staff review team found that participants place the utmost importance on past experiences and lessons learned from events, including the lessons learned from historical blackouts and other significant events related to the viability of restoration plans, and currently share information through restoration working groups and other means. However, the team is concerned that entities that have not experienced blackout conditions may not be fully aware of all the additional insight and lessons learned by entities that have experienced significant blackouts, particularly for blackouts and events in other regions. Therefore, the team recommends that applicable entities that have not recently experienced a blackout or other event which impacted, or could have the potential to impact, the viability of their restoration plans reach out to those who have experienced such events, in an effort to continually improve their restoration plans. Entities could benefit from the sharing of experiences across different regions of the country to gain insight into events that may not have occurred locally within a region, including but not limited to:

- Severe flooding and storm impacts on facilities and equipment depended on for system restoration;
- Effects of extreme temperatures, including severe cold weather impacts on facilities and equipment depended on for system restoration; and
- Preparedness training for the above impacts.

5. Observed Practices for Consideration

The joint staff review team found that lessons learned from past major events have been incorporated into participants' emergency response plans and restoration plans where applicable. The joint staff review team observed the following practices and recommends consideration of these by other entities:

- Use of diversely-located frequency measurements, e.g., PMUs for system operator monitoring of frequency (*see* Island Development and Synchronization section above).
- Maintaining large inventories of transmission line and substation equipment, along with establishing storm response plans, closely monitoring forecasted weather conditions, mobilizing equipment, and activating operations and field personnel to expedite restoration (*see* Island Development and Synchronization section above).

- Developing and implementing extreme weather–related procedures, including alert levels and triggers to initiate the request for additional generator reserves, including the distribution or location of additional reserves (through means such as adjusting reserve requirements). Such procedures could also include cancelling upcoming maintenance or limiting planned outages to enhance operational flexibility during severe weather periods.

- Assigning roles and responsibilities across operator desks during system restoration, as well as identifying points of contact during emergency situations dedicated to providing information. This identification of roles and responsibilities allows operators to focus on restoring the system during an emergency.

- Several participants indicated that lessons learned can be sourced from smaller events just as much as from the larger events, and that sharing and analysis of these events can be accomplished, for example, through the ERO Events Analysis Process.[60]

V. Review of Cyber Security Incident Response and Recovery Plans, and Related Standards Assessment

The joint staff review team reliability assessment also included review of cyber security incident response plans and recovery plans for critical cyber assets, along with associated procedures and resources of the participants, to assess their readiness to respond and recover in the event of a cyber security event. This report provides a breakdown of the review by various response and recovery topics. These topics include:

- Resources, Processes, and Tools for Cyber Incident Response and Recovery;

- External Roles, Interrelationships and Coordination;

- Monitoring for and Detection of Cyber Incidents and Triggers for Incident Response;

- Initial Event Response Actions;

---

[60] See http://www.nerc.com/pa/rrm/ea/Pages/EA-Program.aspx.

- Recovery Planning;
- Review and Verification of Incident Response and Recovery Plans;
- Drills and Training Exercises; and
- Improving Cyber Security Response and Recovery Plans Based on Actual Events and Other Feedback.

As noted above, included at the close of each topic is an analysis of the participants' plans against the relevant Critical Infrastructure Protection (CIP) Reliability Standards, to see where improvements in the clarity or efficacy of the standard may be warranted.

A. Resources, Processes, and Tools for Cyber Incident Response and Recovery

1. Summary

The joint staff review team examined the resources, processes, and tools participants plan to deploy in responding to cyber incidents and in recovery of critical cyber assets, as set out in their cyber incident response and recovery plans. The team considered the following areas: (1) enterprise structuring of cyber security policies; (2) deployment of personnel resources, including defining roles and responsibilities; and (3) facilities and tools for response. The joint staff review team generally found the participants' plans and processes for incident response and asset recovery to be thorough. As described below, some larger participants responsible for multiple registered entities are moving toward an enterprise-wide cyber security approach, and implementation of their incident response and recovery plans is supported by full-time dedicated personnel resources. The joint staff review team recommends that cyber security incident response and recovery plans clearly identify who is responsible for asset response and recovery, specifically designating accountability at the cyber asset level (e.g., EMS servers, RTU concentrators, network routers, etc.), and recommends that measures be taken (including considering changes to the Reliability Standards) to address this issue. Although the joint staff review team recognizes that the Reliability Standards addressing resources, processes, and tools for cyber incident response and recovery will have improved clarity once the approved changes in CIP Version 5 become effective, consideration should be given as to whether the Standards as revised address all of the team's concerns in this respect.[61]

---

[61] *See NERC CIP Standards, version 3*, available at http://www.nerc.com/pa/stand/Pages/ReliabilityStandardsUnitedStates.aspx?jurisdiction=

2. Review of Participants' Response and Recovery Plans

a) Enterprise Structuring of Cyber Security Policies

The joint staff review team found that the size of the participant organization influences the enterprise's structuring and approach to cyber security event response and asset recovery. Participants indicated that the larger the organization, the more likely that an enterprise-wide, top down approach is employed. Some of the largest participants use an enterprise-wide security policy to align all internal entities to the security and business goals of the overall organization, and are moving toward an overarching enterprise security plan.[62] In one example, a participant's plan calls for each business unit within the organization to apply the same cyber security incident handling procedures. For these participants, the enterprise-level policies require the operating companies and business units responsible for critical functions to develop and maintain a security incident response and recovery plan.

Although some smaller participants have business functions governed by enterprise security policies, they generally have more autonomous plans with security-related processes owned by an assigned team in each department or business unit. For these smaller organizations, enterprise-level involvement generally takes the form of review and approval of documentation related to the cyber incident response and cyber asset recovery plans and processes.

b) Personnel Resources, Roles, and Responsibilities

Participants indicated that their resource needs for cyber security have grown significantly in the last five years, and that they expect this growth to continue. All of the participants have full-time personnel dedicated to some aspect of cyber security and response, as defined in their cyber response and recovery plans.

United States. The Commission approved modifications to the currently-effective CIP Standards, referred to as the "CIP Version 5" standards. *See Version 5 Critical Infrastructure Protection Reliability Standards* (Order No. 791), 145 FERC ¶ 61,160 2013).

[62] Internal entities could be wholly-owned affiliates, operating companies, member entities, different NERC-registered functional entities, etc.

Most participants maintain a cyber security response team responsible for the analysis and immediate response to cyber incidents, a cyber security response manager responsible for team governance, and a recovery plan owner responsible for document changes. These individuals are generally not the same personnel responsible for asset recovery. The classification and severity of a given event appear to dictate who is required to be involved in a participant's response. Large catastrophic events like a hurricane, which may result in the loss of critical cyber assets as well as physical equipment, may require most of the response groups mentioned, whereas finding a misplaced or unidentified USB device may require a single team response. The participants' plans varied in the level of detail in defining the personnel that need to be deployed for a given event type, and the particular approach taken. In defining cyber security plan roles and responsibilities, the level of detail of the response plans was shaped by several factors, including geography, size and structure of the organization and holdings, IT and network department size and structure, and vendor support required.

Regardless of the size of the organization, all participants assign roles to a plethora of individuals and groups in their established response and recovery plans, including: corporate IT help desk, telecommunications group, dedicated security teams, dedicated forensic teams, IT and technical managers, local law enforcement, support vendors, and other third parties. The joint staff review team found that a few of the review participants have a dedicated team for cyber security event response and asset recovery that works hand in hand with reliability standard compliance teams. However, from review of participants' plans and discussions with participants, the accountability of these individuals and groups was not always clear. Such a lack of clarity as to accountability could, during implementation, introduce confusion and result in reduced efficiency and effectiveness of recovery.

c) Processes and Tools for Response and Recovery

Processes Used for Event Assessment. All participants' plans characterize and classify the types or severity of events that would trigger the execution of a plan, but only some participants attempt to categorize the severity of all known potential threat events. Most of the participants' plans group events into an impact level of one to five, with the impact level dictating the response. To get an accurate assessment of the impact level, all events require response personnel and tools to perform the initial threat analysis. Whether complex or simple, all participants have an escalation process and response tools that require proper use, communication, and availability.

Facilities and Tools for Response. All of the review participants maintain redundant primary critical EMS/SCADA systems with a replicated backup system capable of assuming all functions in a short failover time. All participants have some degree of device redundancy on the primary system, allowing for high availability and quick recovery from a minor event. In addition to the primary and backup systems, all participants maintain some form of testing or development EMS/SCADA system that mimics the primary system, and in an emergency situation can potentially be used as a spare.

While specific EMS/SCADA installations vary, most participants promote the use of secondary systems for recovery before resorting to data media backups. In fact, some participants do not use tape or other removable media for data backups at all, but instead rely on redundancy and hard-wired drive backups for data recovery. The participants that do use a robust tape data backup system use them for emergencies only. For data retention and restoration, the participants' plans identify approaches, hardware, and responsible personnel, with storage area networks and mirrored disk arrays the most popular approach to data restoration. Most participants do not keep a large inventory of replacement hardware, but rely on a third party for replacement of hardware that has failed.

3. Related Standards Assessment

Currently-effective Reliability Standards CIP-008-3 and CIP-009-3 include specific requirements to ensure that entities maintain planned resources for cyber security incident response and recovery of critical cyber assets. The relevant requirements for each standard are as follows:

CIP-008-3:

R1. Cyber Security Incident Response Plan – The Responsible Entity shall develop and maintain a Cyber Security Incident response plan and implement the plan in response to Cyber Security Incidents. The Cyber Security Incident response plan shall address, at a minimum, the following:

R1.1. Procedures to characterize and classify events as reportable Cyber Security Incidents.

R1.2. Response actions, including roles and responsibilities of Cyber Security Incident response teams, Cyber Security Incident handling procedures, and communication plans.

CIP-009-3:

R1. Recovery Plans – The Responsible Entity shall create and annually review recovery plan(s) for Critical Cyber Assets. The recovery plan(s) shall address at a minimum the following:

R1.1. Specify the required actions in response to events or conditions of varying duration and severity that would activate the recovery plan(s).

R1.2. Define the roles and responsibilities of responders.

...

R4. Backup and Restore – The recovery plan(s) shall include processes and procedures for the backup and storage of information required to successfully restore Critical Cyber Assets. For example, backups may include spare electronic components or equipment, written documentation of configuration settings, tape backup, etc.

The joint staff review team found that Reliability Standard CIP-008-3 Requirements R1, R1.1-R1.2, and CIP-009-3 Requirements R1, R1.1, R1.2, and R4 are sufficiently detailed, enabling participants to effectively identify and maintain planned resources, processes and tools for response and recovery in their plans. However, the above requirements are not clear on accountability for assigned roles and responsibilities for response and recovery of critical cyber assets. This accountability is especially important for interconnected cyber systems which may involve several business units within an organization (e.g., accountability for recovering EMS servers, RTU concentrators, network routers, etc.). Lack of clarity about the accountability of assigned personnel could result in confusion and reduced efficiency of recovery during emergencies.[63]

Also, the joint staff review team found that an important element of security monitoring is to require *identification* of events as possible cyber security incidents, a task which has been more directly addressed in the CIP Version 5 Reliability Standards.[64] With that modification, the team otherwise found the requirements listed above to be clear and effective in promoting necessary planning on the resources, processes and tools to be used for cyber incident response and critical cyber asset recovery.

---

[63] The new CIP version 5 standards may address these concerns to some extent, but may not cover all potential areas of concern. *See* CIP-008-5 Requirement R1, part 1.3, and CIP-009-5 Requirement R1, part 1.2.

[64] CIP-008-5 Requirement R1 Part 1.1 requires applicable entities, for their High Impact BES Cyber Systems and Medium Impact BES Cyber Systems, to include one or more processes to identify, classify, and respond to Cyber Security Incidents. Mandatory compliance with the CIP Version 5 Standards will take effect in April 2016 for High and Medium Impact BES Cyber Systems, superseding the currently-effective Version 3 Standards.

4. Recommendations

Response and recovery plan ownership. The joint staff review team recommends that cyber security incident response plans and recovery plans for critical cyber assets specifically designate accountability at the cyber asset level (e.g., EMS servers, RTU concentrators, network routers, etc.). The team recommends that measures be taken (including considering changes to the Reliability Standards) to address this.

B. External Roles, Interrelationships, and Coordination

1. Summary

The joint staff review team examined the external roles, relationships and coordination required by or needed to implement the participants' cyber response and recovery plans. The team considered the following areas: (1) vendors, third-party support, and external dependencies; and (2) communications and relationships with federal and state law enforcement, task forces, and emergency management offices. The joint staff review team found that the participants have well-developed cyber security incident response plans that include communication plans and otherwise define roles and responsibilities with respect to third parties. The participants also have strong working relationships with local offices of the Federal Bureau of Investigations (FBI), Department of Homeland Security (DHS), and other law enforcement agencies for cyber security and emergency response. The joint staff review team did not identify any issues related to these areas, and found the relevant requirements in the Reliability Standard to be clear and effective in addressing these elements of a cyber incident response or critical cyber asset recovery plan.

2. Review of Participants' Response and Recovery Plans

Technical Support and Hardware. Although participants stress autonomy, all use third parties to varying degrees in support of their cyber security efforts, including technical support. All of the review participants maintain contractual and working relationships with system vendors for EMS and SCADA system technical support and hardware replacement. Participants also rely on EMS and SCADA vendors for security patch updates and assessments for those systems. Additionally, participants rely heavily on Windows and Linux operating system vendors for speedy security patch releases and fixes.

About half of the participants maintain some hardware inventory for critical devices, while the others rely on a third party or device redundancy for recovery.[65] Some participants use redundant dedicated telecommunication lines from vendors for high availability.

Cyber Security Monitoring. The participants primarily contract with third parties for penetration testing and security log analysis and alerting. Third parties responsible for security log reviews and event alerting report back to the participants' incident response teams or other responsible personnel identified in the cyber security response plan. Several participants use third parties in their review of policies, procedures, and restoration/recovery plans, with many reviews being a part of compliance with Reliability Standards.

Cyber Security Event Awareness. All of the review participants rely heavily on the Industrial Control Systems Cyber Emergency Response Team as the primary source for cyber security awareness, but they also rely on the Electric Information Sharing and Analysis Center (Electricity ISAC or E-ISAC),[66] vendors, and other outside sources.

External, Federal, and State Relationships. All participants, regardless of size, consider having working security and functional relationships with law enforcement and other outside entities to be important to their cyber security plans. All participants maintain relationships with relevant federal and state law enforcement entities and task forces, with many having dedicated liaisons to foster two way communication and awareness with these and other groups. Participants specifically mentioned having strong relationships with local offices of the FBI, DHS, and other law enforcement agencies. While all participants must have FBI and law enforcement contacts for events that must be reported on Department of Energy Form OE-417,[67] some participants explicitly

---

[65] The participants that do not maintain an inventory have system redundancy as a compensating measure. Apart from EMS and SCADA redundancy, the participants that maintain spare inventory appear to concentrate on network devices (e.g., firewalls, switches, etc.) and storage devices.

[66] The Electricity ISAC was previously named the Electric Sector Information Sharing and Analysis Center, or ES-ISAC, and the relevant CIP Reliability Standards still reference that name and acronym.

[67] The Department of Energy has established mandatory reporting requirements for electric emergency incidents and disturbances in the United States. See http://www.oe.netl.doe.gov/docs/OE-417_Instr-complete120508.pdf.

incorporate these law enforcement agency contacts in their emergency plans and procedures. These agency contacts may also participate in simulated drills and exercises of participants' emergency communication plans.

3. Related Standards Assessment

Currently-effective Reliability Standards CIP-008-3 and CIP-009-3 include specific requirements to ensure that the roles and relationships for cyber security incident response and critical cyber asset recovery are properly defined in the response plan, including any external roles and responsibilities and communication. The applicable requirements for each standard are as follows:

CIP-008-3:

> **R1.** Cyber Security Incident Response Plan — The Responsible Entity shall develop and maintain a Cyber Security Incident response plan and implement the plan in response to Cyber Security Incidents. The Cyber Security Incident response plan shall address, at a minimum, the following: . . .
>
> **R1.2.** Response actions, including roles and responsibilities of Cyber Security Incident response teams, Cyber Security Incident handling procedures, and communication plans.

CIP-009-3:

> **R1.** Recovery Plans — The Responsible Entity shall create and annually review recovery plan(s) for Critical Cyber Assets. The recovery plan(s) shall address at a minimum the following: . . .
>
> **R1.2.** Define the roles and responsibilities of responders.

The joint staff review team found that the participants have well-developed cyber security incident response plans that include communication plans and otherwise define roles and responsibilities with respect to third parties. Participants also have strong working relationships with local offices of the FBI, DHS, and other law enforcement agencies for cyber security and emergency response. For these reasons, the joint staff review team did not identify any clarity or efficacy issues related to relevant requirements in Reliability Standards CIP-008-3 and CIP-009-3.

C. Monitoring for and Detection of Cyber Incidents and Triggers for Incident Response

1. Summary

The joint staff team examined participants' monitoring for and detection of incidents, and triggers for incident response. The team considered the following areas: (1) monitoring methods and tools to detect anomalies and problems; (2) advances in programs, tools, and expertise for monitoring and detection; and (3) triggers requiring a response to a cyber incident. The joint staff review team found that the incident response plans for monitoring and detection of cyber security incidents vary across the range of participants, with the best of the reviewed plans having comprehensive escalation procedures, containing steps for further implementation based upon the complexity and/or depth and breadth of the threat or vulnerability, i.e. that 'escalate' when the threat or vulnerability risk increases, and make use of advanced tools, support, and expertise. Other reviewed plans lack well-defined characterization, assessment, and escalation of events. Therefore, the joint staff review team recommends that measures be taken (including considering changes to the Reliability Standards) to address the use of specialized technical expertise, advanced tools, and levels of security expertise, to improve event monitoring and response. The team also recommends that measures be taken (including considering changes to the Reliability Standards) to require details around the types of events that should trigger a response and what type should be reported.

2. Review of Response and Recovery Plans

a) Monitoring Methods and Tools to Detect Problems

In comparing participants' cyber incident monitoring methods as set out in their respective plans, the joint staff review team found two major areas of system monitoring pertaining to bulk power system reliability: (1) monitoring cyber system performance, and (2) authorized use of critical assets, critical cyber assets, and their supporting cyber systems. The participants apply controls in various ways to help determine whether critical systems have unwanted or unauthorized activity, and use that information to determine how and when to respond.

Participants' system monitoring consists of automated tools used to monitor network traffic, and can be system-based or host device-based. Participants indicated that the monitoring systems can generate a large quantity of data, and that effective data management tools are therefore important for effective monitoring. Participants indicated that a new threat identified for monitoring can have significant ramifications on operations and often requires timely and appropriate response to the threat. Cyber threat events chosen for monitoring can cover a wide range of activities, such as: (1) new or existing services on host devices suddenly being utilized, (2) unusual login times and unsuccessful attempts, (3) abnormal traffic on a network, (4) changes to a file's integrity and attributes, and (5) escalation of administration rights and suspect network behavior (i.e., traffic/protocols outside the norm).[68] Most participants' response plans have established thresholds for a suspected cyber threat event that may trigger a pre-determined response.

One threshold indicator used by participants automatically assesses large quantities of data, and sends notifications to an Incident Response Team member(s) once a pre-determined trigger is met. The alert notification may prompt additional intervention using manual processes, making it necessary for technicians to manually review the data to determine whether the detected activity should be considered suspicious, warranting further response and threat level escalation, or considered a false positive indication.

Every review participant stressed the importance of having round-the-clock coverage to receive and respond to alert notifications. Some participants have established a dedicated security operations center[69] as an in-house cyber incident and threat assessment center. Security operations center technicians perform the initial analysis of any alert and/or detected suspicious activity. Participants that do not use a security operations center model rely on their subject matter experts or use third-party vendors to process alert notifications. In some cases, participants use a hybrid approach: employees perform the task during business hours and a vendor or network operations center provides support for the balance of the time.

Most information used to perform this analysis comes from an intrusion detection system, but intrusion prevention systems are becoming widely implemented within participants' organizations as well. Organizations may also deploy application whitelisting on user devices, permitting only specified activities, interactive access, and specific processes and programs to run.[70] Another emerging trend is the use of behavioral profiles for each

---

[68] From discussions with the participants and their use of different naming conventions, the joint staff review team chose to use the phrase "cyber threat events" to refer to participant-monitored cyber events that are not yet determined to be cyber security incidents or a cyber threat that did not rise to the level of an entity-declared Cyber Security Incident as defined in the CIP standards (e.g., scanning an IT system for a newly discovered threat described by ICS-CERT).

[69] "Security operations center" is a generic name used in this report to describe a dedicated security monitoring operation.

[70] An application whitelist is a list of applications and application components (libraries, configuration files, etc.) that are authorized to be present or active on a system according to a well-defined baseline. *See* National Institute of Standards and Technology, U.S.

authorized user, so that when a known user's activity deviates from typical behavior, that activity is flagged for further analysis.

b) Implementing Advanced Cyber and Physical Threat Programs and Tools

Participants use a number of third-party vendor products in their cyber security and threat detection programs, procedures, and processes. As participants strive to keep abreast of new and evolving cyber and physical threats, they are partnering with third-party cyber security specialist vendors. Several participants have joined cyber security awareness groups sponsored by governmental authorities, and are working with universities that have advanced cyber security programs. Also, participants' cyber security professionals are coming together to form groups or charters with professionals in similar business or operational models, in an effort to keep current with threats specific to their industry. Some participants have hired cyber security professionals with advanced skills and capabilities to develop advanced in-house cyber security operations centers, and plan to partner or extend their services to other entities outside of their NERC functional registration and footprint. The joint staff review team found that participants' cyber security threat detection teams are staffed with personnel from specialized operational functions such as IT and networking groups that together form a larger cyber security incident command.[71]

Participants acknowledged that developing thorough internal control processes is key for mitigating certain types of slow advanced persistent threats, in which a bad actor or actors can penetrate a system and move across networks while elevating existing accounts and access privileges.[72] Development of internal control processes may require

---

Department of Commerce, *NIST Special Publication 800-167 (Draft) Guide to Application Whitelisting,* (August 2014).

[71] The cyber security group may draw from a typical network operations center, IT support group, 24/7 Help Desk, and EMS/SCADA support. Such a group will use system-specific tools such as intrusion detection systems, hardware monitoring, antivirus, network inspection tools, and security information and event management to inspect network traffic, log files, and logon access.

[72] An advanced persistent threat attacks information assets of national security or strategic economic importance through either cyberespionage or cyber-sabotage. These attacks use technology that minimizes their visibility to computer network and individual

actions such as trusted authorization tickets segmenting business groups. Some participants require additional sponsorships for administrative changes to existing or newly-created user authorization account access, and for escalation of privileges and rights. The use of additional sponsorships or similar processes could help prevent an administrative level insider threat or an escalation of rights attack from an advanced persistent threat. Participants' response plans apply a defense-in-depth posture that includes network detection tools and capabilities, such as host-based intrusion detection systems, antivirus, physical access control systems for physical intrusion threats, EMS/SCADA alarms, firewalls, peripheral system alarms and internal notifications to the 24/7 cyber security monitoring centers described above. Some participants have established a centralized logging system for inspecting many of their system software, login access and physical access controls systems. Participants indicated that specialized software tools and systems can aid in inspecting log files and identifying anomalies for large amounts of data, but that the process of human inspection and intervention is still necessary to determine whether a flagged suspicious item is an actual threat or, for example, an employee who exceeded his or her password attempt limit.[73]

c) Triggers for Responding

All of the review participants have dedicated personnel focused on monitoring systems and devices from a cyber security perspective, and it is common for a participant's IT Help Desk to be the initial point of contact for this monitoring. Users who may detect an anomaly within their environment can report issues through established protocols. Participants' monitoring processes include assigning the reported issues a priority level commensurate with their importance, so business systems may not have the same response expectations as a system critical to operational reliability. Regardless of the

---

computer intrusion detection systems. Advanced persistent threats are directed against specific industrial, economic, or governmental targets to acquire or to destroy knowledge of international military and economic importance. Once an advanced persistent threat has entered its target, the attack can last for months or years; that is, it is a "persistent" threat. *See Encyclopædia Britannica Online*, s. v. "advanced persistent threat (APT)", accessed September 01, 2015, http://www.britannica.com/topic/advanced-persistent-threat.

[73] The team noted that some of the participants, through third party provisions, use advanced monitoring tools which automatically collect and compare information to perform wider-area monitoring for detection of cyber security events.

notification method (Help Desk or automated alert), once a possible incident is determined to warrant further analysis, participants' processes involve initiating the appropriate technical support to evaluate the factors and information pertaining to the alarm/report. Recently, some participants implemented specialized 24/7 incident response teams and a hotline for reporting any suspicious activity or anomaly identified in log files, alarms, communication protocols and changes in system performance. The incident response team may be staffed by in-house operational personnel with IT, network and cyber and physical security experience and backgrounds. Once notified, the incident response team can assess an issue and contact support personnel with the necessary specialized expertise in networks, firewalls, EMS/SCADA systems, cyber threats, and communication systems.

Some participants use a matrix table to evaluate, characterize, and determine the type of cyber threat events occurring on their system or being reported in cyber security alerts, notifications, and advisories. The initial assessment determines if a more thorough review is required. The relevant details are routed to the designated response personnel responsible for business units and operational functions that may be impacted. This routing includes any 24/7 third-party vendor support for cyber security detection and prevention systems employed. For some participants, this more thorough review is fulfilled by a cyber security incident response team sometimes referred to as a cyber security operation center. Some participants are partnering with various specialized cyber security vendors, intrusion detection systems, intrusion prevention systems, and services for more in-depth cyber threat event detection and analysis to identify and help classify events.

Participants' response plans commonly include comparing an event with the matrix criteria, table spreadsheet, or other method used to determine the initial risk assessment and impact, and initial cyber threat event level. Participants apply these methods to obtain an accurate initial cyber threat determination level, which will then initiate the response required for that cyber threat level. The incident response team's response to the event may reveal whether a more serious threat criteria level is present and trigger a greater threat cyber incident level. This increase in incident level may also trigger a different response and additional evaluations, mitigating actions, and notifications. Several participants noted that, while initial threat event classifications are critically important, the triggers and threshold criteria for escalating and de-escalating the level of a threat event are equally important and need to be well understood.

3. Related Standards Assessment

Currently-effective Reliability Standard CIP-008-3 and other standards include specific requirements regarding the monitoring of cyber security events and events or actions responsible for triggering a response.[74] For CIP-008-3, these include:

R1. Cyber Security Incident Response Plan — The Responsible Entity shall develop and maintain a Cyber Security Incident response plan and implement the plan in response to Cyber Security Incidents. The Cyber Security Incident response plan shall address, at a minimum, the following:

R1.1. Procedures to characterize and classify events as reportable Cyber Security Incidents.

R1.2. Response actions, including roles and responsibilities of Cyber Security Incident response teams, Cyber Security Incident handling procedures, and communication plans.

The joint staff review team found that the incident response plans and monitoring programs vary across the range of participants. Several more robust plans have comprehensive escalation procedures, use advanced tools and expertise, and maintain third-party support for monitoring and detecting cyber security incidents. Other participants have less robust plans, which are not as well defined regarding the

---

[74] Reliability Standard CIP-007-3 Requirement R6 requires responsible entities to ensure that all Cyber Assets within the electronic security perimeter, as technically feasible, implement automated tools or organizational process controls to monitor system events that are related to cyber security, and requires maintaining logs of system events to support incident response, as required in CIP-008-3.

Reliability Standard CIP-005-3, Requirement R1.5 requires Cyber Assets used in the access control and/or monitoring of a responsible entity's Electronic Security Perimeter(s) to be afforded the protective measures specified in Standard CIP-008-3. This sub-requirement pulls network communication devices responsible for protecting the electronic security perimeter into the security monitoring requirements of CIP-007 Requirement R6.

The team did not address CIP-006-3 Requirement R2.2, which refers to CIP-008-3, because the sub-requirement pertains to physical security perimeters and mechanisms.

characterization, assessment, and escalation of events, due to a lack of expertise, advanced tools, and third-party support.[75] Use of advanced tools, expertise, and third-party support is not required by CIP-008-3 or by any other relevant Version 3 CIP Reliability Standard. Also, as described above in the section "Planned Resources, Processes, and Tools for Response and Recovery," the joint staff review team observed participants with processes that also include *identifying* events as possible cyber security incidents. CIP-008-3, R1.1 requires procedures to characterize and classify events as reportable cyber security incidents, but does not require identification of the types of possible triggering events as such.

As noted above, this important element is addressed more directly in the CIP Version 5 Standards, which requires each responsible entity to have processes to identify Cyber Security Incidents.[76] Also as described above, in striving to keep abreast of and respond to new and evolving cyber and physical threats, participants recognize the importance of utilizing cyber security technical expertise and advanced tools. The team recognizes that

---

[75] Recognizing the benefits of the use of advanced resources and expertise, some participants employ extensive monitoring programs, while a few rely heavily on complex escalation procedures, advanced tools and third-party support. For the latter participants, the lack of a more streamlined process may introduce room for error.

[76] Reliability Standard CIP-008-5, Requirement R1, Table R1 Part 1.1 requires each responsible entity to have a process(es) to identify Cyber Security Incidents.

The Nuclear Regulatory Commission (NRC) recently adopted a cyber security event reporting rule which specifies, for nuclear licensees and licensee applicants, the kinds of cyber security events that must be reported and the time frame for reporting (from one hour to twenty-four hours depending on the type of event). For example, the NRC requires licensees to notify the NRC within one hour after discovery of a cyber attack that adversely impacted safety-related or important-to-safety functions, security functions, or emergency preparedness functions (including offsite communications), or that compromised support systems and equipment resulting in adverse impacts to safety, security, or emergency preparedness functions within the scope of 10 CFR § 73.54 (Protection of Digital Computer and Communication Systems and Networks). The rule also requires licensees to notify the NRC within four hours of a cyber attack that could have caused an adverse impact to the above, and defines the kinds of events that require notification within eight hours or twenty-four hours. *See Cyber Security Event Notifications,* NRC-2014-0036, 80 Fed. Reg. 67264-01 (Nov. 2, 2015).

CIP Version 5 does not specifically require the use of these; however, through entities' implementation of CIP Version 5, additional insight may be gained to aid in considering future changes to the Reliability Standards to address these important cyber security areas.

4. Recommendations

Use of technical expertise and advanced tools. The joint staff review team has concluded that cyber event monitoring and response would be greatly improved by expanding the use of cyber security technical expertise and advanced technical tools, and recommends that measures be taken (including considering changes to the Reliability Standards) to address the use of these tools to improve cyber event monitoring and response. In considering such measures, it may be appropriate to allow for some experience with CIP versions 5 and 6. In addition, the team recommends that such measures clarify that these advanced tools and resources should be employed in a manner that does not negate the benefits by making the cyber security event monitoring process more cumbersome or unnecessarily burdensome.

Require details on types of cyber security events that should trigger response and reporting. The joint staff review team also recommends that measures be taken (including considering changes to the Reliability Standards) that address the need for cyber security incident response plans to include details around the types of cyber events that should trigger a response (e.g., EMS or SCADA outage, communications network outage, etc.), and what types should be reported. While the team recognizes that CIP version 5 will require responsible entities to have processes to identify cyber security incidents, consideration should be given as to whether any additional clarification or improvements are needed once some experience is gained with CIP version 5.

D. Initial Event Response Actions

1. Summary

The goal of initial cyber event response analysis is to assess whether a given cyber alert or activity warrants further action. Initial event response analysis is a critical step in the response process. In its review of participants' initial event response actions, the joint staff review team examined: (1) triage, (2) bulk power system impact determination, (3) escalation methods and protocols, and (4) event data gathering and containment.

As described below, the joint staff review team did not identify any clarity or efficacy issues with the pertinent Reliability Standards with respect to initial event response actions. The review team recommends that entities consider use of hybrid systems that use a combination of both automation and human analysis as part of initial event analyses.

2. Review of Participants' Response Plans

a) Triage

Most participants use multiple levels or tiers of organizational response for cyber security events, often using a "triage" approach to determine whether further action is warranted and by whom. The triage approach generally includes implementing policies and procedures that address event classification, escalation, responsibilities for response, and reporting obligations.

Some participants have 24/7 dedicated cyber response teams (e.g., incident response teams) with expertise in identifying and tracking cyber threats across the enterprise, while others partner with an existing cyber security service or third-party vendor service. Most participants maintain response teams comprised of employees pulled from business units such as IT, networks, and EMS SCADA to form their incident response team as needed. The incident response team will initially analyze a cyber threat event to determine the degree of response required to address it. For example, if a recently discovered vulnerability identified by the United States Computer Emergency Readiness Team for a particular device (firmware, software version, and release) exists on a participant's system(s), that participant will require a level of response and triage from the incident response team that mitigates the risk and exposure to the entity and the bulk power system.

b) Bulk Power System Impact Determination

The initial steps of an event assessment are focused on the critical systems and locations affected by (or affecting) an event, and on determining the necessary expertise required to assist in mapping out the next steps. All of the review participants employ this approach for their initial response to potential events.

The participants each conduct an in-depth review to assess the potential for a given cyber threat or anomaly to impact the bulk-power system, which includes an analysis of anomalies detected through log inspections and evaluation. This review is typically performed through automation, but the majority of the review participants employ a hybrid system, using a combination of both automation and human analysis. Systems and tools used for this analysis include intrusion detection systems, intrusion prevention systems, inspecting system logs, networking traffic analyses, and other analytical tools. The analysis includes a review of available patches for systems and devices and, if a specific issue or vulnerability is being considered, specialized tools or processes may be used. The joint staff review team believes this hybrid approach to event assessment enhances the industry's ability to respond to cyber security incidents.

c) Escalation Protocols and Methods

All participants rely on a tiered approach to escalation protocols. In a tiered approach to incident response, the event is handled at the lowest tier possible so that the entity does not waste resources when an event does not warrant the full incident response team. One participant uses a four-tiered approach, in which attempts are made to contain and mitigate the event within each tier before moving on to the next higher tier. Various tools for mitigation, detection, monitoring, and forensics are used at each tier with more sophisticated tools used in the higher tiers.

d) Event Data Gathering and Containment

Some of the participants use an enterprise operations center[77] to help with their event analysis and containment of an event while others employ numerous processes and procedures within different departments or functional areas. As noted above, the designated response teams assess the potential impact on EMS or SCADA system availability and take steps to manage that impact.

To help limit the potential propagation of a given cyber threat and to allow for identification of threat sources, all participants use a number of internal security measures and practices. The majority of participants use full time employees where possible for positions that include some level of access to cyber systems. This practice is particularly important for sensitive positions as it limits the entity's exposure to outsider threats and the potential for further propagation of an event. Some of the participants will retain the records of an employee's access to critical cyber systems for up to one month following his or her departure. This record retention practice is mainly for documentation and maintaining a paper trail. Otherwise, supervisors update and revoke access privileges within five days of an employee's departure. One participant takes a different approach to monitoring systems and devices, monitoring every device that is on the network. This approach limits employee lists to a minimum and helps enable the detection of rogue devices that do not belong on the network.

3. Related Standards Assessment

While Reliability Standard CIP-008-3 does not dictate a particular form for initial event response and analysis, it does require applicable entities to have a plan that addresses

---

[77] An enterprise operations center is a group of dedicated employees that review and analyze network traffic data looking for anomalies and/or potential threats to the enterprise network.

"response actions" to cyber incidents, which necessarily includes some form of initial analysis and triage:

CIP-008-3:

R1. Cyber Security Incident Response Plan - The Responsible Entity shall develop and maintain a Cyber Security Incident response plan and implement the plan in response to Cyber Security Incidents. The Cyber Security Incident response plan shall address, at a minimum, the following:

...

R1.2. Response actions, including roles and responsibilities of Cyber Security Incident response teams, Cyber Security Incident handling procedures, and communication plans.

Though the requirements are broadly written and require little in the way of criteria defining an adequate cyber security incident response plan, the joint staff review team found that the participants' plans address, in detail, initial event response actions. All use a tiered approach for triage, and their plans include implementing policies and procedures that address event classification, escalation, responsibilities for response, and reporting obligations. The team's review of participants' plans did not reveal any concerns with the clarity and efficacy of the associated Reliability Standards for these areas, particularly given the approved changes to the standards that will become effective as part of the CIP Version 5 Standards.[78]

4. Observed Practices for Consideration

Some participants' plans include a hybrid system for determining bulk power system impact or threat from a cyber incident. The joint staff review team considers hybrid systems that use a combination of both automation and human analysis to be a beneficial practice for consideration by the industry.

---

[78] Reliability Standard CIP-008-5 includes several new provisions, including: Requirement R1.1, which requires one or more processes to identify, classify, and respond to Cyber Security Incidents; Requirement R1.3, which addresses the roles and responsibilities of Cyber Security Incident response groups or individuals, and Requirement R1.4, which addresses incident handling procedures for Cyber Security Incidents.

E. Recovery Planning

1. Summary

It is crucial that entities have effective recovery plans for critical cyber assets in response to events. The joint staff review team accordingly reviewed the participants' recovery plans and associated testing practices, by examining the stages and processes of participants' recovery plans.

The joint staff review team found that participants' recovery plans for critical cyber assets address the stages and processes of recovery planning included in Reliability Standard CIP-009-3. Participants' plans for recovery from events have well-established strategies for staffing, logistics, emergency facilities, and communications methods, described in detail below. However, the team found assumptions in some participants' recovery plans that could risk a timely recovery. The joint team recommends that measures be taken (including considering revisions to the Reliability Standards) to ensure that recovery plans do not include or implicitly rely on any major inventory assumptions (e.g., assumptions of hardware being available without measures to ensure availability) for critical cyber assets that could significantly affect prompt recovery of critical cyber assets. These measures would mitigate the potential risk of delayed recovery resulting from such assumptions.

2. Review of Participants' Recovery Plans

Participants' recovery plan scenarios are categorized by the severity of the actual event or an anticipated event such as severe weather. Participants' plans for actual and anticipated events have well-established strategies for staffing, logistics, and emergency facilities. Critical to their restoration and recovery efforts are reliable communication protocols and backup communication systems throughout their organizations, departments, business units, groups and personnel identified in the recovery plan. Some of the participants have incorporated related lessons learned into their response and recovery plans, with provisions for an extended loss of communications due to extended power outages, loss of telecom services (landlines, Voice Over Internet Protocol, and mobile), corporate e-mail services, etc.

All participants' response and recovery plans identify the key contact personnel (and backups) by name and department, and owners of the plans. The participants' recovery plans include the plan's objectives and goals at the highest level, and become more granular and specific by the classification of assets or primary business functions. As the recovery plans become more granular and specific by department and function, the plan specifics are to be implemented by designated top level department personnel. Changes to specific recovery plans for most participants require approvals from personnel responsible for that asset, who are generally department heads, and final approvals from the personnel responsible for the entire recovery plan.

Some participants' cyber asset recovery plans enlist corporate IT support and network operations center support because these groups often operate around-the-clock and overlap with many departments and systems, including critical assets, critical cyber assets, and non-critical cyber assets. Specialized groups and owners of a critical asset or critical cyber assets may have sole jurisdiction and ownership of their physical and cyber assets and will determine the level of response required and recovery procedures in their recovery plan.

Under the participants' recovery plans, the first personnel to respond to a given problem or event involving a critical cyber asset will start with troubleshooting their operational systems, business systems and supporting systems. As the cause of the problem becomes more evident, the group enlists other groups and individuals as needed and as identified in their response and recovery plan, including IT support, network support, and vendor support, for both software systems and hardware systems.

As better information becomes available, the scope of the actual and potential threat is assessed and initial response and recovery plans are activated. Participants indicated that it may require days, weeks, or longer to determine the actual root cause of a given cyber event and its impact on various assets types and systems. Specific hardware and software components and specific business functions affected may trigger escalation to a greater severity threat categorization or de-escalate into a lesser response category, as the affected resources may (or may not) be critical assets, critical cyber assets, or non-critical cyber assets that are otherwise important for operations and business functions. There may also be interdependence on vendor response and assistance for critical cyber asset EMS and network systems and supporting business systems.

Participants' plans include procedures for varying levels of loss or degradation of critical assets and critical cyber assets and supporting non-critical cyber asset systems, software, and hardware components. The recovery plans detail a number of response levels for potential events ranging from total physical loss of a critical asset facility, operational loss of a critical cyber asset facility, EMS control center and SCADA loss or degradation, loss of business processes, to the loss of server or network components, including switches, routers, firewalls, and remote terminal units.

The participants' plans vary regarding the back-up computer hardware or other equipment inventory assumptions used for their asset recovery methods. Some participants rely on vendors in part for recovery of their critical cyber assets, but do not necessarily take into account that a particular vendor may need to supply equipment to multiple entities during a large scale event, or otherwise take into account interdependent or common-mode failure scenarios.

3. Related Standards Assessment

Reliability Standard CIP-009-3 requires applicable entities to have processes and procedures in place to recover, backup, and restore critical cyber assets, as follows:

CIP-009-3:

R1. Recovery Plans – The Responsible Entity shall create and annually review recovery plan(s) for Critical Cyber Assets. The recovery plan(s) shall address at a minimum the following:

R1.1. Specify the required actions in response to events or conditions of varying duration and severity that would activate the recovery plan(s).

R1.2. Define the roles and responsibilities of responders.

...

R4. Backup and Restore — The recovery plan(s) shall include processes and procedures for the backup and storage of information required to successfully restore Critical Cyber Assets. For example, backups may include spare electronic components or equipment, written documentation of configuration settings, tape backup, etc.

As noted above, the participants' plans vary regarding the back-up computer hardware or other equipment inventory assumptions used for their asset recovery methods, and the above Reliability Standard requirements allow significant variance in how an entity can recover from a cyber event. For example, some participants rely on vendors, in part, for recovery of their critical cyber assets. In a large scale event, a particular vendor may need to supply equipment to multiple entities. The assumptions may not take into account interdependent or common-mode failure scenarios, which can create the need for multiple entities to recover multiple critical cyber assets from the same vendor(s). Other assumptions may compound this risk, including assumptions regarding availability of spare components from backup facilities or offices that may not be available when needed during the event, and assumptions regarding telecommunication services (cellular and landlines) and e-mail services, which may not be available when needed. Reliance on the assumption that vendors will have the equipment available without some contractual or other guarantee, or otherwise maintaining on-site inventory of vital hardware, could result in a significant delay in asset recovery.

As industry moves toward a more virtual environment, having an effective action plan to restore critical cyber assets is essential. From its review of the participants' critical cyber asset recovery plans, the joint staff review team found that the Reliability Standards need to be clarified in order to effectively support reliability by requiring entities to eliminate any major assumptions incorporated in their recovery plans and procedures which could significantly affect prompt recovery of critical cyber assets. The joint staff review team realizes that it is not possible to eliminate all equipment inventory or availability assumptions, but reliance upon vendors for inventory management may impose significant risk unless availability and timely delivery of replacement hardware is written into contracts. Other such assumptions should also be avoided in order to improve the entity's response to events.

4. Recommendations

Recovery plan inventory assumptions risk. The joint staff review team recommends that measures be taken (including considering changes to the Reliability Standards) to eliminate, to the extent possible, "inventory assumptions" in cyber asset recovery plans that could significantly affect prompt recovery of critical cyber assets. For example, entities may assume that hardware from external sources or other third-party vendor support needed for recovery of critical cyber assets will be available, without necessarily having measures to ensure availability. Likewise, entities may not consider interdependent or common-mode failure scenarios, which can create the need to recover multiple critical cyber assets concurrently from the same vendors.

F. Review and Verification of Incident Response and Recovery Plans

1. Summary

The joint staff review team examined how participants verify the viability of their cyber incident response and critical cyber asset recovery plans, including their periodic reviews of the plans and testing of plans and associated facilities and resources.

As described below, the joint staff review team found that the participants' response and recovery plans address in detail confirmation of the viability of the plans by testing the plan facilities and resources. The joint staff review team nevertheless concludes, for reasons discussed below, that all applicable entities should consider having an independent third party review their cyber incident response and critical cyber asset recovery plans to ensure they are thorough and reliable.

The joint staff review team also observed certain practices that appear to enhance the participants' cyber incident response and recovery planning and testing, and recommends that applicable industry entities consider implementing these approaches in their own recovery plans. Observations by the joint staff review team are detailed below.

2. Review of Participants' Response and Recovery Plans

a) Periodic Reviews of Response and Recovery Plans

Most participants review their cyber security incident response and critical cyber asset recovery plans in-house, with few participants undertaking independent reviews. Of the few participants that use third-party reviewers, only two have multiple independent companies review their plans. One participant, in addition to using independent reviewers, also performs a bi-annual internal CIP sufficiency review to better ensure that it can handle potential cyber security incidents. Those participants who have independent reviews expressed that while these reviews may not in all cases be superior to an in-house review, an independent review of a recovery plan can provide an unbiased perspective and validation of the plan. In addition, these participants indicated that an independent review can provide added value and expertise, and incorporate industry best practices, particularly if the reviewer has the capability and experience of reviewing many industry-wide plans, information and data (i.e., can provide a more comprehensive perspective).

b) Testing of Response and Recovery Plans

The ways in which participants test their response and recovery plans are specific to each participant. Almost all participants use real world events that have either occurred to the participant or to other entities in setting up their testing or exercise scenarios. As described further below in the Drills and Training Exercises section, participants' incident response and recovery plan testing predominantly consists of tabletop paper drills.

The increasing sophistication of cyber security events is driving entities to scrutinize their recovery plans, and evaluate through testing and exercises whether existing recovery resources for critical cyber assets are adequate. Effective recovery plans consider and plan for both small and large impact scenarios.

c) Testing of Recovery Resources

Participants generally test their ability to recover critical cyber assets and associated recovery resources during their back-up control center drills. Some participants' tests are limited to staff traveling to the backup center and powering up the backup resources. Other participants conduct drills for a complete site loss of cyber assets, such as complete loss of a control center or forced site evacuation leading to the transfer of operations to an alternate control center, with some operating for an extended length of time (e.g., greater than 24 hours). Through discussions with participants, the joint staff review team found that exercises involving the actual transfer of control center operations to an alternate site for a period of time are more realistic tests of the functionality of recovery resources than a simple power up of backup control center operations. The drill or actual evacuation event can and often does reveal unknown issues or problems at the alternate site's

SCADA EMS system. Moreover, by running exercises from an alternate control center system for an extended length of time, entities can better evaluate support issues and needs at the alternate site, including the logistics of extended site transfers and the peripheral system needs for running operations at the alternate control center for extended periods of time.

In addition to high impact scenarios, participants often conduct recovery exercises of event scenarios of lesser impact, both in size and scope. These lesser impact scenarios are important, because from a risk perspective, such events are more likely to occur than large catastrophic events. Recovery exercises and scenarios may include a single system loss, network system interruptions, hardware server loss, or loss of a functional system component that can disrupt normal operations and critical business systems.

The joint staff review team found that some participants employ virtualization software to facilitate recovery. Virtualization software products and virtualization technologies can aid servers, workstations and other cyber assets in recovery from unrecoverable hardware disk crashes, corrupted software systems or components, and workstation terminals. In addition to tape backups, a few participants are also using a type of virtual backup referred to as a "golden image" for their critical servers and software components and for network devices like switches and routers.[79] A golden image can significantly reduce the restoration time required to build from a new hardware device. Device restoration is much faster from an imaged software and file system than restoration from files on disk or tape drives.

Discussions with the review participants revealed that conducting certain recovery tests on a live production system or the backup or alternate system is not advisable. This is due to the fact that additional risks may be introduced into the recovery system or facility (e.g., EMS server replacement) that could jeopardize the functionality of the production system or the backup system. For instance, unknown problems with the recovery device could propagate to other production system critical cyber assets (e.g., EMS/SCADA) and prevent the original device from being restored. To this end, participants have installed fully representative test systems for their CCAs and EMS SCADA control systems. Using such a test system, often referred to as a quality assurance system, an entity can

---

[79] In network virtualization, a "golden image" is an archetypal version of a cloned disk that can be used as a template for various kinds of virtual network hardware. The golden image is a master image from which copies can be used to provide a consistent process for creating a disk image. The use of golden images in cloud computing solutions can provide consistency for rebuilding hard drives for recovery or pushing out updates across various virtual machine desktops.

perform full functional testing and restoration and recovery exercises on a system identical to its production environment. In addition, the quality assurance system can also provide spare components in an emergency situation.

Participants expressed that they are continually improving and increasing the availability and redundancy of the EMS control center's systems for operations and business continuity. Advances in both computer EMS software systems and communications between primary control center and alternate control center (such as hot standby and heartbeat) are increasing operational availability and lessening system recovery times.[80] Advances in restoration techniques, virtualization software techniques, and disc imaging are decreasing hardware restoration times compared to restoration from disk and/or tape backup media.[81]

The joint staff review team observed that many participants have spare hardware servers available for testing the recovery of failed servers, switches and firewall components. Some participants have pre-configured hardware servers available as spares. Restoration from a virtualized image in recovery has reduced recovery times significantly compared to restoration from disk or tape drive. However, the actual recovery media entities use, whether virtualized image, disk or tape backup, depends on their particular systems, the amount of data being recovered, and the cost of the solution employed. Participants mentioned that, for emergency situations, having the option of using spare components from identical and redundant systems can shorten restoration and recovery time. EMS servers can be imaged from the alternate EMS system servers or a representative test system such as the quality assurance system.

---

[80] An EMS 'Hot Standby' is a primary EMS and a fully functionally redundant backup EMS system, configured in a constant state of readiness for a quick and seamless takeover if the currently configured primary EMS system's functionality deteriorates or becomes unavailable. In the 'heartbeat' communication process, the currently configured backup EMS continually monitors the health of the current primary EMS's critical processes and functionality. If the currently configured backup EMS (in hot-standby mode) detects a signal or flag that the health and functionality of the Primary EMS is lost or deteriorating, it will start the process of taking over as the primary EMS with complete SCADA functionality.

[81] A disk image is a copy of the entire contents of a storage device, such as a hard drive, DVD, or CD. The disk image represents the content exactly as it is on the original storage device, including both data and structure information. A disk image of a hard drive may be saved as a virtual hard disk.

3. Related Standards Assessment

Reliability Standard CIP-008-3 includes requirements pertaining to review and testing of incident response plans, and Reliability Standard CIP-009-3 includes a requirement for testing backup media essential to critical cyber asset recovery, as follows:

CIP-008-3:

R1. Cyber Security Incident Response Plan – The Responsible Entity shall develop and maintain a Cyber Security Incident response plan and implement the plan in response to Cyber Security Incidents. The Cyber Security Incident response plan shall address, at a minimum, the following:

...

R1.5. Process for ensuring that the Cyber Security Incident response plan is reviewed at least annually.

R1.6. Process for ensuring the Cyber Security Incident response plan is tested at least annually. A test of the Cyber Security Incident response plan can range from a paper drill, to a full operational exercise, to the response to an actual incident.

CIP-009-3:

R1. Recovery Plans – The Responsible Entity shall create and annually review recovery plan(s) for Critical Cyber Assets. The recovery plan(s) shall address at a minimum the following:

R1.1. Specify the required actions in response to events or conditions of varying duration and severity that would activate the recovery plan(s).

R1.2. Define the roles and responsibilities of responders.

R5. Testing Backup Media – Information essential to recovery that is stored on backup media shall be tested at least annually to ensure that the information is available. Testing can be completed off site.

Consistent with these requirements, the participants' plans address the need for periodic review and testing of cyber incident response and critical cyber asset recovery plans. However, while the Reliability Standards currently require annual review of plans and approval by a senior manager or delegate, they do not require plan review by an independent party. As noted above, the joint staff review team concluded from review of the plans and discussion with the participants that independent review of policies, processes, and technical mechanisms included in cyber incident response and critical cyber asset recovery plans can provide an unbiased, more comprehensive perspective. This independent review approach is similar in purpose to other independent reviews required under the Reliability Standards, including reliability coordinator review of a transmission operator's restoration plan under Reliability Standard EOP-005-2. The joint staff review team notes that the projected entity resources needed for conducting an independent review are expected to be similar to those needed for an in-house review or an audit. The team also notes that many of the participants are already employing third-party reviews for compliance review, and that some use third parties for technical best practice reviews. Moreover, many of the participants have established close working relationships with third parties to help stay abreast of developments on cyber threats and prevention approaches, including information received from other electrical sector entities and local government agencies, which information is used as part of these participants' internal reviews.

Notably, under new Reliability Standard CIP-014-1 (Physical Security), applicable entities are required to have an unaffiliated third party verify their required risk assessments identifying critical transmission stations and substations, i.e., those that if rendered inoperable or damaged could result in widespread instability, uncontrolled separation, or cascading within an interconnection. Given that cyber environments are similarly unique, numerous and complex, and create a kaleidoscope of threat and vulnerabilities that will demand unique responses, the joint staff review team believes that the industry as a whole could benefit from independent review of responsible entities' cyber incident response and critical cyber asset recovery plans.

4. Recommendations

Independent review of cyber security response and recovery plans. The joint staff review team recommends that recovery plans for critical cyber assets and cyber security incident response plans be reviewed by an independent authority or third party for the purpose of supporting thoroughness and technical reliability, using a trusted or otherwise qualified third party to ensure a proper security review.

5. Observed Practices for Consideration

In evaluating participants' response and recovery planning reviews and testing, the joint staff review team observed the following practices and recommends consideration of these by other relevant entities:

- Some participants perform exercises or drills that involve the actual transfer of control center operations to an alternate site for a period of time, to test the functionality of the recovery resources. This practice provides a more realistic test of response and recovery readiness as compared to only powering up the backup resources to test their functionality. The drill or actual evacuation event and verification of functionality of recovery resources can and often does reveal unknown issues or problems at the alternate site's SCADA EMS system.

- Some participants perform exercises or drills that require failover to their backup control centers for drills for more than just a few hours. This practice tests whether support systems and other support resources that are needed to run from the backup are readily available and remain available, and allows personnel to become familiar with running from the backup center instead of the primary center. Entities running exercises from their alternate control center system for an extended length of time can better assess support issues and needs, such as the logistics of extended site transfers and the peripheral systems needed for running operations at the alternate control center for extended periods of time.

- Some participants use a type of virtual backup referred to as a golden image for their critical servers and software components and for network devices like switches and routers. This practice can significantly reduce the restoration time required to rebuild and implement hardware that replaces affected hardware during the recovery process, versus utilization of disks or tape storage. Further, reliance upon identical assets used in support environments (e.g., off-line development system assets) to recover the EMS/SCADA production environment can have some drawbacks, due to less-frequent usage and/or software updates.

G. Drills and Training Exercises

1. Summary

The joint staff review team examined how participants' cyber incident response and critical cyber asset recovery plans address drills and training exercises. The team found that the participants' plans require periodic testing or exercising of the response and recovery plans, including testing of backup communications systems and other backup systems used in the plans, typically exercised in the form of tabletop exercises or actual drills. The joint staff review team found that participation in full operational exercises and other more complex simulations provides greater insight into the viability of a given cyber response and recovery plan, and appears to be necessary to develop robust recovery and response plans. Further, participants that have participated in regional tests/exercises which incorporated interdependencies have developed more robust recovery and restoration plans than those that only perform tabletop exercises. The joint staff review team recommends that entities consider, as a best practice, conducting full operational exercises or other more complex simulations of their cyber incident response and critical cyber asset recovery plans, including testing for interdependencies and other vulnerabilities.[82]

2. Review of Participants' Response and Recovery Plans

a) Regional Cyber and Physical Recovery Exercises

One participant held a voluntary exercise simulating a focused cyber and physical attack on the functional entities in its footprint. Another participant engaged in an exercise that the Federal Emergency Management Agency (FEMA) performed in its region, focused more on severe natural disaster conditions (i.e., earthquake, mudslides, tornadoes, hurricanes, etc.), but which extended to recovery of critical cyber assets. This specific exercise was the first conducted by FEMA, but with the success of this exercise, the participant indicated that it is expected to become an annual exercise.

Another review participant enrolls in its reliability coordinator's restoration drill, which involves every entity within the reliability coordinator's footprint. The participant then performs its own large scale exercise, including hypothetical toxic fumes with evacuation of facilities and a complete loss of communications.

In addition, some of the larger participants have held a wide area testing scenario for their footprint, and have invited neighboring utilities to participate in these events. The joint staff review team observes that conducting wide area testing scenarios is a worthwhile practice that the industry should consider, especially for entities with large footprints.

While the three largest participants in the review conduct or participate in regional exercises which involve several entities arranging simulations of cyber or physical attacks, the remaining participants do not perform larger scale exercises that include their neighbors. However, some participants are made aware of their neighbors' exercises so that they can determine how to coordinate with them.

All participants' system operators participate in semi-annual training, in which they review processes and approaches to responding to larger-scale events that may include cyber attacks. In most cases, the training also provides points of contact for outside

[82] The team also notes that testing of operating plans to address loss of control center functionality, conducted pursuant to EOP-008-1, Requirement R7, may be designed to include aspects of testing of and training on entities' required cyber response and recovery plans, thereby providing the necessary information on interdependencies and vulnerabilities.

agencies and groups that would be involved with large scale or severe events. While some participants have well-established relationships and processes for interfacing with outside groups, others, with less well-defined relationships, have determined that better communication with outside participants is necessary. Many participants also include corporate Incident Response Team members in training exercises on simulated cyber or physical attacks, including training on coordination with outside entities. This practice is especially helpful when a liaison is needed with governmental entities.

Telecommunication infrastructure availability is a particular concern for some participants, as it is critical to their cyber incident response or critical cyber asset recovery plans. In order to test the viability of telecommunications, and ensure personnel readiness for cyber or physical attacks, participants deploy telecommunications support personnel to critical locations during all exercises. Some participants are also considering use of other means of communication, such as mobile radios and emergency-only email portals. In addition, some participants send employees to visit and man their telecom operations center during normal operations to allow employees to test their ability to reach the centers in emergency situations.

3. Related Standards Assessment

Reliability Standards CIP-008-3 and CIP-009-3 include specific requirements addressing drills and exercises to test the viability of a responsible entity's cyber incident response and critical cyber asset recovery plans. The relevant requirements for each standard are as follows:

CIP-008-3:

R1. Cyber Security Incident Response Plan – The Responsible Entity shall develop and maintain a Cyber Security Incident response plan and implement the plan in response to Cyber Security Incidents. The Cyber Security Incident response plan shall address, at a minimum, the following: …

R1.2. Response actions, including roles and responsibilities of Cyber Security Incident response teams, Cyber Security Incident handling procedures, and communication plans. …

R1.6. Process for ensuring the Cyber Security Incident response plan is tested at least annually. A test of the Cyber Security Incident response plan can range from a paper drill, to a full operational exercise, to the response to an actual incident.

CIP-009-3:

R2. Exercises — The recovery plan(s) shall be exercised at least annually. An exercise of the recovery plan(s) can range from a paper drill, to a full operational exercise, to recovery from an actual incident. …

R5. Testing Backup Media — Information essential to recovery that is stored on backup media shall be tested at least annually to ensure that the information is available. Testing can be completed off site.

Requirement R1.6 of CIP-008-3 requires applicable entities to test their cyber security incident response plan at least annually. Similarly, requirement R2 of CIP-009-3 requires applicable entities to conduct an annual exercise of their critical cyber asset recovery plan. However, these tests or exercises can take the form of a tabletop exercise or paper drill, which may not address the possible circumstances associated with an actual crisis. Tabletop exercises alone do not, in most cases, identify the potential flaws or omissions in the response and recovery plans being tested. By contrast, the joint staff review team found that participation in full operational exercises and other more complex simulation drills provides much greater insight into the viability of a given cyber response and recovery plan, and appears to be necessary to develop robust recovery and response plans.

Participants who took part in regional exercises reported that the exercises were beneficial, resulted in increased situational awareness, and have often led to changes in existing recovery plans and strategies. Each participant stated that engagement in one or more of the exercises and simulations increased their knowledge and awareness of the challenges in responding to a cyber security incident or a cyber or physical attack, resulting in some form of improvement to their recovery plan, notification process, departmental procedures, or communication procedures. In contrast, the team found that a tabletop exercise has limited value and typically does not involve multiple, simultaneous events or issues escalating in severity and duration. Moreover, the team found that tabletop exercises generally do not provide the same opportunity to identify areas for improvement as compared to more complex simulations, and therefore may not result in improvements to the cyber incident and critical cyber asset recovery plans.[83]

---

[83] Notably, under Reliability Standard EOP-005-2, Requirement R6, an applicable entity must verify that its system restoration plan accomplishes its intended function through analysis of actual events, steady state and dynamic simulations, or other testing. The team found this required verification, along with reliability coordinator review and approval of plans, to be an important element in ensuring that entities develop adequately detailed and thorough system restoration plans.

4. Recommendations

Exercises of response and recovery plans using paper drills. The joint staff review team observed that participation in full operational exercises and other more complex simulations provides greater insight into the viability of a given cyber response and recovery plan, and believes that participation in such exercises by the industry is valuable for developing robust recovery and response plans. The joint staff review team recommends that applicable entities participate in exercise scenarios and simulations structured to gain insight into the viability of cyber response and recovery plans (*i.e.* beyond paper drills and tabletop exercise), including testing for interdependencies and other vulnerabilities.

H. Improving Cyber Security Response and Recovery Plans Based on Actual Events and Other Feedback

1. Summary

The joint staff review team reviewed how participants incorporate feedback and lessons learned from actual cyber security and critical cyber asset recovery events, as well as feedback from other sources regarding the viability of the plans. The joint staff review team found that participants have varying levels of specificity in their processes and procedures for implementing improvements to their cyber security response and recovery plans, including improvements based on lessons learned from actual events. The joint staff review team recommends that further study be conducted about actions being taken by entities when the testing or implementation of their response and recovery plans during actual events reveals the need or opportunity for improvements to the plan.[84] In addition, the study should examine and identify best practices with regard to the types of plan improvements made from entities' analyses of actual cyber events and/or testing. Such information could reveal the need or opportunity for improvements to other entities' response and recovery plans and be a valuable component of a continuous improvement process.

---

[84] The joint staff review team recognizes that CIP version 5 includes requirements for testing and updating cyber response and recovery plans, but the study could provide additional insight as to how these requirements are working and whether they might be improved.

2. Review of Participants' Response and Recovery Plans

a) Actual Cyber Security Response Events

Most participants that the joint team interviewed were fortunate not to have had an actual cyber security incident that impacted their EMS and SCADA system operations. Some participants have never experienced a cyber threat event that included declaring a cyber security incident involving their critical cyber assets. The joint staff review team believes it is especially important to prepare for cyber security events and incidents by exercising cyber security incident scenarios and participating in drills and exercises that test response and recovery plans and procedures. In this manner, feedback from implementing the plans can drive continuous improvement. Participants indicated, and the joint team agrees, that it is far better to find a flaw in the plans through testing or drills than to discover the issue during an actual event.

b) Actual Critical Cyber Asset Recovery Events

Although not precipitated by cyber threat events, some participants have experienced actual events requiring implementation of their critical cyber asset recovery plans. One of the more common events leading to participants' use of such a recovery plan involves the partial or entire loss of EMS and SCADA systems, which are typically classified as critical cyber assets since they are critical components of bulk power system operations.[85]

Some participants have experienced a complete site loss of their EMS SCADA systems due to extreme weather events (e.g., tornado, fire, floods).[86] Most participants indicated that NERC's Lessons Learned documents analyzing the many EMS and SCADA recovery events have been helpful in improving their critical cyber asset recovery plans,

---

[85] EMS systems, SCADA functions, associated hardware and software, networks, communication systems and supporting systems are a large part of the critical cyber assets that must be addressed in an applicable entity's recovery plan and its objective of restoring control center bulk power system operations.

[86] Response and recovery plans typically include scenarios that address varying levels of loss and interruption of the EMS SCADA system, along with recovery plans and procedures for the mobilization of personnel and activation of alternate control centers. Major disruptions to an EMS and control center operations include loss and unavailability of the EMS system processes, server hardware, or communications, and network availability issues.

by revealing, among other things, interdependencies between cyber assets and systems that were not previously known.[87]

c) Process for Improving Plans

Participants' cyber security response and recovery plans include varying levels of specificity in their processes and procedures for implementing improvements to those plans, including improvements based on lessons learned or information gained during actual events or during testing and drills. Some participants appear to incorporate a feedback loop process used to assess, critique, and direct improvements and changes in a cyber response or recovery plan's procedures and methods based on testing, actual events, or other new cyber threat information.[88]

Participants shared examples of improvements made to recovery plans from exercises, actual events, or new information. A common area for improving recovery plans is in communication processes and methods during emergency conditions. In the event that corporate email communications are interrupted, a separate private emergency email system can be used. In the event a loss of a telecom carriers' Voice over Internet Protocol or mobile communication, an emergency satellite phone system could be implemented. Among other things found from exercising their plans, participants have implemented new procedures and processes for improving formal notification channels or improving coordination efforts with their neighbor entities and Authoritative Agencies.

All participants allow for feedback from all entities involved in a drill or exercise to make suggestions and recommendations to their response and recovery plan. Some participants stated that in their feedback loop (*i.e.*, the process used to evaluate performance of their cyber response and recovery plans), certain modifications, improvements, and lessons learned gained during drills and exercises may not rise to the level of a significant change, and therefore may not require modification to their cyber resource or cyber recovery plan. In addition, changes and upgrades in equipment and technology may

---

[87] NERC's Events Analysis program includes a process for developing and issuing "Lessons Learned" documents, intended to ensure the timely dissemination of actionable lessons learned from significant bulk power system events. *See* NERC's Lessons Learned website at http://www.nerc.com/pa/rrm/ea/Pages/Lessons-Learned.aspx.

[88] The joint staff review team found that implementation of a feedback loop can help to correct a plan's procedural mishaps, performance issues in the notification process, communications, and recovery procedures.

require changes to specific recovery procedures and techniques for asset recovery, but will not necessarily result in a change in the overall response and recovery plan.

Another significant feedback area discussed by participants is staying current and active with vendor user support groups and partner relationships with entities that have similar systems. Participants indicated that vendor upgrades and fixes to hardware, software systems, and firmware are not always effectively communicated, or the impact of not implementing the upgrade to systems is not clearly understood. The joint staff review team believes that a feedback process that is part of the overall cyber response and recovery plan can allow for continuous improvements, aid in greater situational awareness and readiness, enhance training programs, shorten response times for cyber events, and fine tune recovery strategies and procedures for such events. Following implementation of response and recovery plans due to an actual event, affected entities should conduct a top to bottom analysis of the event, including identifying any lessons learned that could result in improvements to their (or others') cyber security response or critical cyber asset recovery plans. This analysis should include a determination of whether the actual performance of the response and recovery plan during the test or event indicates that modifications, changes to procedures, and additions and changes to the current response and recovery plans and procedures are needed.

3. Related Standards Assessment

Reliability Standards CIP-008-3 and CIP-009-3 include requirements relating to updates to the cyber security response plan and the critical cyber asset recovery plan, and, as to the latter plan, requiring updates to reflect any changes or lessons learned as a result of an exercise or an actual incident.

CIP-008-3:

R1. Cyber Security Incident Response Plan – The Responsible Entity shall develop and maintain a Cyber Security Incident response plan and implement the plan in response to Cyber Security Incidents. The Cyber Security Incident response plan shall address, at a minimum, the following: …

R1.4. Process for updating the Cyber Security Incident response plan within thirty calendar days of any changes. …

R1.6. Process for ensuring the Cyber Security Incident response plan is tested at least annually. A test of the Cyber Security Incident response plan can range from a paper drill, to a full operational exercise, to the response to an actual incident.

CIP-009-3:

R2. Exercises — The recovery plan(s) shall be exercised at least annually. An exercise of the recovery plan(s) can range from a paper drill, to a full operational exercise, to recovery from an actual incident.

R3. Change Control — Recovery plan(s) shall be updated to reflect any changes or lessons learned as a result of an exercise or the recovery from an actual incident. Updates shall be communicated to personnel responsible for the activation and implementation of the recovery plan(s) within thirty calendar days of the change being completed.

The joint staff review team found that participants have varying levels of specificity in their processes and procedures for implementing improvements to their existing plans, including improvements and/or updates based on exercises or actual events. The joint staff review team examined the requirements regarding updating plans and actual events, and concluded that currently none require applicable entities to employ a feedback loop or continuous improvement process to ensure that cyber security response and recovery plans are up to date. Although the Reliability Standards do require updating critical cyber asset recovery plans based on lessons learned during testing or during an actual event, these updates tend to be administrative changes in nature (e.g., updating documentation such as personnel contact information) versus including the identification of more substantive plan improvements, most likely due to the fact that drills are typically tabletop exercises. Moreover, the Reliability Standards do not require updating the cyber incident response plan or the critical cyber asset recovery plan whenever new information is acquired that could improve the plans. For recovery and response plans to be effective, they must mimic real life scenarios, be applied to production-like systems, and improve with the ever changing technology.[89]

4. Recommendations

Gain further understanding of response and recovery plan updating following testing or actual cyber events. The joint staff review team recommends that a study be conducted to better understand the associated plan improvements made by entities where testing or an actual cyber event reveals the need or opportunity for improvements to a

[89] The team notes that CIP-008-5 Requirement R2, Part 2.2 requires a responsible entity to document any deviations from the written plan that occurred during a response to an incident or an exercise, but does not require an action plan to complete a feedback loop in response to a deviation.

response and recovery plan. This study would support a better understanding of the effectiveness and existence of continuous improvement processes. In addition, the study should examine and identify best practices with regard to the types of plan improvements made from entities' analyses of actual cyber events and/or testing. Such information could reveal the need or opportunity for improvements to other entities' response and recovery plans and be a valuable component of a continuous improvement process.

VI. Appendix 1– Joint Staff Review Team

Federal Energy Regulatory Commission:
Daniel Bogle
Kenneth Githens
Norris Henderson
David Huff
Gilbert Lowe
Raymond Orocco-John
Thomas Reina
Judith Sciullo
Michelle Veloso

North American Electric Reliability Corporation:
Stephen Crutchfield
Tom Hofstetter
Robert Kenyon
Darrell Moore
Katherine Street
Jim Stuart

Northeast Power Coordinating Council, Inc.:
John J. Mosier
Ralph Rufrano

ReliabilityFirst Corporation:
John Idzior
Jeffrey Mitchell

SERC Reliability Corporation:
Steve Corbin
David Greene
Bill Peterson

Western Electricity Coordinating Council:
Darren Nielsen
Tim Reynolds

VII. Appendix 2 –Request Letter for Participation in Reliability Assessment

Request for Participation in Reliability Assessment

Commission staff, in collaboration with NERC and the Regional Entities, is initiating a voluntary review of recovery and restoration plans for selected registered entities. The purpose of this joint staff review is to assess and verify the electric utility industry's bulk power system recovery and restoration planning, and to test the efficacy of the relevant Reliability Standards in achieving or maintaining reliability. The joint staff review is focused on supporting entities in ensuring reliable restoration from reliability events and reviewing the adequacy of the Reliability Standards; it is not a compliance and enforcement initiative.

Recent reliability events, including weather-driven events (e.g., Superstorm Sandy, February 2011 Southwest cold weather rolling blackouts), bulk power system disturbances (e.g., September 2011 Arizona-Southern California Blackout, 2008 Florida Blackout, 2003 Northeast Blackout) and possible cyber/physical attacks have highlighted the potential to cause widespread adverse effects on the bulk power system. Effective system recovery and restoration plans are essential to facilitate a quick and orderly recovery in the aftermath of such events.

The primary objective of this joint staff review is to assess entities' plans for restoration and recovery, and verify how the Reliability Standards support them. To accomplish this objective, the joint staff review will:

- Gather information via outreach with a representative sample of selected entities with significant bulk power system responsibilities.
- Understand the overall state of restoration plans by comparing and contrasting their content, scope and interrelationships.
- Assess the clarity of the Reliability Standards in supporting the adequacy and efficacy of restoration and recovery plans.
- Identify good industry practices or make recommendations to ensure that effective restoration and recovery plans are in place to support reliability.

As an entity with bulk power system significance and broad interrelationships that may impact restoration planning, we are requesting [ENTITY]'s participation in this review. Additionally, other registered entities with interrelated reliability functions that impact, or are impacted by, [ENTITY]'s restoration plan may also be asked to participate in order to achieve comprehensive review of the wider area restoration capabilities.

The focus on the recovery and restoration plan review will be based on the reliability intent of three Reliability Standards:

EOP-005-2 System Restoration Plans from Blackstart Resources[1]
CIP-008-3 Cyber Security — Incident Reporting and Response Planning
CIP-009-3 Cyber Security — Recovery Plans for Critical Cyber Assets

Specifically, documents and information to be requested during the entity outreach, depending on their applicable functions, will include:

- Reliability Coordinator approved restoration plan
- Procedures for deploying blackstart resources
- Selected results of the most recent analysis of actual events, steady state and dynamic simulations, and testing that the restoration plan accomplishes its intended function, including any restoration strategies used to facilitate restoration for recent disturbances or the deployment of blackstart resources
- Existing notes or recommendations recorded as a result of the most recent annual exercise or from an actual incident. Also, any Reliability Coordinator feedback or analysis of last year's system restoration drills, exercises or simulations, as dictated by the particular scope of the drills, exercises, or simulations that were conducted [2]

---

[1] The assessment of EOP-005 will also consider the NERC report "Standards Independent Experts Review Project; An Independent Review by Industry Experts." Located at and accessed April 1, 2014: http://www.nerc.com/pa/Stand/Standards%20Development%20Plan%20Library/Standards_Independent_Experts_Review_Project_Report.pdf

[2] The provided documents may be informative on how other activities required by the above or related Reliability Standards are accomplished (e.g. EOP-006-2 – System Restoration Coordination). In some cases, other information as it relates to the above or

- Cyber Security Incident Response Plan
- Recovery Plan(s) for Critical Cyber Assets

The joint review will also assess entities' reports or recommendations from major events to understand the effectiveness of their recovery and restoration plans following an actual implementation. Reports developed after actual events like Hurricane Sandy, the September 2011 Arizona-Southern California Blackout, the Derecho storms in the Midwest and Mid Atlantic in 2012, and the 2014 Polar Vortex are also requested in order to better put response, recovery and restoration plans into the context of overall reliability efforts. The joint staff review will also use any public or private reports that have already been produced in these areas.

In addition to the information specified above, entities are encouraged to provide any further information or documents that may be helpful in explaining their recovery and restoration planning.

This collaborative assessment by the Commission, NERC and the Regional Entities is an important step in protecting reliability by gauging the electric utility industry's level of preparation for a major event and the ability to recover quickly and efficiently. In anticipation of [ENTITY]'s participation, we thank you and will work closely with you to ensure this project is conducted as a partnership with minimal disruption to your organization. I or my staff will call you at your earliest convenience to provide greater detail and answer any questions or concerns that you may have about this joint staff review.

---

other Standards may be requested later, as needed, in order to have a complete understanding of the applicable entity's restoration and recovery processes.

VIII. Appendix 3 – Standards and Requirements Assessed

EOP-005-2 – System Restoration from Blackstart Resources

In accordance with the scope of the review, Requirements assessed included the restoration plan-related requirements, as well as any requirements that support how the applicable entities test the effectiveness of their plans. These Requirements are listed below.

Reliability Coordinator-Approved Restoration Plan:[1]

R1. Each Transmission Operator shall have a restoration plan approved by its Reliability Coordinator. The restoration plan shall allow for restoring the Transmission Operator's System following a Disturbance in which one or more areas of the Bulk Electric System (BES) shuts down and the use of Blackstart Resources is required to restore the shut down area to service, to a state whereby the choice of the next Load to be restored is not driven by the need to control frequency or voltage regardless of whether the Blackstart Resource is located within the Transmission Operator's System. The restoration plan shall include:

R1.1. Strategies for system restoration that are coordinated with the Reliability Coordinator's high level strategy for restoring the Interconnection.

R1.2. A description of how all Agreements or mutually agreed upon procedures or protocols for off-site power requirements of nuclear power plants, including priority of restoration, will be fulfilled during System restoration.

R1.3. Procedures for restoring interconnections with other Transmission Operators under the direction of the Reliability Coordinator.

---

[1] See Appendix 2 - Request Letter for Participation in Reliability Assessment, which provides the scope of review.

R1.4. Identification of each Blackstart Resource and its characteristics including but not limited to the following: the name of the Blackstart Resource, location, megawatt and megavar capacity, and type of unit.

R1.5. Identification of Cranking Paths and initial switching requirements between each Blackstart Resource and the unit(s) to be started.

R1.6. Identification of acceptable operating voltage and frequency limits during restoration.

R1.7. Operating Processes to reestablish connections within the Transmission Operator's System for areas that have been restored and are prepared for reconnection.

R1.8. Operating Processes to restore Loads required to restore the System, such as station service for substations, units to be restarted or stabilized, the Load needed to stabilize generation and frequency, and provide voltage control.

R1.9. Operating Processes for transferring authority back to the Balancing Authority in accordance with the Reliability Coordinator's criteria.

R4. Each Transmission Operator shall update its restoration plan within 90 calendar days after identifying any unplanned permanent System modifications, or prior to implementing a planned BES modification, that would change the implementation of its restoration plan.

R4.1. Each Transmission Operator shall submit its revised restoration plan to its Reliability Coordinator for approval within the same 90 calendar day period.

R13. Each Transmission Operator and each Generator Operator with a Blackstart Resource shall have written Blackstart Resource Agreements or mutually agreed upon procedures or protocols, specifying the terms and conditions of their arrangement. Such Agreements shall include references to the Blackstart Resource testing requirements.

Selected results of the most recent analysis of actual events, steady state and dynamic simulations, and testing that the restoration plan accomplishes its

intended function, including any restoration strategies used to facilitate restoration for recent disturbances or the deployment of blackstart resources:[2]

R6. Each Transmission Operator shall verify through analysis of actual events, steady state and dynamic simulations, or testing that its restoration plan accomplishes its intended function. This shall be completed every five years at a minimum. Such analysis, simulations or testing shall verify:

R6.1. The capability of Blackstart Resources to meet the Real and Reactive Power requirements of the Cranking Paths and the dynamic capability to supply initial Loads.

R6.2. The location and magnitude of Loads required to control voltages and frequency within acceptable operating limits.

R6.3. The capability of generating resources required to control voltages and frequency within acceptable operating limits.

R9. Each Transmission Operator shall have Blackstart Resource testing requirements to verify that each Blackstart Resource is capable of meeting the requirements of its restoration plan. These Blackstart Resource testing requirements shall include:

R9.1. The frequency of testing such that each Blackstart Resource is tested at least once every three calendar years.

R9.2. A list of required tests including:

R9.2.1. The ability to start the unit when isolated with no support from the `BES or when designed to remain energized without connection to the remainder of the System.

R9.2.2. The ability to energize a bus. If it is not possible to energize a bus during the test, the testing entity must affirm that the unit has the capability to energize a bus such as verifying that the breaker close coil relay can be

---

[2] See Appendix 2 - Request Letter for Participation in Reliability Assessment, which provides the scope of review.

energized with the voltage and frequency monitor controls disconnected from the synchronizing circuits.

R9.3. The minimum duration of each of the required tests.

Existing notes or recommendations recorded as a result of the most recent annual exercise or from an actual incident. Also, any Reliability Coordinator feedback or analysis of last year's system restoration drills, exercises or simulations, as dictated by the particular scope of the drills, exercises, or simulations that were conducted:[3]

R10. Each Transmission Operator shall include within its operations training program, annual System restoration training for its System Operators to assure the proper execution of its restoration plan. This training program shall include training on the following:

R10.1. System restoration plan including coordination with the Reliability Coordinator and Generator Operators included in the restoration plan.

R10.2. Restoration priorities.

R10.3. Building of cranking paths.

R10.4. Synchronizing (re-energized sections of the System).

R12. Each Transmission Operator shall participate in its Reliability Coordinator's restoration drills, exercises, or simulations as requested by its Reliability Coordinator.

R18. Each Generator Operator shall participate in the Reliability Coordinator's restoration drills, exercises, or simulations as requested by the Reliability Coordinator.

CIP-008-3 — Cyber Security — Incident Reporting and Response Planning

---

[3] See Appendix 2 - Request Letter for Participation in Reliability Assessment, which provides the scope of review.

In accordance with the scope of the review, Requirements assessed included the cyber security incident response plan-related requirements, as well as any requirements that support how the applicable entities test the effectiveness of their plans. These Requirements are listed below.

Cyber Security Incident Response Plan:[4]

R1. Cyber Security Incident Response Plan — The Responsible Entity shall develop and maintain a Cyber Security Incident response plan and implement the plan in response to Cyber Security Incidents. The Cyber Security Incident response plan shall address, at a minimum, the following:

R1.1. Procedures to characterize and classify events as reportable Cyber Security Incidents.

R1.2. Response actions, including roles and responsibilities of Cyber Security Incident response teams, Cyber Security Incident handling procedures, and communication plans.

R1.3. Process for reporting Cyber Security Incidents to the Electricity Sector Information Sharing and Analysis Center (ES-ISAC). The Responsible Entity must ensure that all reportable Cyber Security Incidents are reported to the ES-ISAC either directly or through an intermediary.

R1.4. Process for updating the Cyber Security Incident response plan within thirty calendar days of any changes.

R1.5. Process for ensuring that the Cyber Security Incident response plan is reviewed at least annually.

R1.6. Process for ensuring the Cyber Security Incident response plan is tested at least annually. A test of the Cyber Security Incident response plan can range from a paper drill, to a full operational exercise, to the response to an actual incident.

CIP-009-3 — Cyber Security — Recovery Plans for Critical Cyber Assets

[4] See Appendix 2 - Request Letter for Participation in Reliability Assessment, which provides the scope of review.

In accordance with the scope of the review, Requirements assessed included the critical cyber asset recovery plan-related requirements, as well as any requirements that support how the applicable entities test the effectiveness of their plans. These Requirements are listed below.

Recovery Plan(s) for Critical Cyber Assets:[5]

CIP-009-3:

R1. Recovery Plans — The Responsible Entity shall create and annually review recovery plan(s) for Critical Cyber Assets. The recovery plan(s) shall address at a minimum the following:

R1.1. Specify the required actions in response to events or conditions of varying duration and severity that would activate the recovery plan(s).

R1.2. Define the roles and responsibilities of responders.

R2. Exercises — The recovery plan(s) shall be exercised at least annually. An exercise of the recovery plan(s) can range from a paper drill, to a full operational exercise, to recovery from an actual incident.

R3. Change Control — Recovery plan(s) shall be updated to reflect any changes or lessons learned as a result of an exercise or the recovery from an actual incident. Updates shall be communicated to personnel responsible for the activation and implementation of the recovery plan(s) within thirty calendar days of the change being completed.

R4. Backup and Restore — The recovery plan(s) shall include processes and procedures for the backup and storage of information required to successfully restore Critical Cyber Assets. For example, backups may include spare electronic components or equipment, written documentation of configuration settings, tape backup, etc.

R5. Testing Backup Media — Information essential to recovery that is stored on backup media shall be tested at least annually to ensure that the information is available. Testing can be completed off site.

---

[5] See Appendix 2 - Request Letter for Participation in Reliability Assessment, which provides the scope of review.

IX. Appendix 4 – Glossary of Terms Used in Report

Advanced Persistent Threat: A set of stealthy and continuous computer hacking processes, often orchestrated by human(s) targeting a specific entity.

Alternating Current (AC): Current that changes periodically (sinusoidally) with time.

Area Control Error (ACE): The instantaneous difference between a Balancing Authority's net actual and scheduled interchange, plus the instantaneous difference between the interconnection's actual frequency and scheduled frequency and a correction for meter error.

Automatic Generation Control (AGC): A feature of a power system's centralized control system that automatically adjusts generation in a Balancing Authority Area to maintain the Balancing Authority's interchange schedule plus its frequency bias.

Balancing Authority: The responsible entity that integrates resource plans ahead of time, maintains load-interchange-generation balance within a Balancing Authority Area, and supports Interconnection frequency in real time.

Balancing Authority Area: The collection of generation, transmission, and loads within the metered boundaries of the Balancing Authority. The Balancing Authority maintains load-resource balance within this area.

Blackstart Resource: Generating unit and associated equipment with the ability to be started without support from the Bulk Electric System (BES) or is designed to remain energized without connection to the remainder of the BES, with the ability to energize a bus, meeting the transmission operator's restoration plan needs for real and reactive power capability, frequency and voltage control, and that have been included in the transmission operator's restoration plan.

Bulk Electric System (BES): The electrical generation resources, transmission lines, interconnections with neighboring systems, and associated equipment, generally operated at voltages of 100 kV or higher.

Business Continuity Plan: Provides procedures for sustaining mission/business operations while recovering from a significant disruption.

Cascading: The uncontrolled successive loss of system elements triggered by an incident at any location. Cascading results in widespread electric service interruption that cannot be restrained from sequentially spreading beyond an area predetermined by studies.

Cranking Path: A portion of the electric system that can be isolated and then energized to deliver electric power from a generation source to enable the startup of one or more other generating units.

Critical Asset: Facilities, systems, and equipment which, if destroyed, degraded, or otherwise rendered unavailable, would affect the reliability or operability of the Bulk Electric System.

Critical Cyber Asset: Cyber Assets essential to the reliable operation of Critical Assets.

Cyber Security Incident: Any malicious act or suspicious event that compromises, or attempts to compromise, the Electronic Security Perimeter or Physical Security Perimeter of a Critical Cyber Asset; or disrupts, or attempts to disrupt, the operation of a Critical Cyber Asset.

Direct Current (DC): Electric current that is steady and does not change in either magnitude or direction with time. DC is also used to refer to voltage and, more generally, to smaller or special purpose power supply systems utilizing direct current either converted from AC, from a DC generator, from batteries, or from other sources such as solar cells.

Disaster Recovery Plan: Plan that provides procedures for relocating information systems operations to an alternate location.

Distribution Provider: Provides and operates the "wires" between the transmission system and the end-use customer. For those end-use customers who are served at transmission voltages, the transmission owner also serves as the distribution provider. Thus, the distribution provider is not defined by a specific voltage, but rather as performing the distribution function at any voltage.

Extra High Voltage (EHV): Transmission lines with voltages above 765 kV.

Generator Operator: The entity that operates generating unit(s) and performs the functions of supplying energy and Interconnected Operations Services. The generator operator is responsible to have procedures for starting each blackstart resource, in accordance with Reliability Standard EOP-005-2.

Generator Owner: The entity that owns and maintains generating units. Generator owner plant control room personnel also play a role in restoration.

Incident Response Teams: Responsible personnel designated in the cyber security response plan assigned to respond to a Cyber Security Incident or a detected cyber threat event.

Intrusion Detection System: Device or software application that monitors network or system activities for malicious activities or policy violations and produces reports to a management station.

Intrusion Prevention System: Network security appliances that monitor network and/or system activities for malicious activity.

Island, Electrical: An electrically isolated portion of an interconnection. The frequency in an electrical island must be maintained by balancing generation and load in order to sustain operation. Islands are frequently formed after major disturbances wherein multiple transmission lines trip, or during restoration following a major disturbance.

Isochronous Governor Control: An isochronous (or zero droop) governor maintains the same speed regardless of the load, and ensures that the frequency of the electricity generated is constant or flat. Isochronous control mode is used to control frequency in an island during system restoration.

Network Operations Center: One or more locations from which network monitoring and control, or network management, is exercised over a computer, telecommunication or satellite network.

Phasor Measurement Unit (PMU): Device that measures the electrical waves on an electricity grid, using a common time source for synchronization.

Reactive Power: The portion of electricity that establishes and sustains the electric and magnetic fields of AC equipment. Reactive power must be supplied to most types of magnetic equipment, such as motors and transformers. It is also needed to make up for the reactive losses incurred when power flows through transmission facilities. Reactive power is supplied primarily by generators, capacitor banks, and the natural capacitance of overhead transmission lines and underground cables (with cables contributing much more per mile than lines). It can also be supplied by static VAr compensators and other similar equipment utilizing power electronics, as well as by synchronous condensers. Reactive power directly influences system voltage such that supplying additional reactive power increases the voltage. It is usually expressed in kilovars (kvar) or megavars (Mvar), and is also known as "imaginary power."

Regional Entity: An independent, regional entity with delegated authority from NERC to propose and enforce Reliability Standards and to otherwise promote the effective and efficient administration of bulk power system reliability.

Registered Entity: An entity that is a user, owner, or operator of the bulk power system that is generally required to register with NERC.

Regulation: The ability to maintain a quantity within acceptable limits. For example, frequency regulation is the control or regulation of the system frequency to within a tight bandwidth around 60 Hz. Voltage regulation is the control of a voltage level within a set bandwidth. In power systems operations, regulation often refers broadly to changing the output level of selected generators to match changes in system load.

Reliability Coordinator: The entity that is the highest level of authority who is responsible for the reliable operation of the Bulk Electric System, has the Wide Area view of the Bulk Electric System, and has the operating tools, processes and procedures, including the authority to prevent or mitigate emergency operating situations in both next-day analysis and real-time operations. The reliability coordinator has the purview that is broad enough to enable the calculation of Interconnection Reliability Operating Limits, which may be based on the operating parameters of transmission systems beyond any Transmission Operator's vision.

Restoration: The process of returning generators and transmission system elements and restoring load following an outage on the electric system.

Security Information and Event Management: Term for software products and services combining security information management (SIM) and security event management (SEM). This technology provides real-time analysis of security alerts generated by network hardware and applications. It is sold as software, appliances or managed services, and is also used to log security data and generate reports for compliance purposes.

Static VAr Compensators: A combination of shunt reactors and shunt capacitors with switching that is precisely controlled by power electronics to automatically manage reactive power injections and withdrawals from the power system to help maintain proper transmission voltage.

Supervisory Control and Data Acquisition (SCADA): A system of remote control and telemetry used to monitor and control the transmission system.

Synchronize: The process of bringing two electrical systems together by closing a circuit breaker at an interface point when the voltages and frequencies are properly aligned. Also, when generators are brought on-line, they are said to be synchronized to the system.

Synchronous: To be in-step with a reference. The rotor of a synchronous machine, be it a motor or a generator, spins in unison with the power system in terms of frequency.

System Operator: An individual at a control center of a balancing authority, transmission operator, or reliability coordinator, who operates or directs the operation of the bulk electric system (BES) in real-time.

System Restoration Plan: Plan required to allow for restoring the Transmission Operator's System following a Disturbance in which one or more areas of the Bulk

Electric System (BES) shuts down and the use of Blackstart Resources is required to restore the shut down area to a state whereby the choice of the next Load to be restored is not driven by the need to control frequency or voltage regardless of whether the Blackstart Resource is located within the Transmission Operator's System.

Thyristors: Semiconductor devices that act as switches.

Transmission Operator: The entity responsible for the reliability of its "local" transmission system, and that operates or directs the operations of the transmission facilities. The transmission operator is required to have a restoration plan, in accordance with EOP-005-2.

Transmission Owner: The entity that owns and maintains transmission facilities. The transmission owners identified in the transmission operators' restoration plans are required to provide system restoration training to their field switching personnel identified as performing unique tasks associated with the transmission operators' restoration plans, in accordance with EOP-005-2.

Voltage Source Converter: Semiconductor devices that act as switches in the converter but function differently from thyristors. Commutation during the inversion process (DC to AC at the receiving terminal) will take place under all system conditions at the receiving end. This allows a voltage source converter to be used when the system is very weak or blacked out.

X. Appendix 5 - Acronyms Used in Report

AC	Alternating Current
ACE	Area Control Error
AGC	Automatic Generation Control
ALR	Automatic Load Rejection
BES	Bulk Electric System
CIP	Critical Infrastructure Protection Reliability Standards
COM	Communications Reliability Standards
DC	Direct Current
EHV/HV	Extra High Voltage/High Voltage
EOP	Emergency Preparedness and Operations Reliability Standards
EMS	Energy Management System
E-ISAC	Electricity Information Sharing and Analysis Center
FEMA	Federal Emergency Management Agency
FERC	Federal Energy Regulatory Commission
Hz	Hertz
ICCP	Inter-Control Center Communications Protocols
IERP	Independent Experts Review Project
IRO	Interconnection Reliability Operations and Coordination Standards
kV	Kilovolt
MVA	Megavolt Ampere
MW	Megawatt
NERC	North American Electric Reliability Corporation
PER	Personnel Performance, Training, and Qualifications Reliability Standards
PMU	Phasor Measurement Unit
PRC	Protection and Control Reliability Standards
SCADA	Supervisory Control and Data Acquisition
TOP	Transmission Operations Reliability Standards
UFLS	Underfrequency Load Shedding
VAr/MVAr	Volt Ampere reactive/Mega-Volt-Ampere reactive

UNITED STATES OF AMERICA
BEFORE THE
FEDERAL ENERGY REGULATORY COMMISSION

Reliability Standard for Revised	)	
Critical Infrastructure Protection	)	Docket No. RM15-14-000
Reliability Standards	)	

JOINT COMMENTS OF ISOLOGIC, LLC AND THE FOUNDATION FOR RESILIENT SOCIETIES

Submitted to FERC on September 21, 2015

Pursuant to the Federal Energy Regulatory Commission's ("FERC" or "Commission") Notice of Proposed Rulemaking for critical infrastructure protection (CIP) ("CIP NOPR") issued on July 16, 2015, Isologic, LLC and the Foundation for Resilient Societies ("Resilient Societies") respectfully submit joint Comments on the Commission's proposal to approve seven CIP Reliability Standards, to direct the North American Electric Reliability Corporation (NERC) to develop certain modifications to Reliability Standard CIP-006-6, and to develop requirements addressing supply chain management.[1]

Isologic, LLC is a limited liability corporation organized in the State of Maryland in May 2010, with principal office in Edgewater, Maryland. This enterprise engages in cyber protection research and development. Its principal is George R. Cotter.

The Foundation for Resilient Societies, Inc. is a 501(3)(c) non-profit foundation, organized in year 2012, with principal office in Nashua, N.H. Resilient Societies is engaged in research and education to enhance the resiliency of critical infrastructures in the United States and globally.

1 See *Revised Critical Infrastructure Protection Reliability Standards,* Docket No. RM15-14-000; Notice of Proposed Rulemaking, 152 FERC ¶ 61,054 (July 16, 2015), 80 FR 43354-43367 (July 22, 2015).

[NOTE: This docket has been re-typeset for space; pagination is different from the original document.]

Table of Contents

1. Background 1

2. FERC Regulatory Authority 2

3. Voluntary Initiatives 2

4. CIP Standards, Threats and Vulnerabilities 3

5. FERC Study of Grid Cyber Vulnerabilities and Threats 4

6. Recommended Actions on Supply Chain Management 6

7. Protection of Bulk Electric System Communications Networks. 8

7.1 Obvious Vulnerabilities Due to Gaps in Current CIP Standards 8

7.2 Insufficiency of Proposed Steps in CIP NOPR 9

7.3 Need for Lockdown of Grid Attack Surface 10

7.4 Lack of Cyber-Security Protection for Synchrophasor Data Communications 10

7.5 Comprehensive Cyber-Security Protection for Communication Networks 12

8. Initiatives for Enhancement of Regulatory Processes 12

9. FERC/NRC Initiative for Operational Cybersecurity Program 13

10. Concluding Statement and Reasons for Remand 14

List of Figures

Figure 1: Networked Synchrophasors, March 2009 11

Figure 2: Synchrophasor Deployments November 2014 11

1. Background

"We know hackers steal people's identities and infiltrate private e-mail. We know foreign countries and companies swipe our corporate secrets. Now our enemies are also seeking the ability to sabotage our power grid, our financial institutions, and our air traffic control systems. We cannot look back years from now and wonder why we did nothing in the face of real threats to our security and our economy."

President Barack Obama, 2013 State of the Union Address

The vulnerability of critical infrastructure to inadvertent or malicious cyberattack is now widely recognized.[2] But our vulnerabilities did not arise in the immediate past. It has been a decade since the Congress enacted the Energy Policy Act of 2005 that provided FERC authority to protect "communication networks" from cyber incidents and related cyber vulnerabilities.[3]

The Nation's adversaries have convincingly played cyber havoc with poorly protected federal agencies (e.g., Office of Personnel Management (OPM), Internal Revenue Service (IRS), and others), have roamed freely throughout the Defense Industrial Base by downloading priceless R&D for weapons systems, have shown strong capabilities to penetrate critical national infrastructures such as banking, healthcare, and transportation, and have now successfully entered the Grid's[4] critical infrastructure. In this case adversaries are not seeking intellectual property or classified information, but have another objective entirely: insertion of malware should they need to take down the Grid, or to pressure U.S. leadership during international crises. Indeed, Grid vulnerabilities may already be a factor in our hesitant national policies on Russian aggression in Ukraine, responses to Chinese territory grabs in the South China Sea, and negotiations on the Iranian nuclear program. Simply stated, threatened cyber warfare against the U.S. is rampant and already causing negative impacts on foreign policy options and national security.

The United States has no choice. Cybersecurity of the North American Grid must be significantly

2 See the Testimony of Admiral Michael Rogers, Commander, U.S. Cyber Command and Director, National Security Agency, House Permanent Select Committee on Intelligence in Hearing, "Cybersecurity Threats: The Way Forward," November 20, 2014.

Admiral Rogers observed: "the threat of a catastrophic and damaging attack in the United States critical infrastructure like our power or financial networks is actually becoming less hypothetical every day ... Moreover, there are growing reports of attempts to breach the networks and industrial control systems of our electric power operators and critical infrastructure operations."

"... There shouldn't be any doubt in our minds that there are nation-states and groups out there that have the capability to ... enter our systems, to enter those industrial control systems, and to shut down, forestall our ability to operate our basic infrastructure, whether it's generating power across this nation, whether it's moving water and fuel ... So once you're into the system ... it enables you to do things like, if I want to tell power turbines to go offline and stop generating power, you can do that. If I wanted to segment the transmission system so that you couldn't distribute the power that was coming out of power stations, this would enable you to do that. I mean, it enables you to shut down very segmented, very tailored parts of our infrastructure that forestall the ability to provide that service to us as citizens."

3 16 U.S.C. § 824o, Electric Reliability, at sub-section (a)(8) defines "cybersecurity incident" for which reliability standards are authorized as "a malicious act or suspicious event that disrupts, or was an attempt to disrupt the operation of those programmable electronic devices and communication networks including hardware, software, and data that are essential to the reliable operation of the bulk power system."

4 In this filing, we use the term "Grid" for the entire North American electricity system: generation, transmission and distribution. For the FERC and NERC authorities under Section 215 of the Federal Power Act, we use the terms "Bulk Power System" (BPS) and "Bulk Electric System" (BES), respectively.

hardened if the Grid is not to be a major hostage in cyber warfare. Cyber-hardening is also imperative before international terrorist forces reach a level of competency to successfully target the Grid—and this day is not far off given the Grid's dismal state of cybersecurity.

2. FERC Regulatory Authority

The Commission has broad regulatory authority, beyond its duties to set reliability standards under the Federal Power Act. Under Section 201(a) of the Federal Power Act,[5] the Commission has authority to promote interconnection and coordination of the transmission and sale of electricity in interstate commerce. The Commission "may prescribe the terms and conditions of the arrangement to be made ... including the apportionment of cost" and the "compensation or reimbursement reasonably due" to market participants.[6]

The Commission has been, in the main, upheld in appeals relating to the expansion of regional and inter-regional electric markets.[7] The Commission's authority to require "open access" under Section 206 of the Federal Power Act implies also the authority to protect that "open access" from threats to reliable operations that can be foreseen if cyber threats in one entity or one region cause vulnerabilities in another entity or another region of the Bulk Power System.

We ask the Commission to reassess its authority under Section 215 of the Federal Power Act to set reliability standards, in the context of its larger duty to assure a reliable market for the transmission and sale of electric power. Does the Commission thereby have the power to mandate the inclusion of operational cybersecurity programs among all its jurisdictional entities, both as components of mandatory reliability standards and to assure that participants in the market operate in a just and reasonable marketplace? Why should a NERC Registered Entity that seeks to protect its operating assets from cyber-attack be placed at risk by another entity or entities that may be neighbors or more distant, yet bring down electric grid reliability because they have failed to protect their assets, whether assessed as "Low" or "High" Impact, from cyber-attacks that flourish due to weak links in the cyber chain of protection?

Accordingly, we ask that the Commission utilize its broader authority to "provide for an adequate level of reliability of the bulk-power system."[8]

3. Voluntary Initiatives

We acknowledge at the outset of our observations that the electric utility industry has made some progress towards a better cyber-protected bulk electric system. The industry has participated in the development of the National Institute of Standards and Technology (NIST) Framework. It voluntarily reports incidents to the U.S. Department of Homeland Security, though the United States Computer Emergency Readiness Team (U.S. CERT) system. It has developed links with law enforcement agencies, including through the Electricity Sector Information Sharing and Analysis Center (ES-ISAC) system, operated by

---

5 16 U.S.C. §824(a).

6 16 U.S.C. §824(a)(a).

7 See Transmission Access Policy Study Group v. FERC, 225 F.3d 667 (D.C. Cir. 2000), *aff'd sub, nom.* New York v. FERC, 535 U.S. 1, (2002), upholding FERC Order No. 888 and broad FERC regulatory authority consistent with state authority under Section 201(a) of the Federal Power Act.

8 16 U.S.C. § 824*o*(c)(1).

NERC in liaison with the U.S. Department of Energy (DoE) and other federal agencies.[9]

4. CIP Standards, Threats and Vulnerabilities

Section 215 of the Federal Power Act requires the electric utility industry to ensure the reliability of the Bulk Power System (BPS), with industry-wide standards established by the industry itself through an Electric Reliability Organization (ERO), not just engineering standards, but both cybersecurity and physical security standards as well. Cybersecurity standards have been under development and review for over 9 years.[10] And despite overwhelming evidence in industry after industry of successful penetration by criminal, hactivist, terrorist and nation/state adversaries that seize upon the asymmetric advantages of cyberspace, the electric grid relies on NERC CIP standards with questionable effectiveness.

Inevitably, adversarial nation-states in the natural evolution of their strategic doctrine have turned to the North American electric grid as a prime target for offensive cyber weapons, using the tactic of "plausible denial" to obscure penetrations and ultimately, intent. When an adversary develops a cyber attack capability to incapacitate any critical infrastructure, including the North American electric grid, it possesses the ultimate strategic weapon, an advantage useful in many ways. Witness the uncertain reaction of this great nation on cyber penetrations of the OPM and IRS data bases, other entities lacking the means to defend themselves.

In sections 7.4 and 7.5 of this filing, we support FERC's conclusion that Control Center–to–Control Center communications and networks must be secured, but strongly object to exclusion of secure connections to grid facilities other than Control Centers, including critical substations and large generation plants.[11] Vulnerabilities are being ignored. FERC and NERC fail to take into account the porosity of data links, their importance to grid operations, and that microwave and other radio communications are often used in sparsely populated northern latitudes and desert regions.[12]

Protection of "Low Impact" assets is another serious gap. Is the totality of regional "Low Impact" accessible cyber assets less important than a single major generation or transmission facility? Examples of effective system attacks on "Low Impact" assets include Domain Name System (DNS) and Distributed Denial of Service (DDOS) attacks.

Cybersecurity insurance costs for the Grid are increasingly related to threats and vulnerabilities. Lloyds

9 See Frequently Asked Questions about the Electricity Sector Information Sharing and Analysis Center (ES-ISAC). NERC recently sponsored a panel discussion on the role of ES-ISAC in providing top-down support to the bulk electric system. See NERC Electric Reliability Leadership Summit, Washington, D.C., August 25, 2015.

10 FERC Order No.791, November 22,, 2013, 145 FERC ¶ 61,160, provides approval for Critical Infrastructure Protection (CIP) v5/v6 standards subject to several major conditions. Docket RM15-14-000 NOPR provides FERC comments on NERC's response to Order No. 791, and proposed rulemaking.

11 For example, there are approximately 2,500 electric substations within the bulk electric system. In Order 791, para. 42, the Commission observed that "malware inserted via a USB flash drive at a single Low Impact substation could propagate through a network of many substations without encountering a single security control under NERC's proposal." And Low Impact security controls "do not provide for the use of mandatory anti-malware/antivirus protections ..." FERC NOPR RM15-14-000, at Para. 42 (July 16, 2015). In March 2015, for example, a physical break-in at the PG&E Westpark substation near Bakersfield, CA (115 kV/12 kV) resulted in the theft of a SCADA system and extensive damage to alarm systems. See "Experts warn utilities to watch for cyberattacks via substation break-ins," SNL.com, June 2015.

12 See MISO Synchrophasor Overview, Reliability Subcommittee, September 17, 2014.

recently published a white paper[13] aimed at insurance underwriters and reinsurers of primary insurance carriers. This somber analysis of the North American Grid cautions underwriters to be significantly more careful in estimating risk since there is very little empirical evidence of the ability of firms to protect themselves. A deep examination of grid vulnerabilities and threats is, in Lloyds' estimation, overdue, or the game will be lost. Increasing insurance costs provide an additional industry incentive to implement comprehensive cyber protection.

We have observed since 2012 semantic exchange after semantic exchange between FERC and NERC on cybersecurity standards, including deliberate circumlocutions of serious vulnerabilities. Meanwhile, Russian and Chinese cyber experts peer over NERC and FERC's very public shoulders, observing gaps to be exploited. Nowhere is this more evident than in the interminable debates on security of communications networks and on "Low Impact" cyber assets as reflected in FERC Order No. 791.[14] Based on past experience, no rational assessment of the likely outcome of this or following FERC Orders bodes well for the citizens and civic, industrial and national security users of electric power in this great nation.

Short of a national security crisis, the overall responsibility for security of the North American electric grid is shared across FERC, Department of Homeland Security (DHS), Department of Energy (DoE), FERC, and the Nuclear Regulatory Commission (NRC). Only the NRC has developed significant systems and procedures to defend its electric generation facilities, efforts that parallel its very strong structures for physical protection.[15] But we observe the normal federal bureaucratic response to the Russian supply chain penetration[16] of the North American Grid, i.e., "Slip and Slide, Run and Hide." Elsewhere in this filing we argue that there is little chance NERC will be able to effectively develop CIP standards to protect critical facilities from supply chain penetrations; yet that is what the law requires, and therefore the NOPR, proposes.

NERC standards, particularly the much-compromised CIP v5/v6 set, will prove to be hopelessly inadequate by themselves to protect the Bulk Electric System. Federal law, however, does not necessarily prevent national authorities from taking other steps to protect the North American Grid. We believe the major burden for this falls on FERC; however, FERC will need much more effective support from industry and its national agency partners.

5. FERC Study of Grid Cyber Vulnerabilities and Threats

Rather than attempt to categorize cyber assets by risk to the BES, the standards need to be based on a critical analysis of the entire Grid, not just the BES, with vulnerabilities (nodes, data flows, access properties,

13 Lloyds' Emerging Risk Report, Business Blackout: the insurance implications of a cyber attack on the US power grid, Lloyd's and the Centre for Risk Studies, University of Cambridge, July 2015.

14 FERC claimed that "with all BES Cyber Systems being categorized as at least Low Impact," Order No. 791 "offers more comprehensive protection of the bulk electric system." Order No. 791 at para. 2. But CIP v5 "does not require specific controls for Low Impact assets ..." Order No. 791, at Para. 5.

15 The NRC can do little to protect its generation facilities on dependencies for off-site power; this is a FERC/NERC responsibility. Witness the Washington DC power failure on 8 April 2015 when a relatively minor southern Maryland mechanical failure rippled thru the local grid and automatically shut down reactors at Calvert Cliffs nuclear generation site. Note that for some period, the site and the NRC had no way of differentiating this event from a cyber attack. The NRC-announced 45 day study of the event is predictably non-public.

16 "BlackEnergy cyberespionage group targets Linux systems and Cisco routers," IDG News Service, November 4, 2014 7:00 AM PT, Lucian Constantin.

etc.) ranked for meaningful technical standards development, with appropriate penalties for non-compliance. Purveyors of "risk management" often fail to emphasize that there can be no effective risk management without substantial, detailed knowledge of vulnerabilities, and consequent understanding of threats. This information gap characterizes the major failure of the NERC standards and FERC rulemaking processes in-place.

FERC should initiate a major study of cyber vulnerabilities and threats, and with those results, engage the Administration and the Congress, and Canadian authorities in positive steps to secure the North American Grid, over and above NERC-developed industry CIP standards. FERC should ***precede*** this effort by ordering NERC CIP standards that require securing all data communication links, including Control Center – to – Control Center communications, lateral and vertical Reliability Coordinator (RC) communications, substation communications, and network data exchanges. This vulnerability study should be structured by cyber experts and treated as sensitive. Objectives would be:

1. Document the topology of all Grid communications, networks and backups. This includes the technical characteristics of all dedicated point-to-point links, use of virtual private networks (dedicated commercial or shared internet), and all other internet connectivity linked to grid instrumentation, operational functions, and databases. All wireless or other radio or satellite connectivity must be included in the survey. All use of the internet to transmit data to and from unmanned facilities must be documented.[17]

2. Vulnerabilities of communications networks must be documented. Specific security features, where present, should be noted. Methodologies for control of secure communications systems and data required for their use must also be documented. Shared Information Technology (IT) resources which might compromise connectivity must be clear. Presence or absence of access controls, use of wireless and transient devices of any kind, deficiencies of software or other system upgrades must be clearly noted.

3. The use of foreign-developed or designed components and systems that could conceivably be used in supply chain attacks must be documented and justified. This most certainly would require coordination and assistance of vendors.

4. An overall assessment of the Grid topology must then be conducted, against known or expected threats. This assessment must focus on classes and subclasses of facilities and/or functions that support prioritization of nodes, networks, and critical functions for both physical and cybersecurity improvements and rational top-to-bottom security management of the Grid. There must be specific attention to guaranteeing that communications and network facilities supporting critical power interdependencies of nuclear generation facilities are fully secured to include redundant protections.[18]

5. A critical feature of this effort must be a "Red Team" attack on the Grid, short of actual interference with on its operations. This effort should be modelled after Department of Defense (DoD) "Eligible

17 IEEE Spectrum, "Internet-Exposed Energy Control systems Abound," Infracritical presentation at 2014 ICS Cybersecurity Conference. Infracritical researchers found 2.2 million IP addresses for industrial control systems exposed on the Internet.

18 MISO, routinely cautions industry members in internal guidance and regulatory documents: "In the event of a conflict between this document and the MISO Tariff, the MISO Tariff will control, and nothing in this document shall be interpreted to contradict, amend or supersede the MISO Tariff. MISO is not responsible for any reliance on this document by others, or for any errors or omissions or misleading information contained herein." Such language clearly puts the "tariff" above any other regulatory directive on power delivery to nuclear generation sites.

Receiver" exercises.[19] The results should be correlated with the overall Grid assessment outlined above.

6. Procedures must be put in place to maintain and use the databases and other results created by this study and periodically update assessments.
7. All DoE efforts directed to modernization of the North American Grid should take "Lessons-Learned" direction from this study.
8. NIST, other Federal security authorities and appropriate experts from U.S. industry would be the most capable parties to direct the technical constructs of this study.
9. Additional Critical Infrastructure Protection standards should then be modified to capture the results of these efforts.

6. Recommended Actions on Supply Chain Management

We applaud the Commission's decision to address supply chain vulnerabilities for protection of the BES, and compliment the Commission on the careful treatment of the topic in this NOPR. Supply chain vulnerabilities are one of the most difficult areas of cybersecurity since they can affect every industry, most modern civil structures and the Federal government, including its national security systems. The Grid, and certainly the subset that is the BES, is a major critical infrastructure and as we have seen, will be high on the targeting list of any major adversary. Foreign adversaries will place high priority on the creation of attacks on the BES[20] which will give them many options for policy decisions on both preemptive and retaliatory strikes. Supply chain vulnerabilities, whether intentional or opportunistic, are extremely dangerous since they are hard to detect, often difficult to counter, and represent persistent threats since they provide adversaries extended time windows for malice.[21]

Supply chain vulnerabilities are, of course, not new. Users of such systems are seldom aware of the risks they pose. Almost all cases of exploitation come to light very slowly, often requiring cooperative forensics across the security industry that may go on over a long period of time in assessing the full extent of the penetration. It is important to understand that the risks posed are not limited to industrial control systems (ICS) but extend to the full array of computer, communications and network systems, as well as the IT infrastructures supporting operation of the BES, including Control Center functions.[22] Further, many vulnerabilities are often deeply embedded in systems hardware, software, and firmware. In some cases, the vulnerability cannot be examined without destroying the component containing it. And some vendors are reluctant to disclose certain vulnerabilities simply because they are extremely costly to repair. We observe delays when vulnerabilities cross vendor lanes and the cooperation needed to contain the

19 An Essay from the Armed Forces Journal, "Learn Cyber History or Prepare to Repeat it", November 6, 2013.

20 US CERT Alert (ICS-ALERT-14-281-01A) "Ongoing Sophisticated Malware Campaign Compromising ICS (Update A)," Last revised: November 03, 2014. This alert summarized the successful Russian BlackEnergy penetration of several HMI vendors control systems products; these products were inadvertently installed by unsuspecting Grid operators.

21 "At least 700,000 routers given to customers by ISPs are vulnerable to hacking", IDG News Service, March 10, 2015. Many of these routers employ "flawed" firmware provided by a single Chinese company. The majority of attacks on these routers are from Chinese IPs; this is almost certainly a world-wide sophisticated supply chain attack.

22 "Navy Needs New Servers for Aegis Cruisers and Destroyers After Chinese Purchase of IBM Line", U.S. Naval Institute By: Megan Eckstein Updated: May 5, 2015 3:45 PM. Following sale, Lenovo almost immediately preinstalled man-in-the-middle "adware" that hijacks HTTPS traffic on new X86 machines, a supply chain backdoor.

problem is often lacking.[23]

In August 2015, Cisco disclosed that attackers had compromised some of its industrial-scale routers, replacing boot firmware in the operating system. Unlike typical operating system attacks, this attack vector persists across reboots. Significantly, early investigation showed that the United States has more infected routers, sixty-five, than Russia with eleven. Infected routers used by telecommunications providers could give attackers the ability to sniff and modify network traffic, resulting in redirection of users. As electric utilities more often use the shared internet for Control Center–to–Control Center and substation communications, attacks of this kind could be used against the Grid.[24]

Much of what would be required to satisfactorily contain potential damage to the BES simply does not exist. Neither the industry nor its supporting cooperative institutions possess the capability to stay on top of known vulnerabilities let alone work on isolation of new ones. There are no organized cross-industry mechanisms to educate or otherwise support effective supply chain vulnerability management. And in the practical absence of any effective 24/7 operational cybersecurity program for the Grid (BES or beyond), there is simply no mechanism for detecting or isolating or analyzing incidents traceable to supply chain vulnerabilities. For these and many other reasons, there is no comprehensive way NERC can quickly develop effective supply chain vulnerability standards. Should NERC be tasked to do so, we will likely see a repeat of previous NERC standards exercises—fragmented, incomplete, inconsequential multi-year steps leading nowhere. Our strong advice is to limit the NOPR to recommendations for a few "standards requirements" that are immediately necessary:

1. No cyber-related system or component may be installed in the Grid which has been reported by US CERT as being provably vulnerable to supply chain attack, unless the vulnerability has been corrected.

2. Any system or component reported by US Cert as containing an exploitable vulnerability will be immediately removed from operations.

3. Exceptions to (1) or (2) above require waiver by NERC with justification on continued risk to BES reliability.

4. Any hardware or software system developed by or for Grid Responsible Entities will be subject to penetration testing prior to installation in the Grid.

5. An independent 3rd party expert security firm will be used for advice on all purchases of cyber systems known to be developed, manufactured or transported by foreign firms. This applies to all US systems known to include programmable component hardware, software, and/or firmware acquired from foreign firms.

6. ES-ISAC will maintain a repository of supply chain vulnerabilities discovered on the Grid.

---

23 "SDN switches aren't hard to compromise, researcher says", IDG News Service, Aug 5, 2015 6:17 AM PT, Jeremy Kirk. This article reveals that three different network OSes running on Onie, a bare metal switch OS, Switch Light, Cumulus Linux and Mellanox OS permit malware to be installed within Onie's firmware.

24 See "Malware implants on Cisco routers revealed to be more widespread," IDG News Service, September 21, 2015 6:31 AM PT, Lucian Constantin and "Cisco warns customers about attacks installing rogue firmware on networking gear," IDG News Service, August 13, 2015 5:55 AM PT, also by Lucian Constantin.

NERC should establish a Standards Project and associated Standard Drafting Team for management of supply chain vulnerabilities within 90 days to develop these standards under the NOPR. The revised, final NOPR should make clear the Phase 1 standards would be minimal and will be replaced or modified when FERC and NRC complete their study of supply chain vulnerabilities. A final FERC Order should encourage the NERC Standard Drafting Team to consider NIST SP 800-161, "Supply Chain Risk Management Practices for Federal Information Systems and Organizations" (April 2015).

While NERC is developing simple supply chain vulnerability standards, FERC and the NRC should undertake a deeper study of how to protect the Grid (not merely the BES) from supply chain vulnerabilities. A task force would be appropriate for this initiative. Members should include representatives from DoE, DHS, NIST, DoD, and Intelligence Community organizations, representative industry security organizations, appropriate members from the energy industry, and other individuals with strong credentials in this field. Objectives should be to understand the totality of the threat, the specifics of current Grid penetrations, countermeasures that seem appropriate, a proposed program implementable by the industry and regulatory authorities for combatting supply chain vulnerabilities, recommendations to FERC for additional "standards" to be applied to the BES, and recommendations for legislation required to protect the Grid. Here again, the task force should take maximum advantage of the NIST supply chain risk management publication, NIST SP 800-161.

The supply chain vulnerability assessment conducted as above, should be folded into procedures developed for a 24/7 Operational Cybersecurity Program, discussed elsewhere in these comments on the NOPR. This should include procedures for alerting key elements of the BES by the ES-ISAC.

7. Protection of Bulk Electric System Communications Networks.

Given NERC's responsibility as the ERO, we grudgingly accept the need for Commission Orders to NERC on new CIP standards to protect communications networks. Nonetheless, the larger burden on FERC for protecting the United States' dependency on the Grid should be addressed by other means, as discussed elsewhere in these comments.

7.1 Obvious Vulnerabilities Due to Gaps in Current CIP Standards

Over the lengthy period of development of CIP Standards to replace the ineffective CIP v3 set, we have been seriously concerned that the fragmented development of replacement standards has done little or nothing to prevent penetration of the Grid by the nation's adversaries.[25] Nowhere is this more evident

25 We ask the Commission, in extending the scope of its remedial actions beyond a revised NOPR, to consider the Statement for the Record of the Director of National Intelligence (DNI) James R. Clapper, "Worldwide Cyber Threats," September 10, 2015, before the House Permanent Select Committee on Intelligence:

DNI Clapper observed that "Cyber threats to US national and economic security are increasing in frequency, scale, sophistication and severity of impact ... However, the likelihood of a catastrophic attack from any particular actor is remote at this time ... We foresee an ongoing series of low-to-moderate level cyber attacks from a variety of sources over time, which will impose cumulative costs on US economic competitiveness and national security."

"Despite ever-improving network defenses, the diverse possibilities available through remote hacking intrusion, supply chain operations to insert compromised hardware or software, actions by malicious insiders, and mistakes by system users will hold nearly all ICT [Information and Communication Technology] networks and systems at risk for years to come ..."

"Deterrence: Numerous actors remain undeterred from conducting economic cyber espionage or perpetrating cyber attacks. The absence of universally accepted and enforceable norms of behavior in cyberspace has contributed to this situation ..."

than in the NERC dissembling on the most obvious vulnerability of the Bulk Electric System (BES) – security of communications and networks. NERC and the Commission have exhaustively fretted over requirements for "programmable" and "non-programmable" communications and network components, over physical security of "non-programmable components" (cables, wiring, closets), conceptualizing "Electronic Security Perimeters" (these security perimeters are likely to escape inspection in most compliance audits), and by agreeing that a definition of "communications networks" is not necessary since the term is not used in the standards—despite the prominent use of the term ""communications networks" in specific language of the Energy Policy Act (EPA) of 2005.[26] And even CIP 002-5 still contains the ludicrous ***exception*** to required protection of cyber assets, i.e., "Cyber Assets associated with communication networks and data communication links between discrete Electronic Security Perimeters" need not be protected.

7.2 Insufficiency of Proposed Steps in CIP NOPR

While we applaud the Commission's belated efforts to address cyber vulnerabilities, we believe that proposed steps in the CIP NOPR are still woefully insufficient. The Commission should realize that connectivity to and from unsecured BES portals, and for that matter, distribution system assets not covered by NERC CIP standards, provide innumerable opportunities for penetration of the overall communications and networks of the Grid. We are confident that NERC standards alone, however well-devised, cannot prevent a moderately sophisticated take-down (and unrecoverable) attack on the BES.

Defeating defensive measures is certain to be a core tenet of nation-state adversaries in their preparation of cyber space for attack.[27, 28] Making it as hard as possible for potential enemies to be certain of their

"Threat Actors ..."

"Russia. Russia's Ministry of Defense is establishing its own cyber command, which ... will be responsible for conducting offensive cyber activities ... Computer security studies assert that Russian cyber actors are developing means to remotely access industrial control systems (ICS) used to manage critical infrastructures..."

"China. "... Although China is an advanced cyber actor in terms of capabilities, Chinese hackers are often able to gain access to their targets without having to resort to using advanced capabilities ..."

"Iran. ... Iran very likely views its cyber program as one of many tools for carrying out asymmetrical but proportional retaliation against political foe ..."

"North Korea. North Korea is another state actor that uses its cyber capabilities for political objectives ..."

26 Title 16 U.S. Code § 824o .

27 We ask the Commission to include as part of the evidence in Docket RM15-14-000 the June 26, 2015 linked Report of the Group of Governmental Experts on Development in the Field of Information and Telecommunications (ICT), United Nations Document A/70/174, publicly released on July 22, 2015.

The GGE Report on ICT was adopted on August 27, 2015 and referred to the U.N. General Assembly as a proposed Cyber Code of Conduct. The Report recommends, but without any enforcement powers, that "a State should not conduct or knowingly support ICT [Information and Communications Technologies] activity that intentionally damages or otherwise impairs the use and operation of critical infrastructure. States should also take appropriate measures to protect their critical infrastructure from ICT threats. States should not harm the information systems of the authorized emergency response teams of another State or use those teams to engage in malicious international activity. States should encourage the responsible reporting of ICT vulnerabilities and take reasonable steps to ensure the integrity of the supply chain ... UN Doc. A/70/174 (June 2015) at p. 2/17.

Moreover [4] "A number of States are developing ICT capabilities for military purposes. The use of ICTs in future conflicts between States is becoming more likely." [5] "The most harmful attacks using ICTs include those targeted against the critical infrastructure and associated information systems of a State. The risk of harmful ICT attacks against critical infrastructure is both real and serious." UN Doc. A/70/174 (June 2015) at p. 6/17.

Further, [13(i)] "States should take reasonable steps to ensure the integrity of the supply chain so that end users can have confidence in the security of ICT products ..." UN Doc. A/70/174 (June 2015) at p. 8/17.

28 The Commission also has responsibilities to protect the reliability of the bulk electric system in wartime. Under wartime

attack development is the major task before the industry and its regulatory agencies. Equivocation on standards for security of communications and networks should end with Order No. 791.

7.3 Need for Lockdown of Grid Attack Surface

At present, the Grid is largely a loosely-knit and highly porous decentralized set of entities with little or no collective capability for self-protection. Effective security of communications and networks requires that all portals into the Grid be locked down to: (1) prevent a concerted DDOS attack on critical Grid functions, (2) frustrate more sophisticated disruptive attacks that serve to disorient Grid management command and control,[29] (3) prevent manipulation of sensors and their output data streams to frustrate in-place Grid automation systems, and (4) offset clever techniques to interfere with human-machine interfaces (HMI)[30] and their manual functions critical ***at present*** to maintaining balance and reliability of power transmission systems.

7.4 Lack of Cyber-Security Protection for Synchrophasor Data Communications

Widespread synchrophasor deployment is the most important recent advance in operation of the Grid.[31] Moreover, synchrophasor measurements have great potential value in detecting a cyber attack on the Grid. NERC should be aware of this opportunity, but nonetheless has avoided defining the term "communications networks" and therefore evaded protection of synchrophasor communications. And with communications unprotected, synchrophasor data flows represent an enormous security vulnerability to the Grid and therefore to the United States.

The below Figures 1 and 2 show the proliferation of synchrophasor deployments in the North American Grid between 2009 and 2014.

---

conditions, foreign capabilities noted by Adm. Michael Rogers, Commander, Cyber Command, as quoted in Footnote 1, *supra*, pose an existential threat.

29 See Department of Energy "Factors Affecting PMU Installation Costs," October 2014, for a detailed description of the variety of synchrophasor installations supporting Grid management functions.

30 "Siemens Patches Critical SCADA Flaws Likely Exploitable in Recent Attacks," IDG News Services November 27, 2014.

31 NASPI "Representative Data Flows from Transmission Operators to Regional Hubs, Between Regional Coordinators and Between Transmission Operators, Oct 2014.

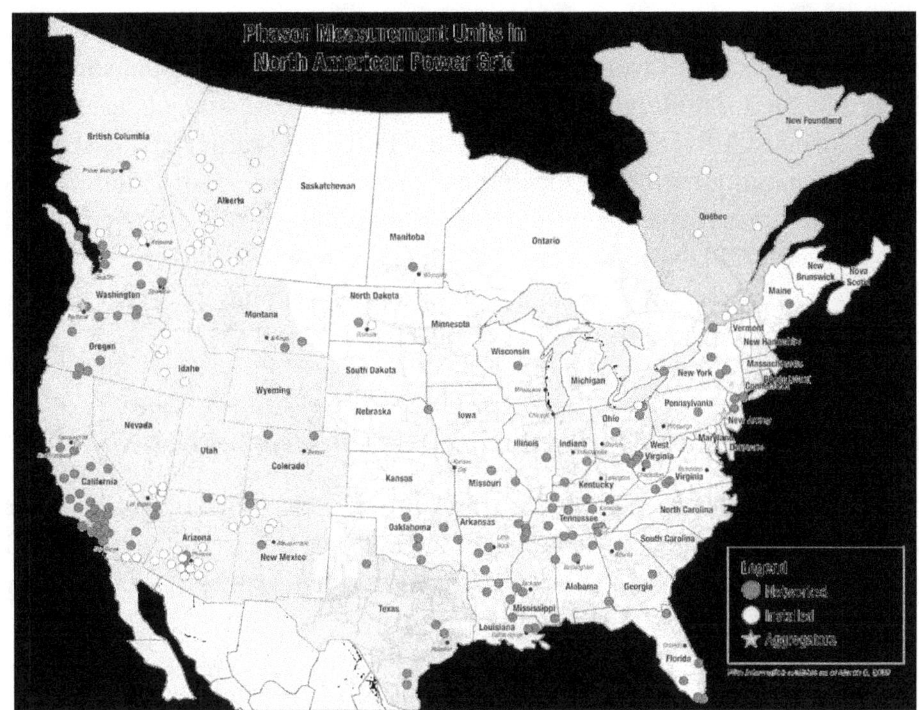

Figure 1: Synchrophasor Deployments, March 2009

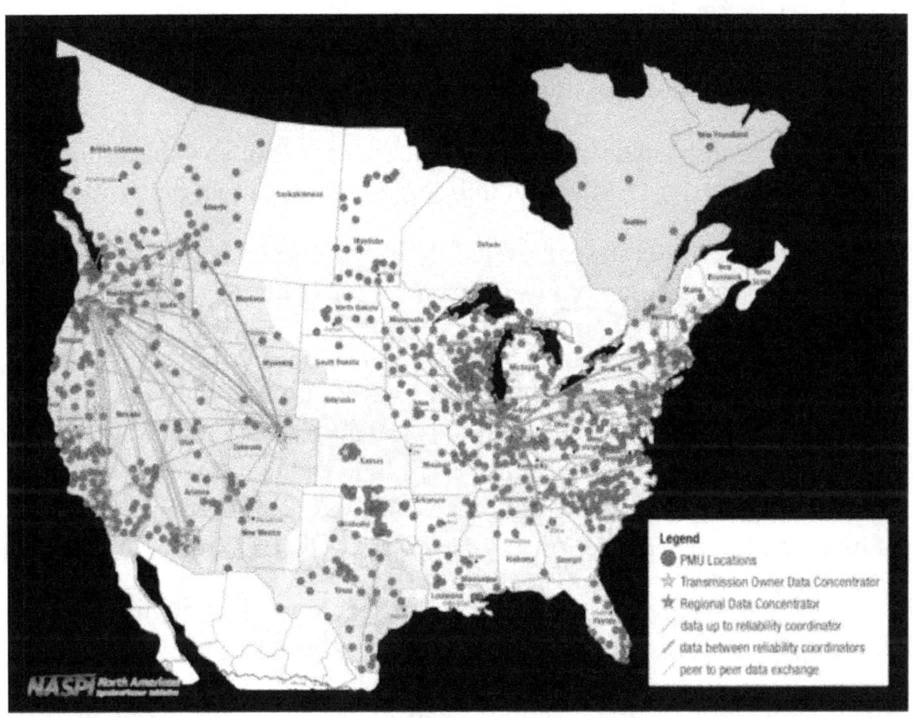

Figure 2: Synchrophasor Deployments November 2014[32]

32 "Using Synchrophasor Data during System Islanding Events and Blackstart Restoration", NASPI Control Room Solutions Task Team Paper, June 2015

We remind FERC that in the workup on CIP v4, never implemented, it was shown that well over 90% of cyber assets were essentially exempt from the draft standards. Most synchrophasor and SCADA instrumentation facilities and their data flows would still be labelled "Low impact" assets, neither identified nor tabulated in the current CIP v5/6. We recognize that FERC had earlier dismissed most synchrophasor data flows (without naming them specifically) as being used for routine maintenance and upkeep. While this may have been reasonable for pilot program installations, the electric utility industry under the North American Synchro Phasor Initiative (NASPI) is far along in dependencies on the highly precise synchrophasor data[33] for wide area management of frequency, phase, current and voltage; (i.e., grid "state" management) on regional, and soon, national levels.

7.5 Comprehensive Cyber-Security Protection for Communication Networks

We urge FERC to require that NERC CIP Standards include ***all*** communications for and networked facilities, not just Control Center–to–Control Center communications. In particular, the NERC CIP Standards should ensure encryption of all synchrophasor/phase measurement unit (PMU) and Supervisory Control And Data Acquisition (SCADA) data flows as a high priority.

8. Initiatives for Enhancement of Regulatory Processes

While deep, long term studies are absolutely essential to reset cybersecurity directions, in the short term FERC and NRC as the regulatory bodies for Grid security should collaborate on an effective set of initiatives to strengthen cybersecurity of the entire Grid, more than just the BES. These steps should include:

1. Create critical professional mass in FERC staffs to understand major cybersecurity vulnerabilities with recommendations to Commissioners on steps to alleviate them.

2. Contract for outside professional support from the security industry to supplement FERC staffs, to take on deep studies of major Grid vulnerabilities, including actual penetrations.

3. Bring pressure to bear on sector specific federal departments (e.g., DoE, DHS, DoD, NIST) to support critical infrastructure protection including R&D of security systems lacking in today's Grid, including secure instrumentation (e.g., synchrophasor/PMU encryptors).

4. Increase pressure on vendor industries to significantly improve the security of their offerings to the energy sector, particularly supply chain dependencies. Zero risks should be tolerated on the most critical Grid components; generators, extra high voltage transformers, switches in critical substations, synchrophasor instrumentation, etc. It may be necessary to establish an industry program of inspection and certification of all such systems containing foreign components.

5. Initiate FERC-NRC collaboration on issues of nuclear power plant interdependencies with the Grid, eliminating exceptions such as the MISO "tariff" rule previously cited.

6. Develop much closer ties to organizations such as the Federal Bureau of Investigation (FBI), National Security Agency (NSA), U.S. Cyber Command, and US-CERT that can advise and assist FERC on cybersecurity programs and processes, and immediately react to a major penetration or an actual attack.

---

33 U.S. Department of Energy, North American Synchrophasor Initiative, Technical Report, Model Validation Workshop, October 2013.

7. Recalibrate the entire NERC standards process to track directly on the lessons learned from the foregoing. We strongly believe that the current High, Medium and Low impact Cyber Asset strategy is misguided; emphasis and priority ought to be based on a functional understanding of actual Grid vulnerabilities identified in the studies recommended in this filing. In the process, insist upon maximum use of federal regulations and guidance that have proven their effectiveness in government use or have been developed for specific CIP domains.[34]

8. Seek legislative reforms when existing legislation is inadequate, including authority to allocate the costs of security initiatives fairly and routinely to energy consumers across the nation.

9. FERC/NRC Initiative for Operational Cybersecurity Program

Unquestionably, the most important single initiative that FERC and the NRC should immediately begin is an Operational Cybersecurity Program for the Grid. Individual firms might have installed 24/7 monitoring for cybersecurity purposes but much, much more is needed, just for minimum cybersecurity "reliability" and situational awareness. There will be the usual industry anti-regulatory, cost and liability objections to the process and discipline that an Operational Cybersecurity Program for the Grid entails. But we point out that the Reliability Coordinator and subordinate reliability designee structures that parallel Grid operations have worked successfully across the Grid. We propose that this recommended Operational Cybersecurity Program be overlaid on the existing 24/7 system for reliability coordination. Much of the new automation capability (for example, synchrophasors) provide installed sensors that can do double-duty on system state and cybersecurity alerting. NASPI already has a number of initiatives along this line.[35] FERC and the NRC should retain oversight of such a program, including penalties for violation of its provisions.

Voluntary information sharing programs by industry such as the current ES-ISAC effort are not enough to ensure operational cybersecurity. Moreover, training to understand the need to protect industrial control systems within the Grid remains woefully inadequate.[36]

The objectives of such an Operational Cybersecurity Program can be stated as follows:

1. To maintain a 24/7 monitoring effort on critical Grid functions and facilities, capable of detecting unauthorized penetration of the Grid or an actual malicious cyber-attack on the Grid.

2. To identify previously-undetected vulnerabilities to senior authorities.

3. To permit overall situational awareness at a local, regional, and national level.

4. To contribute to development of CIP standards for the Grid and support Responsible Entities in standards compliance.

5. To assist in resolving whether Grid anomalies are due to cybersecurity incidents or day-to-day disturbances.

34 See NIST National Cybersecurity Center of Excellence document "Identity and Access Management for Electric Utilities, SP 1800-2a." This is a seminal guidance document on a critical vulnerability of the Grid.

35 See NASPI October 14/15 2015 Chicago Illinois Conference Agenda.

36 See the Comments submitted by Applied Control Solutions, LLC (Joe Weiss) in this Docket on September 11, 2015. Mr. Weiss has compiled at least 60 incidents of electric sector cyber hazards, including five that have directly impacted grid operations.

6. To contribute to more effective FERC and NRC regulatory activities.
7. To provide important information to industry and DoE in Grid modernization.
8. To partner with National Cyber Indications and Warning Centers on Critical Infrastructure Protection.
9. To participate in defense of the Grid in times of national crises.

This Operational Cybersecurity Program should be organized under FERC/NRC/NERC leadership using Reliability Coordinators to help structure the program. Reliability Coordinators could identify Responsible Entities at several levels within NERC Regions, recommend establishment of 24/7 watch "centers" at appropriate critical nodes, and recommend operational features (e.g., reporting procedures) for effective "situational awareness". The Critical Energy Infrastructure Information (CEII) and NRC Safeguards Information (SGI) process could be adapted for handling sensitive details. As a final comment, we do not believe such a FERC/NRC initiative would impinges on NERC's responsibilities; it would be an operational initiative, not a standards effort. Without such a program, the Grid will simply persist in having uncoordinated and fragmented cybersecurity standards implementation by over 2000 industrial organizations—and therefore will remain an extremely attractive target for the nation's adversaries.

10. Concluding Statement and Reasons for Remand

In this comment, we have attempted to propose a comprehensive vision for better cyber protection of the U.S. electric grid. Grid protection initiatives should include voluntary actions, FERC study of grid cyber vulnerabilities and threats, enhancement of regulatory processes, an operational cybersecurity program, and better mandatory standards set by NERC and approved by FERC. While mandatory standards are not enough, we also recognize that any remand to NERC in the current rulemaking must relate to deficiencies in current or proposed standards. We therefore summarize reasons for remand below.

In the nine years since FERC selected NERC as the designated Electric Reliability Organization, NERC has persistently delayed standards-based protection of "communication networks," despite specific legislative requirements in the Section 215 of the Federal Power Act. Not unexpectedly, unprotected communication links—especially internet-reliant links—have allowed foreign adversaries to embed malware within the U.S. electric grid. The open testimony before Congress of the Commander, U.S. Cyber Command and Director, National Security Agency, on vulnerabilities of grid industrial control systems provides independent, official corroboration of the impact of FERC and NERC's inaction on implementing federal law as written.

Protecting routers, switches, and SCADA used in communications networks from supply chain vulnerabilities is another aspect of protecting against "cyber-incidents" as required by Section 215 of the Federal Power Act. Protecting just new equipment with supply chain certifications will not remedy pervasive vulnerabilities now present within the electric grid, nor will protecting just new equipment comply with the specific requirements of Section 215.

To achieve the mandates of Section 215 of the Federal Power Act, the only prudent course of action is for the Commission to remand all or substantially all of the CIP v5/v6 proposed standards to NERC for comprehensive protection of "communications networks," including networks outside "Electronic Security Perimeters" and to also remand for supply chain protection of existing electric grid components, including those components used in "communications networks."

To exclude from mandatory protection facilities judged by NERC to be "Low Impact" is illogical and imprudent, because Low Impact facilities are linked to High Impact facilities; and because Low Impact Facilities networked together can cause High Impact outages. As one public utility has noted in this docket, electric grid substations are integral components of Control Center networks—this requires that all components be protected, not just some. Incomplete protection of communication links to substations can cause Control Centers to lose visibility of critical assets, obviating their control effectiveness.[37] The NERC standard should be remanded to include "Low Impact" assets.

Respectfully submitted by:

George R. Cotter, Isologic LLC

Thomas S. Popik, Chairman

William R. Harris, Secretary,
for the

Foundation for Resilient Societies
52 Technology Way
Nashua, NH 03060-3245
www.resilientsocieties.org

37 See Comment of PNM Resources, Inc., filed in Docket RM15-14-000 on August 24, 2015, observing that "FERC may not be aware that some of these [Control Center] links actually traverse packet systems on a mesh network that includes substations. Requiring physical protection for the entire mesh may prompt Entities with such a configuration to move the backup link to the Entity's Control Center. Such a move would be adverse to the availability of bulk electric system data between the Control Centers ..."

This essay was part of a complete issue focused on EMP and extreme space weather that DomesticPreparedness.com published in June 2016 named Risk. The journal covered the EMP SIG meetings at the NOAA Space Weather Prediction Center's Space Weather Workshop in CO April 27, 2016. EMP SIG members contributed a number of articles that are being reprinted in these proceedings. That complete issue can be found here: http://www.domesticpreparedness.com/pub/docs/DPJJune16.pdf .

Balancing Risk: Understanding & Preparing for Catastrophes

By Catherine L. Feinman

Space weather, nuclear, and catastrophic natural disasters are just lying in wait for the right combination of conditions. Although it is not possible to plan specifically for every type of threat—imaginable and unimaginable—it is necessary to weigh the risks associated with various threats and take sufficient actions to mitigate the devastating effects.

On 27 April 2016, William (Bill) Murtagh, assistant director for space weather at the Office of Science and Technology Policy, Executive Office of the President, addressed leaders from government and industry to share updates on the National Space Weather Strategy and Action Plan and the related tasks and subtasks that are now being assigned to various federal agencies. The strategy urges all community stakeholders to plan and exercise for long-term regional and nationwide blackouts, which would have profound implications for business continuity and disaster planning. Successful mitigation requires a higher level of local community sustainability, especially in lifeline infrastructures such as power, communications, water and sewer, healthcare, emergency management, and law enforcement. High-impact incidents may make it unlikely for outside help to arrive within four days. Forty or 400 days may be more likely.

Developing a National Strategy

Drivers behind the new strategy include: societal and economic impacts; and implications related to losing technology, electric power, and GPS because of the nation's reliance on such technology. After 10 months of development by space weather scientists, preparedness professionals, and policy subject matter experts, the National Space Weather Strategy and Action Plan were introduced in October 2015. However, Murtagh acknowledged that developing the strategy and action plan was the easy part, implementing them will be much more difficult. Therefore, the following six goals were created:

- Establish benchmarks with actionable numbers for space weather events such as induced geo-electric fields, ionizing radiation, ionospheric disturbances, solar radio bursts, and upper atmospheric expansion

- Enhance response and recovery capabilities
- Improve protection and mitigation efforts, such as public information templates for warning messaging
- Improve assessment, modeling, and prediction of impacts on critical infrastructure such as vulnerability assessments of communication systems
- Enhance space weather services through advancing understanding and forecasting
- Increase international cooperation (international effects make it different than typical disasters; effects on magnetic field affect places around the world)

Following Murtagh's presentation, a panel of subject matter experts shared their perspectives from various disciplines on how the strategy and plan will affect business/government continuity and disaster planning.

Healthcare Impacts

Terry Donat, M.D., is the co-chair of the InfraGard's Electromagnetic Pulse Special Interest Group (EMP SIG) Health Advisory Panel and described healthcare as "an indoor sport," with the objective of keeping people healthy. Although space weather may not seem to have a direct effect on healthcare, there are vulnerabilities such as the industry's reliance on infrastructure, technology, fixed facilities, electronic medical records access, and global supply lines. He stated that limited contingent communication between facilities is a significant vulnerability that is often overlooked as circumstances leading to its use are deemed unlikely. As such, it is critical that healthcare providers understand how space weather could affect them directly.

He described two coordinating entities: the sector coordinating council of the U.S. Department of Homeland Security and U.S. Department of Health and Human Services, which released its sector-specific plan in May 2016 to specifically include space weather; and the joint commission on health coordination, which mandates a 96-hour power supply in many areas or aligned with each state's regulations for healthcare facilities. However, space weather is not included in the risk assessment vulnerabilities, and the secondary and tertiary effects are not addressed. The joint commission and other medical organizations, particularly those engaging healthcare executives would help promulgate any necessary information for planning.

Donat explained that, although this problem is unfamiliar to many in the healthcare industry, some measurable factors could be used to inform healthcare workers—for example, threatening space weather conditions that could lead to greater impact. Murtagh reminded attendees that, before the 2010 volcanic eruption in Iceland, many in Europe were unaware of any local threat.

Mental Health Impacts

Dr. Judith Boch is a clinical psychologist at the University of Colorado, Colorado Springs, Trauma Health and Hazards Center and addressed the psychological effects of warnings and public messaging. She emphasized that people are a resource that require psychological resilience. If people are not functioning, then the physical infrastructure around them does not have as much significance. Preparedness and resilience for the human factor requires conservation of intangible resources, such as self-efficacy, problem-solving ability, social support systems, and knowledge. Individuals are part of a bigger system, so systemic issues require support in order to sustain the ability to garner existing or remaining resources for survival. A smartphone is not needed in order to help someone else.

Although the mental health sector cannot inoculate people, it can help them prepare for such events by practicing in advance with small- and large-scale exercises. She pointed out that human resilience needs a clear way to approach the space weather issue and psychology has a lot that it can contribute to the resilience discussion. For example, when telling the public about space weather, the warning information needs to be clear, with a plan attached, with expectations of the public and of the agencies. As people adjust to new ideas and concepts, the topic becomes part of the language and the mitigation efforts a way of life.

Private Sector Impacts

In the private sector, Daniel Gregory is the chief executive officer for Pos-En, a microgrid developer. Over time, the perspectives on space weather effects have changed, with a lot of infrastructure now being protected against non-extreme events. He pointed out that people make 20-year investments, so they need to design the infrastructure to withstand 20 years of potential threats. Unfortunately, he added, many people do not adequately plan for electromagnetic and cosmic events, even though such incidents can have devastating consequences.

He suggested addressing issues with the public in a cost-effective manner in order to be meaningful. With technology that can protect the electric grid, reboot equipment, and provide sufficient back-up power when hit with an electromagnetic pulse, he said the problem is solvable. Attendee William Kaewert suggested finding ways to raise public awareness of this critical issue by leveraging popular media such as climate change activists have since the 1990s to force the spending of billions to avert a potential catastrophe far in the future. In contrast, no monies have been spent to protect the U.S. power system from EMP and space weather, the effects of which would be immediate and castastrophic. The destructive power of each these effects is massive, and has been measured and documented.

Public Sector Impacts

El Paso County Colorado Commissioner Peggy Littleton addressed the space weather issue from a public sector perspective. Her "You're On Your Own" (YOYO) message helps her county better prepare for any possible catastrophic incidents. She encourages communities in her area to remember that they are their own first responders. Ways to do this include an annual zombie run with educational tools – for example, what to do for power, communication, medication, connection to family and friends, etc.

Other programs used by community members in her area include: Lighthouse Prime, which is a community-driven initiative to establish communication within the community; and SNAP (Seattle Neighbors Actively Prepare (SNAP). For catastrophic long-term incidents, resilient communities cannot become too dependent on first responders and military; businesses, organizations, and individuals must find ways to continue operations despite infrastructure failures. Littleton mentioned the Great American Campout and school initiatives (engaging parents through their children) as ways to get preparedness to the local level and help community members learn how to prepare.

Military Impacts

Captain Arthur Glynn is the Command Center Director for NORAD and USNORTHCOM, where aerospace defense is a daily concern because space weather affects the sensors protecting North America and the ability to respond to threats such as missiles, cyberthreats, etc. With the primary mission of protecting the homeland, the U.S. Department of Defense (DOD) has developed a keen ability to plan well, especially for crisis action planning. This ability could be used to facilitate planning efforts within communities.

With solar weather being a major concern, Glynn mentioned some solutions that the DOD is currently implementing, such as: cyber-secure smart grids; and islanding of military bases, with renewable energy, diversified resources, and back-up generators. Lessons should be taken from the DOD's experience to apply such best practices to civilian areas. It is difficult to determine the exact implications of a long-term outage that lasts months or more.

As the chief executive officer for Jaxon Engineering and Maintenance, which performs electromagnetic pulse (EMP) testing, Randy White has observed significant changes and solutions develop over the years – for example, from missile warnings and communication technology to the ability to shoot missiles out of the sky. Being able to survive an EMP, solar flare, or other incident that disables the electric grid for long periods requires that equipment and technology be properly protected.

Public Policy Impacts

William Harris, J.D., is the attorney and a director of the Foundation for Resilient Societies and chair of the EMP SIG Legal Advisory Panel. As an international lawyer specializing in arms control, nuclear non-proliferation, energy policy, and continuity of government, Harris described the cost-recovery opportunities by hardening the electrical grid and implementing standards to require protected transformers. By providing more robust opportunities for cost recovery, more regulatory authority, and more reliable environmental impact statements, communities will become more resilient.

From a legal perspective on healthcare, Harris mentioned how recent studies have shown the effects of solar weather on health hazards, but public policy is challenged because some health studies show increased cardiac and blood pressure visits to emergency rooms during solar storms; whereas other studies show reduced cardiac stress indicated by patient-worn cardiac monitors. Meta-analysis could provide better statistics for space weather consequences—for example, to determine whether there would be an influx of patients with cardiac and other health issues when power is lost during solar storms.

Legislation enacted in 2015 includes:

- The Fixing America's Surface Transportation Act (FAST Act, H.R. 22, Public Law 114-94), which vests emergency powers in the President and Secretary of Energy, pertinent to grid restoration during solar storms; and
- Section 1089 of the National Defense Authorization Act for FY2016 (S. 1356, P.L. 114-92), which re-establishes the Congressional Commission on Electromagnetic Pulse (EMP) and requires assessment of both manmade and solar-derived electromagnetic pulse hazards.

A Global Challenge

Working closely with the United Nations for global standards on space weather and creating a national risk assessment that includes space weather, the U.S. federal government and other planning partners are introducing new space weather legislation to inform the private sector on what actions need to be taken. The new National Space Weather Strategy and Action Plan help communities build actionable items to address significant threats. Even though some people may not care about space weather, they should care about the consequences of space weather. Emergency response planning requires dependable predictions in order to make sound decisions. The global challenge now is to inform policy makers and leaders of partner nations to coordinate an international strategy through federal and nonfederal stakeholder collaborations.

In This Issue

Charles Manto leads this issue of the *DomPrep Journal* with an article about how the new strategy and action plan have created a historic shift in emergency preparedness planning—from short-term to long-term. Joshua Sparber and Benjamin Dancer, who attended the Space Weather Conference in April, each share some key takeaways from the event: space weather models, effective research tools, and electrical systems protection. Dana Goward shares some concerns about electric grid and global positioning system vulnerabilities, with a suggestion for a possible solution.

The final two articles in this issue address other incidents that could have similarly devastating consequences for critical infrastructure, leading to long-term recovery efforts. Jerome Kahan shares a follow-up article to expand on issues concerning nuclear proliferation in the Middle East. Arthur Glynn closes the issue with a looming threat that will cause severe destruction in its path if communities are not fully prepared. Space weather incidents, human-caused disasters, and natural hazards all have the ability to cause long-term consequences, so planning efforts must begin now.

Catherine Feinman joined Team DomPrep in January 2010. As the editor-in-chief, she works with subject matter experts, advisors, and other contributors to build and create relevant content. With more than 25 years of experience in publishing, she heads the DomPrep Advisory Committee to facilitate new and unique content for today's emergency preparedness and resilience professionals. She also holds various volunteer positions, including emergency medical technician, firefighter, and member of the Media Advisory Panel of EMP SIG (InfraGard National Members Alliance).

This essay was part of a complete issue focused on EMP and extreme space weather that DomesticPreparedness.com published in June 2016 named Risk. The journal covered the EMP SIG meetings at the NOAA Space Weather Prediction Center's Space Weather Workshop in CO April 27, 2016. EMP SIG members contributed a number of articles that are being reprinted in these proceedings. That complete issue can be found here: http://www.domesticpreparedness.com/pub/docs/DPJJune16.pdf.

This article by Charles Manto can be found at this website address: http://www.domesticpreparedness.com/First_Responder/Emergency_Management/Space_Weather_-_A_Historic_Shift_in_Emergency_Preparedness/ This article is reprinted with permission from the IMR Group, Inc., publisher of DomesticPreparedness.com, the DPJ Weekly Brief, and the DomPrep Journal. The IMR, Inc. offers no guarantees as to the accuracy of any information presented, but encourages all readers to use IMR, Inc. programs primarily as a resource to facilitate their own research.

Space Weather: A Historic Shift in Emergency Preparedness

By Charles (Chuck) Manto

For the first time since the demise of the civil defense program of the Cold War, the federal government has made one of the most significant modifications to its emergency preparedness message. A three-day emergency kit is no longer sufficient to prepare for emerging threats, whether coming from Earth or from space.

Instead of implying that U.S. communities can always count on being rescued from any disaster in four days—requiring three days of food and water to stay comfortable—the implication now is that local communities might not always receive assistance for a much longer period of time. The source of this change is the new Space Weather Strategy and Action Plan, which were released on 29 October 2015 and explained again at an April 2016 Space Weather Workshop.

The NOAA Space Weather Prediction Center hosted its annual Space Weather Workshop in Broomfield, Colorado, the week of 27 April 2016. This year was special because of the intimate involvement of the White House Office of Science and Technology Policy, which provided keynote addresses and the 2015 promulgation of the National Space Weather Strategy and Action Plan.

Of special interest to the emergency management community is the second of six goals of the strategy that contain the following four elements:

1. "*Complete an all-hazards power outage response and recovery plan:* for extreme space weather events and the long-term loss of electric power and cascading effects on other critical infrastructure sectors;
2. Other low-frequency, high-impact events are also capable of *causing long-term power outages on a regional or national scale.*
3. The plan must *include the Whole Community* and enable the prioritization of core capabilities.
4. *Develop and conduct exercises* to improve and test Federal, State, regional, local and industry-related

space weather response and recovery plans: Exercising plans and capturing lessons learned enables ongoing improvement in event response and recovery capabilities."

Historic Shift in Emergency Preparedness

Long-term national outages of power and other infrastructures that depend on them—including water, sewer, communications, and healthcare institutions—could mean that the entire county might undergo a catastrophe and might not be able to quickly mobilize resources to help many communities. So, unlike the cases of Hurricanes Katrina or Sandy, where help could come within a week or so, help might not arrive in 40 days, or even 400 days.

As awful as that sounds, there might actually be good news for the preparedness community. In the past, having a three-day supply of food and water to some may have seemed a waste of time since they would be rescued in four days anyway. Placing the day-to-day normal disruptions in the context of something much larger and very likely to occur during their lifetimes may motivate them to reconsider this strategy and take greater responsibility to be resilient.

An extreme space weather event—like, the 1859 Carrington event—could create a continental-wide disaster according to the National Strategy, and has a 6 to 12 percent likelihood of occurring every decade. That is a significant likelihood for such a calamitous occurrence. Including high-impact threats in overall disaster planning scenarios provides a sense of importance and immediacy that should compel the whole community to get involved, rather than simply hoping for someone to rescue them.

Preparedness Rhetoric—Beware of Pre-Traumatic Stress Syndrome

Merely discussing high-impact threats can also have a parallelizing effect if not approached properly. When someone is faced with an overwhelming scenario from any high-impact threat, it may seem easier to give up and not even think about the problem let alone begin to plan or take action to mitigate it. The response that causes a shutdown could result from emotional, financial, legal, or political reasons. A new phrase "pre-traumatic stress disorder" could be used to describe the bias many people have toward bad news. For this reason, it is important to provide a sense of hope while conveying a high-impact threat message. It is best to do that within the first minute of conveying the message to avoid having the listener minimize or block the entire conversation.

It helps to mention this problem upfront when discussing plans and conducting workshops or tabletop exercises. It also helps to warn the participants that a given planning exercise might be intended to push the plan to a failure point. In this way, the plan—rather than the participants—fails, which can be an important step in the process for improving plans.

Although the Space Weather Strategy focuses on space weather, it mentions that there are other high-impact threats that can cause similar disasters—for example, manmade electromagnetic threats, cyberattacks, or coordinated physical attacks. When presented together for consideration as an all-hazards collection, the likelihood of any one event in the collection occurring is significantly higher. In the case of natural disasters, preparation can improve the odds of passing through the disaster with less loss. However, in the case of manmade threats, being at least partially prepared could lessen the likelihood of experiencing the disaster in the first place, since bad actors will be tempted to go after easier targets.

Conversely, not taking basic precautions can actually increase the odds of a problem. Having unprotected critical infrastructure is like leaving a sign in front of a house asking people not to come in and take the pile of cash from the table, and then leaving the door open.

Military Warning About Manmade Electromagnetic Threats

In October 2015, the Defense Threat Reduction Agency (DTRA) made a similar request for help in its Small Business Innovation Research Program request for proposal (RFP). In the RFP, DTRA declared that electromagnetic pulse from small nuclear devices detonated in the upper atmosphere or attacks by high-powered microwave devices driven in panel vans could render civilian power grids inoperable where restoration may take weeks, months, or "may not be possible."

This Department of Defense (DOD) RFP further states that, "Such methods should aide in the development of islanding at DOD sites to ensure survivability to *geographically large [electromagnetic] threats.* These methods may also be applied to the *commercial sector and other areas of the government: hospitals, civilian infrastructure, businesses, etc.*" The declaration of DOD is clear: military bases and the civilian institutions that they depend on are overwhelmingly vulnerable to their own near complete reliance on unprepared civilian power grids. The DOD has already started the process of changing by developing distributed energy systems under their own control. However, due to minimal funding, these systems rarely have power storage such as batteries to provide resilience in case of a grid collapse and are even less likely to have EMP protection. This RFP signals a change in this overdependence on extremely vulnerable systems.

This RFP also goes beyond operational and disaster recovery procedures that try to indefinitely work around the problem of no power. Instead, it calls for practical EMP-protected microgrids as a solution to the problem. As a result, this RFP provides hope that individual institutions and communities can in fact protect themselves as part of a prepared whole community approach. However, it also reinforces the message that local communities must become more resilient and not merely wait to be rescued.

High-Impact Threat Assessment & Planning Support

In May 2016, the U.S. Department of Homeland Security (DHS) published Regional Resiliency Assessment Program findings that show similar vulnerabilities of the entire cyber industry (internet, telecommunications facilities, and data centers) to power grid vulnerabilities such as EMP. Two of its key findings include:

- "Data center and content providers may not have a pathway to contribute to resilience efforts and/or communicate criticality during an emergency"—conducting a workshop for the data center community would facilitate communication needs and access to critical resources.
- "Data centers and network providers should consider electromagnetic pulse (EMP) and radio frequency (RF) generator effects in developing resilience and protective measures plans"—conducting additional workshops would facilitate information sharing for EMP/RF threats, protection, and risk management.

In December 2015, the InfraGard National EMP Special Interest Group (SIG) presented copies of its Triple Threat Power Grid Exercise program to leadership of DHS, FEMA, and the National Guard. This program takes an all-hazards approach to high-impact disasters by looking at cyber, space weather, and manmade EMP threats that could result in a 1-, 3-, or 12-month national power outage. Since then, emergency management leaders at the local, state, and federal levels have begun to conduct these high-impact threat workshops nationwide.

InfraGard EMP SIG leaders in the private sector are working with the DHS Office of Infrastructure Protection and National Guard leaders to assist states and local governments in hosting workshops and tabletop exercises to test their disaster and recovery plans in light of these high-impact threats. Local protective security advisors of DHS, EMP SIG, or other InfraGard leaders can be contacted to obtain copies of the program including read-ahead material, a PowerPoint presentation, and customized versions of the facilitator's guide.

Later in 2016, the EMP SIG is expecting to release a "cookbook" of planning assistance that can help communities improve their plans. DomPrep readers are welcome to participate in these discussions and tabletop exercises by checking the DomPrep Calendar for events such as the tabletop exercise of the Society for Disaster Medicine in Maryland in late July, the InfraGard National Congress in Orlando, Florida, the week of September 12, the INCOSE/NASA/InfraGard Energy Tech Conference tabletop exercise in Cleveland on November 2, and the EMP SIG sessions of the Dupont Summit on December 1-2.

__Charles (Chuck) Manto is chief executive officer of Instant Access Networks LLC, a consulting and research and development firm that produces independently tested solutions for EMP-protected microgrids and equipment shelters for telecommunications networks and data centers. He received six patents in telecommunications, computer mass storage, and EMP protection and has another one pending for a smart microgrid controller. He is a senior member of the IEEE and founded and leads InfraGard National's EMP SIG. He can be reached at cmanto@stop-EMP.com__

This essay was part of a complete issue focused on EMP and extreme space weather that DomesticPreparedness.com published in June 2016 named Risk. The journal covered the EMP SIG meetings at the NOAA Space Weather Prediction Center's Space Weather Workshop in CO April 27, 2016. EMP SIG members contributed a number of articles that are being reprinted in these proceedings. That complete issue can be found here: http://www.domesticpreparedness.com/pub/docs/DPJJune16.pdf.

Preparing for Everything Under the Sun

By Josh Sparber

The Space Weather Conference in Broomfield, Colorado, on 25-29 April 2016 focused on improving space weather models and exploring more diverse and effective research tools. Current U.S. policy has shifted in favor of more research and funding, which can only be accomplished through better cooperation between the public and private sectors.

A bright future awaits those whose interest is modeling solar events and the vulnerability of technology to solar events. Attention from the White House has stimulated increased funding, awareness of a need for improving modeling capabilities, and the need for joint government to industry partnerships. Latest policy updates are the Space Weather Research and Forecasting Act (S2817), the October 2015 National Space Weather Action Plan, and the October 2015 National Space Weather Strategy. The commercial sector will assist with increased research and technological operations, including small businesses conducting independent research. The United States Geological Survey has also been given a $1.7M increase in funding.

Status of Space Weather Research

The conference had a strong focus on increasing insight into the current sunspot Cycle 24, the sunspot cycle the sun has been undergoing since 2009. The year 2020 is the anticipated beginning of Cycle 25. Each morning, a prediction of the day's space weather was advanced. The subjects of many of the talks were about advancements in Cycle 24 event predictability. These predictions would include sunspot counts, size and speed of coronal mass ejections and solar winds, Earth-based magnetic and electric fields, interactions of Earth and solar fields, and geoelectric sensing for individual power grid stations.

The following solar-scanning satellites will become available in the near future. The National Oceanic and Atmospheric Administration's (NOAA) Deep Space Climate Observatory (DSCOVR) will be operational in June 2016. DSCOVR will be NOAA's first deep space observatory, a "space buoy" with a National Space and Aeronautics Administration (NASA) four-megapixel charge-coupled device camera. This satellite, deploying a million miles out, will enable scientists to assess better data on how solar storms impact Earth's ozone, aerosols, cloud height and phase, vegetation properties, surface hotspots and Ultraviolet radiation.

NASA's Solar Probe Plus should launch in July 2018. This satellite's ultimate 3.7 million mile proximity from the sun should at least assist existing instrumentation to gather better solar data. Geostationary Operational Environmental Satellite-R Series (GOES-R), a collaborative program of NOAA and NASA, has a launch date of 13 October 2016. Through a United States-Taiwan partnership, Constellation Observing System for Meteorology, Ionosphere, and Climate (COSMIC-2) is being readied; however, funding is not yet available. These latter two satellites will more intensively gather actionable data on solar events affecting Earth's weather.

Other satellites study the solar "architecture" that leads to solar effects. The Solar and Heliospheric Observatory (SOHO), launched in 1995, now shares this duty with at least two other satellites. The Solar Terrestrial Relations Observatory (STEREO) employs two satellites to stereoscopically image coronal mass ejections, and the Solar Dynamics Observatory (SDO) takes helioseismological measurements.

Many government agencies are involved in data gathering and preparing for Earth weather and space weather extremes: NOAA, the United States Geological Survey, the 557th Air Force Weather Wing, Federal Energy Regulatory Commission, the National Energy Reliability Corporation, the National Research Laboratory, the Air Force Research Laboratory, the European Space Agency, and the Advanced Physics Laboratory of Johns Hopkins University. NOAA is now planning memorandums of understanding (MOUs) between itself and NASA, and itself and the National Science Foundation. These MOUs should help mold the give-and-take relationship between research and research's operational implementation: research to operations; and operations to research.

On 28 April 2016, international speakers also described the space weather programs of Japan, Korea, Mexico, Australia, and Ethiopia—all of which are unique in their own ways – and contributed an assortment of viewpoints and measurements to the existing pool of information. One good example is the Australian Energy Market Operators (AEMOs) that protect critical Australian power infrastructure. AEMOs have installed geomagnetic-induced current sensors to help quantify impending solar storm risks to existing Australian power grids.

Another example is the Japanese development of realistic 3-D models of solar flares and a whole atmosphere model called the Ground-to-topside model of Atmosphere and Ionosphere for Aeronomy (GAIA) model. There was also room for a discussion of the historical development of solar observation. The changes in sunspot counting through history and the history of the founding of the NOAA center in Boulder were also highlighted. A side room was reserved for an international crowd of investigators and researchers to display their projects. There were over 60 posters and displays from sources of worldwide scope. One display showed an iPhone application with a 3-D display of an Earth Aurora space weather model.

Touchstones of Progress

A salient point of the conference was the search for a "money chart," a depiction of the danger of a space weather event so crystal clear that decision makers could consider it "actionable." A Hurricane Sandy depiction was an example, showing the landfall pattern within a tolerance band. Space weather instruments could piggyback on a larger satellite. These "CubeSats" would save the fuel and the expense of a separate satellite. Retired Admiral Thad Allen (who led the Coast Guard's efforts following Hurricane Katrina and currently serves as executive vice president for Booz Allen Hamilton), retired Admiral Conrad

Lautenbacher (now chief executive officer for GeoOptics), and others debated pathways to the space weather "enterprise." Conference takeaways included:

- Improving data will evolve understanding from space weather into space climate.
- A part of future capability should be "now-casting" as well as forecasting.
- Joint government/academic/commercial ventures will contribute significantly to scientific advances in the United States during times of limited federal budgets.
- The White House is concerned with space weather and will assist worthy projects.
- Successfully addressing highest impact concerns involves international partnerships.
- Better models will enable more research pushes from users (operations to research).
- Models are being linked together for a more comprehensive view of effects.

The NOAA Boulder Space Weather Prediction Center

The tour of the NOAA Boulder Space Weather Prediction Center (David Skaggs Center) features a downstairs monitoring center, several floors of offices, and a large suspended spherical projector upstairs, named Science on a Sphere. The glassed off solar weather monitoring control room hosts multiple rows of multicolored depictions of daily solar behaviors, with controls underneath. In the visitor area, a large screen transmits these same dynamics. On weekdays, monitors check portions of solar activity and clone information received into daily reports. On weekends, weekly reports are compiled.

In a darkened room, Science on a Sphere displays dozens of projected Earth, sun, planetary, and exo-planetary scenarios in color. These displays result from the projection automation of four computer-controlled cameras. A small sampling of Earth displays were 3-D mappings on the sphere of earthquakes, auroras, hurricanes, wind directions, ocean currents and acidity, carbon dioxide parts per million, worldwide disease vectors, and global air traffic (see Figures 1 and 2).

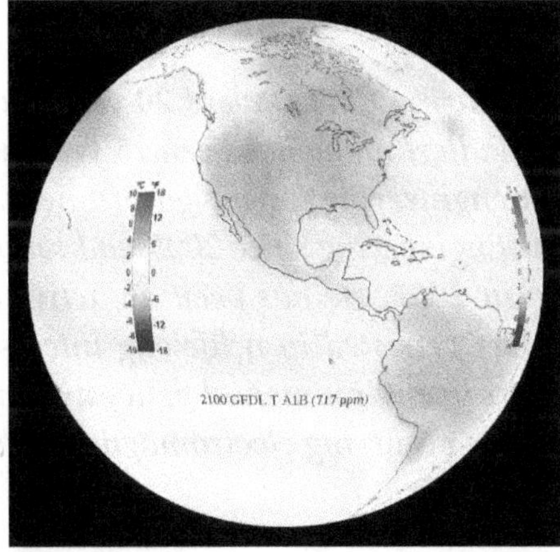

Fig. 1. Worldwide carbon dioxide effects on temperature. (*Source:* NOAA, 2010)

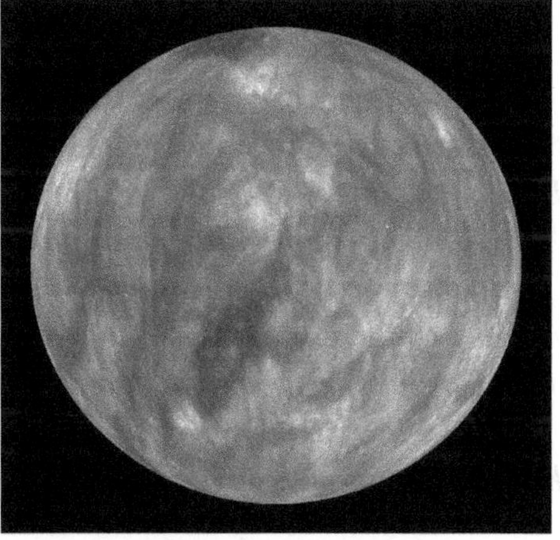

Fig. 2. Composite six-year map of solar active regions based on SOHO extreme ultraviolet observations. (*Source:* NOAA, 2010)

These depictions can time-sequence into the future, rotate on the sphere in any direction, overlay latitude-longitude, and juxtapose attribute labels on the sphere itself—for example, the greatest height of a tsunami based on predictions of the expanding wave. Solar projections capture solar flares, coronas, coronal mass ejections, magnetic fluxes, sunspots, and coronal holes. Downstairs, a large 2-D touch screen mimics these 3-D projections. The 3-D projections can be rotated in any direction by a continuous motion of two fingers placed on the screen itself.

Radiation Experimentation

During the banquet on 27 April 2016, Earth to Sky Calculus, a group of young students led by instructor Dr. Tony Phillips, showed how inexpensive (crowd-funded) experiments could form the basis for careers that will be needed in future space science and meteorology by agencies like NOAA. Their experiments yielded exceptional results for understanding the space weather/Earth weather nexus. As an example of an of this type of experiment, this group examined how solar radiation effects on the Earth and Earth life forms can be measured relatively simply and with a minimum of expense. Using a minimal set of instruments, they send weather balloons into the ionosphere, monitor them, and retrieve them after the balloons burst and the lunch box instrumentation gondolas parachute back to Earth.

They have also traveled by plane, comparing radiation dosages to their instrumentation per height and latitude-longitude. By coordinating their work with microbiologists, the group measured the effects of upper atmosphere radiation on single-celled Halobacteria and yeast. Although the uppermost ionospheric reach of the balloons (before bursting and falling) is similar to the surface of Mars, organisms with a radiation-resistant genotype heartily survived the whole journey. Under radiation, Halobacteria, fortunately also turn a yellow color, which grows darker with greater radiation dosage. A future research avenue could be providing visible checks on air pilot radiation dosage. In the future, it may be critical to know the extent of radiation dosage pilots receive during solar storms and solar flares.

Josh Sparber, after receiving Air Radar training in the U.S. Marine Corps, spent 20 years in the electronics industry. After earning an MSEE from Cal State Fullerton, he has been 15 years with the U.S. Department of Defense, and is involved in system engineering for the government. He has been a member of the International Council on Systems Engineering since 2005 and received his Certified System Engineering Professional certification in 2007. He has been an active and interested member of various environmental groups and has persisted in a lifelong interest in environmental issues. He has just recently completed a masters in environmental policy and management, in which his thesis centers on the impact of naturally occurring electromagnetic pulses on the U.S. national grid and a search for possible system solutions.

This essay was part of a complete issue focused on EMP and extreme space weather that DomesticPreparedness.com published in June 2016 named Risk. The journal covered the EMP SIG meetings at the NOAA Space Weather Prediction Center's Space Weather Workshop in CO April 27, 2016. EMP SIG members contributed a number of articles that are being reprinted in these proceedings. That complete issue can be found here: http://www.domesticpreparedness.com/pub/docs/DPJJune16.pdf.

Space Weather & Electrical Grid – GPS the Weakest Link

By Dana Goward

Among the many important, yet weak, satellite signals that can be disrupted by space weather, the Global Positioning System (GPS) is undoubtedly the most important and the weakest. Two recent public discussions have highlighted the challenges this poses for the national electrical grid, both today and going forward.

In March 2016, the MITRE Corporation released information on "Smart Grid Use of GPS Time." The posted presentation stated that, "(The) Power Grid has a vital dependence on precise time for:

- Time-stamping of operational data (e.g., supervisory control and data acquisition – SCADA)
- Wide area situational awareness
- Synchronization of operations
- Grid management and control
- System and asset protection"

MITRE's most recent effort built upon a 2013 paper that specifically enumerated all the places where GPS timing information was used by electrical grids such as fault detection, substation control, and in SCADA systems.

In May 2016, Alison Silverstein of the North American SynchroPhasor Initiative (NASPI) spoke to the National Space-Based Positioning, Navigation, and Timing (PNT) Advisory Board. As part of a presentation with Netinsight, Silverstein downplayed the use of GPS time in the grid today, but indicated that it was going to be increasingly important going forward. She did acknowledge, though, that the grid is critically dependent on telecommunications and information technology networks, and that those networks are critically dependent on GPS time. Several examples of problems with the GPS signal, such as the January 2016 SVN23 anomaly (when half of the GPS satellites' time transmissions were off by 13.7 microseconds) and faulty installations—and their impact on electrical grids—were discussed. Going forward, she said, the electric power industry would be looking for more reliable, stable sources for wide-area synchronized time signals.

At the same PNT Advisory Board meeting, Andy Proctor, lead for satellite navigation at Innovate UK and chair of the UK Government PNT Group, mentioned that his country was keeping eLoran on the air to provide timing for critical infrastructure and other uses. The idea is to combine eLoran, GPS, and Europe's Galileo satellite navigation and timing system signals as a way of providing trusted time to electrical, telecommunications, and information technology networks.

Pairing eLoran and GPS signals has currency in the United States as well. In December 2015, Department of Defense Deputy Secretary Robert Work and Department of Transportation Deputy Secretary Victor Mendenz told the U.S. Congress that the administration would build an eLoran timing system to help protect critical infrastructure. Unfortunately, little appears to have been done so far.

Dana A. Goward is president of the Resilient Navigation and Timing Foundation, chairman of the Association for Rescue at Sea, and a retired Coast Guard captain. He also is retired from the federal Senior Executive Service.

This essay was part of a complete issue focused on EMP and extreme space weather that DomesticPreparedness.com published in June 2016 named Risk. The journal covered the EMP SIG meetings at the NOAA Space Weather Prediction Center's Space Weather Workshop in CO April 27, 2016. EMP SIG members contributed a number of articles that are being reprinted in these proceedings. That complete issue can be found here: http://www.domesticpreparedness.com/pub/docs/DPJJune16.pdf.

Electrical Systems & 21st Century Threats

By Benjamin Dancer

One leading researcher shares his insights into the existential threats that the electrical infrastructure faces. He proposes that a superhighway with electrical systems protected at multiple points is not only feasible, but it could help reduce carbon emissions, build electromagnetic resilience, and address major space weather events that could threaten the life and health of human populations.

Alexander MacDonald, the former director of the National Oceanic and Atmospheric Administration's (NOAA) Earth System Research Laboratory, gave a talk on 27 April 2016 entitled, "Create a 21st Century Electric System," at the Space Weather Workshop in Broomfield, Colorado. The workshop was co-sponsored by the NOAA Space Weather Prediction Center, the National Science Foundation Division of Atmospheric and Geospace Sciences, and the National Aeronautics and Space Administration Heliophysics Division.

MacDonald's talk was a presentation of a study that he published on 25 January 2016, in the journal Nature Climate Change. He laid out a plan to build an underground, high-voltage direct current (HVDC) superhighway by 2030 that could, "solve the two greatest problems the U.S. faces: the threat of human-induced climate change and the threat of massive homeland destruction from electromagnetic pulse (EMP) or solar ejections." This HVDC superhighway would be the backbone of a multi-point protected electric system (see Figure 1).

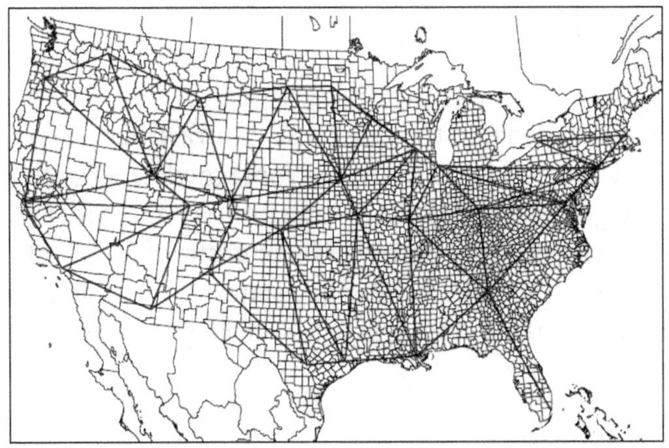

Fig. 1. An HVDC "Superhighway for Electrons." (*Source:* Alexander MacDonald, 2016)

Reducing Carbon Emissions

A NOAA supercomputer was used to calculate future electrical costs, demand, generation, and transmission scenarios. The computer also crunched billions of bits of data from the country's weather history. The study concluded that an HVDC transmission infrastructure that uses converter stations to move power to the present AC distribution points could reduce energy loss and allow weather-driven renewable resources to supply 70 percent of the nation's electricity without raising costs to the consumer. Moreover, implementing the plan would dramatically cut greenhouse gas emissions from power production.

"Our research shows a transition to a reliable, low-carbon, electrical generation and transmission system can be accomplished with commercially available technology and within 15 years," MacDonald said.

Because energy is available in the United States from solar or wind sources (see Figure 2), the solution to the problem of intermittent renewable generation is to ramp up the renewable energy generation system to keep pace with the scale of the country's weather systems.

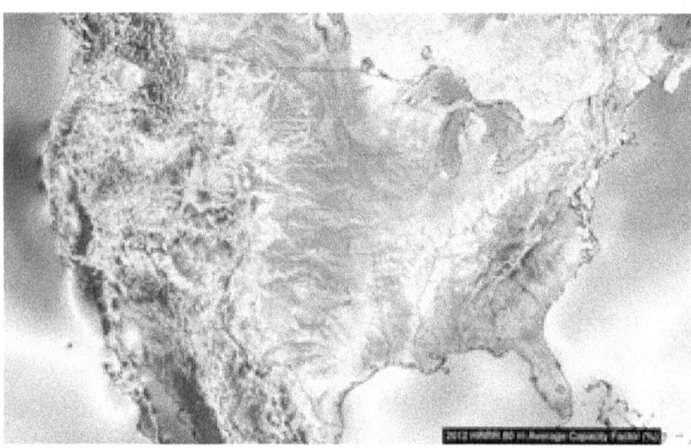

Fig. 2. A high-resolution map based on NOAA weather data shows a snapshot of wind energy potential across the United States in 2012. (*Source:* NOAA, 2016)

One of the greatest obstacles to a transition to renewable energy has been the problem of storage. A large and strategically placed renewable generation system solves the problem of energy storage by allowing electrons generated in one part of the country to be immediately transmitted to any other. This type of efficient transmission is not possible with the current infrastructure.

When describing the HVDC grid, MacDonald employed the analogy of the interstate highway system, which revolutionized the U.S. economy in the 1950s. "With an 'interstate for electrons,' renewable energy could be delivered anywhere in the country while emissions plummet," he said.

Building Electromagnetic Resilience

Citing the Report of the Commission to Assess the Threat to the United States from Electromagnetic Pulse (EMP) Attack, MacDonald concluded that the U.S. civilian infrastructure is essentially unprotected from an EMP attack. An EMP weapon could create an electromagnetic energy field that would damage critical components of the electric grid, including large transformers and supervisory control and data acquisition (SCADA) systems. The potential for damage is so catastrophic that an EMP attack poses an existential threat to the United States. "Today there are rogue states with specially designed EMP nuclear weapons," MacDonald said. "Tomorrow these weapons will be available to terrorists and any small state with a grudge and a balloon."

Addressing Major Space Weather Events

MacDonald also cited space weather as a potential threat. A coronal mass ejection the size of the 1859 Carrington event, or bigger, could pose a threat to the United States on the same scale as an EMP attack. The Carrington event was a massive solar storm that caused telegraph systems all over the world to malfunction or fail.

The design of the proposed underground HVDC network would include protecting it from EMP. The cable system would be protected by an outer sheath that would function much like a faraday cage (a grounded metal screen surrounding a piece of equipment to shield it from electromagnetic influences). Moreover, the cable itself would be grounded at every junction. The converter stations would also be EMP protected.

"An important point," MacDonald said, "is that a protected HVDC network by itself is not enough. We would also have to take action to protect the backbone of the AC distribution system – maybe 5 percent of the AC system would have shields and very rapid response circuit breakers."

In a subsequent interview with the author on 25 May 2016, MacDonald described the importance of protecting the homeland from EMP. There are several mechanisms of destruction that have the potential to disable the power grid, including cyberattack, EMP attack, and space weather. When asked which of the threats caused him the most anxiety, MacDonald said EMP.

Over the course of the last 100 years, society has unwittingly evolved to become dependent on a vulnerable critical infrastructure. People 100 years ago did not need electricity to feed the population or to provide clean drinking water. Today they do. An EMP attack could disable the power grid and create an economic catastrophe on a scale unprecedented in human history. A sophisticated attack has an enormous potential for human casualties. Dr. Peter Pry, a member of the Congressional EMP Commission and executive director of the Task Force on National and Homeland Security, testified before a congressional subcommittee in 2014 that "a nuclear EMP attack ... could kill 9 of 10 Americans."

"The general idea is that a limited set of electric transmission and distribution should be protected," MacDonald said. "In the most likely scenario, where a large geographic area is impacted (e.g., the 10 most western U.S. states by a bomb above the Pacific off of California), there would be a large area with no services – no food, no heat, no fuel, no ability to provide medical or law enforcement, no drinking water, etc. Social breakdown occurs when large numbers of people become desperate and fight among themselves."

MacDonald concluded, "The idea of a robust system is that, if electricity can be made available at key points, services can be available that enable the larger, unharmed part of the U.S. to deliver aid. On the other hand, if the airports and gas stations aren't functional, and the ability to keep public roads, food, and water available are gone in a large area, the disorder and violence can become self-feeding fairly rapidly."

Benjamin Dancer, who works as an advisor and mentor at a Colorado high school, is the author of the literary thriller Patriarch Run, which explores the vulnerability of the nation's critical infrastructure to cyberattack. He also writes about education, parenting, sustainability, and national security. More about the author and the vulnerable critical infrastructure can be found at BenjaminDancer.com.

This essay was part of a complete issue focused on EMP and extreme space weather that DomesticPreparedness.com published in June 2016 named Risk. The journal covered the EMP SIG meetings at the NOAA Space Weather Prediction Center's Space Weather Workshop in CO April 27, 2016. EMP SIG members contributed a number of articles that are being reprinted in these proceedings. That complete issue can be found here: http://www.domesticpreparedness.com/pub/docs/DPJJune16.pdf.

Cascadia Catastrophe—Not If, But When

By Arthur Glynn

A 9.0-magnitude earthquake off the Washington and British Columbia coast along the 700-mile Cascadia Subduction Zone (CSZ)—followed by a tsunami with 90-foot or more wave surges in some areas—is possible based on geological factors and historical accounts. Communities in and around the CSZ, and those with interconnected waterways, need to be prepared for the inevitable.

The last time an earthquake of this magnitude occurred in the region dates back to 1700, when a tsunami reported in Japan was attributed to CSZ seismic activity. According to the U.S. Geological Survey (USGS), movement of the Earth's tectonic plates where the Juan de Fuca plate intersects the Pacific plate generates enough pressure that, when the plates become immovable, the pressure builds until it is released via a very large earthquake (see Figure 1). Historical modelling indicates this occurs approximately every 200-500 years.

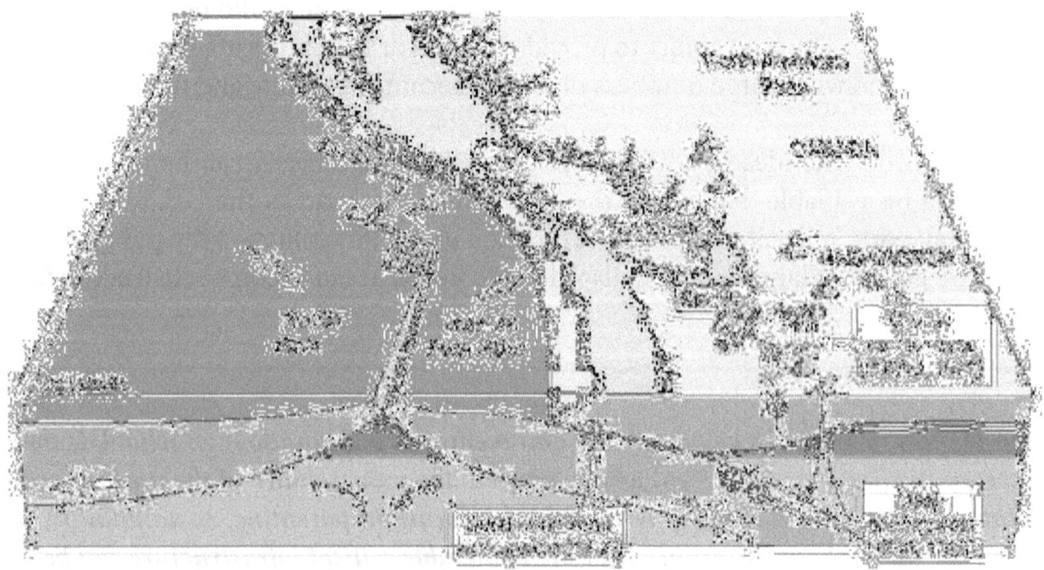

Fig. 1. Map of the Cascadia Fault and related past earthquakes. *Source:* USGS (2000).

Although tsunami warnings along the British Columbia, Washington, and Oregon coastal regions would help people respond appropriately and move to higher ground, inland waterways such as Puget Sound and the Columbia River basin may be less aware of the threat. A comparable scenario was presented 7-10 June 2016 for the Cascadia Rising exercise, which was sponsored by: the Federal Emergency Management Agency (FEMA); Region 10; Washington Military Department, Emergency Management Division; Oregon Military Department, Office of Emergency Management; Idaho Military Division, Idaho Bureau of Homeland Security; United States Department of Defense U.S. Northern Command (USNORTHCOM); United States Department of Defense U.S. Transportation Command (USTRANSCOM); and FEMA National Preparedness Directorate-National Exercise Division, and the Office of Response and Recovery.

Disasters in the Pacific

Recent devastating earthquakes around the Pacific "Ring of Fire" – for example, Indonesia in 2004 (9.1 magnitude, nearly 228,000 fatalities), Chile in 2010 (8.8 magnitude, more than 500 fatalities), Japan in 2011 (9.0 magnitude, more than 18,000 fatalities)—make planning for such an event within North America a necessity. Such earthquakes, coupled with 34 active volcanos (of 40 globally) along the Ring of Fire, mean that everyone in the public and private sectors should be prepared for a cataclysmic earthquake and resulting devastation.

Under FEMA Director Craig Fugate's guidance, FEMA has instituted a permanent catastrophic planning effort to stabilize a catastrophic event within the first 72 hours. This is reflected in FEMA's mission statement, "To support our citizens and first responders to ensure that as a nation we work together to build, sustain and improve our capability to prepare for, protect against, respond to, recover from and mitigate all hazards." However, this does not mean that the federal government will be able to rescue everyone, which is why FEMA emphasizes that everyone following a disaster is a first responder. Due to shear logistical challenges, the success (or failure) of surviving a catastrophic event during the first 72 hours may depend on citizens within the local community.

It may take some time for federal resources to be made available to ease the agony of a catastrophic earthquake that the Cascadia Fault may produce. The normal response following a catastrophic incident is an outpouring of emotional support from across the country and from allies. Promises of support are made, but frankly, it is up to first responders (remember, everyone is a first responder) to take care of themselves individually. Then, and only then, can the individual be part of the greater response.

Preparing Communities With Actionable Plans

Although people have little control over what happens to them, they have complete and utter control of how they respond to what happens to them. Even in the most unlikely events, people can prepare for the unknown, be resilient, and set themselves up to be survivors. The determination to be a survivor, as an individual or a community, is probably the greatest factor determining success or failure—especially during the first 72 hours after initial impact.

When a catastrophic earthquake occurs along the CSZ, whole communities will be cut off from the rest of civilization for a significant amount of time. Lines of communication and electrical transmission cables will be severed; bridges and critical infrastructure destroyed; water contaminated; and access to life-sustaining supplies will be severed. Even large cities—for example, Vancouver, British Columbia; Seattle, Washington; and Portland, Oregon—accustomed to having on-demand resources, will find themselves completely without basic needs. At first, there will be shock and dismay with a consensus sentiment of "How could this happen to us?" Those who have access to communications will hear of outside efforts to help them, but it will take time. And time, is a commodity in short supply.

There will be patience for a little while, especially for those who are not severely affected. For those who are significantly impacted, there may be anger against first responders or those who are less affected. Anger then could escalate toward local, state, and federal civic leaders regarding the delay in disaster response. Even when FEMA and the federal government (including the military) are doing everything within their power to respond, delays are inevitable. Questions will arise about the nation's vast military capability and its response. Although the Department of Defense (DOD)—through USNORTHCOM and USTRANSCOM—has Defense Support of Civil Authorities (DSCA) and logistical support missions respectively, its primary mission is Homeland Defense. In times of major disaster, the nation's adversaries—both state sponsored and nonstate sponsored—have opportunities to take advantage of a perceived national weakness and may initiate attacks to further cripple the United States and gain geopolitical advantage. It is here that the DOD must focus its attention and critical resources during times of national emergency. Though the DOD will do everything in its power—repurposing its vast capabilities to save lives, mitigate human suffering, and prevent great property loss during a national emergency—recent budget cut backs have forced the DOD to focus on their primary mission, homeland defense.

The Next "Big One"

Throughout history, mankind has had many opportunities to mitigate potential disasters but, unless they occur frequently, it is human nature to unknowingly accept an exceedingly high level of risk and be oblivious to warning signs. The nation has received many such warnings in the form of earthquakes near the 9.0 magnitude along the Ring of Fire in Indonesia, Chile, and Japan. Nobody knows where the next 9.0-magnitude earthquake will strike, but the Cascadia Rising exercise was a great first step in preparation for such events. However, the question remains, "Will we be ready when the 'Big One' hits or was this merely an exercise in futility?"

CAPT Arthur Glynn, U.S Navy (Retired), passionate for effective leadership and national defense, has devoted his professional life to ensuring that the American way endures, whether while in uniform or while developing technologies to bolster technological advantages and mitigating national level threats. He recently retired from the United States Navy, where he served as a North America Aerospace Defense Command (NORAD) and United States Northern Command (USNORTHCOM) command center director. Previously, he served as a Navy emergency preparedness liaison officer to USNORTHCOM and FEMA District VIII. He has served as: president/chief executive officer of a manufacturing firm; technology, mergers, and acquisition consultant; emergency management advisor; and financial advisor. He is currently an independent consultant to industry, helping to protect the U.S. critical infrastructure.

Society for Disaster Medicine and Public Health

About Us

The Society for Disaster Medicine and Public Health is a forum for health professionals to collaborate on issues related to the advancement of the discipline of disaster health.

Members participate in the creation of policies and programs that work toward global health security before, during, and after disasters. Members receive a free online subscription to Disaster Medicine and Public Health Preparedness (DMPHP), the Society's official peer-reviewed journal advancing the science and practice of disaster medicine and public health.

Our Vision

The SDMPH is dedicated to the promotion and advancement of excellence in disaster medicine and public health across a broad global, multiprofessional membership. The SDMPH will provide members with resources, organizational structure, and the means to sustain interprofessional interaction and discourse related to the discipline of disaster medicine and public health.

Our Mission

The mission of the SDMPH is to advance and promote excellence in education, training and research in disaster medicine and public health across all phases of the disaster cycle for all potential health system responders based on sound educational principles, scientific evidence and best clinical and public health practices.

SDMPH ● 11300 Rockville Pike, Suite 1000, Rockville, MD 20852
● 718-916-9758 ● info@sdmph.org ● www.sdmph.org

EDITORIAL

Disaster Medicine and Public Health Preparedness: A Discipline for All Health Professionals

James J. James, MD, DrPH, MHA; Georges C. Benjamin, MD, FACP;
Frederick M. Burkle Jr, MD, MPH, DTM, FAAP, FACEP; Kristine M. Gebbie, DrPH, RN;
Gabor Kelen, MD; Italo Subbarao, DO, MBA

Individuals and populations exposed to natural and human-caused disasters confront myriad social, physical, psychological, environmental, and economic conditions that affect health. Lessons learned from Hurricane Katrina (2005), the Haitian earthquake (2010), and other large-scale disasters consistently demonstrate that such events disproportionately affect the most vulnerable members of society, including children, elderly people, and minority populations. Minimizing adverse health outcomes requires cooperative efforts that cross traditional boundaries of health specialties, professions, and nationalities. Health professionals are on the front lines when dealing with injury and disease every day, whether natural or man made.

There are a wide variety of disasters ranging from localized events to large-scale public health emergencies. To respond effectively, health professionals, regardless of specialty or area of expertise, require a fundamental understanding of the disaster management system and the ways in which various health-related roles are integrated to protect health and respond to disease or injury. In a disaster or public health emergency (PHE), health professionals have an obligation to protect and preserve the health, safety, and security of their patients, families, and communities, as well as themselves. All health disciplines should be knowledgeable about the range of illnesses and injuries that may arise and how their particular expertise facilitates effective response. In addition, all must be able to recognize the general features of disasters and PHEs and be knowledgeable about their impact on the population, how to report a potential public health event, and where to access pertinent information as required. Most disaster events are on a scale that communities, whether in the developed or developing world, can manage well. Consequences are usually limited to direct injuries and deaths. In particular, large-scale PHEs place unprecedented demands on the existing public health infrastructure and system that may increase overall morbidity and mortality. PHEs require an added degree of coordination, cooperation, and collaboration between the clinical workforce and public health authorities.

DEFINING THE KNOWLEDGE BASE FOR DISASTER MEDICINE AND PUBLIC HEALTH PREPAREDNESS

It is recognized within the discipline of disaster medicine and public health preparedness (DMPHP) that there are distinct principles and practices across the health and social sciences that provide a foundation for doctrine, education, training, and research within the public health and health care sectors. (DMPHP includes all health professions and specialties, including but not limited to allied health, dentistry, emergency medical services, environmental health, epidemiology, hazardous materials response, medicine, mental health, nursing, pharmacology, public health, toxicology, and veterinary medicine.) Previous definitions have been proposed, but despite their relevance, they have not achieved widespread consensus. To distinguish DMPHP from other health disciplines and professions, a modified definition is proposed that recognizes the essential integration of clinical and public health science and practice into the emergency response system:

DMPHP is defined as the study and collaborative application of sound scientific principles, practices, and standards by multiple health professions for the prevention, mitigation, management, and rehabilitation of injuries, illnesses, and other problems that affect the health, safety, and well-being of individuals and communities in disasters and public health emergencies.

Strong impetus for more focused attention to education, training, and research in DMPHP was provided by Homeland Security Presidential Directive 21 (HSPD-21)[3] and 3 recent consensus reports.[4-6] HSPD-21 specifically calls for the establishment of a discipline that recognizes the unique principles in disaster-related medicine and public health; provides a foundation for the development and dissemination of doctrine, education, training, and research in this field; and better integrates private and public entities into the disaster health system. As precedent for this new discipline, HSPD-21 cites the evolution of the specialty of emergency medicine due to recognition of the special considerations of emergency patient care. HSPD-21 endorses similar action directed to disaster-related public health and medicine, which merits the establishment of a separate formal discipline.

Although DMPHP draws from multiple other fields, to be recognized and embraced as a distinct academic discipline, it must be differentiated by its own unique and distinctive essentials. This can be accomplished through description of an identifiable philosophy for the discipline, a sound conceptual framework, a unique core body of knowledge, and acceptable methodological approaches for the pursuit and development of knowledge in the field.[7] Just as the discipline of biochemistry and its accompanying journals once evolved from the interests of individual experts in organic chemistry, zoology, botany, and other fields, and the discipline of genomics evolved from the interests of individual biochemists, geneticists, pharmacologists, and others, it is envisioned that the discipline of DMPHP will evolve similarly, in response to proper input and nurturing from experts with diverse clinical and public health backgrounds.[8] DMPHP can be seen as a "composite" discipline requiring integrated multidisciplinary study and research to meet its goals.

Proficiency in DMPHP requires knowledge and skills beyond those typically acquired in clinical and public health training and practice, and must encompass unique competencies. The delivery of optimal care in a disaster relies on both clinical and public health expertise, and depends on a common understanding of each health professional's role in the broader emergency management system. To be considered proficient in DMPHP, individuals must demonstrate common mastery of defined essentials in this field. Certain backgrounds (such as may be found in subspecialties within medicine, public health, and nursing, among others) may have further differentiated skills that can be applied effectively in specific disaster events.

EMERGENCY MANAGEMENT ASPECTS OF DMPHP

To prepare for a disaster or PHE, health professionals should learn the essential elements of community and institutional disaster plans, as well as federal and local incident command. Plans should include assessment and characterization of surge capacity assets in the public and private health response sectors, and the extent of their potential assistance in an emergency. Health responders must be knowledgeable about institutional, community, and regional response systems and their respective roles within those structures, including policies and procedures for mobilizing and integrating civilian, military, and other response resources and assets. Health responders also require knowledge of administrative regulations, safety and security issues, systems engineering, decontamination protocols, forensics, use of personal protective equipment, evacuation procedures, continuity planning, and utilization of public information and communication networks.

CLINICAL ASPECTS OF DMPHP

In a disaster, clinicians should be prepared to apply and adapt their usual practices and behaviors, as appropriate, to the recognition, diagnosis, triage, and treatment of seriously injured or ill people, with limited situational awareness and resources. They may be required to apply their accustomed clinical skill set to the assessment and management of people of all ages under a variety of scenarios. At times, they may be called upon to fill nonclinical response functions such as moving patients during a hospital evacuation. Although clinicians specializing in DMPHP should have a universal core knowledge and skill set, understanding the limitations of one's individual clinical capabilities is equally important.

Clinicians and other health responders need to be familiar with medical and mental health implications of the spectrum of disasters and PHEs and recognize that people may have been exposed to nonconventional agents as the source of unusual presentations. This requires competence in identifying the health consequences and treatment of exposure to biological, chemical, radiological, nuclear, and incendiary agents. In a mass casualty situation, health system responders may need to take personal histories, conduct physical examinations, and manage injured or ill people in potentially hazardous environments with limited medical supplies and equipment while maintaining situational awareness. They should be prepared to follow appropriate diagnostic procedures to confirm or refute possible etiologies, and in some cases begin treatment based solely on symptoms and signs. Implementation of safety and protection principles to prevent harm to themselves and others is critical, as is sensitivity to the diagnostic and treatment plans for psychological and behavioral as well as physical trauma.

All health responders should know the ethical and legal structures that govern response to disasters and PHEs, while maintaining the highest possible standards of care under extreme conditions. This encompasses their rights and responsibilities to protect themselves and treat others (including those with potentially contagious diseases), with consideration of issues such as professional liability, worker protection and compensation, licensure, and privacy.

PUBLIC HEALTH ASPECTS OF DMPHP

There are many health system responders who are not clinicians that need to demonstrate proficiency in public health preparedness and response. Although they may not be involved directly in casualty assessment and treatment, the work of these responders is critical to meeting the health needs of affected populations. Actions and interventions that must be considered following the onset of a disaster or PHE include health monitoring and surveillance; outbreak investigation; isolation and quarantine; population-based triage; mass sheltering and feeding; vector control; environmental monitoring; ensuring the safety of food and water supplies; responder and health care worker protection; basic sanitation and hygiene; countermeasure stockpiling, distribution, and dispensing; and management of mass fatalities. This requires basic knowledge of descriptive and analytical epidemiology, laboratory science, environmental and occupational health, infection control, nutrition, effective communication practices and the social sciences.

Health professionals who have direct roles in disaster response should be able to support surveillance efforts and explain the rationale and procedures for case reporting. The basics of risk communication and health messaging will be essential for communicating with affected individuals, their families, and the media regarding exposure risks and potential preventive measures. Finally, just like clinicians, public health responders should know the moral, ethical, and legal issues that are relevant to the management of affected populations and communities and the basic legal framework for public health. They should be familiar with ethical principles that underlie decision making in disasters, such as those impacting allocation of scarce resources.

DEVELOPING CORE CURRICULA AND TRAINING PROGRAMS IN DMPHP

Recent disasters and terrorist events have increased federal interest and attention for the integration of DMPHP into clinical and public health education. In 2006, passage of the Pandemic and All-Hazards Preparedness Act (PAHPA; PL 109-417) created important opportunities to build upon and standardize disaster preparedness education through various programs at the federal, state, and local levels.[9] PAPHA called for the development of integrated, interdisciplinary, and consistent public health and medical disaster response curricula, which would be available to health professionals and health professional schools. Section 304 of the Act states that "the Secretary of Health and Human Services (HHS), in collaboration with the Secretary of Defense, and in consultation with relevant public and private entities, shall develop core health and medical response curricula and training by adapting applicable existing curricula and training programs to improve responses to public health emergencies."

In 2007, HSPD-21 called for federal interagency action and cooperation to ensure that core public health and medical curricula and training developed pursuant to PAHPA address the needs to improve individual, family, and institutional public health and medical preparedness and to develop a mechanism to coordinate public health and medical disaster preparedness and response core curricula and training across executive departments and agencies, to ensure standardization and commonality of knowledge, procedures, and terms of reference within the federal government that also can be communicated to state and local government entities, academia, and the private sector.

To lead federal efforts for the development and delivery of core curricula and training related to medicine and public health in disasters, HSPD-21 specifically calls for the establishment of an academic joint program for disaster medicine and public health, housed at a National Center for Disaster Medicine and Public Health, at the Uniformed Services University of the Health Sciences. The HHS and Department of Defense are required to carry out respective civilian and military missions within this program. In 2009, federal directives aimed at education and training in disaster medicine and public health began to be addressed by the Federal Education and Training Interagency Group. The Group, as authorized under PAHPA, serves as a coordinating body for the delineation of core competencies and education and training standards across federal departments and agencies, as well as state and local government entities, academia, and the private sector in relation to public health emergency and disaster response. The primary charge of this group is to identify and implement a national strategy for the education and training of health professionals in disaster-related medicine and public health. The recently released National Health Security Strategy further emphasizes the importance of professional training, competencies, and standards to help ensure the attainment and maintenance of proficiency by the disaster response workforce.[10]

In 2009, the American Medical Association (AMA) House of Delegates adopted policy calling for formal education and training in DMPHP to be incorporated in all medical school and residency programs.[11] This initiative includes integration of core curricula and training programs to provide a consistent learning experience for physicians-in-training and other students in the health professions. Such training requires consensus on competencies and learning objectives to ensure that course content is based on a well-defined and testable body of knowledge, skill set, and methodology.

To prepare health professionals to respond appropriately and to assist professional schools and continuing education programs to meet this challenge, various organizations and universities have developed competencies for health professionals and other emergency responders.[12-20] To date, many of these efforts have been limited primarily to individual specialties or targeted professionals. This has resulted in a lack of definitional uniformity across professions with respect to education, training, and best practices, thus limiting the establishment of DMPHP at an operational level. To better integrate competencies across all health specialties and professions, a consensus-based educational framework and competency set was published from which educators could devise learning objectives and curricula in DMPHP that are tailored to the needs of all health professionals.[21] The framework includes the delineation of 7 core learning domains and 19 core competencies (Table), as well as 73 specific subcompetencies targeted at 3 broad health personnel categories. A learning matrix also was developed to allow disaster health educators and accreditation entities to incorporate the competencies at any desired proficiency level.

The DMPHP educational framework identifies 3 broad, yet distinct, personnel categories to encompass all health professionals: informed workers/students, practitioners, and leaders. Personnel are expected to perform at different levels of proficiency depending on their experience, professional role, level of education, or job function across the core competencies and subcompetencies. The framework allows for all health professions to be represented in each category, and recognizes the diversity of expected job functions and educational requirements for each health profession involved in disaster prevention, mitigation, response, and recovery. The health personnel categories establish increasing standards for each core competency. The proposed competency set and educational framework were endorsed by the National Disaster Life Support Education Consortium in May 2008. (The Consortium is an unincorporated association jointly sponsored by the AMA and National Disaster Life Support Foundation, Inc, convened by the AMA. It consists of 75 professional organizations and distinguished individuals with interest and expertise in disaster medicine and public health preparedness, as well as experts in professional education and curriculum development, all of whom participate on a voluntary basis.)

Although this vision has been endorsed by many, the implementation is not clear. Decisions about exactly which competencies form the common core for all members of all professions considered to be health professions have not been made. Work that is under way to meet the PAHPA mandate for public health education, for example, does not presume that all public health workers will possess the skills to diagnose individual patient conditions or initiate individual therapies. Similarly, it is unlikely that all licensed physicians and nurses will be expected to have the skills to diagnose and mitigate contamination of a municipal water supply. All of these need a common base that is respectful of all contributions to health and maximizes the efficiency of the health contribution to community readiness, response, and recovery. The DMPHP educational framework provides the best effort to date to facilitate decisions about how best to proceed.

If DMPHP is to be a recognized discipline, then a core standard curriculum must be defined and mastery demonstrated by all who wish to be acknowledged as proficient or "specialist" in this field. Anything less perpetuates the insular "silo" approach that continues today. Specific subcompetencies appropriate for public health practitioners, or certain medical and nursing practitioners, must be considered in addition to the core competencies, however they are defined.

BUILDING THE DMPHP RESEARCH BASE

The effects of conventional disasters and PHEs can be studied through well-established clinical and epidemiological research methods. Such information is critical for adaptation of preparedness, response, and recovery plans. To ensure a sound evidence base for DMPHP, continued research is needed to elu-

TABLE

Core Competencies for All Health Professionals in DMPHP[21]

Competency Domain	Core Competencies
1.0 Preparation and Planning	1.1 Demonstrate proficiency in the use of an all-hazards framework for disaster planning and mitigation.
	1.2 Demonstrate proficiency in addressing the health-related needs, values, and perspectives of all ages and populations in regional, community, and institutional disaster plans.
2.0 Detection and Communication	2.1 Demonstrate proficiency in the detection of and immediate response to a disaster or public health emergency.
	2.2 Demonstrate proficiency in the use of information and communication systems in a disaster or public health emergency.
	2.3 Demonstrate proficiency in addressing cultural, ethnic, religious, linguistic, socioeconomic, and special health-related needs of all ages and populations in regional, community, and institutional emergency communication systems.
3.0 Incident Management and Support Systems	3.1 Demonstrate proficiency in the initiation, deployment, and coordination of national, regional, state, local, and institutional incident command and emergency operations systems.
	3.2 Demonstrate proficiency in the mobilization and coordination of disaster support services.
	3.3 Demonstrate proficiency in the provision of health system surge capacity for the management of mass casualties in a disaster or public health emergency.
4.0 Safety and Security	4.1 Demonstrate proficiency in the prevention and mitigation of health, safety, and security risks to yourself and others in a disaster or public health emergency.
	4.2 Demonstrate proficiency in the selection and use of personal protective equipment at a disaster scene or receiving facility.
	4.3 Demonstrate proficiency in victim decontamination at a disaster scene or receiving facility.
5.0 Clinical/Public Health Assessment and Intervention	5.1 Demonstrate proficiency in the use of triage systems in a disaster or public health emergency.
	5.2 Demonstrate proficiency in the clinical assessment and management of injuries, illnesses, and mental health conditions manifested by all ages and populations in a disaster or public health emergency.
	5.3 Demonstrate proficiency in the management of mass fatalities in a disaster or public health emergency.
	5.4 Demonstrate proficiency in public health interventions to protect the health of all ages, populations, and communities affected by a disaster or public health emergency.
6.0 Contingency, Continuity, and Recovery	6.1 Demonstrate proficiency in the application of contingency interventions for all ages, populations, institutions, and communities affected by a disaster or public health emergency.
	6.2 Demonstrate proficiency in the application of recovery solutions for all ages, populations, institutions, and communities affected by a disaster or public health emergency.
7.0 Public Health Law and Ethics	7.1 Demonstrate proficiency in the application of moral and ethical principles and policies for ensuring access to and availability of health services for all ages, populations, and communities affected by a disaster or public health emergency.
	7.2 Demonstrate proficiency in the application of laws and regulations to protect the health and safety of all ages, populations, and communities affected by a disaster or public health emergency.

cidate the clinical and public health effects of specific disasters; analyze risk factors for adverse social and health effects; and provide for investigation of the effectiveness of clinical and public health interventions and various types of disaster assistance, and the long-term influence of relief operations on the restoration of predisaster conditions. New or modified research tools may be needed to facilitate discoveries in DMPHP.

Dedicated textbooks and peer-reviewed journals, such as *Disaster Medicine and Public Health Preparedness*, are being published to provide the scientific basis and framework for research, education, and training in this field. Additional venues for scholarly discourse in DMPHP include numerous conferences and symposia that have been convened in the United States and abroad. In December 2009, the AMA, in conjunction with the HHS Office of the Assistant Secretary for Preparedness and Response, sponsored the Third National Congress on Health System Readiness. The conference was attended by more than 500 public and private sector health professionals. In February 2010, the National Association of County and City Health Officials held the Fourth Annual Public Health Preparedness Summit, which was attended by approximately 2000 health professionals. In May 2010, the annual Integrated Medical, Public Health, Preparedness and Response Training Summit, sponsored by HHS, was convened as a forum for conducting training, sharing information, and networking among various national organizations involved in preparing for and responding to disasters and public health emergencies. International conferences include the Asia-Pacific Conference on Disaster Medicine as well as meetings sponsored by the World Association for Disaster and Emergency Medicine, the International Society for Disaster Medicine, and the World Health Organization.

Continued validation of principles and practices in DMPHP through sound scientific methods and evidence is fundamental, urgently needed, and essential. Research is needed for the design and evaluation of process and performance measures, educational modalities (eg, lectures, simulations, drills, exercises), and clinical and public health interventions, as well as for the translation of research into improvements in disaster medicine and public health practice. To be meaningful, best practices and performance benchmarks must be evaluated in the context of where these will really be required, in realistic scenarios that involve a community's entire emergency management system, operating as required under the National Response Framework and compliant with the National Incident Management System.

ESTABLISHING THE DISCIPLINE OF DMPHP—THE TIME IS NOW

DMPHP seeks to engage all health professions in efforts to prepare for, respond to, and recover from disasters and PHEs. Because DMPHP relies on the amalgamation of knowledge about health issues affecting individuals and populations in a disaster or public health emergency, it does not belong to any single specialty, profession, or discipline—it belongs to all. It is not simply an extension of dentistry, medicine, nursing, mental health, pharmacy, or a branch of public health. Rather, the discipline extends to all health care and public health professionals whose expertise supports the health-related capacity of emergency response systems. DMPHP is unique in that it can be considered a secondary discipline of all health professionals, as they seek to fulfill professional and societal obligations to patients, populations, and communities in a disaster or public health emergency.

Education and training in DMPHP should be integrated as a basic element of lifelong learning for all clinical and public health professionals. Considering the relevance of this field for all health professionals, schools and entities responsible for the training, continuing education, credentialing, and certification of health professionals should incorporate cross-cutting competencies in DMPHP into curricula at the undergraduate, graduate, and postgraduate levels. Mechanisms must be developed to coordinate public health and clinical disaster preparedness and response education in the public and private sectors to ensure standardization and commonality of knowledge, procedures, and terms of reference.

Core curricula and training programs are needed to provide a consistent learning experience for all health professionals. Developing such curricula presents a daunting challenge—disasters, terrorism, and public health emergencies can occur in multiple scenarios, with diverse clinical and public health outcomes, many of which are not addressed in current health professional education. Certainly, DMPHP topic areas must be relevant to the roles they will play and be reasonably attainable, considering time and financial resources. Despite the challenges of integrating new content into existing health professional curricula, the risk of not doing so can no longer be ignored.

DMPHP is more than just clinical care and public health. There are also major elements of politics, economics, social sciences, and logistics that must exist to plan and respond effectively. DMPHP professionals provide care, leadership, and community guidance throughout all phases of a disaster. They serve to interface with public safety and emergency management personnel, government agency officials, legislators, and the media, and facilitate coordination of private and public sector disaster response assets. As colleagues of a formally recognized discipline, DMPHP professionals can provide the essential expertise and leadership to facilitate the integration of the clinical and public health sectors as well as civilian-military coordination that forms a resilient national disaster health system.

A new organizational entity that has the committed resources to provide comprehensive, dedicated leadership and support for the promotion and advancement of this field is needed to provide the structure and means for sustaining multiprofessional interaction and discourse in DMPHP, with a broad membership. As the umbrella organization for DMPHP, this entity could develop and foster mechanisms to coordinate public health and clinical disaster preparedness and response core curricula and training across professions. As envisioned, the mission of this organization would be to achieve and promote excellence in education, training, and research related to DMPHP for all health professionals based on sound educational principles, scientific evidence, and best clinical and public health practices. To fulfill this mission and realize its desired impact, this new organization would support a membership dedicated to formalized, lifelong learning in DMPHP with a shared vision to create a network of personnel who are ready, willing, and able to meet the health and safety needs of all ages and populations affected by disasters and public health emergencies.

This editorial was unanimously endorsed by all DMPHP Board members in attendance at the annual Editorial Board meeting on April 28, 2010.

Acknowledgments

The authors gratefully acknowledge James Lyznicki, MS, MPH, and Edbert Hsu, MD, MPH, for their contributions to this commentary.

1. Brown RKB. Disaster medicine. What is it? Can it be taught? *JAMA*. 1966; 197:133-136.
2. Gunn SWA, Masellis M. The scientific basis of disaster medicine. *Ann MBC*. 1992;5:51-55 http://www.medbc.com/annals/review/vol_5/num_1/text/vol5n1p51.htm. Accessed March 12, 2010.
3. Homeland Security Presidential Directive 21 (HSPD-21). Public Health and Medical Preparedness. Washington, DC: The White House. October 18, 2007. http://fas.org/irp/offdocs/nspd/hspd-21.htm. Accessed March 12, 2010.
4. Institute of Medicine Report Series on the Future of Emergency Care in the U.S. Health System: (a) *Emergency Care for Children: Growing Pains*. (b) *Emergency Medical Services at the Crossroads*. (c) *Hospital-Based Emergency Care: At the Breaking Point*. Washington, DC: National Academies Press; 2006.
5. American Medical Association, American Public Health Association. *Improving Health System Preparedness for Terrorism and Mass Casualty Events: Recommendations for Action. A Consensus Report of the AMA/APHA Linkages Leadership Summit*. Chicago: American Medical Association; 2007. http://www.ama-assn.org/ama1/pub/upload/mm/415/final_summit_report.pdf. Accessed March 12, 2010.
6. Institute of Medicine. *Research Priorities in Emergency Preparedness and Response for Public Health Systems: A Letter Report*. Washington, DC: National Academies Press; 2008. http://www.nap.edu/catalog.php?record_id=12136. Accessed March 12, 2010.
7. Cameron-Traub E. An evolving discipline. In Gray C, Pratt R, eds. *Towards a Discipline of Nursing*. Melbourne, Australia: Churchill Livingstone; 1991.
8. James JJ, Subbarao I, Lanier WL. Improving the art and science of disaster medicine and public health preparedness. *Mayo Clin Proc*. 2008;83: 559-562.
9. Pandemic and All-Hazards Preparedness Act (PAHPA); 2006. http://www.hhs.gov/aspr/opsp/pahpa/index.html. Accessed March 12, 2010.
10. US Department of Health and Human Services (HHS). *National Health Security Strategy of the United States of America*. Washington, DC: HHS; 2009. http://www.hhs.gov/aspr/opsp/nhss/nhss0912.pdf. Accessed March 12, 2010.
11. Policy H-295.868; AMA Policy Database. As cited in Report 15 of the Council on Medical Education. *Education in Disaster Medicine and Public Health Preparedness During Medical School and Residency Training*. CME Report 15 (A-09). Chicago: American Medical Association; 2009.
12. Association of Schools of Public Health. *Public Health Preparedness and Response Core Competency Development Project*. http://www.asph.org/document.cfm?page=1081. Accessed March 12, 2010.
13. Medical Reserve Corps. *MRC Core Competencies Matrix*. Washington, DC: Office of the Surgeon General; 2007. http://www.medicalreservecorps.gov/File/MRC%20TRAIN/Core%20Competency%20Resources/Core_Competencies_Matrix_April_2007.pdf. Accessed March 12, 2010.
14. Hsu EB, Thomas TL, Bass EB, et al. Healthcare worker competencies for disaster training. *BMC Med Educ*. 2006;6:1-9 http://www.biomedcentral.com/1472-6920/6/19. Accessed March 12, 2010.
15. Barbara JA, Macintyre AG, Shaw G, et al. *VHA-EMA Emergency Response and Recovery Competencies: Competency Survey, Analysis, and Report*. Washington, DC: Institute for Crisis, Disaster, and Risk Management, The George Washington University; 2005.
16. Hospital Core Competency Sub Committee and Health, Medical, Hospital, and EMS Committee Florida State Working Group. *State of Florida Recommended Core Competencies & Planning/Mitigation Strategies for Hospital Personnel*; 2004. http://www.emlrc.org/pdfs/disaster2005presentations/HospitalDisasterMgmtCoreCompetencies.pdf. Accessed March 12, 2010.
17. *Educational Competencies for Registered Nurses Responding to Mass Casualty Incidents*. Nashville: International Nursing Coalition for Mass Casualty Education; 2003. http://www.aacn.nche.edu/Education/pdf/INCMCECompetencies.pdf. Accessed March 12, 2010.
18. Center for Public Health Preparedness, Columbia University Mailman School of Public Health and the Center for Health Policy, Columbia University School of Nursing, Greater New York Hospital Association, The Commonwealth Fund. *Emergency Preparedness and Response Competencies for Hospital Workers*. New York: Center for Health Policy, Columbia University School of Nursing; 2003. http://www.ncdp.mailman.columbia.edu/files/hospcomps.pdf. Accessed March 12, 2010.
19. Center for Health Policy, Columbia University School of Nursing. *Bioterrorism and Emergency Readiness: Competencies for All Public Health Workers*. New York: Center for Health Policy, Columbia University School of Nursing; 2002. https://www.train.org/Competencies/btcomps.pdf. Accessed March 12, 2010.
20. American College of Emergency Physicians NBC Task Force. *Developing Objectives, Content, and Competencies for the Training of Emergency Medical Technicians, Emergency Physicians, and Emergency Nurses to Care for Casualties from Nuclear, Biological, or Chemical (NBC) Incidents: Final Report*. Washington, DC: Department of Health and Human Services, Office of Emergency Preparedness; 2001.
21. Subbarao I, Lyznicki JM, Hsu EB, et al. A consensus-based educational framework and competency set for the discipline of disaster medicine and public health preparedness. *Disaster Med Public Health Preparedness*. 2008; 2:57-68.

POWERING THROUGH

From Fragile Infrastructures

To Community Resilience

Electromagnetic Pulse Special Interest Group

Purpose: A Call to Action

Be Prepared for a Nation-wide Long Term Electric Grid Failure—Weeks, Months or Years

- Planning for consequence management
- Mobilizing whole of community
- Create Resilient Community Islands

Suggestions and Resources for Sectors

- Electrical Power
- Water/Wastewater
- Food
- Communications
- Transportation
- Financial services
- Medical and Public Health
- Security
- Sector Interdependencies

Suggestions and Resources for Key Actors

- Citizens
- Urban Communities
- Rural Communities
- Non-government Organizations
- Universities and Community Colleges
- Businesses
- Supply Chains
- Emergency Organizations
- Local and State Governments
- National Guard
- Federal Government

To Obtain the Book

- Contact: www.empcenter.org

POWERING THROUGH
FOR HIGH IMPACT THREATS TO THE POWER GRID
VERSION 1.0

A Preparedness Initiative by the InfraGard EMP SIG		
Title Page & Foreword		
Section	**Title**	**Description**
Executive Summary	Executive Summary	States purposes; identifies authors; links five grid threats (EMP, solar storms, cyber-attacks, physical attack & radio frequency weapons) to equipment and functions at risk (see color chart)
Chapter I	Introduction	Purpose of the guide; intended audience; high-level walk-through of the chapters and their content
Chapter II	Power Grids and Interdependencies	Composition and importance of the grid; complex dependencies and interdependencies and other U.S. Critical Infrastructure Sectors; grid vulnerabilities; projections relating to cascading and catastrophic impacts of long-duration grid failures.
Chapter III	U.S. National Framework for Disaster Preparedness	Background on the overall constitutional, statutory and policy framework for disaster preparedness and response
Chapter IV	Enhancing Nationwide Grid Preparedness	Suggested approaches and specific actions for enhancing national grid preparedness
Chapter V	Enhancing Preparedness of Other Critical Infrastructures	Importance of blackout preparedness across selected critical infrastructure sectors; general approaches to enhancing preparedness; suggested actions for several particular sectors
Chapter VI	Resilient Communities using an Island Concept	Emerging concepts for building resilience that would promote individual survival, community recovery, and societal stability
Chapter VII	Key Preparedness Actors	Identification of key actors in "whole of nation" grid preparedness, along with their respective roles and potential to contribute
Chapter VIII	Response and Recovery in Long-Duration Nationwide Grid Failures	Recommended approaches and specific actions related to post-event response and recovery efforts in grid failures ranging from weeks to months or even years
Chapter IX	Conclusions	A call to action for preparedness & a discussion of Version 2.0
Appendix 1	Glossary	Definitions for key terms used in the guide
Appendix 2	Planning and Response: Frameworks and Tools	Overview and crosswalk of different planning systems
Appendix 3	Preparedness Maturity Model	A maturity model for evaluating preparedness
Appendix 4	Cybersecurity and Industrial Control / SCADA Systems	Discussion of how cyber attacks target the grid
Appendix 5	Blackstarting the Power Grid	A plan for restarting the grid after a failure
Appendix 6	Supply Chain Resiliency	An examination of the supply chain and its ability to respond
Appendix 7	International Disaster Assistance	FEMA vision for U.S. incorporation of international disaster assistance
Appendix 8	Resilience Legislative	The highlights of legislative action federally and in the states with their current plans.
Appendix 9	Reference Material & Organizational Contacts	Select references used in developing the guide and appropriate resources for further research on preparedness
Appendix 10	Contributor Biographies	Brief background on key contributors to the guide

The following bibliographical materials have been consolidated into the planning guide from the EMP SIG named, Powering Through.

APPENDIX 9: Reference Material & Organizational Contacts

1. InfraGard EMP SIG: Triple Threat Power Grid Exercise. (Published Nov 2015 with a companion Power Point Presentation and a facilitator's guide via controlled circulation) The back cover provides a quick overview of the key issue regarding space weather from their own goal #2 and the link to "other high impact threats". We focus on the concept that this is the first time that the federal government has been willing to admit that they won't always be able to rescue us by day 4 after a disaster. In the case of high-impact events it may be day 40 or day 400. The introduction provides an overview to the history of the space weather issue and the related EMP issue brought up the by the DoD DTRA RFP asking for a systematic approach to providing EMP protected microgrids for military bases and the entities they need off base. The book also has a section for "read ahead" material for an exercise that also provides an overview of the threats. Both the read-ahead material and the day of the event material include bibliographies for further reading. See Amazon – High Impact Threats
2. InfraGard EMP SIG: Planning Resilience for High-Impact Threats: This conference proceedings (Dec 2015) provides verbatim transcripts of the previous conference (2014) speakers with hyperlinks to the recorded video presentations from Sen. Ron Johnson, Congressman Bartlett and other federal, state, local officials and leaders from the private sector. It also includes additional publications and official documents that arose since the last conference. In this way, someone who comes to the conference gets an overview of what happened the year before and key items since then so they can get up-to-date quickly. Key documents include the Space Weather Strategy, the DTRA RFP for EMP protected microgrids, and the formation documents of a major engineering association, the International Council on Systems Engineering, that just created the equivalent of the EMP SIG in their association called the Critical Infrastructure Protection and Recovery Working Group. A hyperlinked bibliography is also available. See Amazon – High Impact Threats
3. InfraGard EMP SIG: Mitigating High-Impact Threats to Critical Infrastructure: This previous conference proceedings (Dec 2014) has great shelf life since it covers the 2013 presentations by Bill Murtagh of NOAA covering the super solar storm near miss of 2012 just prior to the publications of his articles that led to his assignment as lead of the White House group organizing the National Space Weather Strategy and Action Plan. It also featured a presentation by Scott McBride of Idaho National Laboratory. This was the briefer public presentation on tests of live power grids at

their location (greater than 600 sq. miles) that included Emprimus equipment with live transformers. We provided a full day of private briefings for DOE officials who were banned from travel during the sequestration. The bibliography included 80 links to rare materials on EMP including the economic impact assessment I commissioned with support of the EMP Commission. See Amazon – High Impact Threats

4. InfraGard EMP SIG: High-Impact Threats to Critical Infrastructure: This was the first conference proceedings (published Dec 2013) covering the second conference (2012) after our initial work at the National Defense University where we instigated the first comprehensive DoD evaluation of a nationwide collapse of infrastructure due to space weather that included participation from the White House, DoD, DOE, DHS, National Governors Association, counties, cities, FERC, NOAA, FBI, InfraGard and others from the private sector. This conference included first hand discussion of the Fukushima disaster from Mr. Karakawa of Japan. See Amazon – High Impact Threats

2015 National Space Weather Strategy
-White House National Science & Technology Council

The National Space Weather Strategy (Strategy) and the accompanying National Space Weather Action Plan (Action Plan) together seek to enhance the integration of existing national efforts and to add important capabilities to help meet growing demands for space-weather information.

October 2015

https://www.whitehouse.gov/sites/default/files/microsites/ostp/final_nationalspaceweatherstrategy_20151028.pdf

Alternate Funding for Disaster Recovery A Guide to Available Federal Programs and Funding Resources

Prepared by: Witt O'Brien's

Washington, DC 20005

The purpose of this guide is to provide basic information about programs of assistance available to individuals, businesses, and public entities after a disaster. These programs help individuals cope with their losses, and affected businesses and public entities restore their structures and operations.

February 2013

http://www.wittobriens.com/external/content/document/2000/1883986/1/Witt-O'Brien's-Alternate-Disaster-Funding_Feb2013.pdf

Agricultural Disaster Assistance

Congressional Research Service (CRS)

Megan Stubbs
Specialist in Agricultural Conservation and Natural Resources Policy
April 14, 2016
The federal crop insurance program offers subsidized policies designed to protect crop producers from unavoidable risks associated with adverse weather and weather-related plant diseases and insect infestations.
https://www.fas.org/sgp/crs/misc/RS21212.pdf

City Resilience Index 2014
Rockefeller Foundation
April 2014
Risk assessments and measures to reduce specific foreseeable risks will continue to play an important role in urban planning.
In addition, cities need to ensure that their development strategies and investment decisions enhance, rather than undermine the city's resilience.
https://www.rockefellerfoundation.org/app/uploads/City-Resilience-Framework1.pdf

City Resilience Index 2015
Rockefeller Foundation
December 2015
This report provides a holistic articulation of city resilience, structured around four dimensions and 12 goals that are critical for the resilience of our cities.
https://www.rockefellerfoundation.org/report/city-resilience-framework

Cloud Computing Strategy, Department of Defense, Chief Information Officer
July 2012
http://dodcio.defense.gov/Portals/0/Documents/Cloud/DoD%20Cloud%20Computing%20Strategy%20Final%20with%20Memo%20-%20July%205%202012.pdf

Common Cybersecurity Vulnerabilities in Industrial Control Systems
Industrial Control Systems Cyber Emergency Response Team
Trent Nelson, Project Manager, Idaho National Laboratory;
May Chaffin, Cyber Researcher, Idaho National Laboratory
May 2011
Retrieved February 20, 2016

Correlated and compiled in this report are vulnerabilities from general knowledge gained from DHS CSSP assessments and
Industrial Control Systems Cyber Emergency Response Team (ICS-CERT) activities describing the most common types of cybersecurity vulnerabilities as they relate to ICS.
https://ics-cert.us-cert.gov/sites/default/files/recommended_practices/DHS_Common_Cybersecurity_Vulnerabilities_ICS_2010.pdf

Critical Infrastructure Protection
Federal Agencies Have Taken Actions to Address Electromagnetic Risks, but Opportunities Exist to Further Assess Risks and Strengthen Collaboration
United States General Accountability Office (GAO)
March 2016
http://www.gao.gov/products/GAO-16-243

Cross-sector emergency planning for water providers and healthcare facilities
Publications: Journal - American Water Works Association
Welter, Gregory; Bieber, Steven; Bonnaffon, Heidi; Deguida, Nicholas; Socher, Myra
Issue Date: January 2010
Volume / Number: 102, Number 1
Water professionals can best serve their communities by reaching out to their local emergency management agencies and hospital associations and work with them to create successful and cost-effective emergency plans.
http://www.awwa.org/publications/journal-awwa/abstract/articleid/23327.aspx

Defense Support of Civil Authorities
Joint Publication 3-28
July 31, 2013
This publication provides overarching guidelines and principles to assist commanders and their staffs in planning, conducting, and assessing defense support of civil authorities (DSCA)
http://www.dtic.mil/doctrine/new_pubs/jp3_27.pdf

EIS (Electric Infrastructure Protection Council) "Electric Infrastructure Protection (EPRO®) Handbook II Volume 2 – Water"

This two-volume work, EPRO Handbook II, provides options to strengthen the resilience of two especially vital infrastructure components against Black Sky power outages: the water sector, with its uniquely essential services, and the natural gas infrastructure on which power generation increasingly depends. EPRO II provides a detailed range of preparedness options that infrastructure owners and operators can consider and adapt to help meet their own system specific needs. The Handbook also offers a methodology for developing Black Sky playbooks to help these sectors guide emergency operational planning and resilience initiatives.
July 18, 2016
http://eiscouncil.org/App_Data/Upload/7f41c325-654e-4c67-be3d-6941645f4485.pdf

EIS (Electric Infrastructure Protection Council) "Electric Infrastructure Protection (EPRO®) Handbook II Volume 1 – Resilient Fuel Resources for Power Generation in Black Sky Events"
Infrastructure Protection
July 18, 2016
This two-volume work, EPRO Handbook II, provides options to strengthen the resilience of two especially vital infrastructure components against Black Sky power outages: the water sector, with its uniquely essential services, and the natural gas infrastructure on which power generation increasingly depends. EPRO II provides a detailed range of preparedness options that infrastructure owners and operators can consider and adapt to help meet their own system specific needs. The Handbook also offers a methodology for developing Black Sky playbooks to help these sectors guide emergency operational planning and resilience initiatives.
http://www.eiscouncil.com/App_Data/Upload/149e7a61-5d8e-4af3-bdbf-68dce1b832b0.pdf

Electromagnetic Pulse (EMP) Attack: A PREVENTABLE Homeland Security Catastrophe
The Heritage Foundation 2008
By Jena Baker McNeill and Richard Weitz, Ph.D.
http://www.heritage.org/Research/Reports/2008/10/Electromagnetic-Pulse-EMP-Attack-A-Preventable-Homeland-Security-Catastrophe

Electromagnetic Pulse Threats To U.S. Military and Civilian Infrastructure [H.A.S.C. No. 106–31]
Testimony to the House Committee on Homeland Security

Hearing before the Military Research and Development Subcommittee
of The Committee on Armed Services House of Representatives
One Hundred Sixth Congress, 1999
http://commdocs.house.gov/committees/security/has280010.000/has280010_0f.htm

EMP: America's Achilles Heel
Hillsdale College
Frank J. Gaffney
President, Center for Security Policy
June 2005 • Volume 34, Number 6
http://imprimis.hillsdale.edu/emp-americas-achilles-heel/

Energy Sector-Specific Plan (SSP)
U.S. Department of Homeland Security
2015
The 2015 Energy SSP is closely aligned with the National Infrastructure Protection Plan 2013: "Partnering for Critical Infrastructure Security and Resilience" (NIPP 2013) and the joint national priorities, which were developed in collaboration by representatives from all critical infrastructure sectors, including Energy.
https://www.dhs.gov/sites/default/files/publications/nipp-ssp-energy-2015-508.pdf

Executive Order 13636, Improving Critical Infrastructure Cybersecurity
The White House Executive Order
February 2013
https://www.whitehouse.gov/the-press-office/2013/02/12/executive-order-improving-critical-infrastructure-cybersecurity

Grid Assurance
2011
Strategic Resiliency Solutions Company
http://www.gridassurance.com

Guest Editorial: Emerging Technologies
IEEE JOURNAL ON SELECTED AREAS IN COMMUNICATIONS, VOL. 34, NO. 3, 457
Shuguang Cui, Fellow, IEEE, John S. Thompson, Fellow, IEEE, Tomohiko Taniguchi, Fellow, IEEE,

Latif Ladid, Member, IEEE, Jie Li, Senior Member, IEEE, Andrew Eckford, Senior Member, IEEE,
and Vincent W.S. Wong, Fellow, IEEE
March 2016
http://ieeexplore.ieee.org/stamp/stamp.jsp?arnumber=7430394

Guidance on Diagnosis and Treatment for Healthcare Providers
Radiation Emergency Medical Management (REMM)
US Department of Health and Human Services
http://www.remm.nlm.gov

Guide to Industrial Control Systems (ICS) Security
National Institute of Standards and Technology
May 2011
U.S. Department of Homeland Security
Retrieved February 20, 2016
http://nvlpubs.nist.gov/nistpubs/SpecialPublications/NIST.SP.800-82r2.pdf

Healthcare and Public Health Sector-Specific Plan
U.S. Department of Homeland Security
May 2016
The Sector's integrated approach to managing all-hazards risks to HPH (Healthcare/Public Health) critical infrastructure and the HPH workforce includes several key components:
•Identifying and preparing for a range of potential threats and hazards;
•Reducing the vulnerabilities of identified critical assets, systems, and networks, including those associated with critical internal and out-of-sector dependencies and interdependencies;
•Mitigating the potential impacts to and enabling the timely restoration of critical infrastructure as a result of emergencies that do occur; and
•Adapting to changing conditions to withstand and rapidly recover from disruptions due to emergencies, irrespective of the causal factors.
http://www.phe.gov/Preparedness/planning/cip/Documents/2016-hph-ssp.pdf

Health Physics Society
General guidance for rad emergency response
http://www.hps.org

High Altitude Electromagnetic Pulse (HEMP) and High Power Microwave (HPM) Devices: Threat Assessments Congressional Report Services 2004 – Updated August 2008
Clay Wilson
Specialist in Technology and National Security
Foreign Affairs, Defense, and Trade Division
http://www.fas.org/man/crs/RL32544.pdf

How to Protect Your Computer
Federal Bureau of Investigation
Harp, D., & Gregory-Brown, B.
June 2015
Retrieved February 20, 2016
https://www.fbi.gov/scams-safety/computer_protect

How to stay safe and help others after a nuclear explosion
Citizen web page, Ventura County, California.
http://vchca.org/nuclear-educational-campaign/information-for-community-members

HSPD 8 "National Preparedness"
Department of Homeland Security
December 13, 2003
This directive establishes policies to strengthen the preparedness of the United States to prevent and respond to threatened or actual domestic terrorist attacks, major disasters, and other emergencies by requiring a national domestic all-hazards preparedness goal, establishing mechanisms for improved delivery of Federal preparedness assistance to State and local governments, and outlining actions to strengthen preparedness capabilities of Federal, State, and local entities.
http://fas.org/irp/offdocs/nspd/hspd-8.html

Implementation of resilient production systems by production control
Robust Manufacturing Conference (RoMaC 2014)
Institute of Ergonomics, Manufacturing Systems and Automation, Otto von Guericke University, Universitätsplatz 2, Magdeburg 39106, Germany
Matthias Heinicke*

Procedia CIRP 19 (2014) 105 – 110
2212-8271 © 2014 Elsevier B.V.
Selection and peer-review under responsibility of the International Scientific Committee of "RoMaC 2014"
http://www.sciencedirect.com/science/article/pii/S2212827114006295

LDS Preparedness Manual
http://thesurvivalmom.com/wp-content/uploads/2010/08/LDS-Preparedness-Manual.pdf
January 1, 2011

Lighthouse Prime
Training and support for neighbors interested in sustaining good communications and resilient communities in times of trouble.
www.LightHousePrime.com

Lights Out
David Crawford
2010, Post-EMP event fiction book. Detailed description of the event and post-event society
http://www.lightsoutsaga.com/saga.html

Lights Out: A Cyberattack, A Nation Unprepared, Surviving the Aftermath
Ted Koppel, Host of ABC's Nightline
October 2015
http://tedkoppellightsout.com

Mother Earth News
For local sustainability
motherearthnews.com

National Alliance for Radiation Readiness
NARR seeks to address the problems of limited visibility for radiation preparedness, confusion about roles and responsibilities in a radiological incident among partners, and the need for robust tools for practitioners in the field.
http://www.radiationready.org

NARUC Regional Mutual Assistance Groups: A Primer

Created under the National Council on Electricity Assurance program, a project of the National Association of Regulatory Utility Co
Miles Keogh, Sharon Thomas, NARUC Grants & Research
November 2015
Electric utilities across the country have been providing mutual aid to each other during emergencies for years. One strategy for communicating and coordinating information as well as tangible resources needed on a wider scale is to use regional mutual assistance groups (RMAGs). This paper explains what an RMAG is, identifies some of the reasons why they are a central mechanism for assuring electric grid reliability and resilience of the power system, and offers suggestions for how we can take a great idea and make it even stronger and better.
https://pubs.naruc.org/pub/536E475E-2354-D714-5130-C13478337428

National Space Weather Action Plan
White House National Science and Technology Council
October 2015
The National Space Weather Strategy (Strategy), released concurrently with this National Space Weather Action Plan (Action Plan), details national goals for leveraging existing policies and ongoing research and development efforts regarding space weather while promoting enhanced domestic and international coordination and cooperation across public and private sectors.
https://www.whitehouse.gov/sites/default/files/microsites/ostp/final_nationalspaceweatheractionplan_20151028.pdf

NCFRP, REPORT 30: Making U.S. Ports Resilient as Part of Extended Intermodal Supply Chains
Transportation Research Board of the National Academies
2014
TRB's National Cooperative Freight Research Program (NCFRP) Report 30: Making U.S. Ports Resilient as Part of Extended Intermodal Supply Chains focuses on identifying and elaborating on the steps needed to coordinate freight movements through ports in times of severe stress on existing operating infrastructures and services.
http://onlinepubs.trb.org/onlinepubs/ncfrp/ncfrp_rpt_030.pdf

National Infrastructure Protection Plan (NIPP), the second edition of which was published by DHS

NIPP 2013: Partnering for Critical Infrastructure Security and Resilience
December 2013
In February 2013, the President issued Presidential Policy Directive 21 (PPD-21), "Critical Infrastructure Security and Resilience", which explicitly calls for an update to the National Infrastructure Protection Plan (NIPP). This update is informed by significant evolution in the critical infrastructure risk, policy, and operating environments, as well as experience gained and lessons learned since the NIPP was last issued in 2009.
https://www.dhs.gov/sites/default/files/publications/NIPP%202013_Partnering%20for%20Critical%20Infrastructure%20Security%20and%20Resilience_508_0.pdf

National Planning System
U.S. Department of Homeland Security
February 2016
This document contains an overview of the National Planning System and includes:
- The Planning Architecture, which describes the strategic, operational, and tactical levels of planning and planning integration; and
- The Planning Process, which describes the steps necessary to develop a comprehensive plan, from forming a team to implementing the plan.
http://www.dhsem.state.co.us/sites/default/files/National_Planning_System_20151029.pdf

National Strategy for Global Supply Chain Security Implementation Update
US Government
January 2013
United States Government's policy to strengthen the global supply chain to protect the welfare and interests of the American people and to enhance our Nation's economic prosperity.
https://www.whitehouse.gov/sites/default/files/national_strategy_for_global_supply_chain_security.pdf

Neighborhood Preparedness Consortium (NPC)
Directory of Resources,
George Washington University
The Neighborhood Preparedness Consortium (NPC) is a private-public emergency management partnership in the greater

Foggy Bottom neighborhood in Washington, DC comprised of local, regional, federal, international, corporate and non-profit partners. For information, email: npc@gwu.edu
https://campusadvisories.gwu.edu/neighborhood-preparedness-consortium
https://campusadvisories.gwu.edu/partnerships

NERC Reliability Standard BAL-002
SERC Regional Criteria Contingency Reserve Policy
SERC Reliability Corporation
October 28, 2014
This document establishes standard terminology and minimum requirements governing the amount, availability, distribution, and activation of Contingency Reserve within the SERC Region in conformance with NERC requirements and provides documentation of current Contingency Reserve practices of SERC member systems (Balancing Authorities).
http://www.serc1.org/docs/default-source/program-areas/standards-regional-criteria/regional-criteria/serc-reg-criteria_contingency-reserve-policy-rev-6-(10-28-14).pdf?sfvrsn=2)

Mutual assistance between companies during emergencies takes several forms. During more routine emergencies (e.g., high loads due to extreme weather or loss of a large unit), neighboring NERC Balancing Authorities (BAs) have emergency assistance provisions (are required in NERC Reliability Standards) for capacity and energy. Since these are bi-lateral agreements between private companies, I know of none that post that information publicly. In addition to one-to-one assistance, a number of entities have joined together to form Reserve Sharing Groups (RSGs), which allow sharing of emergency assistance among multiple companies. Several do not post information publicly, but an example that does is the Northwest Power Pool
http://www.nwpp.org/our-resources/NWPP-Reserve-Sharing-Group

NERC Special Report: Spare Equipment Database (SED) System
August 2011 Report
The SED is primarily a tool to be developed, populated, and managed by participating organizations to facilitate timely communications between those needing long-lead time equipment damaged in a HILF event and those equipment owners who may be able to share
existing equipment being held as spares by their organization.
http://www.nerc.com/docs/pc/sedtf/SEDTF_Report_Draft_PC_Meeting_2.pdf

NIMS ICS Unified Command and Control

Keeping "Pollution Catastrophe" off Katrina's resume' of tragic consequences

CDR Roger Laferriere, U.S. Coastguard Deputy Sector Commander, Honolulu, Hawaii
Mr. Tracy Long, Security/Emergency Response Advisor, Chevron Pipe Line Company
Mr. Greg Guerriero, Incident Commander, Shell Oil Products U.S.
http://uscgproceedings.epubxp.com/i/85793-win-2006-07/27

NIST Draft Cyber-Physical Systems Framework Available for Public Comment

... the draft framework is intended to describe foundational concepts and provide a methodology for understanding, designing, and building CPS that can work with one another. As such, it is hoped that the framework will provide a basis upon which tools, standards, and applications can be based.

October 8, 2015
https://pages.nist.gov/cpspwg/
https://www.ansi.org/news_publications/news_story.aspx?menuid=7&articleid=d00342ab-e834-4062-8e8e-3beb4f102db5

One Second After

William R. Forstchen
2009, Historical Fiction on a worst-case EMP scenario
http://www.onesecondafter.com

Planning Guidance for Response to a Nuclear Detonation – 2nd Edition

National Security Staff, Interagency Policy Coordination Subcommittee for Preparedness & Response to Radiological and Nuclear Threats
June 2010
http://www.remm.nlm.gov/PlanningGuidanceNuclearDetonation.pdf
Thorough and easier to read. 1.5MB

Planning Resilience for High-Impact Threats to Critical Infrastructure: Conference Proceedings InfraGard National EMP SIG Sessions at the 2014 DuPont Summit

Charles L. Manto, Stephanie Lokmer
November 10, 2015
The InfraGard National Electromagnetic Pulse Special Interest Group (EMP SIG) was formed in July 2011 for the purpose of sharing information about catastrophic threats to

our nation's critical infrastructure. Those threats include extreme space weather, manmade EMP, cyber-attacks, coordinated physical attacks and pandemics.
https://www.amazon.com/Planning-Resilience-High-Impact-Critical-Infrastructure/dp/1633912612/ref=sr_1_1?s=books&ie=UTF8&qid=1472007279&sr=1-1&keywords=Planning+Resilience+for+High+Impact

Presidential Policy Directive (PPD) 8 entitled "National Preparedness"
March 30, 2011
PPD-8 is aimed at strengthening the security and resilience of the United States through systematic preparation for the threats that pose the greatest risk to the security of the nation, including acts of terrorism, cyber-attacks, pandemics, and catastrophic natural disasters.2011
https://www.dhs.gov/presidential-policy-directive-8-national-preparedness

Presidential Policy Directive (PPD) 21
The White House
Critical Infrastructure Security and Resilience
February 12, 2013
The Presidential Policy Directive (PPD) on Critical Infrastructure Security and Resilience advances a national unity of effort to strengthen and maintain secure, functioning, and resilient critical infrastructure.
https://www.whitehouse.gov/the-press-office/2013/02/12/presidential-policy-directive-critical-infrastructure-security-and-resil

Press Release by DTRA,(DoD) "Accelerating Society-wide EMP Protection of Critical Infrastructure and Microgrids"
The Defense Threat Reduction Agency (DTRA)/SCC announces the beginning of a Small Business Innovation Research (SBIR) contract with Instant Access Networks, LLC (IAN) and its subcontractors as of March 28, 2016 entitled, "Accelerating Society-wide EMP Protection of Critical Infrastructure and Micro-grids".
June 24, 2016
http://highfrontier.org/wp-content/uploads/2016/07/DTRA-IAN-Press-Release-June-24-2016.pdf

Protect Your Computer From Viruses, Hackers, and Spies. Retrieved February 20, 2016
State of California Department of Justice
Office of the Attorney General
Stouffer, K., Pillitteri, V., Lightman, S., Abrams, M., & Hahn, A.
May 2015
https://oag.ca.gov/privacy/facts/online-privacy/protect-your-computer

Radiation Resilient City Initiative
UPMC Center for Health Security
A Local Planning Tool to Save Lives Following a Nuclear Detonation
http://www.radresilientcity.org

Report of the Commission to Assess the Threat to the United States from Electromagnetic Pulse (EMP) Attack Presidential Commission 2004
http://www.empcommission.org/docs/empc_exec_rpt.pdf

Report of the Commission to Assess the Threat to the United States from Electromagnetic Pulse (EMP) Attack
Critical National Infrastructures
Commission Members
Dr. John S. Foster, Jr.
Mr. Earl Gjelde
Dr. William R. Graham (Chairman)
Dr. Robert J. Hermann
Mr. Henry (Hank) M. Kluepfel
Gen Richard L. Lawson, USAF (Ret.)
Dr. Gordon K. Soper
Dr. Lowell L. Wood, Jr.
Dr. Joan B. Woodard
April 2008
This report presents the results of the Commission's assessment of the effects of a high altitude electromagnetic pulse (EMP) attack on our critical national infrastructures and provides recommendations for mitigation.
http://www.empcommission.org/docs/A2473-EMP_Commission-7MB.pdf

Revised Critical Infrastructure Protection Reliability Standards

156 FERC ¶ 61,050
UNITED STATES OF AMERICA FEDERAL ENERGY REGULATORY COMMISSION 18 CFR Part 40
[Docket No. RM15-14-002; Order No. 829]
Issued July 21, 2016
The Federal Energy Regulatory Commission (Commission) directs the North American Electric Reliability Corporation to develop a new or modified Reliability Standard that addresses supply chain risk management for industrial control system hardware, software, and computing and networking services associated with bulk electric system operations. The new or modified Reliability Standard is intended to mitigate the risk of a cybersecurity incident affecting the reliable operation of the Bulk-Power System.
https://www.ferc.gov/whats-new/comm-meet/2016/072116/E-8.pdf

Selling the apocalypse
Congress Blog feed
The Hill
By David Stuckenberg
May 10, 2016
http://thehill.com/blogs/congress-blog/homeland-security/279165-selling-the-apocalypse

Society of Nuclear Medicine & Molecular Imaging
http://www.snm.org
Radiation Injury Treatment Network
http://ritn.net

Southeastern Electric Exchange
SEE is a non-profit, non-political trade association of investor-owned electric utility companies
For storm restoration efforts (primarily distribution-related, but can also include transmission and generation if impacted), mutual assistance organizations (most notable the Southeastern Electric Exchange, or SEE) have been in place for decades. Started in the 1930's, the name is a bit misleading today in that participants range from New Mexico to New York and Florida to parts of Michigan and Illinois.
http://www.theexchange.org/aboutus.html

Spilling the Beans

Chris Vickery, MacKeeper
January 26, 2016
Retrieved February 20, 2016
https://mackeeper.com/blog/post/185-spilling-the-beans

Strategies, Protections, and Mitigations for the Electric Grid from Electromagnetic Pulse Effects
U.S. Department of Energy
Idaho National Laboratory
January 2016
This report identifies known grid impacts from EMP threats, effectiveness, and potential costs of known mitigations, areas for government and private partnerships in better protecting the electric grid, and gaps in knowledge and protection strategy.
https://inldigitallibrary.inl.gov/STI/INL-EXT-15-35582.pdf

Strengthen Your Financial Preparedness for Disasters and Emergencies
Emergency Financial First Aid Kit (EFFAK)
FEMA
September 2015
Includes a checklist of important documents and forms to compile your relevant information:

- Household Identification
- Financial and Legal Documentation
- Medical Information
- Household Contacts

http://www.fema.gov/media-library-data/1441313659987-38b0760a58131b871d494ddacbf52b6e/EFFAK_2015_508.pdf

Supervisory Practices Regarding Banking Organizations and their Borrowers and Other Customers Affected by a Major Disaster or Emergency
Board of Governors of the Federal Reserve System
Washington, D.C. 20551
Division of Banking
Supervisor and Regulation
Division of Consumer and Community Affairs
March 29, 2013
[This document is a highlight of] the supervisory practices that the Federal Reserve can employ when banking organizations and their borrowers and other customers are affected by a major disaster or emergency. Major disasters include hurricanes, tornadoes, floods, earthquakes, blizzards, and other natural catastrophes, as well as fires and explosions

https://www.federalreserve.gov/bankinforeg/srletters/sr1306.pdf

The Great Campout
National Wildlife Foundation
http://www.nwf.org/Great-American-Campout.aspx

The National Preparedness Goal
FEMA, US Department of Homeland Security
September 2015
https://www.fema.gov/media-library/assets/documents/25959

The State of Security in Control Systems Today
SANS Institute
Office of the Attorney General
Retrieved February 20, 2016
https://www.sans.org/reading-room/whitepapers/analyst/state-security-control-systems-today-36042

The Transition Handbook: From Oil Dependency to Local Resiliency
Rob Hopkins
http://transitionwhatcom.org/docs/transition_handbook_1_the_head.pdf
https://www.amazon.com/Transition-Handbook-Dependency-Local-Resilience/dp/0857842153#reader_0857842153

Transferability of Self-Healing Principles to the Recovery of Supply
Procedia CIRP 19 (2014) 14 – 20
ScienceDirect
Robust Manufacturing Conference (RoMaC 2014)
Network Disruptions – The Case of Renesas Electronics
Marie Brüninga*, J. Henning Buchholzb, Julia Bendula
http://www.sciencedirect.com/science/article/pii/S2212827114006398

Triple Threat Power Grid Exercise: High Impact Threats Workshop and Tabletop Exercises Examining Extreme Space Weather, EMP and Cyber Attacks
Charles Manto, Dr. George Baker III, Terry Donat MD, David Hunt
October 30, 2015
This InfraGard National Electromagnetic Pulse Special Interest Group (EMP SIG) exercise package facilitates discussions, planning and preparation for catastrophic events involving

the electrical grid and the cascading impacts to other critical infrastructure and the community.
https://www.amazon.com/Triple-Threat-Power-Grid-Exercise/dp/1633912493

Understand A Major National Security Threat - Improvised Nuclear Device
National Security Staff, Interagency Policy Coordination Subcommittee for Preparedness & Response to Radiological and Nuclear Threats
June 2010
http://www.youtube.com/watch?v=gxb9rg4MQgk

Understanding and Mitigating Catastrophic Disruption and Attack
Denise M.B. Masi, Eric E. Smith, and Martin J. Fischer
Analysts have amassed much data that points to vulnerabilities in telecommunications and cybersecurity.
Examining past natural disasters and major attacks can provide valuable insight into mitigating new ones.
https://blackboard.angelo.edu/bbcswebdav/institution/LFA/CSS/Course%20Material/BOR4301/Readings/UnderstandingAndMitigating.pdf

US Department of Homeland Security
http://www.ready.gov

Water Sector Resilience Final Report and Recommendations
June 2016
The National Infrastructure Advisory Council (NIAC)
NIAC provides the President of the United States with advice on the security and resilience of the critical infrastructure sectors and their functional systems, physical assets, and cyber networks.
NIAC was asked to 1) assess security and resilience in the Water Sector, 2) uncover key water resilience issues, and 3) identify potential opportunities to address these issues.
https://www.dhs.gov/sites/default/files/publications/niac-water-resilience-final-report-508.pdf

What If the Biggest Solar Storm on Record Happened Today?: Repeat of 1859 Carrington Event would devastate modern world, experts say
National Geographic
Richard A. Lovett, for National Geographic News
March 4, 2011
http://news.nationalgeographic.com/news/2011/03/110302-solar-flares-sun-storms-earth-danger-carrington-event-science

Emergency Transportation Operations			
Publication	**Publication Number**	**EDL Number**	**Contact**
Highway Evacuations in Selected Metropolitan Areas: Assessment of Impediments (HTML, PDF 3.7MB)	FHWA-HOP-10-059	N/A	Kimberly.Vasconez@dot.gov
FHWA's Emergency Transportation Operations Publications Series Presents: The Best of Traffic Incident Management, Traffic Planning for Special Events and Evacuation & Disaster Planning (CD)	FHWA-HOP-10-053	N/A	Kimberly.Vasconez@dot.gov
Good Practices in Transportation Evacuation Preparedness and Response: Results of the FHWA Workshop Series (HTML, PDF 748KB)	FHWA-HOP-09-040	N/A	Kimberly.Vasconez@dot.gov
Evacuating Populations With Special Needs - Routes to Effective Evacuation Planning Primer Series (HTML, PDF 21MB)	FHWA-HOP-09-022	N/A	Kimberly.Vasconez@dot.gov
Information Sharing Guidebook for Transportation Management Centers, Emergency Operations Centers, and Fusion Centers (HTML, PDF 3MB)	FHWA-HOP-09-003	N/A	Kimberly.Vasconez@dot.gov
Operational Concept - Assessment of the State of the Practice and State of the Art in	FHWA-HOP-08-020	N/A	Kimberly.Vasconez@dot.gov

Evacuation Transportation Management (HTML, PDF 923KB)			
Interview and Survey Results: Assessment of the State of the Practice and State of the Art in Evacuation Transportation Management (HTML, PDF 660KB)	FHWA-HOP-08-016	N/A	Kimberly.Vasconez@dot.gov
Literature Search for Federal Highway Administration - Assessment of the State of the Practice and State of the Art in Evacuation Transportation Management (HTML, PDF 2.4MB)	FHWA-HOP-08-015	N/A	Kimberly.Vasconez@dot.gov
Technical Memorandum for Federal Highway Administration on Case Studies - Assessment of the State of the Practice and State of the Art in Evacuation Transportation Management (HTML, PDF 1.4MB)	FHWA-HOP-08-014	N/A	Kimberly.Vasconez@dot.gov
Using Highways For No-Notice Evacuations - Routes to Effective Evacuation Planning Primer Series (HTML, PDF 20.9MB)	FHWA-HOP-08-003	N/A	Kimberly.Vasconez@dot.gov
Best of Public Safety and Emergency Transportation Operations CD	FHWA-JPO-08-037	14417	Kimberly.Vasconez@dot.gov
Common Issues in Emergency Transportation Operations Preparedness and Response: Results of the FHWA Workshop	FHWA-HOP-07-090	N/A	Kimberly.Vasconez@dot.gov

Series (HTML, PDF 1.1MB)			
Best Practices in Emergency Transportation Operations Preparedness and Response: Results of the FHWA Workshop Series (HTML, PDF 567KB)	FHWA-HOP-07-076	N/A	Kimberly.Vasconez@dot.gov
Communicating With the Public Using ATIS During Disasters: A Guide for Practitioners (HTML, PDF 2.3MB)	FHWA-HOP-07-068	14339	Kimberly.Vasconez@dot.gov
Managing Pedestrians During Evacuation of Metropolitan Areas (HTML, PDF 396KB)	FHWA-HOP-07-066	N/A	Kimberly.Vasconez@dot.gov
Routes to Effective Evacuation Planning Primer Series: Using Highways During Evacuation Operations for Events with Advance Notice (HTML, PDF 2.8MB)	FHWA-HOP-06-109	N/A	Kimberly.Vasconez@dot.gov
Transportation Evacuation Planning and Operations Workshop	FHWA-HOP-06-076	14184	Laurel.Radow@dot.gov
Coordinating Military Deployments on Roads and Highways: A Guide for State and Local Agencies (HTML, PDF 1.6MB)	FHWA-HOP-05-029	N/A	Kimberly.Vasconez@dot.gov
What Have We Learned About Intelligent Transportation Systems? Chapter 2: What Have We Learned About Freeway, Incident and Emergency Management and Electronic Toll Collection? (PDF 116KB)	FHWA-OP-01-006	13318	Kimberly.Vasconez@dot.gov
Intelligent Transportation	FHWA-JPO-	6326	ITSPUBS@dot.gov

Systems Field Operational Test Cross-Cutting Study: Emergency Notification and Response (PDF 273KB)	99-033		
Faster Response Time, Effective Use of Resources – Integrating Transportation and Emergency Management Systems (PDF 929KB)	FHWA-JPO-99-004	6874	ITSPUBS@dot.gov
Speeding Response, Saving Lives – Automatic Vehicle Location Capabilities for Emergency Vehicles (PDF 823KB)	FHWA-JPO-99-003	6866	ITSPUBS@dot.gov
Enhancing Public Safety, Saving Lives – Emergency Vehicle Preemption (PDF 1.12MB)	FHWA-JPO-99-002	6871	ITSPUBS@dot.gov
Effects of Catastrophic Events on Transportation Systems Management and Operations: Howard Street Tunnel Fire Baltimore City	Web publication only	13754	Kimberly.Vasconez@dot.gov
Effects of Catastrophic Events on Transportation Systems Management and Operations: Northridge Earthquake January 17, 1994	Web publication only	13775	Kimberly.Vasconez@dot.gov
Effects of Catastrophic Events on Transportation Systems Management and Operations: Cross-Cutting Study	Web publication only	13780	Kimberly.Vasconez@dot.gov

Organizational Contacts

InfraGard https://www.infragard.org/

National Governors Association (NGA) http://www.nga.org/cms/home.html

National Association of Counties (NACO) http://www.naco.org/

National Emergency Management Association (NEMA) http://www.emacweb.org/

Department of Homeland Security Protective Security Advisor Program (PSA)

https://www.dhs.gov/protective-security-advisors

Volunteers Active in Disasters (VOAD) http://www.nvoad.org/ (See the following table of those organizations who are members of VOAD)

American Red Cross http://www.redcross.org/mo2s

American Logistic Aid Network (ALAN) http://alanaid.org/

FEMA Region I http://www.fema.gov/region-i-ct-me-ma-nh-ri-vt

FEMA Region II https://www.fema.gov/region-ii-nj-ny-pr-vi-0

FEMA Region III http://www.fema.gov/region-iii-dc-de-md-pa-va-wv

FEMA Region IV https://www.fema.gov/region-iv-al-fl-ga-ky-ms-nc-sc-tn

FEMA Region V https://www.fema.gov/region-v-il-mi-mn-oh-wi

FEMA Region VI https://www.fema.gov/region-vi-arkansas-louisiana-new-mexico-oklahoma-texas

FEMA Region VII http://www.fema.gov/region-vii-ia-ks-mo-ne

FEMA Region VIII https://www.fema.gov/region-viii-co-mt-nd-sd-ut-wy

FEMA Region IX https://www.fema.gov/fema-region-ix-arizona-california-hawaii-nevada-pacific-islands

FEMA Region X https://www.fema.gov/region-x-ak-id-or-wa

National Members: 62		State/Territory Members: 56	
Adventist Community Services All Hands Volunteers Alliance of Information and Referral Systems Amateur Radio Relay League American Bible Society American Red Cross Americares Billy Graham Rapid Response Team Brethren Disaster Ministries Buddhist Tzu Chi Foundation Catholic Charities Churches of Scientology Disaster Response Church World Service Convoy of Hope Cooperative Baptist Fellowship Direct Relief Disciples of Christ Episcopal Relief and Development Feeding Americas Feed the Children Habitat for Humanity International Headwater Disaster Relief Heart to Heart International HOPE AACR Hope Coalition America Hope Force International HOPE Worldwide Humane Society of the U.S ICNA Relief USA Institute for Congregational Growth and Trauma	International Medical Corp International Orthodox Christian Charities Islamic Relief USA Latter-Day Saint Charities Link2Health Solutions (Disaster Distress Helpline Luther Disaster Response Mennonite Disaster Service Mission to North America NESHAMA Association of Jewish Chaplains The National Baptist Convention USA Nazarene Disaster Response Disaster Operation Blessing Points of Light Partnerships with Native Americans Presbyterian Disaster Assistance Rebuilding Together Samaritans Purse Save the Children Southern Baptist Convention St. Bernard Project Team Rubicon The Jewish Federation of North America The Salvation Army Society of St Vincent De Paul ToolBank United Church of Christ United Methodist Committee on Relief United Way Worldwide World Renew	Alabama VOAD Alaska VOAD American Samoa VOAD Arizona VOAD Arkansas VOAD California VOAD Colorado VOAD Connecticut VOAD Delaware VOAD District of Columbia VOAD Florida VOAD Georgia VOAD Guam VOAD Hawaii VOAD Idaho VOAD Illinois VOAD Indiana VOAD Iowa VOAD Kansas VOAD Kentucky VOAD Louisiana VOAD Maine VOAD Maryland VOAD Massachusetts VOAD Michigan VOAD Minnesota VOAD Mississippi VOAD Montana VOAD	Missouri VOAD Nebraska VOAD Nevada VOAD New Hampshire VOAD New Jersey VOAD New Mexico VOAD New York VOAD North Carolina VOAD North Dakota VOAD Northern Mariana Islands VOAD Ohio VOAD Oklahoma VOAD Oregon VOAD Pennsylvania VOAD Puerto Rico VOAD Rhode Island VOAD South Carolina VOAD South Dakota VOAD Tennessee VOAD Texas VOAD United States Virgin Islands VOAD Utah VOAD Vermont VOAD Virginia VOAD Washington VOAD West Virginia VOAD Wisconsin VOAD Wyoming VOAD

Table 1: National Volunteer Organizations Active in Disasters

RESILIENCY ASSESSMENT
Ashburn, Virginia

WHAT IS THE RRAP?

The Regional Resiliency Assessment Program (RRAP) is a cooperative, non-regulatory assessment program implemented to examine the resilience of critical infrastructure and systems through regional analysis. The program, led by the Department of Homeland Security (DHS) Office of Infrastructure Protection (IP), addresses a range of hazards that could have significant consequences, both regionally and nationally.

Each RRAP typically involves data gathering and analytical effort, followed by continued technical assistance to support resilience building. RRAPs can incorporate various components, including voluntary facility vulnerability assessments, targeted studies and modeling, first responder capability evaluations, subject matter expert workshops, and other valuable information-exchange forums. The RRAP process involves the following steps. >>>

The Office of Infrastructure Protection leads the national effort to mitigate risk to, strengthen the protection of, and enhance the all-hazard resilience of the Nation's critical infrastructure.

Data Centers Depend on Electricity and Water for Cooling

STEP 1 **Assess critical infrastructure on a regional level, focusing on threats, vulnerabilities, and consequences from an all-hazards perspective**

STEP 2 **Identify dependencies, interdependencies, and cascading effects** by developing awareness of how this system can be disrupted through facility site assessments, cyber assessments, one-on-one interviews, and facilitated meetings with a wide range of stakeholders

STEP 3 **Identify gaps in resilience** through additional research and analysis aimed at improvements in stakeholder planning and preparedness

STEP 4 **Assess capabilities to protect critical infrastructure** and to establish, communicate, and implement alternatives for commuters and transit providers

STEP 5 **Coordinate efforts to enhance resilience** by developing Resilience Enhancement Options to be implemented by State and local stakeholders with the support of DHS IP

WHY ASHBURN, VIRGINIA?

The cluster of data centers in the greater Ashburn area in Loudoun County serves as the primary global Internet traffic hub on the East Coast owing to the presence of a major Internet exchange point (IXP). With the unique concentration of both fiber and power, on average, 50 to 70 percent of all Internet traffic flows through the greater Ashburn-area data centers.[1,2] Greater Ashburn-area data centers provide primary and secondary continuity and backup for the information technology (IT) infrastructure of Federal, State, and local governments. The protection of greater Ashburn-area network and data infrastructure assets is critical to the continuity of operations at governmental agencies and private companies that, in turn, supply day-to-day services to critical utilities and the public.

[1] Garber, K., 2009, "The Internet's Hidden Energy Hogs: Data Servers," *U.S. News and World Report*, March 24, http://www.usnews.com/news/energy/articles/2009/03/24/the-internets-hidden-energy-hogs-data-servers, accessed August 31, 2015.

[2] O'Connell, J., 2014, "Data Centers Boom in Loudon County, but Jobs Aren't Following," *The Washington Post*, January 17, http://www.washingtonpost.com/business/capitalbusiness/data-centers-boom-in-loudoun-county-but-jobs-arent- following/2014/01/17/b4a704c8-7f0e-11e3-93c1-0e888170b723_story.html, accessed August 31, 2015.

ASHBURN, VIRGINIA
RRAP PARTICIPANTS

- U.S. Department of Homeland Security (DHS)
  - National Protection and Programs Directorate
    - Office of Cyber and Infrastructure Analysis
    - Office of Cybersecurity and Communications
      - National Coordinating Center for Communications
      - National Cyber Security Division
    - Office of Infrastructure Protection
      - Infrastructure Information Collection Division
      - Protective Security Coordination Division Sector Outreach and Programs Division
  - U.S. Immigration and Customs Enforcement
- Federal Bureau of Investigation

- Northern Virginia Regional Intelligence Center
- Virginia Council of Governments
- Virginia Department of Emergency Management
- Virginia Office of Public Safety and Homeland Security
- Virginia Information Technology Agency
- Virginia State Police

- Fairfax County Police Department
- Fairfax Water
- Loudoun County Fire and Rescue
- Loudoun County Office of Economic Development
- Loudoun County Office of Emergency Management
- Loudoun County Sheriff's Department
- Loudoun Water

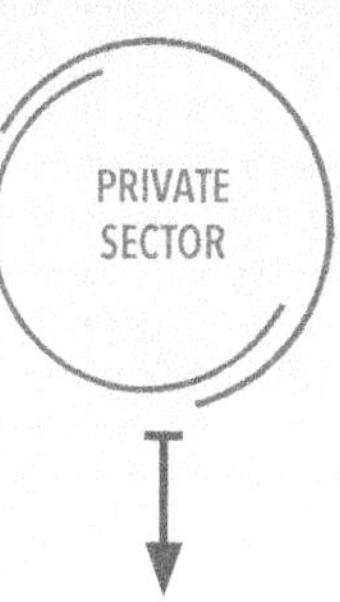

- Tier 1 Internet Service Providers
- Internet Exchange Providers
- Major Content Providers
- Content Delivery Networks
- Fuel Providers
- Logistics Providers
- Electric Utility Providers
- Optical Fiber Providers
- Telecom Providers
- Financial Service Providers

100+
Data
Centers
Belmont
Ashburn
Broadlands
Countryside
Sugarland Run
Dulles Town Center
Sterling
Loudoun Valley Estates
Brambleton
Washington Dulles International Airport
Reston
VA 267
VA 643
VA 621
VA 607
VA 640
VA 7
VA 659
VA 606
US 50
SR 228
SR 606
SR 665
33
6
14
20
6
6
2

KEY FINDINGS
RESILIENCE ENHANCEMENT OPTIONS

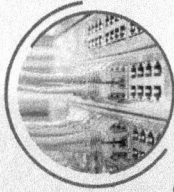

Centralized Internet exchange points (IXPs) pose the largest single points of failure for Internet resilience.

A study should be conducted that simulates the outage of a centralized peering facility. Such a simulation would include modeling the traffic and Transmission Control Protocol congestion during an outage of IXP facilities in the greater Ashburn area. The study should also include workshops to establish a common vocabulary and taxonomy that would benefit both customers and providers in communicating and evaluating resilience needs.

Lack of transparency in both network and data center infrastructure makes adequate resilience planning difficult.

The Virginia Office of Public Safety and Homeland Security, in concert with private sector stakeholders, should coordinate a workshop on the development of cloud and data center service taxonomies and assessments. Such taxonomies and/or assessments should allow for equal comparison of resilience features across providers and empower customers by creating honest competition.

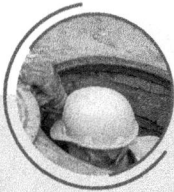

Law enforcement does not have adequate training or information to recognize suspicious activity around unprotected fiber routes and vaults.

Law enforcement personnel should engage industry stakeholders to facilitate training and education on fiber routes and suspicious activity. This training should include locations that are particularly vulnerable or exposed, or that have a larger-than-average possibility of leading to consequences. It should also address how and when to approach maintenance personnel, and how to confirm that they are authorized to work in a given area.

Data center and content providers do not have a way to contribute to resilience efforts and/or communicate criticality during an emergency.

A workshop should be conducted for the data center community so that all parties can communicate their needs for connections and access to emergency operation center (EOC) resources and for communication pathways during an emergency. Through information sharing and analysis centers, data center and content providers should become more involved and obtain greater access to EOC resources.

KEY FINDINGS
RESILIENCE ENHANCEMENT OPTIONS

Transportation infrastructure and trucking companies supplying diesel fuel may not be able to adequately support the data center community in an extended power interruption.

An in-depth study of the region's fuel supply chain should be conducted specifically to address concerns relating to fuel supplies and the ability to deliver available fuel during an extended power outage. Such a study should include best- and worst-case scenarios for road conditions after a disaster. Data center and network providers should be present at the State's annual exercise within the region. Their participation would improve decision makers' understanding concerning the critical roles of data centers in response efforts.

Data center and network providers have not adequately factored radio frequency (RF) weapons and associated electromagnetic pulse (EMP) effects into their resilience and protective measures planning.

Workshops should be conducted to support Ashburn-area data centers in their efforts to improve resilience to EMP effects, including RF weapons. A process should be established to keep data centers and local law enforcement up to date on new EMP/RF threats. Discussion about whether and to what extent EMP protection should be deployed should be a standard risk management topic for data centers.

Communication and education efforts between data center providers and fire department personnel as a part of resilience planning are inadequate.

Data centers and emergency services personnel should work together to arrange training and education sessions to help ensure that fire department personnel are aware of distinctive data center needs and environments. Data centers should consider installing radio infrastructure that operates on frequencies used by fire and police department radios to assist with operations during emergency situations.

RELATED ACTIVITIES – EMP AWARENESS SEMINARS

In May 2015, an EMP and RF Devices awareness seminar was conducted in conjunction with the Ashburn RRAP.

DHS guest speaker Kevin Briggs provided a brief overview of the threats associated with nuclear EMP and solar super storms and identified their potential impact on data centers. He also provided a briefing on his DHS report entitled EMP Protection Guidelines for Data Centers.

FBI InfraGard EMP Special Interest Group (SIG) guest speakers included Dr. George Baker, Dr. Paul R. Hayes and Chuck Manto. Drs. Baker and Hayes provided an overview of intentional electromagnetic interference threats, and Mr. Manto presented on an economic model of EMP protection, a civilian approach to an EMP rating system calibrated from U.S. military standards, examples of mitigation methods, and the role of the InfraGard EMP SIG to request additional information.

For more information about the Ashburn, Virginia, RRAP, contact **PSCDOperations@hq.dhs.gov**

For more information visit:
http://www.dhs.gov/publication/rrap-fact-sheet

www.ingramcontent.com/pod-product-compliance
Lightning Source LLC
Chambersburg PA
CBHW061737210326
41599CB00034B/6705

* 9 7 8 1 6 3 3 9 1 4 2 9 2 *